Smail Djebali

Algebraic Topology

Also of Interest

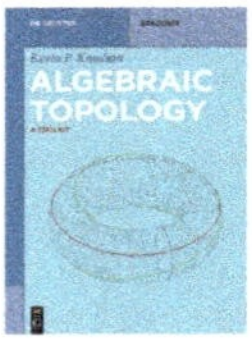

Algebraic Topology
A Toolkit
Kevin P. Knudson, 2024
ISBN 978-3-11-101481-4, e-ISBN (PDF) 978-3-11-101485-2

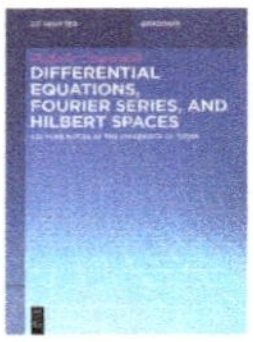

Differential Equations, Fourier Series, and Hilbert Spaces
Lecture Notes at the University of Siena
Raffaele Chiappinelli, 2023
ISBN 978-3-11-129485-8, e-ISBN (PDF) 978-3-11-130252-2

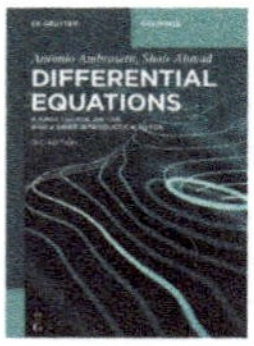

Differential Equations
A first course on ODE and a brief introduction to PDE
Shair Ahmad, Antonio Ambrosetti, 2023
ISBN 978-3-11-118524-8, e-ISBN (PDF) 978-3-11-118567-5

Topics in Complex Analysis
Dan Romik, 2023
ISBN 978-3-11-079678-0, e-ISBN (PDF) 978-3-11-079681-0

Advanced Mathematics
An Invitation in Preparation for Graduate School
Patrick Guidotti, 2022
ISBN 978-3-11-078085-7, e-ISBN (PDF) 978-3-11-078092-5

Smail Djebali

Algebraic Topology

Constructions, Retractions, and Fixed Point Theory

DE GRUYTER

Mathematics Subject Classification 2020
Primary: 54A05, 55M20, 55Q05; Secondary: 54B17, 54C15, 55N40

Author
Prof. Dr. Smail Djebali
Department of Mathematics
Faculty of Sciences
Imam Mohammad Ibn Saud
Islamic University
P. O. Box 90950
Riyadh 11623
Saudi Arabia

Laboratoire "Théorie du point Fixe et Applications"
École Normale Supérieure
Kouba (Algiers)
Algeria
ibdjebali@imamu.edu.sa

ISBN 978-3-11-151736-0
e-ISBN (PDF) 978-3-11-151738-4
e-ISBN (EPUB) 978-3-11-151778-0

Library of Congress Control Number: 2024941706

Bibliographic information published by the Deutsche Nationalbibliothek
The Deutsche Nationalbibliothek lists this publication in the Deutsche Nationalbibliografie; detailed bibliographic data are available on the Internet at http://dnb.dnb.de.

Cover image: estherpoon / iStock / Getty Images Plus
Typesetting: VTeX UAB, Lithuania

www.degruyter.com

For my deceased parents and brothers

Introduction

On the borderline of topology and algebra, algebraic topology is the study of topological spaces through algebraic methods. This book introduces some of the most important tools employed in algebraic topology: the theory of homotopy (homotopy of paths and homotopy of loops), the theory of retraction (retract sets and retraction maps), the higher homotopy groups and the homology groups. After a thorough description of the methods related to these theories, Brouwer's fixed- point theorem for the disc in the plane and the general finite-dimensional case are presented. The infinite-dimensional case concerning Schauder's fixed-point theorem is investigated by an approximation approach. The book offers applications of the fixed-point theory to the solvability of some differential and integral equations. Then further fixed-point theorems are proved, including the Borsuk–Ulam theorem. The degree of mappings between spheres of Euclidean space is discussed as an application of the homology groups of the sphere.

Actually, the book explores the theory and various applications of the basic methods employed in algebraic topology. A simple mathematical approach is used throughout to perform several standard constructions of topological spaces by means of a large number of examples as well as to introduce the theory of homotopy and the theory of homology. The book is primarily designed for Bachelor, Master's students and first-year PhD students since it is accessible to any reader with a good understanding of the general topology and metric spaces. Thus, the book is aimed to serve as an introductory graduate course in algebraic topology. It is also suitable for advanced undergraduate students who aspire to easily grasp some new concepts in algebraic topology. Indeed, the presentation adopted in this book relies essentially on point-set topology rather than on homotopy theoretic arguments or homological algebra theory. The book consists of twelve chapters followed by one chapter of Appendices:

1. Background in topology
2. Quotient topology
3. Topological constructions
4. The fundamental group
5. Covering maps and lifting maps
6. Fundamental groups of the circle and the sphere
7. Borsuk's theory of retracts
8. Fundamental groups of some surfaces
9. Higher homotopy groups
10. Elements of homology theory
11. Fixed-point theorems
12. Applications
13. Appendices

The first three chapters are usually covered by any textbook on point-set topology. Chapters 4 to 7 are traditionally delivered in advanced topology courses. The remaining chapters introduce algebraic topology to Master students in mathematics. The layout of the chapters is as follows. In order to understand most mathematical concepts developed in the book, Chapter 1 and Chapter 2 provide some background notions in point-set topology, in particular, all those properties of the quotient topology. These are reasonably self-contained chapters. Further properties of particular subsets of topological spaces

https://doi.org/10.1515/9783111517384-201

are suggested as exercises at the end of sections of Chapter 1. Students with a modest knowledge of the general topology can benefit from the first two chapters. Senior students at more advanced graduate levels willing to get insight into deeper topics can use this book as a first course in algebraic topology, skipping Chapter 1 and Chapter 2.

Chapter 3 is a follow-up of the first two chapters. In this chapter, the quotient topology is applied to construct several familiar topological spaces. Those spaces will serve as fundamental examples for the theoretical results offered in the subsequent chapters. Most textbooks in topology devote some chapters to the construction of new topological spaces. We recommend [1, 5, 27].

The concept of the fundamental group was first introduced in 1895 by French mathematician Henri Poincaré in his influential paper and seminal work "Analysis Situs" [78]. The definition of the fundamental group is based on both the homotopy theory and the group theory. The fundamental group is vital in building up algebraic topology by connecting analysis with algebra. The algebraic properties of some groups provide useful features of the topological structures of some relevant spaces. Chapter 4 introduces the concept of homotopy of paths and describes in detail the main properties of the fundamental group. In particular, the analytical properties of some fundamental groups are discussed in the second part of this chapter. The definitions and some properties of contractible sets and simply connected spaces, which arise in some complex analysis formulas are introduced from the homotopy theory.

Before we tackle the computation of some fundamental groups, we investigate the main topological properties of covering maps and lifting maps in Chapter 5. Covering maps are used to show the existence and uniqueness of the lifting map. The case of a loop, a path and then the general case are discussed separately.

The key target of Chapter 6 is the computation of the fundamental groups of the unit circle and the sphere S^n, for $n \geq 2$. The property of the nonretraction of the circle is derived from Section 6.1 and Section 6.3. In this chapter, the extension problem of maps defined only on the unit circle to the whole disk is discussed in Section 6.2 as a prelude to the formulation of the Borsuk–Ulam theorem. In Section 6.4, the degree of a map is defined. Finally, five equivalent versions of the Borsuk–Ulam theorem are proved for the sphere S^2 of Euclidean space.

The theory of retracts of Karol Borsuk (retract sets and retraction maps) is intimately related to the problem of the extendability of continuous functions. Fully developed in Chapter 7, the theory is exploited to prove that the sphere is not a retract of the ball. The process involved in the proof makes use of the theory of the fundamental group. The theory of retraction initiated in [8, 9] is relevant to the fixed-point property of some topological spaces. For instance, Brouwer's fixed- point theorem for the disc in the plane is shown using the nonretraction of the circle S^1. Chapter 6 and Chapter 7 pave the way to the next one.

The computation of the fundamental groups of some surfaces is reported in Chapter 8 with an emphasis on some familiar surfaces. One can further find the description of the fundamental groups of some particular sets, including star-convex sets and AR spaces. Some of the homeomorphisms considered in Chapter 1 will help in the computation of the fundamental groups of some surfaces through the Seifert–van Kampen theorem (Section 8.1). We will also appeal to the planar representations of the surfaces. In Section 8.2, we discuss the torus, the projective plane, the projective space, the Möbius band (introduced in 1858 by August Ferdinand Möbius) and the Klein bottle (invented by Felix Klein in 1882). The sphere, the torus and the projective plane are involved in the classification of the compact surfaces. The theory of the fundamental group is completely developed in Chapter 8.

The general theory of the fundamental group is presented for higher dimensions in Chapter 9. The extension, which first appeared in 1935, is owed to Witold Hurewicz, 40 years after Poincaré's Analysis Situs [78] appeared. The Freudenthal suspension theorem due to Hans Freudenthal helps in computing many higher homotopy groups, especially those of the sphere. Since the theory of homotopy is sufficiently investigated in Chapter 4, only some notions and results specific to the dimension greater than one are discussed in Chapter 9, more precisely the Abelian higher fundamental group $\pi_n(X, x_0)$ and in particular the higher homotopy groups of the sphere S^k in Euclidean space.

Chapter 10 describes the main properties of the simplicial homology, the singular homology and the relative homology. This chapter is independent of the previous chapters. The homology groups of the ball, the sphere, the quotient space and some other manifolds are computed. The bulk of this chapter builds up these theories either by a direct construction (in the case of the simplicial homology and the singular homology) or by an axiomatic treatment based on Eilenberg–Steenrod axioms (in the case of the relative homology). A second course taking up the homology theory in more detail is strongly recommended for those students working in algebraic topology, homological algebra, differential topology or differential geometry. In this chapter, the focus is rather on the proofs of some results and on some applications, which make use of the Mayer–Vietoris exact sequence (Section 10.4). The chapter ends with the definitions of the Euler and Betti numbers involved in the classification of compact surfaces.

The interplay between the homotopy theory and the retraction theory developed in Chapter 8 and Chapter 9 is crucial in order to understand Chapter 11, which explores some fixed-point theorems as first applications. Using the nonretraction of the sphere S^{n-1}, we put the stress on one of the numerous proofs of Brouwer's fixed-point theorem in Euclidean space $\mathbb{R}^n$. Some equivalent forms of this theorem and alternative proofs are first given in this chapter. In 1911, the mathematician Luitzen Egbertus Jan Brouwer established this fixed-point theorem, named after him. Further equivalent versions of Brouwer's fixed-point theorem are then proved. They include the Borsuk–Ulam theorem and the Lyusternik–Schnirelmann theorem established in 1947 jointly by Lazar Aronovich Lyusternik and Lev Genrikhovich Schnirelmann. The more general version of Brouwer's fixed-point theorem is Schauder's fixed-point theorem valid in any normed space. Schauder's fixed-point theorem was proved by Juliusz Schauder in 1930. The proof relies on a Schauder approximation result, namely Lemma A.7.1. In this chapter, we explain the difference between the finite- and the infinite-dimensional cases. In particular, we show that in the infinite-dimensional case, unlike the finite-dimensional case, the ball does retract on its boundary. In order to establish the retraction of the sphere in the infinite-dimensional case, we have appealed to a nice result due to V. L. Klee (1956) and Dugundgi's extension theorem (1951). Chapter 11 ends with a discussion of the Lefschetz fixed-point theorem, Granas fixed-point theorem and some of its variants.

Finally, some applications of the fixed-point theory and the homology theory are reported in Chapter 12. Nine equivalent formulations of Brouwer's fixed-point theorem are summarized in this chapter. This theorem is also used to prove the Perron–Frobenius theorem, a fundamental theorem of algebra. Schauder's fixed-point theorem is employed to discuss the solvability of some differential and integral equations, which are set on infinite-dimensional spaces such that the spaces of continuously differentiable functions. More precisely, we prove the existence of local and global solutions to a class of first-order initial value problems and a nonlinear second-order boundary value problem. Then some applications of homology theory are proposed. The concept of the degree of a mapping acting between spheres is first discussed as a generalization of the winding number introduced in Section 6.2 in the bidimensional case. After computing the degree of reflection map and the antipodal map, the hairy ball theorem is proved. The Borsuk–Ulam theorem has many applications in fixed-point theory, graph theory and game theory (Nash equilibria). The last section of Chapter 12 is devoted to this important theorem. The classical formulation of this theorem was proved in 1933 by Borsuk in Lviv, Poland (now Ukraine). Brouwer has first connected the homotopy theory and the homology theory by proving that two maps are homotopic if and only if they have same degree, a result established in Section 12.5.

In the Appendices, we have collected some auxiliary notions from basic functional analysis (pasting lemma, extension and approximation lemmas, compactness criteria, Minkowski's functional...), general algebra and linear algebra (homomorphism, cyclic group, free group, free product of groups, free product with amalgamation,...). This is material that we have referred to throughout the book. We hope this makes the book self-contained.

To make easier searching topics in the book, the subject index entries are arranged in as follows: "Theorem, Jordan" or "Theorem, Ascoli" or "Lemma, pasting" or "Lemma, Stampacchia" or "map, compact" or "map, projection" or "space, connected" or "space, broom" and so on.

The book aims at providing an application-oriented presentation of some fundamental concepts in algebraic topology by using merely point-set topology. We have refrained from reporting long and heavy technical proofs of some advanced theorems, which can be found in most specialized textbooks. Some theorems are briefly reported with emphasis on the applications, for instance, the proofs of the classification of compact surfaces, the Hurewicz theorem, the Mayer–Vietoris theorem are omitted. The Seifert–van Kampen theorem is proved in a special situation. For the same reason, we have missed the treatment of some concepts such that the cell homology and the computation of the fundamental groups of graphs and polyhedra. The interested readers can consult the rich literature, e. g., [27, 51, 67, 71, 91, 99].

We do not claim any originality of the results or the proofs presented herein. All the material can be found elsewhere and we have tried to indicate the original sources of the various results, which we have learned. For each result presented, the oldest references are cited whenever possible. However, an effort has been made to make a clear presentation of the results trying to reach an almost self-contained teaching document. Most parts of the book have been prepared during the health confinement and the lockdown resulting from the Covid-19 pandemic (2020–2022). Of course, the treatment may suffer some defects. We welcome constructive criticism and helpful suggestions for future improvement of the material of this book.

At the origin of the idea of the preparation of this book, our ulterior motivation was really to derive some known fixed-point theorems from the homotopy and homology theory and then discuss the solvability of some differential equations. For many years, we have been teaching the retract theory and the homotopy theory for Master's degree students at École Normale Supérieure (Kouba, Algiers). The course deals with the topological degree theory and its applications in fixed-point theory and differential equations. We have found the homotopy and homology theories so fascinating that more sophisticated material has been prepared and developed giving birth to this book. The literature is abundant of textbooks in algebraic topology. For those readers willing to delve deeper into the advanced theories, additional results can be found among the suggested references. One hundred six references and sources cited and used in this textbook are reported in the bibliography. Most of them have inspired many parts of the book.

To facilitate the readability of the book, we have strived to present in details and simplicity all topics covered. For some theorems and propositions, we propose more than one proof. A profusion of examples and applications of the results presented in the book are proposed to illustrate the theory. With such a desire to make the book more oriented toward applications, 280 worked examples, 80 illustrative figures and around 70 diagrams (commutative diagram or composition of maps) have been included to clearly demonstrate the theory and highlight the topological concepts investigated in this book. All figures have been drawn using Geogebra free software. The commutative diagrams were drawn using Latex and tikz associated tools or using free editors: (https://tikzcd.yichuanshen.de/) or (https://q.uiver.app/). The main fundamental groups (homology groups, respectively) computed in this book are summed up in tables drawn in Chapter 8 and Chapter 10, respectively.

To enrich the content of the results presented, all chapters or sections are brightened up with practical exercises of different and varying degrees of complexity. For this purpose, 320 exercises have been carefully selected. They are kept within the scope of the chapters. Unfortunately, no detailed answers are provided in this first edition of the book. Writing all the solutions might take much more time. However, we have indicated hints to some questions and we have referred to the original sources for others, whenever available. Students can work out the proposed questions and come up with solutions on their own.

I wish to thank some of my colleagues and former PhD students for reading several parts of the book. I quote late Karima Ait-Mahiout, Saadia Benchabane, Bilal Boulfoul, Toufik Moussaoui, Aboubakr-Khaled Sadallah (ENS-Kouba, Algiers, Algeria), Karima Mebarki (University of Bejaia, Algeria), Zakir Ahmed, Faryad Ali, Said El Manouni (IMSIU, Riyadh, KSA).

I am grateful to all those who showed me support during the preparation of the book.

Special thanks go to Professor Ranis Ibragimov, Acquisitions Editor, who believed in the book project and the editorial board of "De Gruyter Textbook" for accepting to include the book in the series. Also, I would like to thank De Gruyter VTeX production team for excellent proofing and careful checking of the textbook.

Riyadh, January 2020–September 2024 Pr S. Djebali (djebali@hotmail.com)

Contents

Notation and symbols

$\mathbb{N} = \{1, 2, \ldots, \ldots\}$	set of positive natural numbers
$\mathbb{Z}$	set of integers
$\mathbb{Q}$	set of rational numbers
$\mathbb{R}$	set of real numbers
$\mathbb{R}^n$	n-dimensional real Euclidean space
$\mathbb{C}$	set of complex numbers
$f(A)$	image of set the A
$f^{-1}(A)$	inverse image of the set A
$f_{\|A}$	restriction of f to A
$\widetilde{f}$	extension of f
$B(a, R)$	closed ball centered at a with radius R
$\mathring{B}(a, R)$	open ball centered at a with radius R
B^n	unit closed ball in $\mathbb{R}^n$
$\mathring{B}^n$	unit open ball in $\mathbb{R}^n$
S^n	unit sphere in $\mathbb{R}^{n+1}$)
$S^{n-1} = \partial B^n$	boundary of B^n
$\mathcal{N}_x$	Neighborhood of the point x
$\mathrm{d}(x, A)$	distance from a point x to the set A
$\operatorname{diam} A$	diameter of the set A
$\mathring{A}$,	interior of the set A
$\overline{A}$	closure of the set A
$\mathrm{Co}(A)$	closure of the convex hull of the set A
i, j	inclusion map
Id	identity map
γ	path connecting two points
$f * g$	product of paths
$\overline{h}$	induced map
h_*	induced homomorphism
r	retraction map
AR	absolute retract
$\sim$	equivalence relation
$[x]$	class of x
$X/_\sim$	quotient space defined by an equivalence relation
X/A	quotient space defined by a set
$\pi : X \longrightarrow X/A$	canonical projection map
$C(X)$	cone over X
$\Sigma(X)$	suspension over X
$X * Y$	join of spaces X and Y
$G_1 * G_2$	free product of groups G_1 and G_2
$X \vee Y$	wedge sum of X and Y
$X \wedge Y$	smash product of X and Y
Z_f	adjunction space along f
C_f	mapping cylinder of f
K_f	mapping cone of f
$C = S^1 \times [0, 1]$	bounded cylinder
$C = S^1 \times [0, \infty)$	unbounded cylinder
$\mathbb{T}^2$	torus

https://doi.org/10.1515/9783111517384-202

$\mathbb{M}^2$	Möbius band
$C(\mathbb{M}^2)$	Core of the Möbius band
$\partial(\mathbb{M}^2)$	boundary of the Möbius band
$M\#N$	connected sum of manifolds M and N
$M^\circ = M \setminus \mathring{B}^n$	hollowed n manifold
$\widehat{X}$	covering space
F	Fiber
$p : \widehat{X} \longrightarrow X$	covering map, fiber bundle
(X, A) $(A \subseteq X)$	pair of spaces
$p : (X, A) \longrightarrow (Y, B)$	map of pairs
$\pi_1(X)$	fundamental group of X
$\pi_n(X)$	nth higher homotopy group of X
$H_n(X)$	nth homology group of X
$H_n(X, A)$	nth relative homology group of the pair (X, A)
∂	boundary operator
$\omega(f)$	winding number of f
$\deg f$	degree of the map f
$\simeq$	homotopic to (maps)
$\simeq$	homotopically equivalent (spaces)
$\cong$	isomorphic to (groups)
$\cong$	homeomorphic to (spaces)
$\equiv$	identically equal to (maps)
$X \setminus Y$	set-theoretic difference
$\times$	product
$f \times g$	product of functions f and g
$\oplus$	direct sum, product
Z_n	n-cycle
B_n	n-boundary
C_n	n-chain
$\triangle_p$	simplicial complex
σ	simplicial map
$\chi(K)$	Euler number of K
$\beta_n(X)$	Betti number of X
$\Lambda(f)$	Lefschetz number of a f
$\|x\|$	absolute value of the real number x
$\|z\|$	modulus of the complex number z
$\\|x\\|$	norm of the vector x
$(X, \\|\cdot\\|)$	real Banach space with a norm $\\|\cdot\\|$
$\mathscr{C}(K, E)$	space of continuous functions on a compact space K with values in a space E, endowed with norm $\\|f\\| = \sup\{\\|f(x)\\|_E : x \in K\}$
$L^p(\Omega, E)$	(Banach) Lebesgue space of Bochner measurable functions with pth summable power with norm $\\|f\\|_p = (\int_\Omega \\|f\\|_E^p d\mu)^{1/p}$
FPP	fixed-point property
$\mathrm{Fix}(f)$	set of fixed points of f

List of Figures

https://doi.org/10.1515/9783111517384-203

1 Background in topology

This first chapter offers an exposition of some fundamental concepts in topology and begins with a reminder of the topology and the continuity of maps (Section 1.1 and Section 1.2). We assume a basic knowledge of metric spaces and their properties. Then we present the essential properties of compact spaces and compact maps in Section 1.3. The definition of the fundamental group (Chapter 4) relies on the definitions of a connected space and a path-connected space (Section 1.4). Some homomorphisms are discussed in the last section (Section 1.5), where several examples are provided. Homeomorphic spaces are used throughout the book not only in the construction of some topological spaces but in the computation of some fundamental groups as well. The main properties of topological spaces are well developed in most textbooks in topology, e. g., [6, 17, 18, 25, 61, 66, 76, 103].

1.1 Topological space

1.1.1 Open and closed sets

Definition 1.1.1. Let X be a nonempty set and $\mathscr{T}$ a family of subsets of X. The pair $(X, \mathscr{T})$ is called a topological space if the following three conditions are satisfied:

(1) X and the empty set $\emptyset$ belong to $\mathscr{T}$.

(2) If $\{O_\alpha\}_{\alpha\in A} \subseteq \mathscr{T}$, then $\bigcup_{\alpha\in\Lambda} O_\alpha \in \mathscr{T}$ for any set Λ. We say that $\mathscr{T}$ is closed under arbitrary unions.

(3) If $\{O_i\}_{1\le i\le N} \subseteq \mathscr{T}$, for some $N \in \mathbb{N}$, then $\bigcap_{1\le i\le N} O_i \in \mathscr{T}$. We say that $\mathscr{T}$ is closed under finite intersections.

The family $\mathscr{T}$ is called *a topology* and elements of $\mathscr{T}$ are called *open sets*. The complement of an open set is called a *closed set.*

Unless a specified mention of the topology $\mathscr{T}$, a topological space $(X, \mathscr{T})$ will be merely denoted X and called space for brevity.

Definition 1.1.2. A subset of a space X is called *clopen* if it is closed and open in X.

Example 1.1.3. (1) The topology $\mathscr{T} = \{X, \emptyset\}$ is called the trivial topology.

(2) If $\mathscr{T}$ consists of all subsets of X, then $\mathscr{T}$ is called the discrete topology and X is the discrete space.

Example 1.1.4. A metric space (E, d) is a topological space, where the topology is the collection of all open balls.

Recall the following.

https://doi.org/10.1515/9783111517384-001

Definition 1.1.5. A *metric space* is a pair (E, d), where $d : E \times E \longrightarrow [0, +\infty)$ satisfies the three axioms:
(a) $d(x, y) = 0$ if and only if $x = y$;
(b) $d(x, y) = d(y, x)$ for all $x, y \in E$ (symmetry);
(c) $d(x, y) \leq d(x, z) + d(z, y)$ for all $x, y, z \in E$ (triangle inequality).

Definition 1.1.6. A *normed space* $(X, \|\cdot\|)$ is a vector space X over a field K and a map

$$\|\cdot\| : X \longrightarrow [0, +\infty),$$

which satisfies the three axioms:
(a) $\|x\| = 0$ if and only if $x = 0$;
(b) $\|\lambda x\| = |\lambda|\|x\|$ for all $x, y \in X$ and $\lambda \in K$;
(c) $\|x + y\| \leq \|x\| + \|y\|$ for all $x, y, z \in X$ (triangle inequality).

A normed space is a metric space with metric $d(x, y) = \|x - y\|$.

Example 1.1.7. Let $X = \{\frac{1}{n}, n = 1, 2, \dots\}$ be a subset of the real line endowed with the standard topology. Then X is a discrete space for the subspace topology, since for every $n = 1, 2, \dots$, there exists $0 < \varepsilon < \frac{1}{n(n+1)}$ such that $\{\frac{1}{n}\} = X \cap (\frac{1}{n} - \varepsilon, \frac{1}{n} + \varepsilon)$. Hence, each element of X is an open set. Note however that the space $Y = X \cup \{0\}$ is not discrete since $\{0\}$ is not an open set.

Definition 1.1.8. Let $(X, \mathscr{T})$ be a space, $U \subseteq X$ and $x \in X$. U is said to be a neighborhood of x if there exists an open set $O \in \mathscr{T}$ such that $x \in O \subseteq U$. We write $U \in \mathscr{N}_x$.

Let $A \subset X$ be a nonempty subset of a space X.

Definition 1.1.9. (1) A point $x \in A$ is called an interior point if $A \in \mathscr{N}_x$. The set of all interior points of A, called the interior set, is denoted $\mathring{A}$.
(2) A point $x \in X$ is called an adherent point of the subset A if every neighborhood of x meets the set A. The set of all adherent points, called the closure of A, is denoted $\overline{A}$.
(3) A point $x \in X$ is called a boundary point of A if it is adherent to A but not an interior point of A. The set of all boundary points of A, called the boundary of A, is denoted ∂A. Thus,

$$\partial A = \overline{A} \cap (X \setminus \mathring{A}) = \overline{A} \cap \overline{(X \setminus A)}.$$

The following characterization follows from the definition.

Proposition 1.1.10. (1) *The set A is open if and only if $A = \mathring{A}$;*
(2) *A is closed if and only if $A = \overline{A}$.*

1.1.2 Separation theorems

Definition 1.1.11. We say that X is a T_1 space if for every distinct points $x, y \in X$, there exists an open subset $U \subset X$ such that $x \in U$ and $y \notin U$.

Proposition 1.1.12. *The following statements are equivalent:*
(1) *X is a T_1 space;*
(2) *For all $x \in X$, $\{x\} = \bigcap\{U : U \in \mathcal{N}_x\}$;*
(3) *For all $x \in X$, $\{x\}$ is a closed set.*

Proof. (1) $\Longrightarrow$ (2): For every $x \in X$, $\{x\} \subset \bigcap\{U : U \in \mathcal{N}_x\}$. Conversely, if $y \in \bigcap\{U : U \in \mathcal{N}_x\}$ and $y \neq x$, then by Assumption (1) there exists an open set U such that $x \in U$ and $y \notin U$, which is a contradiction. Hence, (2) holds.

(2) $\Longrightarrow$ (1): Let $y \neq x$ be two elements of X. Then $y \notin \bigcap\{U : U \in \mathcal{N}_x\}$. As a consequence, there exists $U \in \mathcal{N}_x$ such that $y \notin U$, proving (1).

(1) $\Longrightarrow$ (3): Let $x \in X$ and $y \neq x$. By assumption, there exists an open set U such that $y \in U$ and $x \notin U$. Hence, $U \subset X \setminus \{x\}$, proving that $X \setminus \{x\}$ is an open set.

(3) $\Longrightarrow$ (1): Let $y \neq x$ be two elements of X. Then $y \in U = X \setminus \{x\}$, which is open; whence (1). □

Definition 1.1.13. A space X is called *Hausdorff* (or satisfies the T_2-axiom) if every pair $x, y \in X$ have disjoint open neighborhoods.

Proposition 1.1.14. *The following statements are equivalent:*
(1) *X is a T_2 space;*
(2) *For every distinct points $x, y \in X$, there exists a subset $U \in \mathcal{N}_x$ such that $y \notin \overline{U}$;*
(3) *For all $x \in X$, $\{x\} = \bigcap\{\overline{U} : U \in \mathcal{N}_x\}$.*

Proof. (1) $\Longrightarrow$ (2): Let $x \neq y$ be two distinct points of X. By (1), there exist $U \in \mathcal{N}_x$ and $V \in \mathcal{N}_y$ such that $U \cap V = \emptyset$. Then $U \subset X \setminus V$ implies $\overline{U} \subset X \setminus V$ since V is open. Hence, $V \subset X \setminus \overline{U}$ and so $y \notin \overline{U}$.

(2) $\Longrightarrow$ (3): For $x \in X$, let $y \in \overline{U}$ for all $U \in \mathcal{N}_x$. If $y \neq x$, then by assumption there exists $U \in \mathcal{N}_x$ such that $y \notin \overline{U}$, a contradiction. The inclusion $\{x\} \subseteq \bigcap\{\overline{U} : U \in \mathcal{N}_x\}$ is obvious.

(3) $\Longrightarrow$ (1): Let $x \neq y$ be two elements of X. Then there exists $U \in \mathcal{N}_x$ such that $y \notin \overline{U}$, i. e., $y \in V = X \setminus \overline{U}$ with $U \cap V = \emptyset$. □

Remark 1.1.15. (1) By Exercise 1, Section 1.3, T_2 is equivalent to the closedness of the diagonal $\Delta = \{(x, x) \in X \times X\}$ in the product topology.
(2) In Proposition 1.3.12, we will prove that in a Hausdorff space a compact K and a point $x \neq\in K$ can be separated.

Clearly, every T_2 space is a T_1 space. Every metric space is T_2. Product of T_2 spaces is a T_2 space. A subspace of a T_2 is a T_2 space.

Definition 1.1.16. A space X is *a regular space* (or a T_3 space) if X is a T_1 space and for every closed subset $A \subset X$ and $x \notin A$, x and A can be separated by disjoint open sets, i. e., there are open sets U and V such that $x \in U, A \subset V$ and $U \cap V = \emptyset$.

Clearly, every T_3 space is a T_2 space.

Proposition 1.1.17. *A space X is a regular space if and only if for every $x \in X$ and every neighborhood $V \in \mathcal{N}_x$, there exists a neighborhood $U \in \mathcal{N}_x$ such that $\overline{U} \subset V$.*

Proof. (1) Let X be a regular space and $V \in \mathcal{N}_x$ an open neighborhood of x. Then, $C = X \setminus V$ is a closed subset which does not contain x. By assumption, there exist open sets U_1 and U_2 such that $x_1 \in U_1, C \subset U_2$, and $U_1 \cap U_2 = \emptyset$. Hence, $U_1 \subset X \setminus U_2 \subset X \setminus C = V$. Taking the closure sets,

$$\overline{U_1} \subset \overline{X \setminus U_2} = X \setminus U_2 \subset X \setminus C = V.$$

(2) Assume that the condition of the proposition holds. Let $x \in X$ and $A \subset X$ a closed subset that does not contain x. Then $V = X \setminus A$ is an open neighborhood of x. By hypothesis, there exists a neighborhood $U \in \mathcal{N}_x$ such that $\overline{U} \subset V$. Let $U_1 = U \ni x$ and $U_2 = X \setminus \overline{U}$. Then $X \setminus V \subset X \setminus \overline{U}$, which implies $A \subset U_2$ and $U_1 \cap U_2 = \emptyset$. □

Definition 1.1.18. A space X is said to be *a normal space* (or a T_4 space) if it is T_1 and for every two disjoint closed sets A and B. There exist two disjoint open sets U and V containing A and B, respectively.

Clearly, every T_4 space is a T_3 space. Hence,

$$T_4 \Longrightarrow T_3 \Longrightarrow T_2 \Longrightarrow T_1.$$

Proposition 1.1.19. *A space X is normal if and only if for every closed subset $A \subset X$ and every open subset $V \subset X$ containing A. There exists an open set $U \subset X$ such that $A \subset U \subset \overline{U} \subset V$.*

Proof. (1) Let X be normal, $A \subset X$ a closed subset and $V \subset X$ an open subset containing A. Then the closed sets A and $B = X \setminus V$ are separated, i. e., there exist disjoint open sets U_1 and U_2 such that $A \subset U_1$ and $B \subset U_2$. Hence, $U_1 \subset X \setminus U_2 \subset V$, which implies $\overline{U_1} \subset \overline{X \setminus U_2} = X \setminus U_2 \subset V$.

(2) Conversely, assume that the condition of the proposition holds and let A, B be two disjoint closed subsets. For the closed set A and the open set $V = X \setminus B$, which contains A, there exists an open set $U \subset X$ such that

$$A \subset U \subset \overline{U} \subset V = X \setminus B.$$

We conclude that the open sets $U_1 = U$ and $U_2 = X \setminus \overline{U}$ are disjoint and contain A and B, respectively. □

An example of a normal space will be provided by a compact Hausdorff space (see Corollary 1.3.14).

Definition 1.1.20. A space X is said to be *a completely regular* or a Tychonoff space if it is T_1 and for all $x \in X$ and every closed subset $A \subset X$, which does not contain x, there exists a continuous function (see Definition 1.2.1) $f : X \longrightarrow [0,1]$ such that $f(x) = 0$ and $f(y) = 1$ for all $y \in A$.

Definition 1.1.20 may be expressed equivalently as the following.

Proposition 1.1.21. *A space X is completely regular if and only if for all $x \in X$ and every open neighborhood $U \in \mathcal{N}_x$. There exists a continuous function $f : X \longrightarrow [0,1]$ such that $f(x) = 0$ and $f(y) = 1$ for all $y \in X \setminus U$.*

Tychonoff space is also called $T_{3\frac{1}{2}}$ space because it can been inserted between T_4 and T_3 according to the following implications:

$$T_4 \Longrightarrow T_{3\frac{1}{2}} \Longrightarrow T_3 \Longrightarrow T_2 \Longrightarrow T_1. \tag{1.1}$$

Indeed, if X is a Tychonoff space, then for any closed subset A of X and every $x \notin A$, the open sets $U_1 = f^{-1}([0,1/2)) \ni x$ and $U_2 = f^{-1}((1/2,1]) \supset A$ are disjoint, then separate the set A and the point x. Hence, $T_{3\frac{1}{2}}$ implies T_3. By Urysohn's lemma (Theorem A.4.1), T_4 implies $T_{3\frac{1}{2}}$. The letter T comes from the German word "trennung," which means separation.

Furthermore, these spaces relate to compact and paracompact spaces (see Section 1.3) according to the diagram in Figure 1.1, Section 1.3.

1.1.3 Limit and limit point

Definition 1.1.22. We say that a sequence $(x_n)_n$ *converges to a limit* $x \in X$, if for all $U \in \mathcal{N}_x$, there exists $n_0 \in \mathbb{N}$ such that $x_n \in U$ for all $n \geq n_0$.

Definition 1.1.23. We say that $x \in X$ is a *limit point* of a sequence $(x_n)_n$, if for all $U \in \mathcal{N}_x$ and all $n \in \mathbb{N}$, there exists $n_0 \geq n$ such that $x_{n_0} \in U$.

Remark 1.1.24. The limit points of a sequence $(x_n)_n \subset X$ form the set

$$\mathscr{A} = \bigcap_{n \in \mathbb{N}} \overline{\mathscr{A}_n},$$

where $\mathscr{A}_n = \{x_n, x_{n+1}, \ldots\}$.

Remark 1.1.25. Every limit of a sequence is a limit point of the sequence. Generally, a limit point is not unique as the real sequence $x_n = (-1)^n$ shows.

However, the following result is easy to check.

Proposition 1.1.26. *In a Hausdorff space, every sequence has at most one limit.*

Proposition 1.1.27. *A point $x \in X$ is a limit point of a sequence $(x_n)_n$ if and only if x is a limit of some subsequence $(x_{n_k})_k$.*

Proof. Let $x \in X$ be a limit point of a sequence $(x_n)_n$. Then, for every open neighborhood $U \in \mathcal{N}_x$ of x and all $n \in \mathbb{N}$, there exists some positive integer $n_0 \geq n$ such that $x_{n_0} \in U$. Since n is arbitrary, in the definition of a limit point, set $n = n_0$. Then there exists some positive integer $n_1 \geq n_0$ such that $x_{n_1} \in U$. Continuing this way, we construct a sequence $(x_{n_k})_k \subset U$ such that for all $k \geq 0$, $x_{n_k} \in U$ that is, $\lim_{k\to\infty} x_{n_k} = x$. The converse is easy to check. □

We now consider the case of a metric space (E, d).

Definition 1.1.28. A sequence $(x_n)_n \subset E$ is said to be *a Cauchy sequence* if for every positive ε, there exists a positive integer n_0 such that for all $p, q \geq n_0$, and we have $d(x_p, x_q) < \varepsilon$.

Remark 1.1.29. (1) Every Cauchy sequence is bounded.
(2) Every convergent sequence is a Cauchy sequence. But the converse is not true.

Recall that the diameter of a set A is $\operatorname{diam}(A) = \sup\{d(x,y),\ x,y \in A\}$.

Proposition 1.1.30. *Let $(x_n)_n$ be a sequence of a metric space E. Then*

$$(x_n)_n \text{ is a Cauchy sequence} \quad \text{if and only if} \quad \lim_{n\to\infty} \operatorname{diam}(\mathcal{A}_n) = 0.$$

Proof. A sequence $(x_n)_n$ is a Cauchy sequence if and only if for every positive real number ε, there exists a positive integer n_0 such that $d(x_k, x_{k'}) < \varepsilon$ for all k, $k' \geq n_0$. This is equivalent to the following statements written symbolically:

$$\begin{aligned}
& \forall\, \varepsilon > 0,\ \exists\, n_0 \in \mathbb{N}: \quad \sup_{k,k' \geq n_0} d(x_k, x_{k'}) < \varepsilon \\
\Leftrightarrow\quad & \forall\, \varepsilon > 0,\ \exists\, n_0 \in \mathbb{N}: \quad \operatorname{diam}(\mathcal{A}_{n_0}) < \varepsilon \\
\Leftrightarrow\quad & \forall\, \varepsilon > 0,\ \exists\, n_0 \in \mathbb{N},\ \forall\, n \geq n_0: \quad \operatorname{diam}(\mathcal{A}_n) \leq \operatorname{diam}(\mathcal{A}_{n_0}) < \varepsilon \\
\Leftrightarrow\quad & \lim_{n\to\infty} \operatorname{diam}(\mathcal{A}_n) = 0,
\end{aligned}$$

because $\mathcal{A}_n \subseteq \mathcal{A}_{n_0}$ for all $n \geq n_0$. □

Corollary 1.1.31. *If a Cauchy sequence in a metric space (E, d) has a limit point x, then x is the limit of the sequence.*

Proof. By Remark 1.1.24, $x \in \mathcal{A}$, which means $x \in \bigcap_{n\geq 1} \overline{\mathcal{A}}_n$. Let $(x_n)_n$ be a Cauchy sequence. By Proposition 1.1.30, $\lim_{n\to\infty} \operatorname{diam}(\mathcal{A}_n) = 0$, i. e., for all $\varepsilon > 0$, there exists $n_0 \in \mathbb{N}$ such that $\operatorname{diam}(\mathcal{A}_n) < \varepsilon$ for all $n \geq n_0$. Since $\operatorname{diam}(\overline{\mathcal{A}}_n) = \operatorname{diam}(\mathcal{A}_n)$ and x_n, x belong to $\overline{\mathcal{A}}_n$, we write symbolically

$$\forall \varepsilon > 0,\ \exists n_0 \in \mathbb{N}, \quad d(x, x_n) \le \operatorname{diam}(\overline{\mathscr{A}_n}) \le \varepsilon, \quad \forall n \ge n_0.$$

Therefore, $\lim_{n\to\infty} d(x, x_n) = 0$. □

Definition 1.1.32. (1) A metric space (E, d) is said to be *complete* if every Cauchy sequence is convergent.
(2) A Banach space is a complete normed space.

Example 1.1.33. Every discrete space is complete.

1.1.4 Subspace and product topology

Definition 1.1.34. Let X be a set $\mathscr{B}$ a collection of subsets of X. We say that $\mathscr{B}$ is *a basis* if it satisfies the following properties:
(1) For every $x \in X$, there exists $B \in \mathscr{B}$ such that $x \in B$.
(2) For all B_1, B_2 in $\mathscr{B}$ and $x \in B_1 \cap B_2$, there is $B_3 \in \mathscr{B}$ such that $x \in B_3 \subset B_1 \cap B_2$.

Theorem 1.1.35. *Given a basis $\mathscr{B}$, define a collection $\mathscr{T}$ of subsets of X by $U \in \mathscr{T}$ if for every $x \in U$, there exists $B_x \in \mathscr{B}$ such that $x \in B_x \subset U$. Then $\mathscr{T}$ is a topology, called the topology generated by the basis $\mathscr{B}$.*

Proof. (1) Clearly, X and $\emptyset$ belong to $\mathscr{T}$.

(2) Let $(U_\alpha)_{\alpha\in J}$ be an indexed family of elements of $\mathscr{T}$ and $U = \bigcup_{\alpha\in J} U_\alpha$. For every $x \in U$, there exists $\alpha \in J$ such that $x \in U_\alpha$. Since $U_\alpha \in \mathscr{T}$, there exists $B_x \in \mathscr{B}$ such that $x \in B_x \subset U_\alpha \subset U$, proving that $U \in \mathscr{T}$.

(3) Let $(U_i)_{1\le i\le N}$ be a finite family of elements of $\mathscr{T}$ and $U = \bigcap_{1\le i\le N} U_i$. For every $x \in U$, $x \in U_i$ for all $1 \le i \le N$. Hence, there exist $(B_i)_{1\le i\le N} \in \mathscr{B}$ such that $x \in B_i \subset U_i$ for all $i \in [1, N]$. By definition of the basis, there exists some $B_0 \in \mathscr{B}$, such that $x \in B_0 \subset \bigcap_{1\le i\le N} U_i \subset U$, proving that $U \in \mathscr{T}$. □

The converse of Theorem 1.1.35 is easy to check.

Theorem 1.1.36. *Let $\mathscr{B} \subset \mathscr{T}$ be a subfamily of the topology $\mathscr{T}$ such that for all $U \in \mathscr{T}$ and every $x \in U$, there exists $B_x \in \mathscr{B}$ such that $x \in B_x \subset U$. Then $\mathscr{B}$ is a basis for the topology $\mathscr{T}$.*

Let $(X, \mathscr{T}_X)$ be a topological space and $Y \subseteq X$ a nonempty subset. The proof of the following theorem is immediate.

Theorem 1.1.37. *The collection $\mathscr{T}_Y = \{U \cap Y,\ U \in \mathscr{T}_X\}$ is a topology, called* the subspace topology.

Theorem 1.1.38. *Let $\mathscr{B}_X$ be a basis for the topology $\mathscr{T}_X$. Then the collection $\mathscr{B}_Y = \{B \cap Y,\ B \in \mathscr{B}_X\}$ is a basis for the topology $\mathscr{T}_Y$.*

Proof. (1)(a) Let $x \in Y \subseteq X$. Since $\mathscr{B}_X$ is a basis for $\mathscr{T}_X$, there exists $B \in \mathscr{B}_X$ such that $x \in B$. Let $B' = B \cap Y$. Then $B' \in \mathscr{B}_Y$ and $x \in B'$.

(b) Let B_Y^1, B_Y^2 be in $\mathscr{B}_Y$ and $x \in B_Y^1 \cap B_Y^2$. Then there exist B_X^1, B_X^2 in $\mathscr{B}_X$ such that $B_Y^1 = B_X^1 \cap Y$ and $B_Y^2 = B_X^2 \cap Y$. Since $x \in B_X^1 \cap B_X^2$ and $\mathscr{B}_X$ is a basis for $\mathscr{T}_X$, there exists $B_3 \in \mathscr{B}_X$ such that $x \in B_3 \subset B_X^1 \cap B_X^2$. Let $B_3' = B_3 \cap Y$. Then $x \in B_3' \subset B_Y^1 \cap B_Y^2$. Therefore, $\mathscr{B}_Y$ is a basis.

(2) Let $V \in \mathscr{T}_Y$. Then there exists $U \in \mathscr{T}_X$ such that $V = U \cap Y$. Every x in V belongs to U and Y. Since $\mathscr{B}_X$ is a basis for $\mathscr{T}_X$, there exists $B \in \mathscr{B}_X$ such that $x \in B \subset U$. Let $B' = B \cap Y$. Then $B' \in \mathscr{B}_Y$ and $x \in B' \subset U \cap Y = V$. Hence, $\mathscr{B}_Y$ is a basis for the topology $\mathscr{T}_Y$. □

Let $(X, \mathscr{T}_X)$ and $(Y, \mathscr{T}_Y)$ be topological spaces.

Theorem 1.1.39. *The collection $\mathscr{B}_{X\times Y} = \{U \times V,\ U \in \mathscr{T}_X,\ V \in \mathscr{T}_Y\}$ is a basis in $X \times Y$.*

Proof. (1) For all $(x,y) \in X \times Y$, there exist $B_x \in \mathscr{B}_X$ and $B_y \in \mathscr{B}_Y$ such that $(x,y) \in B_x \times B_y \subset X \times Y$.

(2) Let B_1, B_2 be in $\mathscr{B}_{X\times Y}$. Then there exist $U_i \in \mathscr{T}_X$, $V_i \in \mathscr{T}_Y$, $i = 1,2$ such that $B_1 = U_1 \times V_1$ and $B_2 = U_2 \times V_2$. For $(x,y) \in B_1 \cap B_2$, we have $x \in U_1 \cap U_2$ and $y \in V_1 \cap V_2$. By definition of the topologies $\mathscr{T}_X$ and $\mathscr{T}_Y$, there exist $B_x^i \in \mathscr{B}_X$ and $B_y^i \in \mathscr{B}_Y$, $i = 1,2$ such that $x \in B_x^i \subset U_i$ and $y \in B_y^i \subset V_i$ for $i = 1,2$. Let $B_3 = (B_x^1 \times B_y^1) \cap (B_x^2 \times B_y^2) \subset B_1 \cap B_2$. Finally, $(x,y) \in B_3$ completes the proof. □

Definition 1.1.40. The topology generated by the basis $\mathscr{B}_{X\times Y}$ is called the product topology on the Cartesian product $X \times Y$.

The following result shows that fewer elements $\mathscr{T}_X \times \mathscr{T}_Y$ form a basis for the product topology. The proof is skipped.

Theorem 1.1.41. *Let $\mathscr{B}_X$ and $\mathscr{B}_Y$ denote bases for the topologies $\mathscr{T}_X$ and $\mathscr{T}_Y$, respectively. The collection $\mathscr{B} = \{B_1 \times B_2,\ B_1 \in \mathscr{B}_X,\ B_2 \in \mathscr{B}_Y\}$ is a basis for the product topology $\mathscr{T}_{X\times Y}$.*

We end this section with a result on the subspace product topology.

Theorem 1.1.42. *Let $(X, \mathscr{T}_X)$ and $(Y, \mathscr{T}_Y)$ be topological spaces and $A \subset X$, $B \subset Y$ two subspaces. Then the product topology on $A \times B$ is the same as the topology inherited from the product topology $\mathscr{T}_{X\times Y}$.*

Proof. Let $\mathscr{B}_1 = \{U \times V,\ U \in \mathscr{T}_A,\ V \in \mathscr{T}_B\}$ be the basis for the product topology in $X \times Y$ and $\mathscr{B}_2 = \{(U \times V) \cap (A \times B),\ U \in \mathscr{T}_A,\ V \in \mathscr{T}_B\}$ the basis for the subspace topology. To prove that $\mathscr{B}_1 = \mathscr{B}_2$, just observe that $(A \times B) \cap (U \times V) = (A \cap U) \times (B \cap V)$. Also, for every $U \in \mathscr{T}_A$ and $V \in \mathscr{T}_B$, there exist $U_X \in \mathscr{T}_X$ and $V_Y \in \mathscr{T}_Y$ such that $U = U_X \cap A$ and $V = V_Y \cap B$. Hence,

$$U \times V = (U_X \cap A) \times (V_Y \cap B) = (U_X \times V_Y) \cap (A \times B),$$

proving that the topologies are the same. □

Exercises (Section 1.1)

1. Describe a topology on the space X in the following cases:
(1) X consists of the straight lines parallel to the x-axis in Euclidean plane $\mathbb{R}^2$;
(2) X is the collection of all circles of the plane;
(3) X is the collection of all circles of the plane, which have centers on the x-axis.

2. Prove that the collection $\mathcal{T}$ of subsets of the real line $U \subseteq \mathbb{R}$ such that $\mathbb{R} \setminus U$ is either empty or an infinite set is not a topology.

3. Prove that the collection $\mathcal{T}$ of subsets of the positive integers $U \subseteq \{1, 2, \dots\}$ such that $\mathcal{T}$ consists of the empty set, the set of positive integers and all finite subsets, is not a topology.

4. Prove that

$$X \setminus \overline{A} = \overset{\circ}{\widehat{X \setminus A}} \quad \text{and} \quad X \setminus \mathring{A} = \overline{X \setminus A}.$$

5. Prove Proposition 1.1.10.

6. Prove Proposition 1.1.26.

7. Prove that:
(1) $\mathring{A} \cap \partial A = \emptyset$ and $\overline{A} = \mathring{A} \cup \partial A$;
(2) A closed $\iff \partial A \subseteq A \iff \partial A = A \setminus \mathring{A}$;
(3) A open $\iff \partial A = \overline{A} \setminus A$.

8. Let A, B be disjoint subsets of a space X with A open set. Prove that $A \cap \overline{B} = \emptyset$.

9. A point $x \in X$ is an accumulation point of a subset $A \subset X$ if every neighborhood of x meets the set A in a point other that x. The set of all accumulation points is called the derived set and denoted A'. Prove the following:
(1) $\overline{A} = A' \cup A$;
(2) $(A \cup B)' = A' \cup B'$ and $(A \cap B)' = A' \cap B'$;
(3) A closed $\iff A' \subseteq A$;
(4) If X is a Hausdorff space, then A' is a closed set.

10. Let (X, d) be a metric space. Prove that $x \in A'$ if and only if $d(x, A \setminus \{x\}) = 0$.

11. Show that in a Hausdorff space, every finite subset is closed.

12. Let A, B be subsets of a space X. Prove the following:
(1) $\mathring{A} \cup \mathring{B} \subseteq \overset{\circ}{\overbrace{A \cup B}}$ and $\mathring{A} \cap \mathring{B} = \overset{\circ}{\overbrace{A \cap B}}$;
(2) $\overline{A \cup B} = \overline{A} \cup \overline{B}$ and $\overline{A \cap B} \subseteq \overline{A} \cap \overline{B}$;
(3) $\overline{A} \setminus \overline{B} \subseteq \overline{A \setminus B}$.

13. Let $A \subseteq X$ and $B \subseteq Y$ be two subsets.
(1) Prove that:
  (a) $\overline{A \times B} = \overline{A} \times \overline{B}$;
  (b) $\overset{\circ}{\overbrace{A \times B}} = \mathring{A} \times \mathring{B}$.
(2) Deduce that $\partial(A \times B) = (A \times \partial B) \cup (\partial A \times B)$.

14. Let $B = B(x_0, R)$ and $\mathring{B}$ be a closed ball and an open ball of a normed space X. Prove that:
(1) $\mathring{B} = \text{Interior}(B)$;
(2) $B = \overline{\mathring{B}}$.

15. Given an example of a metric space (X, d) where:
(1) the closure of an open ball is not a closed ball;
(2) the radius is not half the diameter.

16. Let (X, d) be a metric space.
(1) Prove that every closed subset $C \subset X$ can be written as a countable intersection of open sets (C is called a G_δ set). More precisely,

$$C = \bigcap_{n=1}^{\infty} \Theta_n, \quad \text{where } \Theta_n = \bigcup_{x \in C} B(x, 1/n).$$

(2) Deduce that every open subset $\mathcal{O}$ can be written as a countable union of closed sets ($\mathcal{O}$ is called an F_σ set).

17. (1) Prove that from any real sequence, one can extract a monotonic subsequence. (*Hint:* consider the set of peaks given by $E = \{m \in \mathbb{N} : \; x_n \leq x_m, \text{ for all } n \geq m\}$.)
(2) Deduce that every bounded real sequence has a convergent subsequence (Bolzano–Weierstrass theorem).

18. (1) Let (X, d) be a metric space, $x \in X$ and $(x_n)_n \subset X$ a sequence such that every subsequence of $(x_n)_n$ has a sub-subsequence that converges to x. Prove that $\lim_{n\to\infty} x_n = x$. (*Hint:* One may consider $\liminf_{n\to\infty} d(x_n, x)$ and $\limsup_{n\to\infty} d(x_n, x)$.)
(2) Let $(x_n)_n$ a bounded real sequence with only one limit point x. Prove that $\lim_{n\to\infty} x_n = x$. (*Hint:* One may use part (1) both with Exercise 17(2).) By a counterexample, show that the result no longer holds if $(x_n)_n$ is unbounded.

1.2 Continuous map

Let $(X, \mathscr{T})$ and $(Y, \mathscr{T}')$ be two spaces and $f : X \longrightarrow Y$ a map.

Definition 1.2.1. (1) The map f is said to be continuous at a point $x \in X$ if for every open neighborhood V of $f(x)$, there is an open neighborhood U of x such that $f(U) \subseteq V$.
(2) The map f is said to be continuous on X if f is continuous at every point $x \in X$.

Next are presented some equivalent formulations of the continuity at a point as well as the continuity over a space. $\overbrace{f^{-1}(V)}^{\circ}$ will denote the interior of the inverse image set $f^{-1}(V)$.

Theorem 1.2.2. *The following statements are equivalent:*
(1) *f is continuous at a point $x \in X$.*
(2) *For every open neighborhood V of $f(x)$, there exists an open set U such that $x \in U \subseteq f^{-1}(V)$.*
(3) *For every set $V \subseteq Y$ such that $f(x) \in \mathring{V}$, we have $x \in \overbrace{f^{-1}(V)}^{\circ}$.*

Proof. The equivalence (1) $\Longleftrightarrow$ (2) follows from the fact that $f(U) \subseteq V$ if and only if $U \subseteq f^{-1}(V)$.

(2) $\Longrightarrow$ (3): Let $V \subseteq Y$ be such that $f(x) \in \mathring{V}$. Since $\mathring{V} \in \mathscr{T}'$, by assumption there exists $U \in \mathscr{T}$ such that $x \in U \subseteq f^{-1}(\mathring{V})$. Moreover, $\mathring{V} \subseteq V$ implies $f^{-1}(\mathring{V}) \subseteq f^{-1}(V)$. Hence, $x \in U \subseteq f^{-1}(V)$, proving that $x \in \overbrace{f^{-1}(V)}^{\circ}$.

(3) $\Longrightarrow$ (2): Let V be an open set containing $f(x)$. By supposition, $x \in \overbrace{f^{-1}(V)}^{\circ}$ and by definition of the interior set, there is some open set $U \subseteq X$ such that $x \in U \subseteq f^{-1}(V)$, proving claim (2). □

Theorem 1.2.3. *The following statements are equivalent:*
(1) *f is continuous on X.*
(2) *For each open set $V \subseteq Y$, $f^{-1}(V)$ is open in X.*
(3) *For each closed set $C \subseteq Y$, $f^{-1}(C)$ is closed in X.*
(4) *For each subset $B \subseteq Y$, $f^{-1}(\mathring{B}) \subseteq \overbrace{f^{-1}(B)}^{\circ}$.*
(5) *For each subset $A \subseteq X$, $f(\overline{A}) \subseteq \overline{f(A)}$.*

Proof. (1) $\Longrightarrow$ (2): Let $V \subseteq Y$ be an open set and let $x \in f^{-1}(V)$, i. e., $f(x) \in V$. By hypothesis: (1) there exists $U \in \mathscr{T}$ such that $x \in U$ and $f(U) \subseteq V$. Thus, $x \in U \subseteq f^{-1}(V)$, proving (2).

(2) $\Longrightarrow$ (1): Let $x \in X$ with $V \in \mathscr{T}'$ and $x \in f^{-1}(V)$. By hypothesis (2), $f^{-1}(V)$ is open in X. Taking $U = f^{-1}(V)$, we get $x \in U$ and $f(U) = V$.

(2) $\Longleftrightarrow$ (3): Note that V open in Y means $C = Y \setminus V$ closed in Y. Since $f^{-1}(C) = X \setminus f^{-1}(V)$, then

$$f^{-1}(V) \text{ open} \quad \Longleftrightarrow \quad f^{-1}(C) \text{ closed},$$

proving the claim.

(2) $\Longrightarrow$ (4): Let $B \subseteq Y$. By hypothesis, $f^{-1}(\mathring{B})$ is open. Since $f^{-1}(\mathring{B}) \subseteq f^{-1}(B)$, then $f^{-1}(\mathring{B}) \subseteq \overbrace{f^{-1}(B)}^{\circ}$ follows from the definition of the interior set $\overbrace{f^{-1}(B)}^{\circ}$.

(4) $\Longrightarrow$ (2): Let $V \subseteq Y$ be an open set. By assumption, $f^{-1}(\mathring{V}) = f^{-1}(V) \subseteq \overbrace{f^{-1}(V)}^{\circ}$. Hence, $f^{-1}(V) = \overbrace{f^{-1}(V)}^{\circ}$ is open.

(3) $\Longrightarrow$ (5): Let $A \subseteq X$ be any subset. Since $\overline{f(A)}$ is closed, by hypothesis $f^{-1}(\overline{f(A)})$ is closed. However,

$$A \subseteq f^{-1}(f(A)) \subseteq f^{-1}(\overline{f(A)}).$$

Since the latter set is closed, we have

$$\overline{A} \subseteq \overline{f^{-1}(f(A))} \subseteq \overline{f^{-1}(\overline{f(A)})} = f^{-1}(\overline{f(A)}).$$

Therefore,

$$f(\overline{A}) \subseteq f(f^{-1}(\overline{f(A)})) \subseteq \overline{f(A)}.$$

(5) $\Longrightarrow$ (3). Let $C \subseteq Y$ be a closed set. By hypothesis, we have

$$f(\overline{f^{-1}(C)}) \subseteq \overline{f(f^{-1}(C))} \subseteq \overline{C} = C.$$

Hence, $\overline{f^{-1}(C)} \subseteq f^{-1}(C)$ and equality holds, proving that $f^{-1}(C)$ is closed. This completes the proof of the theorem. □

A sixth equivalent version of continuity is proposed in Exercise 2 of this section.

Example 1.2.4. Let X, Y be spaces. Then the projection maps given by $\mathrm{pr}_1(x,y) = x$ and $\mathrm{pr}_2(x,y) = y$ are continuous.

Example 1.2.5. Let (E,d) and (E',d') be two metric spaces and $f : E \longrightarrow E'$ a Lipschitz function. Then f is uniformly continuous, hence continuous.

Recall the following.

Definition 1.2.6. Let (E,d) and (E',d') be two metric spaces and $f : E \longrightarrow E'$.

(1) The map f is called *a Lipschitz function* if there exists some positive constant k such that

$$d'(f(x),f(y)) \leq kd(x,y), \quad \text{for all } x,y \in E.$$

(2) The map f said to be *uniformly continuous* if for each positive number ε, there exists a positive number δ such that $d'(f(x),f(y)) \leq \varepsilon$ whenever $(x,y) \in E^2$ satisfy $d(x,y) \leq \delta$.

We end this section with the definition of open and closed map.

Definition 1.2.7. *A closed map* (open map, respectively) is a map $f : X \longrightarrow Y$, which carries closed sets (open sets, respectively) of X into closed sets (open sets, respectively) of Y.

Theorem 1.2.8. *The following statements are equivalent:*

(1) *f is open on X.*

(2) *For each subset $A \subseteq X, f(\mathring{A}) \subseteq \overbrace{f(A)}^{\circ}$.*

Proof. (2) $\Longrightarrow$ (1): Let $U \subseteq X$ be an open set, i. e., $\mathring{U} = U$. Then, by assumption $f(U) = f(\mathring{U}) \subseteq \overbrace{f(U)}^{\circ}$, proving that $f(U) = \overbrace{f(U)}^{\circ}$ is open in Y.

(1) $\Longrightarrow$ (2): Let $A \subseteq X$ and $y \in f(\mathring{A})$. Then $y = f(x)$ for some $x \in \mathring{A}$. Since f is open, so is $f(\mathring{A})$. Since $y \in f(\mathring{A}) \subseteq f(A), y \in \overbrace{f(A)}^{\circ}$, as claimed. □

Exercises (Section 1.2)

1. Let $(X, \mathscr{T})$ be a discrete space and $(Y, \mathscr{T}')$ be any space. Prove that:

(1) any map $f : X \longrightarrow Y$ is continuous,

(2) the result does not hold if $(Y, \mathscr{T}')$ is the discrete space.

2. Prove that $f : X \longrightarrow Y$ is continuous if and only if for every subset $B \subseteq Y$,

$$\overline{f^{-1}(B)} \subseteq f^{-1}(\overline{B}).$$

3. (1) Let X and Y be spaces. Prove that if $f : X \longrightarrow Y$ and $g : Y \longrightarrow Z$ are continuous maps, then the composite $g \circ f : X \longrightarrow Z$ is continuous map.

(2) (a) Prove that if $f, g : X \longrightarrow \mathbb{R}$ are continuous maps, then $\alpha f + \beta g$ is continuous for all scalars α, β.

(b) What about the product $f.g$?

4. Prove that:

(1) the constant map $c : X \longrightarrow y_0 \in Y$ is continuous.

(2) the inclusion map $i : A \subseteq X \hookrightarrow X$ given by $i(x) = x$ for all $x \in A$ is continuous.

5. Let $f : X \longrightarrow Y$ be a continuous map and $A \subseteq X$ dense in X. Prove that $f(A)$ is dense in $f(X)$.

6. Let $f : X \longrightarrow Y$ be a continuous map. Prove that:

(1) $f : X \longrightarrow Y'$ is continuous, whenever $f(X) \subseteq Y' \subseteq Y$ (restricting the range);

(2) $f : X \longrightarrow Y'$ is continuous for all $Y' \supseteq Y$ (expanding the range).

7. Let $(X, \mathscr{T})$ and $(Y, \mathscr{T}')$ be two spaces. A map $f : X \longrightarrow Y$ is said to be *sequentially continuous* at a point $x \in X$ if for all sequence $(x_n)_n \subseteq X$, $\lim_{n\to\infty} f(x_n) = f(x)$ whenever $\lim_{n\to\infty} x_n = x$.

(1) Prove that if f is continuous at x, then it is sequentially continuous at x.

(2) (a) Assume that every point of X has a countable fundamental system of neighborhoods. Prove that if f is sequentially continuous at x, then it is continuous at x.

(b) What about the case of a metric space (E, d)?

8. Let $(X, \mathscr{T})$, $(Y, \mathscr{T}')$ be two spaces and $f : X \longrightarrow Y$ a map. Prove that f is closed if and only if for every subset $A \subseteq X$, $\overline{f(A)} \subseteq f(\overline{A})$.

9. Let $(X, \mathscr{T})$, $(Y, \mathscr{T}')$, $(Z, \mathscr{T}'')$ be spaces and $f : X \longrightarrow Y$, $g : Y \longrightarrow Z$ two maps. Prove that if f and g are open (closed, respectively), then the composite map $g \circ f$ is open (closed, respectively).

10. Let $(X, \mathscr{T})$, $(Y, \mathscr{T}')$, $(Z, \mathscr{T}'')$ be spaces and $f : X \longrightarrow Y$, $g : Y \longrightarrow Z$ two maps. Prove that if $g \circ f$ is open (closed, respectively) and g is a continuous injective map, then f is open (closed, respectively).

11. Let $(X, \mathscr{T})$, $(Y, \mathscr{T}')$, $(Z, \mathscr{T}'')$ be spaces and $f : X \longrightarrow Y$, $g : Y \longrightarrow Z$ two maps. Prove that if $g \circ f$ is open (closed, respectively) and f is a continuous surjective map, then g is open (closed, respectively).

12. Let $f : X \longrightarrow Y$ and $g : X' \longrightarrow Y'$. Define the product map $h : X \times X' \longrightarrow Y \times Y'$ by $h(x,y) = (f(x), g(x))$.

(1) Prove that if f and g are continuous, then h is continuous.

(2) Prove that if f and g are open, then h is open.

13. (1) Let $X = A \cup B$, where the subsets A and B satisfy $A \setminus B \subseteq \mathring{A}$ and $B \setminus A \subseteq \mathring{B}$. Let $f : A \longrightarrow Y$ and $g : B \longrightarrow Y$ be continuous maps such that $f(x) = g(x)$ for all $x \in A \cap B$. Prove that the extended map h defined by

$$h(x) = \begin{cases} f(x), & x \in A, \\ g(x), & x \in B \end{cases}$$

is continuous on X.

(2) By considering the case when A and B are open subsets, deduce the pasting Lemma A.2.1. (*Hint:* see [12, 2.5.11].)

1.3 Compact set and compact map

1.3.1 Compact set

Let X be a space and $Y \subseteq X$.

Definition 1.3.1. (1) An open covering of Y is a collection $(U_i)_{i\in J}$ of open subsets of X such that $Y \subseteq \bigcup_{i\in J} U_i$.
(2) The space Y is called *compact* if every open covering $(U_i)_{i\in J}$ of Y has a finite subcovering, i. e., there are $U_1, U_2, \ldots, U_n$ such that $Y \subseteq \bigcup_{i=1}^{n} U_i$, for some positive integer n.

Taking the complements, we obtain an equivalent definition.

Definition 1.3.2. A space is compact if every family of closed sets $(F_i)_{i\in I}$ with empty intersection has a finite subfamily with empty intersection. Symbolically,

$$\bigcap_{i\in I} F_i = \emptyset \quad \Rightarrow \quad \exists n \in \mathbb{N}\colon \quad \bigcap_{i=1}^{n} F_i = \emptyset.$$

An immediate consequence is the following.

Lemma 1.3.3 (Closed nesting subsets property). *Let X be a compact space and $(C_n)_{n\in\mathbb{N}}$ a sequence of nonempty closed nested subsets, i. e., $C_{n+1} \subset C_n$ for all n. Then $\bigcap_{n\in\mathbb{N}} C_n \neq \emptyset$.*

Proof. On the contrary, assume that $\bigcap_{n=1}^{\infty} C_n = \emptyset$. By compactness of X, there is some $p \in \mathbb{N}$ such that $\bigcap_{n=1}^{p} C_n = \emptyset$, that is to say $C_p = \emptyset$, a contradiction. □

Example 1.3.4. The set of reals $\mathbb{R}$ is not compact since the covering $(-n, n)_{n\in\mathbb{N}}$ has no finite subcovering for otherwise $\mathbb{R}$ would be bounded.

Proposition 1.3.5. *The interval $[a, b]$ is compact in the set $\mathbb{R}$ of real numbers.*

Proof. Let $\mathscr{S}$ be an open covering of $[a, b]$. Denote $E = \{x \in [a, b] : \exists S \in \mathscr{S} : [a, x] \subset S\}$. We claim that $E = [a, b]$.

(1) The set E is nonempty for $a \in E$ and $[a, a] = \{a\}$. Since $a \in [a, b] \subset \bigcup_{S\in\mathscr{S}} S$, there exists some $S \in \mathscr{S}$ such that $[a, a] \in S$.

(2) The set E is bounded for $E \subseteq [a, b]$. Let $c = \sup E$. We claim that $c \in E$. Since $c \in \overline{E} \subseteq \overline{[a, b]} = [a, b]$, $c \in [a, b] \subseteq \bigcup_{S\in\mathscr{S}} S$. Hence, there exists some $S_c \in \mathscr{S}$ such that $c \in S_c$. Since S_c is open in $\mathbb{R}$, there exists $\varepsilon > 0$ such that $(c - \varepsilon, c + \varepsilon) \subseteq S_c$. Since c is the supremum of E, there exists $x \in E$ such that $c-\varepsilon < x \leq c+\varepsilon$. Thus, $x \in (c-\varepsilon, c+\varepsilon) \subseteq S_c$. But $x \in E$ means that $[a, x] \subseteq S'_c$ for some $S'_c \in \mathscr{S}$. Therefore, $[a, c] = [a, x] \cup [x, c] \subseteq S'_c \cup S_c$, and thus $c \in E$.

(3) We have $\sup E = b$. If $c < b$, then there would exist some $d \in (c, c + \varepsilon) \cap [a, b]$ such that

$$[a, d] = [a, c] \cup (c, d] \subseteq [a, c] \cup (c, c + \varepsilon)$$

$$\subseteq [a,c] \cup S_c$$
$$\subseteq (S'_c \cup S_c) \cup S_c.$$

Then $d \in E$ with $c < d$, which contradicts $c = \sup E$. Finally, $c = b \in E$ and $E = [a,b] = \mathscr{S}$. So, $[a,b]$ has a finite subcovering of $\mathscr{S}$. □

Proposition 1.3.6. (1) *If X is compact and Y is closed in X, then Y is compact.*
(2) *If $Y \subseteq X$ is compact Hausdorff, then Y is closed in X.*

Proof. (1) Let $Y \subseteq \bigcup_{i\in J} U_i$ be an open covering of Y. Then $X \subseteq (X\setminus Y) \bigcup_{i\in J} U_i$, where $X\setminus Y$ is open. Since X is compact, there exists a finite subcovering $X \subseteq (X \setminus Y) \bigcup_{k=1}^{N} U_k$, and thus $Y \subseteq \bigcup_{k=1}^{N} U_k$, which is a finite open subcover. Then Y is compact.

(2) Let $Z = X \setminus Y$ and $x \in Z$. Since Y is Hausdorff, for every $y \in Y$, there exist open sets $U_y \in \mathscr{N}(y)$ and $V_y(x) \in \mathscr{N}(x)$ such that $U_y \cap V_y = \emptyset$. The family $(U_y)_{y\in F}$ is a covering of Y. Since Y is compact, it has a finite subcovering $Y \subseteq \bigcup_{i=1}^{n} U_{y_i}$. As a consequence, $x \in \bigcap_{i=1}^{n} V_{y_i} \subseteq X \setminus Y$, proving that the complement $X \setminus Y$ is an open set. □

Remark 1.3.7. In Proposition 1.3.6(2), the Hausdorff property is essential. To see why, consider $\mathbb{R}$ with the finite complement topology. Then any interval $[a,b]$ is compact (see Exercise 6) but $[a,b]$ is not closed.

Theorem 1.3.8. *Let $f : X \longrightarrow Y$ be a continuous map. If X is compact, then $f(X)$ is compact.*

Proof. Let $(V_i)_{i\in J}$ be an open covering of $f(X)$: $f(X) \subseteq \bigcup_{i\in J} V_i$. Since f is continuous, $f^{-1}(\bigcup_{i\in J} V_i)$ is an open set. Moreover, $X \subseteq f^{-1}(\bigcup_{i\in J} V_i) = \bigcup_{i\in J} f^{-1}(V_i)$. Since X is compact, $X \subseteq \bigcup_{i=1}^{n} f^{-1}(V_i)$ for some positive integer n. Hence,

$$f(X) \subseteq f\left(\bigcup_{i=1}^{n} f^{-1}(V_i)\right) = f\left(f^{-1}\left(\bigcup_{i=1}^{n} V_i\right)\right) \subseteq \bigcup_{i=1}^{n} V_i,$$

proving our claim. □

The next result follows from Proposition 1.3.6 and Theorem 1.3.8.

Corollary 1.3.9. *Let X be compact and Y Hausdorff. Then every continuous function $f : X \longrightarrow Y$ is a closed map.*

Theorem 1.3.10. *A finite product of compact spaces is a compact space.*

For the proof, we refer to [76, Theorem 26.7].

Example 1.3.11. Since $[a,b]$ is compact (Proposition 1.3.5), so is the n-cube $\prod_{k=1}^{n}[a_k, b_k]$ of $\mathbb{R}^n$.

Before leaving this subsection, let us go back to the separation theorems (Subsection 1.1.2) and connect them to compactness. A first result is given by the following.

Proposition 1.3.12. *Let X be a Hausdorff space, K a compact subset and $x \in X \setminus K$. Then K and $\{x\}$ can be separated by disjoint open sets.*

Proof. By the Hausdorff property, for every element $y \in K$, there exist disjoint open sets $U_y \in \mathscr{N}_x$ and $V_y \in \mathscr{N}_y$. Since $K \subset \bigcup_{y \in K} V_y$ and K is compact, there exists a finite open subcovering $K \subset V = \bigcup_{1 \le i \le n} V_{y_i}$. Let $U = \bigcap_{1 \le i \le n} U_{y_i}$. Then $U \in \mathscr{N}_x$, $K \subset V$, V is open and $U \cap V = \emptyset$. □

Corollary 1.3.13. *Let X be a Hausdorff space and K_1, K_2 two compact subsets. Then K_1 and K_2 can be separated.*

Proof. By Proposition 1.3.12, for each $x \in K_1$, there exist open sets $U_x \in \mathscr{N}_x$ and $V_x \supseteq K_2$ such that $U_x \cap V_x = \emptyset$. Since $K_1 \subset \bigcup_{x \in K_1} U_x$ and K_1 is compact, there exists a finite open subcovering $K_1 \subset U = \bigcup_{1 \le i \le n} U_{x_i}$. Let $V = \bigcap_{1 \le i \le n} V_{x_i}$. Then V is open, $K_2 \subset V$ and $U \cap V = \emptyset$. □

Corollary 1.3.14. *A Hausdorff compact space X is normal.*

Let us recall the paracompactness of a space, a concept weaker than the Hausdorff compactness. Paracompact spaces were introduced by Jean Dieudonné in 1944.

Definition 1.3.15. (1) A family of sets is said to be a neighborhood-finite refinement if each point of the space has a neighborhood. which intersects the family for at most finitely many elements.
(2) A space is paracompact if it is T_2 and every open covering has a neighborhood-finite refinement.

Remark 1.3.16. (1) By Stone's theorem [96], every metric space is paracompact.
(2) By [25, Theorem 2.2, Chapter VIII], a paracompact space is normal, a result stronger that Corollary 1.3.14.
(3) By [25, Theorem 6.4, Chapter XI], every locally compact space is a Tychonoff space.

Thus, the diagram in Figure 1.1 completes the implications set up in (1.1).

$$
\begin{array}{ccccccccccc}
\text{metric space} & & & & & & & & & & \\
\Downarrow & & & & & & & & & & \\
\text{paracompact} & \Longrightarrow & T_4 & \Longrightarrow & T_{3\frac{1}{2}} & \Longrightarrow & T_3 & \Longrightarrow & T_2 & \Longrightarrow & T_1 \\
\Uparrow & & & & \Uparrow & & & & & & \\
\text{compact Hausdorff} & & \Longrightarrow & & \text{loc. compact Hausdorff} & & & & & &
\end{array}
$$

Figure 1.1: Connection between some topological spaces.

1.3.2 Compactification

Definition 1.3.17. (1) Let X be a compact space and $Y \subset X$ such that $\overline{Y} = X$. Then Y is said to be relatively compact.
(2) If further $X \setminus Y$ consists of a single point, then X is called the *one-point compactification* of Y (or Alexandroff compactification). We denote this $X = Y^\infty$.

We prove that a necessary and sufficient condition for a space Y to have a one-point compactification is that Y is Hausdorff locally compact but not compact (see, e. g., [76, Theorem 29.1]. Recall the following.

Definition 1.3.18. A space is called *locally compact* if every point has a compact neighborhood.

Example 1.3.19. (1) A compact space is trivially locally compact.
(2) Any discrete space is locally compact.
(3) The real line is locally compact but not compact (see Example 1.3.4).
(4) The set of rational numbers is not locally compact.

Proposition 1.3.20. (1) *Let Y be a Hausdorff locally compact but not compact space. Then Y has Hausdorff compact a one-point compactification X.*
(2) *Conversely, if X is a Hausdorff compact one-point compactification of a subspace Y, then Y is a Hausdorff locally compact but not a compact space.*

Proof. (1) Let Y be a Hausdorff locally compact, which is not compact and let $X = Y \cup \{\infty\}$, where $\infty \notin Y$. Define a topology on X by stating that $U \subseteq X$ is open if and only if either U is an open subset of Y or $U = X \setminus K$, where K is compact in Y. Clearly, this defines a topology on X. We can describe the space X.

(a) The space X is Hausdorff. Let x and y be distinct elements in X. If both elements are in Y, then they can be separated by open neighborhoods because Y is Hausdorff. If $x \in Y$ and $y = \infty$, then since Y is locally compact, x has a compact neighborhood. Then there exist an open set U and a compact subset $K \subset Y$ such that $x \in U \subset K$. Hence, $V = X \setminus K$ is open in X and $y \in V$ by definition of ∞. Finally, $U \cap V = \emptyset$.

(b) The space X is compact. Let $(U_i)_{i \in J}$ be an open covering of X. Then $(Y \cap U_i)_{i \in J}$ is an open covering of Y. Since $\infty \in U_{i_0}$ for some $i_0 \in J$, $U_{i_0} \nsubseteq Y$ and so $X \setminus U_{i_0}$ is compact. Moreover, $X \setminus U_{i_0} \subseteq \bigcup (Y \cap U_i)_{i \in J, i \neq i_0}$. Hence, there exists a finite subcovering of the compact set $X \setminus U_{i_0}$. By adding the open set U_{i_0}, we get a finite covering of the space X.

(c) We have $\overline{Y} = X$. It is sufficient to check that $\infty \in \overline{Y}$. Indeed, let $U \in \mathcal{N}_\infty$. Then $K = X \setminus U$ is compact in Y. Since Y is not compact, $K \neq Y$, which implies that $U \cap Y \neq \emptyset$, as claimed.

(2) Conversely, let X be Hausdorff compact. If X is a one-point compactification of a subspace $Y \subset X$, then Y is Hausdorff because X is. To show that Y is locally compact, let $x \in Y$, $U \in \mathcal{N}_x$ and $V \in \mathcal{N}_\infty$ such that $U \cap V = \emptyset$. Then $K = X \setminus V$ is compact

in X. Furthermore, for each $x \in K$, $x \notin V$ and so $x \neq \infty$, which implies $x \in Y$. Hence, $x \in U \subset K \subset Y$. Finally, Y is not compact for otherwise $Y = \overline{Y} = X$, a contradiction. □

Definition 1.3.21. More generally, if Y is a noncompact space, *a compactification* of Y is a compact space X such that Y is homeomorphic to a dense subset of X (see Definition 1.5).

Remark 1.3.22. A sufficient and necessary condition for the subspace Y to have a compactification is that Y is completely regular (see, e. g., [27, Theorem 3.5.1]).

Example 1.3.23. (1) The interval $X = [0,1]$ is a one-point compactification of $Y = [0,1)$.
(2) We have $X = \mathbb{R}^n \cup \{\infty\} \cong S^n$ and $Y = \mathbb{R}^n$ (Proposition 1.5.28).

Remark 1.3.24. (1) Unlike Theorem 1.3.8 and Theorem 1.3.29, the continuous image of a locally compact space is not necessarily locally compact. A simple counterexample is provided by the identity map from a discrete space to the same space with a topology for which it is not locally compact.
(2) However, if $f : X \longrightarrow Y$ is continuous open and X is locally compact, then the image $f(X)$ is locally compact. Indeed, let $y \in f(X)$ and $y = f(x)$, $x \in X$. Since X is locally compact, there are U open and K compact such that $x \in U \subset K$. Then $y \in f(U) \subset f(K)$, where $f(K)$ is compact by Theorem 1.3.8 and $f(U)$ is open since f is an open map.

Notice that locally compact normed spaces have some nice properties.

Proposition 1.3.25. *Let X be a normed space. The following statements are equivalent:*
(1) *The closed unit ball is compact.*
(2) *Every closed ball is compact.*
(3) *The space X has finite dimension.*
(4) *The space X is locally compact.*

Proof. Any closed ball is the image of the closed unit ball under shifting and dilation, which are continuous maps, whereupon follows the equivalence (1) $\Longleftrightarrow$ (2). The equivalence (1) $\Longleftrightarrow$ (3) is Riesz theorem (see, e. g., [84, p. 61]).

(3) $\Longrightarrow$ (4). If X has finite dimension n, then X is homeomorphic to Euclidean space $\mathbb{R}^n$, where every point x has the closed ball $B(x,r)$ as a compact neighborhood for every $r > 0$.

(4) $\Longrightarrow$ (2). If X is locally compact, then every $x \in X$ has a compact neighborhood K, and thus there exists a compact closed ball $B(x, r_0) \subset K$. □

1.3.3 Sequential compactness in metric spaces

Definition 1.3.26. A metric space X is called *sequentially compact* if every sequence in X has a convergent subsequence in X.

Let (X, d) be a metric space and $A \subset X$ a bounded subset.

Definition 1.3.27. (1) Let $\varepsilon > 0$. We say that A has an ε-net if there exists $n_0 = 1, 2, \ldots$ such that $A \subseteq \bigcup_{i=1}^{n_0} \mathring{B}(x_i, r_i)$, where $\{x_1, x_2, \ldots, x_{n_0}\} \subset X$ and $0 < r_i \leq \varepsilon$ for all $i = 1, \ldots, n_0$.

(2) The subset A is totally bounded if A has an ε-net for every positive real number ε.

Theorem 1.3.28. *Every compact space X is sequentially compact. The converse is true if X is a metric space.*

Proof. (1) Let X be a compact space. By applying Lemma 1.3.3 to the sequence of sets $\overline{\mathscr{A}_n} = \overline{\{x_n, x_{n+1}, \ldots\}}$, we obtain that $\mathscr{A} = \bigcap_{n \in \mathbb{N}} \overline{\mathscr{A}_n} \neq \emptyset$, that is to say $(x_n)_n$ has a limit point. Proposition 1.1.27 completes the proof.

(2) Conversely, let (X, d) be a metric space and $(U_i)_{i \in I}$ an open covering of X. We proceed in three steps.

(a) We claim that there exists some $\varepsilon > 0$ such that for all $x \in X$ and all $r \in (0, \varepsilon]$, there exists $i_0 \in I$ such that

$$B(x, r) \subseteq U_{i_0}. \tag{1.2}$$

Arguing by contradiction, assume that for all positive ε, there exist $x \in X$ and $r \in (0, \varepsilon]$ such that for all $i \in I$, $B(x, r) \not\subset U_i$. In particular, for every $n \in \mathbb{N}$, there exist $x_n \in X$ and $r_n \in (0, 1/n]$ such that

$$B(x_n, r_n) \not\subset U_i, \quad \text{for all } i \in I.$$

Let x be a limit point of the built-up sequence $(x_n)_{n \in \mathbb{N}}$. So, there exists $i_0 \in I$ such that $x \in U_{i_0}$. Since U_{i_0} is open, there exists $r_0 > 0$ such that $B(x, r_0) \subset U_{i_0}$. Since x is a limit point of (x_n), there exists $n_0 \in \mathbb{N}$ such that $x_{n_0} \in B(x, r_0/2)$. Hence, $B(x_0, r_0/2) \subset B(x, r_0)$. Indeed, if $d(y, x_{n_0}) < r_0/2$, then

$$d(y, x) \leq d(y, x_{n_0}) + d(x_{n_0}, x) < r_0/2 + r_0/2 = r_0.$$

We deduce that $B(x_{n_0}, r_0/2) \subset A_{i_0}$, which is a contradiction.

(b) Since the metric space X has a limit point, we claim that X is totally bounded, that is to say for every $\varepsilon > 0$, X can be covered by a finite number of balls with diameters less than ε, i. e., symbolically,

$$\exists N_\varepsilon \in \mathbb{N}, \forall i \in \{1, \ldots, N_\varepsilon\}, \exists x_i^\varepsilon \in X, \exists r_i^\varepsilon \in (0, \varepsilon], \quad X \subseteq \bigcup_{i=1}^{N_\varepsilon} B(x_i^\varepsilon, 2r_i^\varepsilon). \tag{1.3}$$

Let $\varepsilon_0 > 0$ and $x_0 \in X$. By contradiction, if X is not totally bounded, then we could pick some $x_1 \in X \setminus B(x_0, \varepsilon)$, then an element $x_2 \in X \setminus [B(x_1, \varepsilon_0) \cup B(x_0, \varepsilon_0)]$, and so on. Recurrently, we construct a sequence $(x_n)_n \subset X$ such that

$$x_{n+1} \notin \bigcup_{k=0}^{n} B(x_k, \varepsilon_0).$$

Hence, $d(x_p, x_q) \geq \varepsilon_0$ for all $p \neq q$. We deduce that the sequence (x_n) is not a Cauchy sequence and has no Cauchy subsequence. As a consequence (x_n) has no limit point, a contradiction.

(c) From (1.3), the space X can be covered by a finite number of balls $(1 \leq i \leq N)$ $B(x_i, r_i)$ such that $x_i \in X$ and $0 < r_i \leq \varepsilon$ for all $1 \leq i \leq N$. This with (1.2) yields

$$X \subseteq \bigcup_{i=1}^{N} B(x_i, r_i) \subseteq \bigcup_{i=1}^{N} U_i,$$

proving the compactness of E. □

As for Theorem 1.3.8, we have the following.

Theorem 1.3.29. *Let $f : X \longrightarrow X'$ be a continuous map. If X is sequentially compact, then $f(X)$ is sequentially compact.*

Proof. Given a sequence $(y_n)_n \subset f(X)$, let $(x_n)_n \subset X$ be such that $f(x_n) = y_n$. Since X is sequentially compact, there exists a subsequence $(x_{n_k})_k \subset X$, which converges to some limit $x \in X$. Since f is continuous, f is sequentially continuous (Exercise 7, Section 1.2). Hence, $\lim_{k\to\infty} y_{n_k} = f(x)$. □

Proposition 1.3.30. *Let X be a space and $(x_n)_n \subset X$ a sequence converging to some limit $x \in X$. Then the subspace $Y = (x_n)_n \cup \{x\}$ is compact, that is to say Y is one-point compactification of the sequence $(x_n)_n$.*

Proof. Let $(U_i)_{i\in J}$ be an open covering of Y. Then there exists $i_0 \in J$ such that $x \in U_{i_0}$. Since $\lim_{n\to\infty} x_n = x$, there is a positive integer n_0 such that $x_n \in U_{i_0}$ for all $n \geq n_0$. As a consequence, there are only finite number of terms of the sequence, which lie outside U_{i_0}, say $\{x_{n_1}, x_{n_2}, \ldots, x_{n_p}\} \subset X \setminus U_{i_0}$ $(p < n_0)$. Therefore, $Y = \bigsqcup_{k=1}^{p} U_{i_k} \cup U_{i_0}$, which is a finite subcovering of $(U_i)_{i\in J}$. □

Proposition 1.3.31. *Let X be a compact space. A sequence is convergent if and only if it has a unique limit point.*

Proof. (1) The unique limit point of a convergent sequence is its limit.

(2) According to Remark 1.1.24, let $\mathscr{A}_n = \{x_n, x_{n+1}, \ldots\}$ and $\mathscr{A} = \bigcap_{n\in\mathbb{N}} \overline{\mathscr{A}_n}$, the set of limit points. Assume that $(x_n)_n$ has only one limit point $x \in E$, i. e., $\mathscr{A} = \{x\}$. Let $U \in \mathscr{N}_x$ be an open neighborhood of x. Then $(\mathscr{A}_n \cap (X \setminus U))_n$ is a decreasing sequence of closed sets with empty intersection. By Definition 1.3.2, there exists $n_0 \in \mathbb{N}$ such that $\mathscr{A}_{n_0} \cap (X \setminus U) = \emptyset$. Hence, $\mathscr{A}_n \cap (X \setminus U) = \emptyset$ for all $n \geq n_0$, i. e., $\mathscr{A}_n \subset U$, proving that $(x_n)_n$ converges to the limit x. □

Theorem 1.3.32. *Let $X \subset \mathbb{R}^n$. Then X is compact if and only if X is closed and bounded.*

Proof. (1) If X is compact, it is closed by Theorem 1.3.6. The union of open balls $\mathring{B}(0,n)$, $n = 1, 2, \ldots$ is an open covering of the compact space X. Hence, $X \subseteq \bigcup_{n=1}^{p} \mathring{B}(0,n)$ for some positive integer p. Hence, X is bounded.

(2) Conversely, let X be a bounded closed subset of $\mathbb{R}^n$. Then there is some $M > 0$ such that $\|x\| \leq M$ for all $x \in X$, i. e., $X \subseteq [-M, M]^n$, which is compact by Example 1.3.11. Since X is closed, it is compact. □

Example 1.3.33. The closed ball $B^n = \{x \in \mathbb{R}^n : \|x\| \leq 1\}$ and its boundary, the sphere $S^{n-1} = \{x \in \mathbb{R}^n : \|x\| = 1\}$ are compact sets.

Proposition 1.3.34. *If a metric space E is compact, then:*
(1) *E is complete;*
(2) *E is totally bounded.*

Proof. (1) Let $(x_n)_n$ be a Cauchy sequence in E. By Theorem 1.3.28, E is sequentially compact. Then Corollary 1.1.31 implies that $(x_n)_n$ has a limit point x. Finally, Corollary 1.1.31 implies that $\lim_{n\to\infty} x_n = x$, proving the completeness of X.

(2) By contradiction, assume that X is not totally bounded. Then there exists $\varepsilon > 0$ such that for every finite subset $S \subset X$, $X \neq \bigcup_{s\in S} B(s,\varepsilon)$. Let $x_0 \in X$, $x_1 \in X \setminus B(x_0,\varepsilon), \ldots, x_n \in X \setminus \bigcup_{i=0}^{n-1} B(x_i,\varepsilon)$. Then the sequence $(x_n)_n$ has no convergent subsequence because $d(x_j, x_k) > \varepsilon$ for all $j \neq k$. This contradicts the compactness of E. □

Remark 1.3.35. Conversely, we can prove that a complete totally bounded metric space is compact, i. e., in metric spaces:

$$\text{compact} \quad \Longleftrightarrow \quad \text{complete and totally bounded.}$$

Actually, in a totally bounded metric space, each sequence has a Cauchy subsequence. We omit the details.

1.3.4 Compact map

Let X, Y be two normed spaces and $\Omega \subseteq X$ an open subset.

Definition 1.3.36. (1) A map $f : \Omega \longrightarrow Y$ is said to be *compact* if the range $f(\Omega)$ is relatively compact in Y (i. e., $\overline{f(\Omega)}$ is compact in Y).
(2) The map f is *completely continuous* if it is continuous and the image of every bounded set is relatively compact.

Remark 1.3.37. (1) Every compact continuous map is completely continuous. The converse holds if Ω is bounded.
(2) A compact map is not necessarily continuous as the function in Figure 1.2 shows, where $f([0,2]) \subseteq [0,1]$.

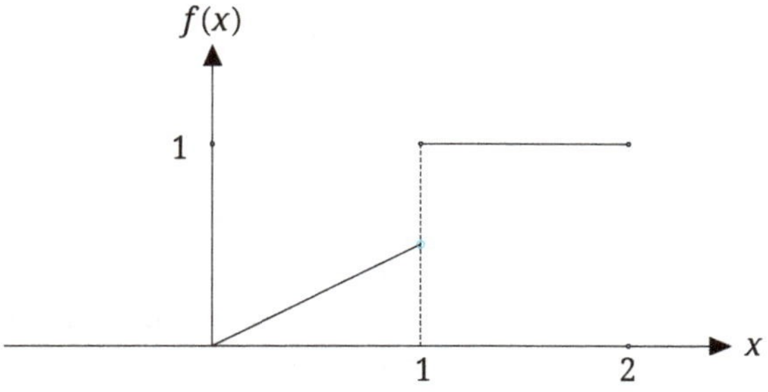

Figure 1.2: The map f is compact but not continuous.

(3) However, a compact linear map is continuous and the converse holds if f has a finite rank, i. e., the dimension of the range space is finite.

(4) If X has a finite dimension, then every endomorphism on X is compact and continuous.

Proposition 1.3.38. *A map $f : X \longrightarrow Y$ is compact if and only if every sequence $(x_n)_n$ in X has a subsequence $(x_{n_k})_k$ such that $(f(x_{n_k}))_k$ converges in Y.*

Proof. (1) Assume that f is compact and let $(x_n)_n \subset X$. Since $(f(x_n))_n$ is relatively compact, by Theorem 1.3.28, there exists a subsequence $(x_{n_k})_k$ such that $(f(x_{n_k}))_k$ converges in Y.

(2) Conversely, let $(y_n)_n \subset f(X)$ and let $(x_n)_n \subset X$ be such that $y_n = f(x_n)$ for all $n \in \mathbb{N}$. By supposition, there is a subsequence $(x_{n_k})_k$ such that $\lim_{k\to\infty} f(x_{n_k}) = y_0 \in Y$. Therefore, $f(X)$ is sequentially compact, hence compact by Theorem 1.3.28. □

Exercises (Section 1.3)

1. Prove that a space X is Hausdorff if and only if the diagonal $\Delta = \{(x,x) \in X \times X\}$ is closed in the product topology.

2. Let $f, g : X \longrightarrow Y$ be two continuous functions and Y a Hausdorff space. Prove that:

(1) the set $\{x \in X : f(x) = g(x)\}$ is closed in X.

(2) Suppose that $f(x) = g(x)$ for all $x \in A$, where A is dense in X. Prove that f and g are identical.

3. (1) Let $f : X \longrightarrow Y$ be a continuous map and Y a Hausdorff space. Prove that the set $\{(x,x') \in X \times X : f(x) = f(x')\}$ is closed in $X \times X$.

(2) Let $f : X \longrightarrow Y$ be an open surjective map such that the set $\{(x,x') \in X \times X : f(x) = f(x')\}$ is closed. Prove that Y is a Hausdorff space.

4. Let $f : X \longrightarrow Y$ be a continuous injection and Y a Hausdorff space. Prove that X is Hausdorff.

5. Let (E,d) be a metric space and $K \subseteq E$ a compact subset. For some constant $r > 0$, define the sets

$$C = \bigcup_{x \in K} B(x,r) \quad \text{and} \quad U = \bigcap_{x \in K} \mathring{B}(x,r).$$

Prove that C is closed and U is open in E. (*Hint:* Show that the map $x \longmapsto \operatorname{dist}(x,K)$ is a Lipschitz function (Definition 1.2.6) and $C = \{y \in E : \operatorname{dist}(y,K) \leq r\}$, where $\operatorname{dist}(y,K) = \inf\{d(y,x) : x \in K\}$ is the distance from y to K, or prove directly that $\overline{C} \subset C$ using the sequential compactness of K.)

6. Given an infinite space X, consider the family of sets defined by

$$\mathscr{T} = \{U \subseteq X : U = \emptyset \text{ or } X \setminus U \text{ is finite}\}.$$

(1) Prove that $\mathscr{T}$ is a topology, called the finite complement topology.
(2) Prove that $(X, \mathscr{T})$ is compact but not Hausdorff.

7. Let X be a space, $x_0 \in X$ and Y a compact space. Suppose that there exists an open set $U \subset X \times Y$ such that $\{x_0\} \times Y \subset U$. Prove *the tube lemma*: There exists an open set $V \subset X$ such that $\{x_0\} \times Y \subset V \times Y \subset U$. (*Hint:* See [76, Lemma 26.8].)

8. Let $\mathrm{pr}_1 : X \times Y \longrightarrow X$ be the projection map given by $\mathrm{pr}_1(x,y) = x$.
(1) Prove that if Y is compact, then pr_1 is a closed map. (*Hint:* See [27, Example 3.1.16].)
(2) Let $X = Y = \mathbb{R}$ be endowed with the standard topology. Prove that pr_1 is an open map but not a closed map.

9. Let $f : X \longrightarrow Y$ be a map, where Y is compact Hausdorff.
(1) Prove that f is continuous if and only if the graph $G_f = \{(x, f(x)),\ x \in X\}$ of f is closed in $X \times Y$.
(2) What about the equivalence when Y is not compact? Discuss the case when $X = \mathbb{R}$ is endowed with the usual topology, $F = \mathbb{R}$ has the discrete topology and f is the identity map.

10. Let $X, Y \subset \mathbb{R}^n$ be two subspaces such that X is compact and Y is closed. Prove that X and Y are disjoint if and only if $\inf\{\|x - y\| : x \in X, y \in Y\} > 0$.

11. Let X and Y be Hausdorff spaces and $f : X \longrightarrow Y$ a continuous map. We say that f is *a proper map* if $f^{-1}(K)$ is a compact subset of X for every compact subset $K \subset Y$.
(1) Prove that a proper map is a closed map.
(2) Assume that f is continuous, surjective, closed and for all $y \in Y, f^{-1}(\{y\})$ is compact.
   (a) Prove that f is a proper map.
   (b) Deduce that X is compact (locally compact, respectively) whenever Y is compact (locally compact, respectively).

(3) (a) If X and Y are finite-dimensional normed spaces, prove that

$$f \text{ is a proper map} \iff$$
$$\text{for every bounded subset } B \subset Y, f^{-1}(B) \text{ is bounded in } X.$$

(b) Let $f : \mathbb{R}^n \longrightarrow \mathbb{R}^n$ be a continuous and coercive map, i. e., $\lim_{\|x\|\to\infty} \|f(x) = \infty$. Prove that f is a proper map.

(4) Let $f : \mathbb{R}^n \longrightarrow \mathbb{R}^n$ be a continuous map. Prove that

$$f \text{ is a homeomorphism} \iff f \text{ is bijective and proper.}$$

12. Let $f : X \longrightarrow Y$ be a proper map. Prove that X is locally compact whenever Y is.

1.4 Connectedness and path-connectedness

1.4.1 Connected sets

Definition 1.4.1. A space X is *connected* if it satisfies one of the following equivalent conditions:

(1) X is not the disjoint union of two nonempty open subsets.
(2) The space X is not the disjoint union of two nonempty closed subsets.
(3) The only clopen subsets are X and $\emptyset$.
(4) Every continuous function from X to $\{0, 1\}$ with the discrete topology is constant.

Definition 1.4.2. When X is the disjoint union of two nonempty open subsets U, V, we say that it is *separated* and $\{U, V\}$ is called *a separation.*

Example 1.4.3. The only connected subspaces in $\mathbb{R}$ are bounded intervals, rays and $\mathbb{R}$ (see Exercise 1).

Example 1.4.4. The set of rational numbers $\mathbb{Q}$ in the real line is separated since we can take $U = (-\infty, \sqrt{2})$ and $V = (\sqrt{2}, \infty)$. We can also see that $\mathbb{Q}$ is disconnected because it is not an interval (Example 1.4.3).

Example 1.4.5. Let $S^1 \subset \mathbb{R}^2$ and $S^2 \subset \mathbb{R}^3$ denote the unit circle of Euclidean plane and the unit sphere of Euclidean space, respectively. By identifying S^1 with the equatorial circle, we can see $S^2 \setminus S^1$ is not connected as union of two disjoint open sets, the northern and the southern open hemispheres.

Example 1.4.6. Let the linear group $\mathrm{GL}_n = \mathrm{GL}_n(\mathbb{R})$ of nonsingular real $n \times n$ matrices with the subspace topology inherited from $\mathbb{R}^{n^2}$. The group GL_n is not connected because it can be written as the union of two disjoint open sets, the sets of matrices with positive and negative determinants, respectively (see Exercise 24).

Theorem 1.4.7. *The continuous image of a connected space is a connected space.*

Proof. Let the diagram

$$X \xrightarrow{\quad f \quad} f(X) \xrightarrow{\quad \phi \quad} \{0,1\}$$

where f and ϕ are continuous. Then, for every $y \in f(X)$, there exists $x \in X$ such that $\phi(y) = \phi(f(x))$. The function $\phi \circ f$ is constant because X is connected. Hence, ϕ is constant. By Definition 1.4.1(4), $f(X)$ is connected. □

Lemma 1.4.8. *Let X be a space and $\{U, V\}$ a separation. If Y is a connected subset of X, then either $Y \subset U$ or $Y \subset V$.*

Proof. Since $\{Y \cap U, Y \cap V\}$ is a separation of Y and Y is connected, either $Y \cap U = \emptyset$ or $Y \cap V = \emptyset$. □

Theorem 1.4.9. *Let $A \subseteq B \subseteq \overline{A} \subseteq X$, where A is connected. Then B is connected.*

Proof. Let U, V be an open separation of B. Then $U = B \cap U'$ and $V = B \cap V'$, where U' and V' are open sets in X. Since $A \subseteq U \cup V$ and A is connected, by Lemma 1.4.8, either $A \subseteq B \cap U'$ or $A \subseteq B \cap V'$. Suppose $A \subseteq B \cap U'$. Then

$$A \cap V' = A \cap B \cap V' \subseteq B \cap U' \cap V' = U \cap V = \emptyset.$$

Consequently, $\overline{A} \cap V' = \emptyset$ for otherwise there is some $x_0 \in \overline{A} \cap V'$. Then $V' \in \mathscr{N}_{x_0}$ and $A \cap V' \neq \emptyset$, which is a contradiction. This shows that B is connected. □

Corollary 1.4.10. *The closure of a connected set is a connected set.*

Remark 1.4.11. The interior of a connected set is not necessarily connected as shows the subspace $X = C_1 \cup C_2$ of Euclidean plane, where

$$(C_1) : (x-1)^2 + y^2 = 1 \quad \text{and} \quad (C_2) : (x+1)^2 + y^2 = 1.$$

Theorem 1.4.12. *The union of connected subspaces whose intersection is nonempty is connected.*

Proof. Let $A = \bigcup_{\alpha \in J} A_\alpha$, where A_α is connected for all $\alpha \in J$ and let $x_0 \in \bigcap_{\alpha \in J} A_\alpha$. Let U, V be a separation of A. Then either $x_0 \in U$ or $x_0 \in V$. Assume that $x_0 \in U$. Since A_α is connected for all $\alpha \in J$, either $A_\alpha \subseteq U$ or $A_\alpha \subseteq V$. Since A_α contains x_0, necessarily $A_\alpha \subseteq U$. Therefore, $A = \bigcup_{\alpha \in J} A_\alpha \subseteq U$, and thus $V = \emptyset$, a contradiction. □

Actually, we have a more precise result.

Theorem 1.4.13. *Let A and B be two connected subspaces such that $\overline{A} \cap B \neq \emptyset$. Then the union $A \cup B$ is a connected space.*

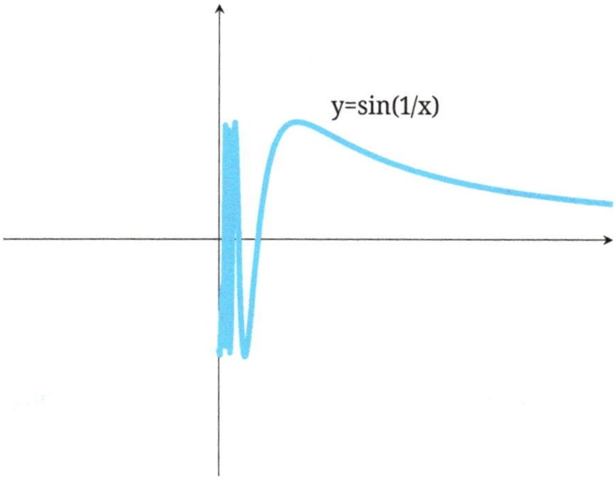

Figure 1.3: Topologist's sine curve.

Proof. If $A \cap B \neq \emptyset$, the claim is Theorem 1.4.12. Suppose that $A \cap B = \emptyset$ and by contradiction $A \cup B$ is not connected. Then there exist disjoint open subsets U, V such that $A \cup B = U \cup V$. Since A is a connected space, either $A \cap U$ or $A \cap V$ is the empty set. Likewise, either one of the sets $B \cap U$ or $B \cap V$ is empty. For instance, let $A \cap U = B \cap V = \emptyset$. Then $U = B$ and $V = A$. Let $a \in \overline{A} \cap B$. Then $a \in U$ and U must intersect A, which is a contradiction. □

Example 1.4.14. Let $f : (0,1] \to \mathbb{R}$ be the topologist's sine function $f(x) = \sin(1/x)$. Set

$$A = \{(x,y) \in \mathbb{R}^2 : y = f(x),\ x \in (0,1]\}, \quad B = \{(x,y) \in \mathbb{R}^2 : x = 0 \text{ and } -1 \leq y \leq 1\}.$$

Since f is continuous, by Theorem 1.4.7, A is connected. The interval B is connected. Since for instance, $(0,0) \in \overline{A} \cap B$, we deduce that the union $A \cup B$ is a connected space (see Figure 1.3).

1.4.2 Path-connected sets

Definition 1.4.15. (1) We say that a continuous function $f : I \to X$ is a *path* from an initial point $x_0 \in X$ to a final (terminal) point $x_1 \in X$ if $f(0) = x_0$ and $f(1) = x_1$ (see Figure 1.4).
(2) A path is called a *loop* if $x_0 = x_1$. In this case, x_0 is called a *base point*.
(3) For some $x_0 \in Y$, the constant map $f(s) = x_0$, $s \in I$ is called the *trivial loop*.

Definition 1.4.16. A space X is *path-connected* if any two points of X can be joined by a path.

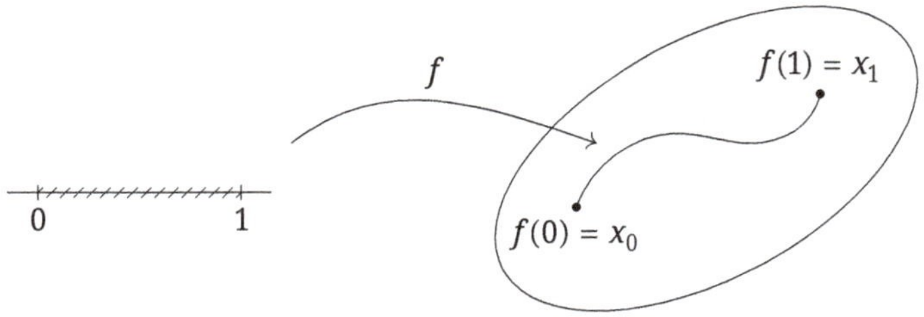

Figure 1.4: Path.

Example 1.4.17. Any convex subset of a vector space is path-connected. A path joining x_0 and x_1 is the straight line $f(t) = tx_0 + (1-t)x_1$. In particular, open and closed balls of Euclidean space are path-connected.

Theorem 1.4.18. *A path-connected space X is connected.*

Proof. Let $\{U, V\}$ be a separation of X and let $x_0 \in U, x_1 \in V$. Then there exists a path γ joining x_0 to x_1. Since γ is continuous, $\{\gamma^{-1}(U), \gamma^{-1}(V)\}$ is a separation of the connected interval $[0,1]$, a contradiction. Therefore, X must be connected. □

Remark 1.4.19. In general, the converse of Theorem 1.4.18 is not true. However, as we will see, a connected open subset of $\mathbb{R}^n$ is path-connected.

Theorem 1.4.20. *The continuous image of a path-connected space is a path-connected space.*

Proof. Let $y_1, y_2 \in f(X)$, where X is path-connected and f continuous. Let $x_1, x_2 \in X$ be such that $y_1 = f(x_1)$ and $y_2 = f(x_2)$ and let γ be a path joining x_1, x_2 in X. Then $f \circ \gamma$ is a path joining y_1 and y_2 in $f(X)$. □

Example 1.4.21. The unit circle S^1 is path-connected though not convex. S^1 is the continuous image of the interval $[0,1]$ under $f(t) = (\cos t, \sin t)$. As a direct proof, if $x, y \in S^1$, then $f(t) = \frac{x\cos(t\frac{\pi}{2})+y\sin(t\frac{\pi}{2})}{\|x\cos(t\frac{\pi}{2})+y\sin(t\frac{\pi}{2})\|}$ is a path joining x to y.

Example 1.4.22. The space $X = \mathbb{R}^{n+1} \setminus \{0\}$ is path-connected for $n \geq 1$. Let $x, y \in X$ and distinguish between two cases.

(1) If the line $f(t) = tx + (1-t)y$, $t \in \mathbb{R}$ does not pass through the origin, then we can take $\gamma = f$ as a path joining x to y, for $t \in [0,1]$.
(2) If the origin belongs to the line f, take any broken-line passing through a third point z and missing the origin, for instance,

$$\gamma(t) = \begin{cases} (1-2t)x + 2tz, & 0 \leq t \leq 1/2, \\ 2(1-t)z - (1-2t)y, & 1/2 \leq t \leq 1. \end{cases}$$

In fact, we have proved that X is polygonally-connected (see Exercise 13).

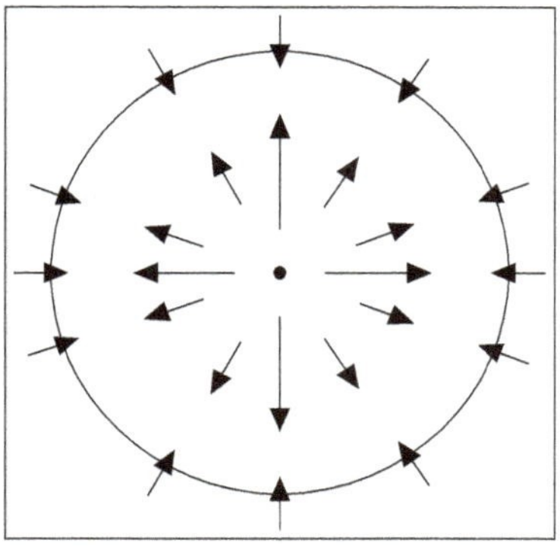

Figure 1.5: Radial retraction.

Example 1.4.23. For $n \geq 1$, the sphere S^n is path-connected as the continuous image of the punctured space $X = \mathbb{R}^{n+1} \setminus \{0\}$ by means of the so-called radial retraction $f(x) = r(x) = x/\|x\|$ (see Figure 1.5).

Remark 1.4.24. The closure of a path-connected set is not necessarily path-connected as the topologist's sine curve shows. The set

$$A = \{(x,y) \in \mathbb{R}^2 : y = f(x),\ x \in (0,1]\}$$

is path-connected since the sine function $f(x) = \sin(1/x)$ is continuous (see Figure 1.3).

In Example 1.4.14, we proved that the union $\mathscr{S} = A \cup B = \overline{A}$ is connected, where

$$B = \{(x,y) \in \mathbb{R}^2 : x = 0 \text{ and } -1 \leq y \leq 1\}.$$

We claim that $\mathscr{S}$ is not path-connected. Otherwise, let γ be a path joining the origin $0 \in B$ to some point $x_0 \in A$ and define the set $E = \{t \in [0,1] : \gamma(t) \in B\}$. Since $0 \in E$, $E \neq \emptyset$. Then $\bar{t} = \sup E \in E$ for $E = \gamma^{-1}(\{0\} \times [-1,1])$ is closed. Hence, $\gamma(\bar{t}) \in B$ and $\gamma(t) \in A$ for all $t \in [\bar{t}, 1]$. Let $x_n = \frac{1}{\frac{\pi}{2}+n\pi}$. Then there exists a sequence of points $t_n \geq \bar{t}$ such that $\gamma(t_n) = (x_n, \sin(1/x_n))$. By continuity of γ, $\lim_{n\to\infty} \gamma(t_n) = \gamma(\bar{t})$ but $(x_n, \sin 1/x_n) = (x_n, (-1)^n)$, a contradiction.

Another way to check that $\mathscr{S}$ is not path-connected is to note that $\mathscr{S}$ is the union of the disjoint path-connected subsets A and B. Hence, $\mathscr{S}$ has two path-components (see the next subsection).

1.4.3 Components and path-components

1.4.3.1 Components

Define a relation $\sim$ in X by $x \sim y$ if there exists a connected subspace containing x and y. Then $\sim$ is an equivalence relation.

Definition 1.4.25. The equivalence class $C(x) = [x]$ of a point x is called *the connected component* (or component for short) of x.

Example 1.4.26. A space X is connected if and only if for every $x \in X$, $C(x) = X$.

Example 1.4.27. The connected components of the product space $\mathbb{Q}\times\mathbb{R}$ are the sets $x\times\mathbb{R}$, where $x \in \mathbb{Q}$.

Theorem 1.4.28. *The components of X are disjoint connected sets whose union is X. Each connected subset intersects only one component.*

Proof. The fact that the components constitute a separation of X follows from the definition of the equivalence class. The last statement is a consequence of the equivalence relation. To prove that a component $C = C(x_0)$ is connected, note that $C = \bigcup_{x\in C} A_x$, where A_x is a connected subset containing x and x_0. Since the intersection $\bigcap_{x\in C} A_x$ contains x_0, the union C is connected by Theorem 1.4.12. □

Theorem 1.4.29. *The components of X are closed sets.*

Proof. The closure $\overline{C}$ of a component C is a connected set by Corollary 1.4.10. By definition of C, $\overline{C} \subseteq C$, hence $\overline{C} = C$. □

Definition 1.4.30. A space X is totally disconnected if $C(x) = \{x\}$ for all $x \in X$.

Example 1.4.31. The space $\mathbb{Q}$ is totally disconnected. The component of a rational number x is the closed set $C(x) = \{x\}$ but it is not open.

1.4.3.2 Path-components

Define a relation $\sim$ in X by $x \sim y$ if there exists a path joining x and y. Clearly, $\sim$ is reflexive and if $\phi(t)$ is a path from x to y, then $\phi(1-t)$ is a path from x to y, proving symmetry of $\sim$. To check that $\sim$ is transitive, let ϕ, ψ be paths from x to y and from y to z, respectively. Then the map γ given by $\gamma(t) = \phi(2t)$, for $0 \le t \le 1/2$ and $\gamma(t) = \psi(2t-1)$, for $1/2 \le t \le 1$ is a path from x to z. Hence, $\sim$ is an equivalence relation.

We have the following.

Corollary 1.4.32. *Let $x_0 \in X$. Then X is path-connected if and only if for every $x \in X$, there exists a path joining x to x_0.*

Corollary 1.4.33. *The union of a collection of path-connected spaces whose intersection is nonempty is path-connected.*

Definition 1.4.34. The equivalence class $C(x) = [x]$ of a point x is called *the path-component*, which contains x.

The following result is similar to Theorem 1.4.28. The proof is left as an exercise.

Theorem 1.4.35. *The path-components of X are disjoint path-connected sets whose union is X. Each path-connected subset intersects only one path-component.*

Remark 1.4.36. Unlike the components, a path-component need not be closed. For example, for the topologist's sine curve (see Remark 1.4.24), the subset $B = \{0\} \times [0,1]$ is closed in $\mathbb{R}^2$, hence closed in $\mathscr{S}$ but not open in $\mathscr{S}$ for the complement $\mathscr{S} \setminus B = A$ is not closed in $\mathscr{S}$. The subset A is open in $\mathscr{S}$ (because the complement $\mathscr{S} \setminus A = B$ is closed) but not closed (for $\overline{A} \neq A$).

Definition 1.4.37. A space X is totally path-disconnected if the path-component of each element $x \in X$ is the one-set element $\{x\}$.

Example 1.4.38. The set of rationals $\mathbb{Q}$ as well as every totally disconnected space are totally path-disconnected.

1.4.4 Local connectedness

Definition 1.4.39. (1) A space X is said to be *locally connected* if every point $x \in X$ has a basis of connected neighborhoods, that is to say for every $U \in \mathcal{N}_x$, there exists an open connected set V such that $x \in V \subseteq U$.

(2) A space X is said to be *locally path-connected* if every point $x \in X$ has a basis of path-connected neighborhoods, i. e., for every $U \in \mathcal{N}_x$, there exists a path-connected open set V such that $x \in V \subset U$.

Example 1.4.40. A discrete space is locally connected.

Example 1.4.41. Each interval of the real line is connected and locally connected.

Example 1.4.42. The set of rational numbers $\mathbb{Q}$ is neither connected nor locally connected.

Example 1.4.43. (1) The union of two disjoint intervals is locally connected but not connected.

(2) Another example is provided by the linear group $\mathrm{GL}_n = \mathrm{GL}_n(\mathbb{R})$ of nonsingular real $n \times n$ matrices (see Exercises 24 and 25).

Example 1.4.44. The topologist's sine curve is connected but not locally connected. Indeed, the origin $(0,0)$ has not a basis of connected neighborhoods.

Remark 1.4.45. Example 1.4.44 shows that:

(1) not every connected space is locally connected. However, a compact space is locally compact;

(2) if every point of X has a connected neighborhood, then X is not necessarily locally connected.

Theorem 1.4.46. *A space is locally connected if and only if the components of every open subset are open sets.*

Proof. (1) Let X be locally connected and $C \subseteq U \subseteq X$, where C is a component and U is open. For each $x \in C$, let V be open, connected and $x \in V \subseteq U$. By definition of the component C, $x \in V \subseteq C$, proving that C is open.

(2) Conversely, let $x \in X$ and $U \in \mathcal{N}_X$ be an open neighborhood of x. Let $C = C(x)$ be the component of x in U. By assumption, C is open. Then take $V = C$ to conclude that X is locally connected. □

Corollary 1.4.47. *Let X be a locally connected space and $U \subseteq X$ an open subset. Then U is locally connected.*

Example 1.4.48. Every open subset $U \subseteq \mathbb{R}^n$ is locally connected.

Corollary 1.4.49. *In a locally connected space X, the connected components are clopen sets.*

We have an analogous result for locally path-connected spaces. The proof is left as an exercise.

Theorem 1.4.50. *A space is locally path-connected if and only if the path-components of every open subset are open sets.*

Remark 1.4.51. (1) Local connectedness and local path-connectedness are topological invariant.
(2) The continuous image of a locally connected space is not locally connected as the identity map $\mathrm{Id} : (\mathbb{Q}, d) \longrightarrow (\mathbb{Q}, |\cdot|)$ shows where d is the discrete metric and $|\cdot|$ the standard metric. However, the continuous image of a locally connected space is locally connected if further the map is either closed or open (see Exercises 20, 21).

Proposition 1.4.52. *The following statements are equivalent:*
(1) *Each path-component is open.*
(2) *Each point has a path-connected neighborhood.*

Proof. Since (1) $\Longrightarrow$ (2) is trivial, assume that (2) holds and let $x \in X$ and C a path-component containing x. By hypothesis, there exists an open path-connected set U containing x. By definition of C, $x \in U \subseteq C$, whereupon (1) follows. □

Theorem 1.4.53. *X is path-connected if and only if X is connected and every point has a path-connected neighborhood.*

Proof. The necessary condition is trivial. So, assume that X is connected and every point has a path-connected neighborhood. Let $C \subseteq X$ be a path-component. By Proposition 1.4.52, C is open and it is closed by Theorem 1.4.29. Since X is connected, $C = X$, proving that X has only one path-component. □

Corollary 1.4.54. *Let X be a connected and a locally path-connected space. Then X is path-connected.*

Proof. Since X is a locally path-connected space, every point has a path-connected neighborhood. The result follows from Theorem 1.4.53. □

Corollary 1.4.55. *Let X be locally connected and $U \subseteq X$ an open subset. Then the components of U are clopen and path-connected.*

Proof. Let C be a component of U. By Corollary 1.4.49, C is open (and closed). Corollary 1.4.47 implies that both U and C are locally connected. Corollary 1.4.54 concludes the proof. □

Exercises (Section 1.4)

1. Let $A \subseteq \mathbb{R}$ be a nonempty set. Prove that

$$A \text{ connected} \iff A \text{ path-connected} \iff A \text{ convex} \iff A \text{ interval.}$$

2. (1) Prove Theorem 1.4.35.
(2) Prove Theorem 1.4.50.

3. Prove that the product of two connected (resp., path-connected, locally connected) spaces is connected (resp., path-connected, locally connected).

4. Let X be a space. Prove that X is connected if and only if the diagonal $\Delta = \{(x, x) \in X \times X\}$ is connected in the product space $X \times X$.

5. Let $B \subset X$ be a subset of a space X and $A \subset X$ a connected subset, which intersects both B and $X \setminus B$. Prove that A intersects the boundary of B.

6. Let $A, B \subset X$ be subsets of a path-connected space X such that $X = \mathring{A} \cup \mathring{B}$. Prove that each path-component of A meets B.

7. (1) Let X and Y be spaces. Prove that X and Y are connected (path-connected, respectively) if and only if $X \times Y$ is connected (path-connected, respectively) for the product topology.
(2) Let X, Y be connected spaces and $A \subset X$, $B \subset Y$ two nonempty connected subsets such that $A \neq X$ and $B \neq Y$. Prove that $(X \times Y) \setminus (A \times B)$ is connected.

8. Let $C \subset X$ be a nonempty clopen connected subset. Prove that C is a connected component.

9. Let A, B two closed subsets (or open subsets) of a space X such that $A \cup B$ and $A \cap B$ are path-connected. Prove that A and B are path-connected.

10. Let $U \subseteq \mathbb{R}^n$ be a nonempty open subset of Euclidean space. Prove that:
(1) A is connected if and only if A is path-connected;
(2) The components of U are clopen and path-connected. (*Hint:* (1) and (2) follow from Corollary 1.4.55.)
(3) The components of U are countable.
(4) What about the number of components when U is further bounded?

11. Let $C \subset \mathbb{R}^2$ be closed and homeomorphic to S^1, i. e., a Jordan curve (see Definition 1.5.5). Show that $\mathbb{R}^2 \setminus C$ is not connected. We say that C separates $\mathbb{R}^2$. This result will be discussed in a more general context in Theorem 10.4.22 and Exercise 21, Chapter 12.

12. Let $C \subset \mathbb{R}^n$ be a countable subset ($n \geq 2$) and $X = \mathbb{R}^n \setminus C$. Show that:
(1) $\overline{X} = \mathbb{R}^n$ (X is dense in $\mathbb{R}^n$);
(2) X is path-connected. (*Hint:* The case $n = 2$ can be easily visualized by connecting two points of the plane through uncountable number of segment lines.)

13. A set $A \subseteq \mathbb{R}^n$ is said to be polygonally connected if any two points a, b of A can be connected by a polygonal path in A, i. e., there exist a finite number of points $x_k \in A$, $(k = 0, 1, \ldots N)$, such that $x_0 = a$, $x_N = b$ and $L(x_{k-1}, x_k) \subseteq A$, for $k = 1, \ldots, N$, where $L(x_{k-1}, x_k)$ is the polygonal line joining x_{k-1} to x_k.
(1) Prove that every polygonally connected subset of $\mathbb{R}^n$ is path-connected.
(2) Let $A \subseteq \mathbb{R}^n$ be an open subset, $a \in A$ and U the set of points $x \in A$, which can be polygonally connected in A to a. Prove that U is open.
(3) Prove that every open connected set in $\mathbb{R}^n$ is polygonally connected.

14. Let X be locally path-connected. Prove that the components and the path-components are the same. (*Hint:* See Corollary 1.4.54.)

15. Let $A \subset X$ be a clopen subset.
(1) Show that A is union of components of X.
(2) Suppose that A is further connected. Deduce that A is a connected component (Exercise 8).

16. Prove that every point has a connected (path-connected) neighborhood if and only if every component (path-connected component) is clopen.

17. Let X have a finite number of components. Prove that each component is clopen.

18. Let X be a compact and locally connected space. Prove that:
(1) X is locally path-connected;
(2) X has a finite number of components.

19. Let $f : \mathbb{N} \longrightarrow \mathbb{R}$ be given by $f(0) = 0$ and $f(n) = 1/n$. Show that f is continuous, $\mathbb{N}$ is locally connected but $f(\mathbb{N})$ is not.

20. Let X be a locally connected space and $f : X \longrightarrow Y$ a continuous and closed (or open) map. Prove that Y is locally connected.

21. Prove that the deleted comb space (see Figure 1.6)

$$Cb = \bigcup_{n=1}^{\infty} \left(\left\{ \frac{1}{n} \right\} \times I \right) \cup (I \times \{0\}) \cup \{(0,1)\}$$

is connected but neither path-connected nor locally connected (see also Example 4.9.3).

22. Prove that the broom space (see Figure 1.7)

$$\left\{ (x_1, x_2) \in \mathbb{R}^2 : \ 0 \leq x_1 \leq 1, \ x_2 = \frac{x_1}{n}, \ n = 1, 2, \ldots \right\} \cup (I \times \{0\})$$

is path-connected but not locally connected (see also Example 4.9.4).

Figure 1.6: Comb space.

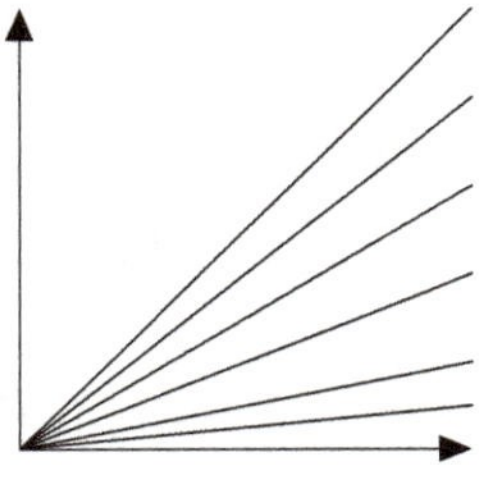

Figure 1.7: Infinite broom space.

23. Consider the topologist's whirlpool defined by $W = A \cup S^1$, where the spiral

$$A = \left\{ \frac{\theta}{1+\theta}\cos\theta, \frac{\theta}{1+\theta}\sin\theta, \ \theta \geq 0 \right\}$$

is given in polar coordinates. Draw W in the plane. Prove that W is connected but not path-connected.

24. Prove that the general linear group of invertible matrices GL_n (see Example 1.4.6) has two components:

$$G^+ = \{A \in \mathrm{GL}(n) : \det A > 0\} \quad \text{and} \quad G^- = \{A \in \mathrm{GL}(n) : \det A < 0\}.$$

(*Hint:* Check that G^+ and G^- are open and homeomorphic sets, then show that G^+ is connected. Recall that the determinant of $n \times n$ matrix is continuous as a polynomial in the n^2 entries.)

25. Prove that GL_n is locally path-connected.

26. Let X and Y be two spaces and $f : X \longrightarrow Y$ a map. f is said to be universal if for any map $g : X \longrightarrow Y$, there exists $x \in X$ such that $f(x) = g(x)$.

(1) Prove that a universal map is surjective.

(2) Let f_1 and $f_1 \circ f_2$ be universal maps. Prove that f_2 is universal.

(3) Prove that if the space X is connected, then every continuous map $f : X \longrightarrow [0,1]$ is universal.

1.5 Homeomorphism

The unit open ball in $\mathbb{R}^n$ will be denoted $\mathring{B}^n$ and the closed ball B^n.

1.5.1 Homeomorphic spaces

Definition 1.5.1. (1) *A homeomorphism* is a bijective map $f : X \longrightarrow Y$, which is bicontinuous (f and f^{-1} are continuous). X and Y are called *homeomorphic spaces.*

(2) $f : X \longrightarrow Y$ is *a local homeomorphism* if for every $x \in X$, there exists a neighborhood $U_x \in \mathcal{N}_x$ of x such that the image $f(U_x)$ is open in Y and $f_{|U_x} : U_x \longrightarrow f(U_x)$ is a homeomorphism.

Remark 1.5.2. If X and Y are homeomorphic spaces with some homeomorphism h, then so are the cut-points $X \setminus \{x\}$ and $Y \setminus \{h(x)\}$ for every $x \in X$. This observation may help to show that two spaces are not homeomorphic.

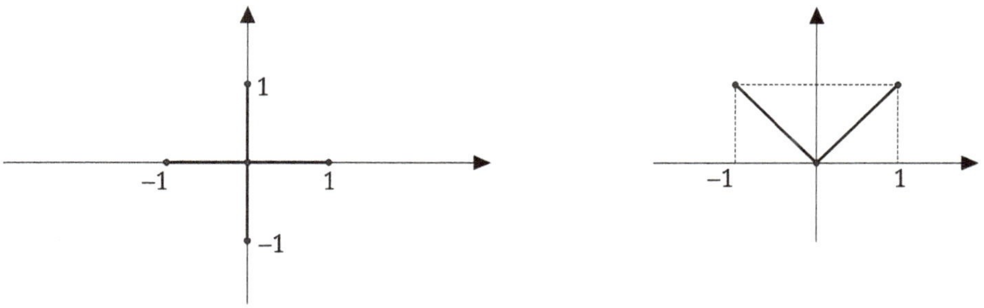

Figure 1.8: Cross figure and Figure ∨ are not homeomorphic.

Definition 1.5.3. (1) Some topological properties such as Hausdorff, compactness, connectedness, path-connectedness, first and second countability are preserved under the homeomorphism. They are called *topological invariant*.
(2) A map $f : X \longrightarrow Y$ is called *embedding* if the subspace $f(X)$ is homeomorphic to X.

Example 1.5.4. The figure ∨ is not homeomorphic to the cross symbol X (see Figure 1.8). Removing the center of the cross yields four path-components. Figure ∨ with one point removed has either one component (in case an end point is removed) or two components if the removed point is an interior point.

Let Y be a space.

Definition 1.5.5. (1) A *curve* is a continuous map $f : [0,1] \longrightarrow Y$.
(2) If $f : [0,1] \longrightarrow Y$ is an embedding, the image $\mathscr{A} = f([0,1])$ is called *an arc* of Y. $\mathring{\mathscr{A}}$ is the interior of the arc.
(3) If $f : S^1 \longrightarrow Y$ is an embedding, where S^1 is the unit circle, then $\mathscr{J} = f(S^1)$ is called *a simple closed curve* or a Jordan curve of Y.
(4) *A curved triangle* is a space homeomorphic to a closed triangular region of the plane. With such a homeomorphism, an edge and a vertex of the curved triangle are homeomorphic images of an edge and a vertex of the triangle, respectively.

Proposition 1.5.6. *Let $\mathscr{A} = f([0,1])$ be an arc of Y. Then either $\mathring{\mathscr{A}} = \emptyset$ or $\mathring{\mathscr{A}} = f(0,1)$.*

Proof. Assume $\mathring{\mathscr{A}} \neq \emptyset$.

(1) For $y \in \mathring{\mathscr{A}}$, there exists an open neighborhood $V \in \mathscr{N}_y$ such that $y \in V \subseteq \mathscr{A} = f([0,1])$. Then there is some $x \in [0,1] : y = f(x) \in \mathring{\mathscr{A}}$, which implies $x \in f^{-1}(\mathring{\mathscr{A}})$. Since f is continuous, we have

$$f^{-1}(\mathring{\mathscr{A}}) \subseteq \overbrace{f^{-1}(\mathscr{A})}^{\circ} = \overbrace{[0,1]}^{\circ} = (0,1).$$

Hence, $x \in (0,1)$ and $y = f(x) \in f(0,1)$, i. e., $\mathring{\mathscr{A}} \subseteq f(0,1)$. In particular, the endpoints 0 and 1 do not belong to $f^{-1}(\mathring{\mathscr{A}})$ because f is injective.

(2) Since $f(0,1)$ is open and $f(0,1) \subseteq f([0,1]) = \mathscr{A}$, we have $f(0,1) \subseteq \mathring{\mathscr{A}} \subseteq f([0,1])$. □

Example 1.5.7. The map $h : \mathbb{R} \longrightarrow (-1,1)$ given by $h(x) = \frac{x}{1+|x|}$ is a homeomorphism whose inverse is $h^{-1}(x) = \frac{x}{1-|x|}$. The map k given by $\frac{1}{1-x} - \frac{1}{x}$ is a also a homeomorphism from $(-1,1)$ to $\mathbb{R}$.

Example 1.5.8. Any open interval (a,b) is homeomorphic to $(-1,1)$ with homeomorphism $h : (-1,1) \longrightarrow \mathbb{R}$ given by $h(x) = \frac{b-a}{2}x + \frac{b+a}{2}$ and inverse $h^{-1}(x) = \frac{b+a-2x}{b-a}$. This is also a homeomorphism between the closed intervals $[a,b]$ and $[-1,1]$.

Example 1.5.9. The map $h : \mathbb{R}^2 \longrightarrow P^+$ given by $h(x,y) = (e^x, y)$ is a homeomorphism, where $P^+ = \{(x,y) \in \mathbb{R}^2 : x > 0\}$ is the right half-plane.

Example 1.5.10. More generally, the subset $\{(x,y) \in \mathbb{R}^2 : y > a\}$ is homeomorphic to the plane $\mathbb{R}^2$ with homeomorphism given by $h(x,y) = (x, \ln(y-a))$ and inverse $h^{-1}(x,y) = (x, a + e^y)$.

Example 1.5.11. By identifying $\mathbb{R}^2$ and the complex plane $\mathbb{C}^2$, the map $h : \mathbb{R}^2 \longrightarrow \mathring{B}^2$ given by $\frac{z}{1+|z|}$ is a homeomorphism of the Euclidean plane onto the open ball.

Example 1.5.12. More generally the n-dimensional open ball $X = \mathring{B}^n = \{x = (x_1, x_2, \ldots, x_n) \in \mathbb{R}^n : \|x\| < 1\}$ and Euclidean space $\mathbb{R}^n$ are homeomorphic with homeomorphism $h(x) = \frac{1}{1-\|x\|}x$ from $\mathring{B}^n$ to $\mathbb{R}^n$. The inverse map is $h^{-1}(x) = \frac{1}{1+\|x\|}x$.

Example 1.5.13. From Example 1.5.12, we deduce that the punctured open ball $\mathring{B}^n \setminus \{0\}$ is homeomorphic to $\mathbb{R}^n \setminus \{h(0)\}$, for some homeomorphism h. By making a shift, $\mathring{B}^n \setminus \{0\}$ is homeomorphic to $\mathbb{R}^n \setminus \{0\}$.

Example 1.5.14. (1) Let $\|x\| = (\sum_{i=1}^{i=n} x_i^2)^{1/2}$ be the Euclidean norm. Then the n-dimensional closed ball $X = B^n = \{x = (x_1, x_2, \ldots, x_n) \in \mathbb{R}^n : 0 \leq \|x\| \leq 1\}$ is homeomorphic to the n-cube

$$I^n = \{x = (x_1, \ldots, x_n) : -1 \leq x_i \leq 1,\ i = 1, 2, \ldots, n\}.$$

The homeomorphisms are

$$h(x) = \frac{\max_{1 \leq i \leq n}(|x_i|)}{\|x\|}x \quad \text{from } I^n \text{ to } B^n$$

and

$$h^{-1}(x) = \frac{\|x\|}{\max_{1 \leq i \leq n}(|x_i|)}x \quad \text{from } B^n \text{ to } I^n$$

for $x \neq 0$ and $h(0) = 0$. The continuity at the origin follows from the standard inequalities

$$\max_{1\le i\le n}(|x_i|) \le \|x\| \le \sqrt{n}\max_{1\le i\le n}(|x_i|).$$

In addition, h maps the boundary

$$\partial I^n = \{x = (x_1, \dots, x_n) : -1 \le x_i \le 1, \text{ for } i = 1, 2, \dots, n \text{ and } x_i = \pm 1, \text{ for some } i\}$$

onto the boundary $S^{n-1} = \partial B^n$. As a consequence,

$$I^n \setminus \partial I^n \cong B^n \setminus S^{n-1} = \mathring{B}^n \cong \mathbb{R}^n \cong (I \setminus \partial I)^n = \mathring{I}^n. \tag{1.4}$$

(2) For example, the closed unit disc $B^2 = \{(x,y) \in \mathbb{R}^2 : x^2 + y^2 \le 1\}$, the unit square $S = \{(x,y) \in \mathbb{R}^2 : |x| \le 1, |y| \le 1\}$ and the closed elliptical region $E = \{(x,y) \in \mathbb{R}^2 : x^2 + (y/2)^2 \le 1\}$ are homeomorphic.

Example 1.5.15. (1) The closed ball B^{m+n} and the product $B^m \times B^n$ are homeomorphic. This follows from Example 1.5.14 and the associativity of the product of intervals $[0,1]$.

(2) The construction of the homeomorphism in Example 1.5.14 shows that the open ball $\mathring{\overline{B^{m+n}}}$ and the interior of the product $B^m \times B^n$ are homeomorphic, too.

Example 1.5.16. (1) Making use of Exercise 13(2), Section 1.1, we obtain the homeomorphism

$$S^{m+n-1} = \partial B^{m+n} \cong \partial(B^m \times B^n) = (S^{m-1} \times B^n) \cup (B^m \times S^{n-1}).$$

For instance, $S^3 \cong (S^1 \times B^2) \cup (B^2 \times S^1)$.

(2) In Exercise 11, this section, we will mention spaces homeomorphic to the solid tori $S^{m-1} \times B^n$ and $B^m \times S^{n-1}$.

Example 1.5.17. Define the figure eight (representing the number eight), denoted $8 \cong C_1 \cup C_2$, to be homeomorphic to the union of the circles

$$C_1 = \{(x,y) \in \mathbb{R}^2 : (x-1)^2 + y^2 = 1\} \quad \text{and} \quad C_2 = \{(x,y) \in \mathbb{R}^2 : (x+1)^2 + y^2 = 1\}.$$

Then the figure eight 8 is not homeomorphic to the symbol $\vee$ (the letter "vee") (see Figure 1.9) represented by the union

$$\vee \cong \{0 \le x \le 1 : y = x\} \cup \{-1 \le x \le 0 : y = -x\}.$$

Indeed, let $h : \vee \longrightarrow 8$ be a possible homeomorphism. Then the space $X = \vee \setminus \{(-1,1),(1,1)\}$ is path-connected. However, three possibilities may occur for the space $Y = 8 \setminus \{h(-1,1), h(1,1)\}$:

(1) The points $h(-1,1)$ and $h(1,1)$ lie on the same circle. So, Y has two components, and thus cannot be homeomorphic to X.

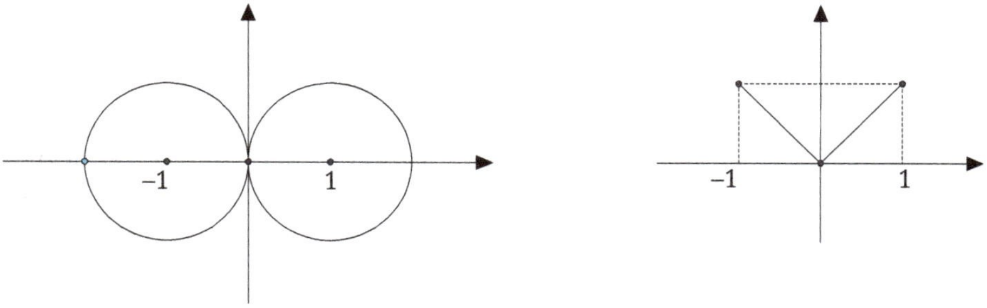

Figure 1.9: Figure 8 and Figure ∨ are not homeomorphic.

(2) The point $h(-1,1)$ lies on one of the circles and $h(1,1)$ is the intersection point. Then Y has three components and again cannot be homeomorphic to X.
(3) The point $h(-1,1)$ lies on one of the circles and $h(1,1)$ is on the other one. In this case, the figure Y has only one component. By removing once again the intersection point from Y, one gets four components. However, removing one point from X yields a figure with two components, wherever the point is removed. Therefore, no homeomorphism can be constructed between X and Y.

Example 1.5.18. In Euclidean space $\mathbb{R}^n$ $(n \geq 2)$, the map $h(x) = \frac{1}{\|x\|^2}x$ is a homeomorphism between the punctured open ball $\mathring{B}^n \setminus \{0\}$ and the hollowed space $\mathbb{R}^n \setminus B^n$, where B^n is the closed unit ball. The inverse map is h itself (h is called an inversion). h is also a homeomorphism between $B^n \setminus \{0\}$ and $\mathbb{R}^n \setminus \mathring{B}^n$ and a homeomorphism between $\mathbb{R}^n \setminus \{0\}$ and itself.

Example 1.5.19. If $h : \mathbb{R} \longrightarrow \mathbb{R}^2$ were a homeomorphism, then it would induce a homeomorphism $h : \mathbb{R} \setminus \{0\} \longrightarrow \mathbb{R}^2 \setminus \{f(0)\}$, which is impossible for $\mathbb{R}^2 \setminus \{f(0)\}$ is path-connected (Example 1.4.22) whereas $\mathbb{R} \setminus \{0\}$ is the union of two open intervals. Thus, connectedness is not preserved. Hence, $\mathbb{R}^2 \not\cong \mathbb{R}$.

Example 1.5.20. The inclusion map $i : A \subset X \hookrightarrow X$ is an embedding but not a homeomorphism.

Example 1.5.21. The space $X = [0,1]$ and the closed upper semicircle $Y = S^1_+$ are homeomorphic via the homeomorphism $h(t) = (\cos \pi t, \sin \pi t)$. The map h is also a homeomorphism from the open interval $(0,1)$ to the open upper semicircle $\mathring{S}^1_+$.

Example 1.5.22. (1) More generally, for $n \geq 1$, the n-dimensional closed ball

$$X = B^n = \{x = (x_1, x_2, \ldots, x_n) \in \mathbb{R}^n : \|x\| \leq 1\}$$

and the closed northern hemisphere (upper cap)

$$Y = S^n_+ = \{x = (x_1, x_2, \ldots, x_n, x_{n+1}) \in S^n : x_{n+1} \geq 0\}$$

are homeomorphic with homeomorphism $h : X \longrightarrow Y$ given by $h(x) = (x, \sqrt{1 - \|x\|^2})$. The inverse map is the projection $h^{-1}(x_1, x_2, \ldots, x_n, x_{n+1}) = (x_1, x_2, \ldots, x_n)$. Notice that h maps the boundary S^{n-1} of B^n onto the boundary S^{n-1} of the hemisphere.

(2) Likewise, $h(x) = (x, -\sqrt{1 - \|x\|^2})$ maps B^n homeomorphically onto the closed southern hemisphere (lower cap).

Example 1.5.23. The closed circular annulus

$$A = A(0; 1, 2) = \{x \in \mathbb{R}^2 : 1 \leq \|x\| \leq 2\}$$

is homeomorphic to the closed cylinder $C = S^1 \times [0, 1]$ with homeomorphism $h : A \longrightarrow C$ given by $h(x_1, x_2) = (x_1/\|x\|, x_2/\|x\|, \|x\| - 1)$. The inverse map is $h^{-1}(y_1, y_2, y_3) = ((1 + y_3)y_1, (1 + y_3)y_2)$.

Example 1.5.24. The unbounded cone

$$C = \{(x, y, z) \in \mathbb{R}^3 : x^2 + y^2 = z^2,\ z \geq 0\}$$

is homeomorphic to the plane $\mathbb{R}^2$ with homeomorphism $h(x, y) = (x, y, \sqrt{x^2 + y^2})$ and inverse $h^{-1}(x, y, z) = (x, y)$; the projection on the plane $z = 0$.

Example 1.5.25. The unbounded cylinder $S^1 \times (0, \infty)$ is homeomorphic to the punctured plane $\mathbb{R}^2 \setminus \{0\}$ with homeomorphisms given in polar coordinates by

$$h(\cos\theta, \sin\theta, r) = (r\cos\theta, r\sin\theta) \quad \text{and} \quad h^{-1}(r\cos\theta, r\sin\theta) = (\cos\theta, \sin\theta, r),$$

where $r > 0$ and $\theta \in [0, 2\pi]$.

Example 1.5.26. The punctured sphere $X = S^n \setminus \{N_p\}$ and Euclidean space $Y = \mathbb{R}^n$ are homeomorphic, where $N_p = (0, 0, \ldots, 0, 1) \in \mathbb{R}^{n+1}$ is the northern pole. The homeomorphism $h : X \longrightarrow Y$ is the stereographic projection defined by (see Figure 1.10):

$$h(x_1, x_2, \ldots, x_{n+1}) = \frac{1}{1 - x_{n+1}}(x_1, x_2, \ldots, x_n)$$

whose inverse is given by

$$h^{-1}(y_1, y_2, \ldots, y_n) = \frac{2}{\|y\|^2 + 1}\left(y_1, y_2, \ldots, y_n, \frac{\|y\|^2 - 1}{2}\right).$$

Remark 1.5.27. Note that the homeomorphism h defined in Example 1.5.26 satisfies

$$\lim_{x \to N_p} \|h(x)\|^2 = \lim_{x_{n+1} \to 1^-} \frac{1 + x_{n+1}}{1 - x_{n+1}} = \infty.$$

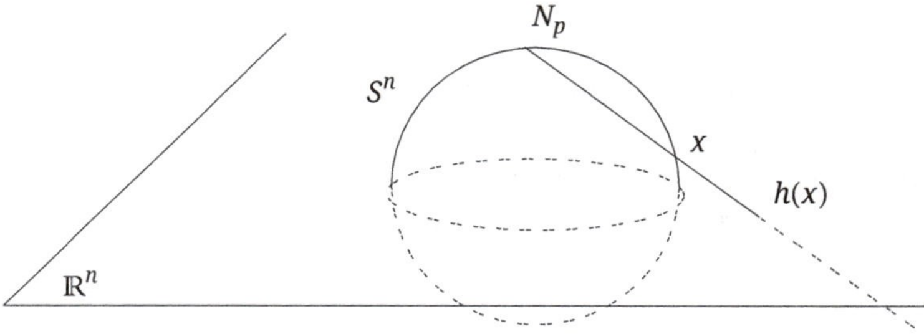

Figure 1.10: Stereographic projection.

This suggests to extend h by continuity to the entire sphere S^n by the map $\tilde{h}$ given by

$$\tilde{h}(x) = \begin{cases} h(x), & x \in S^n \setminus \{N_p\}, \\ \infty, & x = N_p \end{cases}$$

in order to obtain that S^n is homeomorphic to $\mathbb{R}^n \cup \{\infty\}$. This is made more precise by the following compactification result.

Proposition 1.5.28. *The sphere S^n is a one-point compactification of Euclidean space $\mathbb{R}^n$.*

Proof. By Example 1.5.26, the punctured sphere $S^n \setminus N_p$ is homeomorphic to $\mathbb{R}^n$, which is locally compact but not compact. Clearly, $(S^n \setminus N_p) \cup N_p = S^n$ is compact. Given Definition 1.3.17, it remains to show that $S^n \setminus N_p$ is dense in S^n. Let $x \in S^n$. If $x \neq N_p$, then the constant sequence $x_n \equiv x$ converges to x. If $x = N_p$, then the sequence given by $x_n = (0, 0, \ldots, \frac{\sqrt{2n-1}}{n}, 1 - \frac{1}{n})$ lies in $S^n \setminus N_p$ and converges to N_p, proving that $\overline{S^n \setminus N_p} = S^n$. According to Definition 1.3.17, S^n is a one-point compactification of $S^n \setminus N_p$. This with Example 1.5.26 completes the proof. □

1.5.2 Further properties

Remark 1.5.29. (1) The squaring map $f : \mathbb{R} \longrightarrow [0, \infty)$ given by $f(x) = x^2$ is a continuous open surjection, which is not a local homeomorphism.

(2) A continuous bijective map is not necessarily a homeomorphism as the identity $\mathrm{Id}_{\mathbb{R}} : (\mathbb{R}, \mathscr{P}(\mathbb{R})) \longrightarrow (\mathbb{R}, |\cdot|)$ shows, where $\mathscr{P}(\mathbb{R})$ is the discrete topology and $|\cdot|$ the standard topology of the real line, respectively.

However, for bijective maps we have the following result.

Theorem 1.5.30. *Let $f : X \longrightarrow Y$ be a bijective map. The following statements are equivalent:*

(1) *f is a homeomorphism;*

(2) *f is continuous and open (closed);*
(3) *f and f^{-1} are open (closed).*

Proof. If f is a homeomorphism, then f and f^{-1} are continuous maps so (1) implies (2). If (2) holds, then for every open set $V \subseteq Y, f^{-1}(V)$ is open by continuity of f, whence (3). Finally, (3) implies that both f and f^{-1} are continuous, i. e., (1). □

Example 1.5.31. Let f be the map defined by $f(t) = e^{2i\pi t}$.
(1) The function $f : [0,1] \longrightarrow S^1$ is not a homeomorphism. Since $f(0) = f(1)$, it is not injective.
(2) The function $f : (0,1) \longrightarrow S^1$ is not a homeomorphism, because it is not surjective for $a = (1,0) \notin f(0,1)$.
(3) The function $f : [0,1) \longrightarrow S^1$ is not a homeomorphism, because it is not open. Indeed, the positive semicircle $f([0,1/2) = S^1_+$ is not open whereas $[0,1/2) = (-1/2,1/2) \cap [0,1)$ is open in $[0,1)$. In Exercise 21, it is suggested to show that $f : \mathbb{R} \longrightarrow S^1$ is open. In addition, $[1/2,1)$ is closed in $[0,1)$ but $f([1/2,1))$ is not. Hence, f is not a closed map.

Example 1.5.32. (1) It is immediately checked that $f : (0,1) \longrightarrow S^1 \setminus \{a\}$ is a homeomorphism (the general case was examined in Example 1.5.26).
(2) More generally, let $\mathcal{J} = h(S^1)$ be a Jordan curve (Definition 1.5.5), where h is some homeomorphism. For some point $a = (1,0) \in S^1$, the cut-point $\mathcal{J} \setminus \{h(a)\}$ is homeomorphic to $(0,1)$.

Example 1.5.33. Consider the figure eight $8 = C_1 \cup C_2$ (see Figure 1.9 in Example 1.5.17), where C_1 and C_2 are copies of the unit circle. Then the space, which consists in Figure 8 with the points $b = (2,0)$ and $-b = (-2,0)$ removed, is homeomorphic to the open cross symbol $Y = Y_1 \cup Y_2$, where $Y_1 = (-1,1) \times \{0\}$ and $Y_2 = \{0\} \times (-1,1)$ (see Figure 1.11). Indeed, by Example 1.5.21, there are homeomorphisms $h_1 : C_1^+ \setminus \{b,(0,0)\} \longrightarrow X_1$, $h_2 : C_1^- \setminus \{b,(0,0)\} \longrightarrow X_1$, $k_1 : C_2^+ \setminus \{-b,(0,0)\} \longrightarrow X_3$ and $k_2 : C_2^- \setminus \{-b,(0,0)\} \longrightarrow X_4$. For $i = 1,2$, C_i^+ and C_i^- denote the upper half and lower half of the circle C_i, respectively. $X_1 = (-1,0) \times \{0\}$, $X_2 = \{0\} \times (-1,0)$, $X_3 = (0,1) \times \{0\}$ and $X_4 = \{0\} \times (0,1)$ are the four branches of the cross figure. Then the map h given by

$$h(x) = \begin{cases} h_1(x), & x \in C_1^+ \setminus \{b,(0,0)\}, \\ h_2(x), & x \in C_1^- \setminus \{b,(0,0)\}, \\ k_1(x), & x \in C_2^+ \setminus \{-b,(0,0)\}, \\ k_2(x), & x \in C_2^- \setminus \{-b,(0,0)\}, \\ 0, & x = (0,0) \end{cases}$$

is the required homeomorphism.

Proposition 1.5.34. (1) *Every local homeomorphism is continuous and open.*
(2) *A bijective local homeomorphism is a homeomorphism.*

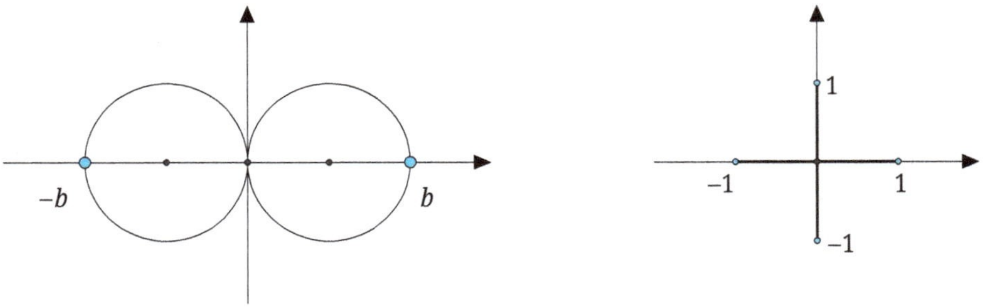

Figure 1.11: Double-punctured Figure 8 and Figure X are homeomorphic.

Proof. (1) (a) Let $f : X \longrightarrow Y$ be a local homeomorphism, $x \in X$ and $V \in \mathscr{N}_{f(x)}$ an open neighborhood of $f(x)$. By definition, there exists $U_x \in \mathscr{N}_x$ such that

$$f_{|U_x} : U_x \longrightarrow f(U_x) \text{ is a homeomorphism.} \tag{1.5}$$

Then $V \cap f(U_x)$ is an open neighborhood of $f(x)$, which is contained in $f(U_x)$. By (1.5), there exists $U'_x \in \mathscr{N}_x \cap U_x$ such that

$$f(U'_x) \subseteq V \cap f(U_x) \subseteq V,$$

proving the continuity of f.

(b) To prove that f is an open map, let $U \subseteq X$ be an open set, $V = f(U)$ and $y_0 \in V$. Then there exists $x_0 \in U$ such that $f(x_0) = y_0$. By assumption, there exists an open neighborhood $U_{x_0} \in \mathscr{N}_{x_0}$ such that

$$f_{|U_{x_0}} : U_{x_0} \longrightarrow f(U_{x_0}) \text{ is a homeomorphism.} \tag{1.6}$$

Let $U'_{x_0} = U \cap U_{x_0}$ be an open neighborhood of x_0 and $V'_{x_0} = f(U'_{x_0})$. Then $x_0 \in U'_{x_0} \subseteq U_{x_0}$. By (1.6), V'_{x_0} is an open set and $y_0 \in V'_{x_0} \subseteq f(U) = V$, proving that V is an open set.

(2) f is continuous and open by (1) and bijective, hence a homeomorphism by Theorem 1.5.30. □

From Proposition 1.5.34, we have the following.

Corollary 1.5.35. *A function $f : X \longrightarrow Y$ is a local homeomorphism if and only if it is open, continuous and locally injective.*

A practical situation is given by the Weierstrass theorem.

Theorem 1.5.36. (1) *Let $f : X \longrightarrow Y$ be a continuous bijective map, where X is compact and Y Hausdorff. Then f is a homeomorphism.*

(2) *Let $f : X \longrightarrow Y$ be a continuous one-to-one map, where X is compact and Y Hausdorff. Then f is an embedding.*

Proof. (1) Let $C \subseteq X$ be a closed subset, hence compact. Since f is continuous, $f(C)$ is compact, hence closed. By Theorem 1.5.30, f is a homeomorphism.

(2) Same reasoning. □

Some further homeomorphisms needed in the sequel are now established. The following two results complement the one in Example 1.5.23.

Theorem 1.5.37. *The open annulus of Euclidean plane*

$$A(0;1,2) = \{x \in \mathbb{R}^2 : 1 < \|x\| < 2\}$$

is homeomorphic to the punctured open ball

$$\mathring{B}^2 \setminus \{(0,0)\} = \{x \in \mathbb{R}^2 : 0 < \|x\| < 1\}.$$

Proof. (1) The map $h(x) = x + r(x)$ is a homeomorphism from $\mathring{B}^2 \setminus \{(0,0)\}$ to $A(0;1,2)$, where r is the radial retraction. Indeed, h is continuous and for $0 < \|x\| < 1$,

$$\|h(x)\| = \left(1 + \frac{1}{\|x\|}\right)\|x\| = \|x\| + 1 \in (1,2).$$

Hence, $h(x) \in A(0;1,2)$. The function h is further injective. Indeed,

$$\begin{aligned} h(x_1) = h(x_2) \quad &\Longleftrightarrow \quad \left(1 + \frac{1}{\|x_1\|}\right)x_1 = \left(1 + \frac{1}{\|x_2\|}\right)x_2 \\ &\Longrightarrow \quad \left(1 + \frac{1}{\|x_1\|}\right)\|x_1\| = \left(1 + \frac{1}{\|x_2\|}\right)\|x_2\| \\ &\Longleftrightarrow \quad \|x_1\| + 1 = \|x_2\| + 1 \\ &\Longleftrightarrow \quad \|x_1\| = \|x_2\| \\ &\Longleftrightarrow \quad x_1 = x_2. \end{aligned}$$

The last equivalence follows from the first one. Finally, h is onto and its inverse is given by

$$h^{-1}(x) = \frac{\|x\| - 1}{\|x\|}x.$$

Indeed, if $y = (1 + \frac{1}{\|x\|})x$, then

$$\|y\| = \|x\| + 1 \quad \Longrightarrow \quad \|x\| = \|y\| - 1 \quad \Longrightarrow \quad x = \frac{1}{1 + \frac{1}{\|y\|-1}}y.$$

Hence, $x = \frac{\|y\|-1}{\|y\|}y$. Clearly, h^{-1} is continuous.

(2) The result can be viewed by using the radial shifting $h(r,\theta) = (r - 1, \theta)$ in terms of the polar coordinates. □

Theorem 1.5.38. (1) *For $n \geq 2$, the function $h(x,t) = tx$ is a homeomorphism from the cylinder $S^{n-1} \times (0,1)$ to $\mathring{B}^n \setminus \{0\}$ and a homeomorphism from $S^{n-1} \times (0,\infty)$ to $\mathbb{R}^n \setminus \{0\}$. The inverse map is $h^{-1}(x) = (\frac{x}{\|x\|}, \|x\|)$. Furthermore, $S^{n-1} \times (0,1]$ and $B^n \setminus \{0\}$ are homeomorphic via the same homeomorphism.*

(2) *The open bounded and unbounded cylinders $S^{n-1} \times (0,1)$ and $S^{n-1} \times (0,\infty)$ are homeomorphic with homeomorphism $h(x,t) = (x, \frac{t}{1+t})$ and inverse $h^{-1}(x,t) = (x, \frac{t}{1-t})$.*

Proof. (1) (a) h is injective: Let (x_1,t_1), (x_2,t_2) be either in $S^{n-1} \times (0,1)$ or in $S^{n-1} \times (0,\infty)$ with $h(x_1,t_1) = h(x_2,t_2)$. Then $t_1x_1 = t_2x_2$ and

$$\|t_1x_1\| = \|t_2x_2\| \iff t_1\|x_1\| = t_2\|x_2\| \iff t_1 = t_2$$

because $x_1 \in S^{n-1}$ and $x_2 \in S^{n-1}$. Hence, $x_1 = x_2$ since $t_1, t_2 > 0$.

(b) h is surjective: Let $y \in \mathring{B}^n \setminus \{0\}$. Then $y = tx$ with

$$t = \|y\| \quad \text{and} \quad x = \frac{y}{\|y\|}.$$

Therefore, $h^{-1}(y) = (\frac{y}{\|y\|}, \|y\|)$.

The difference between the cases of the bounded and unbounded cylinders is that if $t \in (0,1)$, then $\|tx\| = t \in (0,1)$ for all $x \in S^{n-1}$, and thus $tx \in B^n \setminus \{0\}$ but if $t \in (0,\infty)$, then $\|tx\| = t > 0$ for all $x \in S^{n-1}$, which implies that $tx \in \mathbb{R}^n \setminus \{0\}$.

(c) Clearly, h and h^{-1} are continuous.

(2) The map $f(t) = \frac{t}{1+t}$ is a homeomorphism from $(0,\infty)$ to $(0,1)$ with inverse $f^{-1}(y) = \frac{y}{1-y}$. □

Remark 1.5.39. (1) It is easily checked that $h(x) = (\frac{x}{\|x\|}, \|x\|)$ is a homeomorphism from $\mathbb{R}^n \setminus \{0\}$ to $S^{n-1} \times (0,\infty)$.

(2) By Example 1.5.13, we already know that $\mathring{B}^n \setminus \{0\}$ and $\mathbb{R}^n \setminus \{h(0)\}$ are homeomorphic.

Exercises (Section 1.5)

1. Let $f : X \longrightarrow Y$ be a bijective map between two spaces. Prove that the following statements are equivalent:

(1) f is a homeomorphism.

(2) For all $A \subseteq X, f(\overline{A}) = \overline{f(A)}$.

(3) For all $B \subseteq Y, f^{-1}(\overline{B}) = \overline{f^{-1}(B)}$.

2. Prove that the radial retraction $r : \mathbb{R}^{n+1} \setminus \{0\} \longrightarrow S^n$ given by $r(x) = x/\|x\|$ is an open map. Is it a closed map?

3. Let $f : \mathbb{R} \longrightarrow \mathbb{R}$ be a bijection. Show that f is a homeomorphism if and only if f is monotonic (i. e., either increasing or decreasing).

4. Let $f : I \subseteq \mathbb{R} \longrightarrow J = f(I)$ be strictly monotonic, where I is an open set and J is an interval of the real line. Show that f has an inverse map $f^{-1} : J \longrightarrow I$, which is strictly monotonic and continuous.

5. Let $f : I \longrightarrow \mathbb{R}$ be strictly monotonic and continuous, where $I \subseteq \mathbb{R}$ is an open interval. Show that:
(1) $J = f(I)$ is an open interval,
(2) f is an embedding.

6. Let $f : S^1 \longrightarrow \mathbb{R}$ be a continuous map. Show that there exists $x \in S^1$ such that $f(x) = f(-x)$. Hence, S^1 does not embed in the real line. (*Hint:* Make use of $g(x) = f(x) - f(-x)$ both with the connectedness of $g(S^1)$). This is Borsuk–Ulam in dimension 1 (see Theorem 6.4.6 for dimension 2).

7. Let $f : I \longrightarrow \mathbb{R}^n$ be a continuous one-to-one map, where $I \subseteq \mathbb{R}$ is an interval. Show that f is an embedding.

8. For $n \geq 2$, let $f : [0,1] \longrightarrow \mathbb{R}^n$ be a continuous and one-to-one map. Show that $\overset{\circ}{\overbrace{f([0,1])}} = \emptyset$ (the interior of an arc in $\mathbb{R}^n$ is the empty set).

9. Let the product space $X \times Y$ be endowed with the product topology. Show that for each $y \in Y$, the map $\phi : X \longrightarrow X \times \{y\}$ such that $\phi(x) = (x, y)$ is a homeomorphism, where $X \times \{y\}$ is equipped with the subspace topology.

10. Let

$$A = \{(x_1, x_2, x_3, x_4) \in S^3 : x_1^2 + x_2^2 \leq 1/2\}$$

and

$$B = \{(x_1, x_2, x_3, x_4) \in S^3 : x_1^2 + x_2^2 \geq 1/2\}$$

be subsets of the sphere S^3 in $\mathbb{R}^4$. Prove that:
(1) A and B are homeomorphic to *the solid tori* $B^2 \times S^1$ and $S^1 \times B^2$, respectively.
(2) The sphere S^3 is homeomorphic to the union of two solid tori with intersection homeomorphic to a torus. This is an alternative proof to Example 1.5.16.

11. Let

$$A_1 = \{(x, y) \in S^{m+n-1} : \|x\|^2 \geq 1/2\} \quad \text{and} \quad A_2 = \{(x, y) \in S^{m+n-1} : \|x\|^2 \leq 1/2\},$$

where $\|x\|^2 = \sum_{k=1}^{k=m} x_k^2$ and $\|y\|^2 = \sum_{k=1}^{k=n} x_k^2$ for $(x, y) \in \mathbb{R}^{m+n}$.
(1) Show that $h_1 : A_1 \longrightarrow S^{m-1} \times B^n$ and $h_2 : A_2 \longrightarrow B^m \times S^{n-1}$ are homeomorphisms, where

$$h_1(x,y) = \left(\frac{x}{\sqrt{1 - \|y\|^2}}, \sqrt{2}y \right) \quad \text{and} \quad h_2(x,y) = \left(\sqrt{2}x, \frac{y}{\sqrt{1 - \|x\|^2}} \right).$$

(2) Check that $A_1 \cap A_2 \cong S^{m-1} \times S^{n-1}$. Conclude. (*Hint:* See Exercise 10 and [38, Chapitre II, Proposition 1.7]).

12. Let $X \subset \mathbb{R}^2$ be the union of n segments with a common endpoint. Prove that $\mathbb{R}^2 \setminus X$ is homeomorphic to the punctured plane.

13. Prove that the quadric surface in $\mathbb{R}^n$ given by $\sum_{k=1}^{k=n-1} x_k^2 = 1 + x_n^2$ is homeomorphic to $S^{n-2} \times \mathbb{R}$. (*Hint:* Use $h(x_1, x_2, \ldots, x_n) = (x_1/\sqrt{1+x_n^2}, x_2/\sqrt{1+x_n^2}, \ldots, x_{n-1}/\sqrt{1+x_n^2})$.)

14. Prove that the unit circle S^1 and the sphere S^2 are not homeomorphic.

15. Prove that the closed unit disc B^2 and the closed annulus $A(0;1,2)$ are not homeomorphic.

16. Let $h : X \longrightarrow Y$ be a homeomorphism and $A \subset X$. Prove that:

(1) $f(\overline{A}) = \overline{f(A)}$;

(2) $f(\mathring{A}) = \overset{\circ}{\overline{f(A)}}$;

(3) $f(\partial A) = \partial f(A)$.

17. Let $h : X \longrightarrow Y$ and $h' : X' \longrightarrow Y'$ be two homeomorphisms. Prove that the map $H : X \times X' \longrightarrow Y \times Y'$ defined by $H(x, x') = (h(x), h'(x'))$ is a homeomorphism. Conclude.

18. A graph is the union of arcs $(A_\alpha)_\alpha$ such that the intersection of two arcs is either empty or a single one end-point. Prove that a graph is connected if and only if every pair of vertices can be connected by an edge path. (*Hint:* See [76, Lemma 84.6].)

19. Let $f : X \longrightarrow Y$ be a continuous, surjective, open map such for all $y \in Y$, the set $f^{-1}(y)$ is connected. Prove that a subset $B \subseteq Y$ is connected if and only if the inverse image $f^{-1}(B)$ is connected.

20. Let $A \subseteq S^1$ be such that $A \neq S^1$. Prove that A and S^1 are not homeomorphic.

21. Let $f : \mathbb{R} \longrightarrow S^1$ be the map defined by $f(t) = e^{2i\pi t}$. Prove that f is an open map. (*Hint:* See [92, Lemma 14.3.3].)

2 Quotient topology

One of the applications of the quotient topology consists in building up new spaces from old ones. This is the motivation of this chapter. Some general properties of quotient spaces are explored in Section 2.4. Before this, the general notion of the identification maps is discussed in Section 2.1. A situation of particular interest for our purposes concerns the quotient map, which is developed in Section 2.2. Finally, we treat separately the quotient spaces defined by a map (Section 2.3) or by a set (Section 2.5).

2.1 Identification map

Let X and Y be two spaces.

Definition 2.1.1. A surjective map $f : X \longrightarrow Y$ is called *an identification map* if for every $U \subset Y$,

$$f^{-1}(U) \text{ open in } X \quad \Longleftrightarrow \quad U \text{ open in } Y.$$

An identification map is obviously continuous. The following result is easy to check.

Proposition 2.1.2. *Let $B \subset Y$ be an open (or closed) subset and $f : X \longrightarrow Y$ an identification map. Then the restriction $f_{|f^{-1}(B)} : f^{-1}(B) \longrightarrow B$ is an identification map.*

The next result is useful for applications.

Proposition 2.1.3. *If f is surjective, continuous and open (or closed), then f is an identification map.*

Proof. Let $U \subset Y$ be such that $f^{-1}(U)$ is open in X. Since f is onto, $f(f^{-1}(U)) = U$ is open because f is open. The proof is identical if f is a closed map. □

Example 2.1.4. A homeomorphism is an identification map.

Example 2.1.5. The exponential map $p_0 : [0,1] \longrightarrow S^1$ defined by $p_0(t) = e^{2\pi it}$ is clearly surjective, continuous and closed map because $[0,1]$ is compact. Then p_0 is an identification map. However, f is not open since $[0,1/2) = [0,1] \cap (-1/2,1/2)$ is open in $[0,1]$ but its image, which is the upper semicircle with the point $(1,0)$ and without the point $(-1,0)$, is not open.

Corollary 2.1.6. *Let $f : X \longrightarrow Y$ be continuous and has a continuous right-inverse. Then f is an identification map.*

Proof. The map f is onto because it is right-invertible. Let $U \subset Y$ be such that $f^{-1}(U)$ is open in X. Let g be a continuous right-inverse of f. Since g is continuous, $g^{-1}(f^{-1}(U))$ is an open set. Then $g^{-1}(f^{-1}(U)) = (f \circ g)^{-1}(U) = \mathrm{Id}_{|Y}^{-1}(U) = U$ is open, whence the claim. □

https://doi.org/10.1515/9783111517384-002

Example 2.1.7. Let $f : \mathbb{R}^2 \longrightarrow [0,\infty)$ be given by $f(x,y) = x^2 + y^2$. Then f is continuous and $f \circ g = \mathrm{Id}_{|[0,\infty)}$, where $g(t) = (\sqrt{t}, 0)$. Hence, f is an identification map.

Corollary 2.1.8. *Let $f : X \longrightarrow Y$ be continuous and surjective with X compact and Y Hausdorff. Then f is an identification map.*

Proof. Let $C \subseteq X$ be a closed subset and $C' = f(C)$. Since X is compact and f continuous, C and then $f(C)$ are compact sets. Hence, $f(C)$ is closed in Y and the claim follows from Proposition 2.1.3. □

Next, we present an important factorization result.

Theorem 2.1.9. *Let $f : X \longrightarrow Y$ be an identification map and $g : Y \longrightarrow Z$ a map. Then g is continuous if and only if $g \circ f : X \longrightarrow Z$ is continuous.*

Proof. Assume that $g \circ f$ is continuous and let $U \subset Z$ be open in X. Then $(g \circ f)^{-1}(U) = f^{-1}(g^{-1}(U))$ is open in X. Since f is an identification map, $g^{-1}(U)$ is open in Y, as desired. The converse is a consequence of the continuity of the composite of two continuous functions. □

A converse of Theorem 2.1.9 holds (see [25, Theorem 3.3, Chapter VI]).

Theorem 2.1.10. *Let $f : X \longrightarrow Y$ be a continuous surjective map such that for every space Z, any map $g : Y \longrightarrow Z$ is continuous whenever $g \circ f$ is. Then f is an identification map.*

Proof. On the topological space $(Y, \mathscr{T})$, define the topology

$$\mathscr{T}' = \{U \subset Y : f^{-1}(U) \text{ is open in } X\}.$$

Clearly, $\mathscr{T} \subseteq \mathscr{T}'$. Let $p : X \longrightarrow (Y, \mathscr{T}')$ be an identification map and $\mathrm{Id}_{|Y} : (Y, \mathscr{T}) \longrightarrow (Y, \mathscr{T}')$ the identity map. Let $U \subseteq \mathscr{T}'$. Since $(\mathrm{Id}_{|Y} \circ f)^{-1}(U) = f^{-1}(U)$ is open by definition of the topology $\mathscr{T}'$, $\mathrm{Id}_{|Y} \circ f$ is continuous. The condition of the theorem implies that $\mathrm{Id}_{|Y}$ is continuous. For the same reason, $(\mathrm{Id}_{|Y}^{-1} \circ p)^{-1}(U) = p^{-1}(U)$ implies that $\mathrm{Id}_{|Y}^{-1}$ is continuous. Then $\mathrm{Id}_{|Y}$ a homeomorphism between $(Y, \mathscr{T})$ and $(Y, \mathscr{T}')$. Therefore, $\mathscr{T} = \mathscr{T}'$ and f is an identification map. □

The next result is a universal property of identifications.

Proposition 2.1.11. *Let $f : X \longrightarrow Y$ be an identification map and $g : X \longrightarrow Z$ a continuous map. Then there exists a continuous map $h : Y \longrightarrow Z$ such that $h \circ f = g$ if and only if g is constant on each inverse image set $f^{-1}(y)$, $y \in Y$, called the fibers of f (or fibers).*

Proof. We only prove the condition of sufficiency. Define the map $h : Y \longrightarrow Z$ by $h(y) = g(f^{-1}(y))$. Since g is constant on $f^{-1}(y)$, h is well-defined and satisfies $h(f(x)) = g(f^{-1}(f(x))) = g(x)$ for all $x \in X$. To show that h is continuous, let $V \subseteq Z$ be an open set. Since $g^{-1}(V) = f^{-1}(h^{-1}(V))$ and g is continuous, $f^{-1}(h^{-1}(V))$ is an open subset of X. Since f is an identification map, $h^{-1}(V)$ is open in Y. □

Proposition 2.1.11 shows that the map g can be factored as the composite of f and a continuous map h. Next, we take up a particular identification map, namely the quotient map.

2.2 Topology on a quotient space

Let X be a space, $\sim$ an equivalence relation on X and $\pi : X \longrightarrow X/_{\sim}$ the *canonical projection map* defined by $\pi(x) = [x]$, $x \in X$. We are going to define a topology on $X/\sim$, called the quotient topology.

Definition 2.2.1. A set $U \in X/_{\sim}$ is said to be open if $\pi^{-1}(U)$ is open in X. The collection of all open sets U is called *the quotient topology* induced by π and the mapping π is called *a quotient map.*

Theorem 2.2.2. *The quotient topology is a topology.*

Proof. (1) Since $\pi^{-1}(\emptyset) = \emptyset$ and $\pi^{-1}(X/_{\sim}) = X$, $\emptyset$ and $X/_{\sim}$ are open sets in the quotient topology.

(2) Let $(U_i)_{i\in I}$ be a family of open sets in $X/_{\sim}$, i. e., $\pi^{-1}(U_i)$ is open, for each $i \in I$. Since $\pi^{-1}(\bigcup_{i\in I} U_i) = \bigcup_{i\in I} \pi^{-1}(U_i)$, by definition of the quotient topology $\bigcup_{i\in I} U_i$ is open in $X/_{\sim}$. In the same way, since $\pi^{-1}(\bigcap_{i=1}^{i=N} U_i) = \bigcap_{i=1}^{i=N} \pi^{-1}(U_i)$, the intersection of finite open sets is an open set. □

Remark 2.2.3. Clearly, a quotient map is an identification map.

Remark 2.2.4. Since, by definition, $\pi : X \longrightarrow X/_{\sim}$ is continuous, if X is compact (connected, path-connected, respectively), then $X/_{\sim}$ is compact (connected, path-connected, respectively).

Example 2.2.5. (1) In the real line $X = \mathbb{R}$, define the equivalence relation $x \sim y \Longleftrightarrow x - y \in \mathbb{Z}$. Then $[x] = \{x + n,\ n \in \mathbb{Z}\}$. Since for every $x \in \mathbb{R}$, there exists $n \in \mathbb{Z}$ (integer part of x) such that $0 \leq x - n < 1$, $\mathbb{R}/_{\sim} = \{\{\{x\}, x \in (0,1)\}, \{\mathbb{Z}\}\}$. Let $h : (0,1) \longrightarrow \mathbb{R}/_{\sim}$ be the map defined by $h(x) = \{x\}$, i. e., $h = \pi_{|(0,1)}$, where π is the projection map. Then h is an embedding, which is open because $\pi^{-1}(h(U)) = U$ for every open subset $U \subseteq (0,1)$. Since $[0] = [1]$, we conclude that the quotient space $\mathbb{R}/_{\sim}$ is homeomorphic to the open interval $(0,1)$ with the endpoints identified. By Theorem 2.5.11, it is concluded that $\mathbb{R}/_{\sim}$ is homeomorphic to S^1.

(2) An alternative way to see this result is to consider the surjective homomorphism

$$f : (\mathbb{R}, +) \longrightarrow (S^1, \cdot) \quad \text{given by } f(t) = e^{2i\pi t}.$$

Then $\operatorname{Ker} f = \mathbb{Z}$, and thus by the fundamental theorem of homomorphism (Theorem B.2.14), $\mathbb{R}/\mathbb{Z} \cong S^1$, where $\mathbb{R}/\mathbb{Z}$ is a quotient group (Proposition B.2.9).

This is extended to any dimension in Example 2.3.12.

In Example 2.2.5, notice that the quotient space $X/_{\sim}$ is Hausdorff because $Y = S^1$ is. In Exercise 6, a necessary and sufficient condition for the quotient space to be Hausdorff is provided. In Section 2.5, we will consider conditions for a quotient space to be Hausdorff.

Since a quotient map is an identification map, Theorem 2.1.9 takes the following form.

Theorem 2.2.6. *Let Y be a space and $g : X/_{\sim} \longrightarrow Y$ a map. Then g is continuous if and only if $g \circ \pi : X \longrightarrow Y$ is continuous.*

Definition 2.2.7. A subset $A \subset X$ is *saturated* if $\pi^{-1}(\pi(A)) = A$.

Proposition 2.2.8. *The following statements are equivalent:*
(1) *A is saturated;*
(2) *$A = \bigcup_{x \in A}[x]$;*
(3) *There exists a subset $B \subset X/_{\sim}$ such that $A = \pi^{-1}(B)$.*

Proof. (1) $\Longrightarrow$ (2): Assume (1) and let $y \in [x]$ for some $x \in A$, i. e., $\pi(x) = \pi(y)$. Hence, $\pi(y) \in \pi(A)$, and thus $y \in \pi^{-1}(\pi(A))$, which is A by hypothesis. Then $y \in A$. Conversely, for each $y \in A$, $y \in [y]$, which completes the proof of (2).

(2) $\Longrightarrow$ (3): When (2) holds, just take $B = \pi(A)$.

(3) $\Longrightarrow$ (1): This follows from the identity $\pi^{-1}(\pi(A)) = \pi^{-1}(\pi(\pi^{-1}(B))) = \pi^{-1}(B) = A$. □

Proposition 2.2.8 states that a subset A is saturated if and only if all elements equivalent to an element of A belong to A.

2.3 Equivalence relation defined by a map

Let $f : X \longrightarrow Y$ define the equivalence relation $x_1 \sim x_2 \Longleftrightarrow f(x_1) = f(x_2)$ for every $x_1, x_2 \in X$. Consider the map $\overline{f} : X/_{\sim} \longrightarrow Y$ given by $\overline{f}([x]) = f(x)$. The following diagram commutes:

$$\begin{array}{ccc} X & \xrightarrow{f} & Y \\ {\scriptstyle\pi}\downarrow & \nearrow_{\overline{f}} & \\ X/\sim & & \end{array}$$

The map $\overline{f}$ is single-valued and continuous.

Definition 2.3.1. $\overline{f}$ is called the *map induced by* f.

Remark 2.3.2. (1) Since π is an identification map, by Proposition 2.1.11, f is constant on each fiber $\pi^{-1}([x])$, for $x \in X$.

(2) The map $\overline{f}$ is one-to-one since if $\overline{f}([x_1]) = \overline{f}([x_2])$, then by definition $f(x_1) = f(x_2)$. Hence, $x_1 \sim x_2$ and $[x_1] = [x_2]$.
(3) Consequently, if a map $f : X \longrightarrow Y$ defines an equivalence relation, then it can be factored as $f = \overline{f} \circ \pi$, where $\overline{f}$ is injective and π is surjective.

Theorem 2.3.3. *Let $f : X \longrightarrow Y$ be a surjective continuous map. The following statements are equivalent:*
(1) *The induced map $\overline{f}$ is a homeomorphism.*
(2) *For every open subset $U \subset X/_\sim$, f maps the saturated open set $\pi^{-1}(U)$ into an open set in Y.*
(3) *The function f is an identification map.*

Proof. (1) $\Longrightarrow$ (2): Given an open set $U \subset X/_\sim$, by definition of the quotient topology $\pi^{-1}(U)$ is open in X. Since $\overline{f}$ is a homeomorphism, the range $f(\pi^{-1}(U)) = \overline{f}(U)$ is open in Y.

(2) $\Longrightarrow$ (1): Since $\overline{f} \circ \pi = f$ and f is continuous, by Theorem 2.2.6 $\overline{f}$ is continuous. The map $\overline{f}$ is surjective because f is. Then $\overline{f}$ is a bijection. Given an open set U of $X/_\sim$, $\pi^{-1}(U)$ is open in X. By supposition, the range $f(\pi^{-1}(U)) = \overline{f}(U)$ is open in Y. Then $\overline{f}$ is an open map, hence a homeomorphism.

(2) $\Longrightarrow$ (3): Let $V \subset Y$ be open subset. By continuity of f, $f^{-1}(V)$ is open in X. Conversely, let $f^{-1}(V)$ be open in X. From set theory and the continuity of f, we can check that $f^{-1}(V)$ is saturated. Then $\pi^{-1}(\pi(f^{-1}(V)))$ is open in X. By hypothesis (2), $f(\pi^{-1}(\pi(f^{-1}(V))))$ is open in Y. Finally, using again a set theoretic reasoning and that f is onto, one can check that the latter set is nothing but V, proving the claim.

(3) $\Longrightarrow$ (2): Let U be open in $X/_\sim$, i. e., $\pi^{-1}(U)$ is open in X. Since $f^{-1}(f(\pi^{-1}(U))) = \pi^{-1}(U)$, $f^{-1}(f(\pi^{-1}(U)))$ is open in X. Hence, $\pi(f^{-1}(V))$ is open in $X/_\sim$. By hypothesis (3), $f(\pi^{-1}(U))$ is open in Y, as claimed. □

Remark 2.3.4. We can replace "open" by "closed" in the statements of Theorem 2.3.3.

Proposition 2.3.5. *Let $f : X \longrightarrow Y$ be a continuous surjective map and Y is a Hausdorff space. Then the quotient space $X/_\sim$ is Hausdorff, where the relation $\sim$ is defined by f.*

Proof. Let $[x] \neq [x']$ be two distinct elements of the quotient space $X/_\sim$. Then $\overline{f}([x])$ and $\overline{f}([x'])$ are distinct elements in Y, where $\overline{f}$ is the map induced by f. Since Y is Hausdorff, there exist open neighborhoods $V \in \mathcal{N}_{\overline{f}([x])}$ and $V' \in \mathcal{N}_{\overline{f}([x'])}$ such that $V \cap V' = \emptyset$. By continuity of $\overline{f}$, there exist open neighborhoods $U \in \mathcal{N}_{[x]}$ and $U' \in \mathcal{N}_{[x]'}$ such that $\overline{f}(U) = V$ and $\overline{f}(U') = V'$. Since $\overline{f}$ is bijective, it follows that $U \cap U' = \emptyset$, proving that $X/_\sim$ is Hausdorff. □

As immediate consequences of Corollary 2.1.8, Proposition 2.1.3 and Theorem 2.3.3, we have the following.

Corollary 2.3.6. *Let $f : X \longrightarrow Y$ be a surjective continuous open map (or closed map). Then the induced map $\bar{f}$ is a homeomorphism.*

Corollary 2.3.7. *Let $f : X \longrightarrow Y$ be a surjective continuous map with X compact and Y Hausdorff. Then the induced map $\bar{f}$ is a homeomorphism.*

Example 2.3.8. Let $X = [0,1]$ be given with the equivalence relation

$$t_1 \sim t_2 \iff \begin{cases} \text{either } t_1 = t_2 \\ \text{or } t_1, t_2 \in \{0,1\}. \end{cases}$$

Let $f : X \longrightarrow Y = S^1$ be defined by $f(t) = e^{2\pi i t}$. Clearly, $f(t_1) = f(t_2) \iff t_1 \sim t_2$. Since X is compact and Y is Hausdorff, by Corollary 2.3.7, f induces a homeomorphism $\bar{f} : [0,1]/_\sim \longrightarrow S^1$. This result will be directly proved in Theorem 2.5.11 for the case of a general closed ball B^n instead of $[0,1]$.

Example 2.3.9. Let $X = [0,1] \times [0,1]$ be the square with the equivalence relation

$$(r,t) \sim (r',t') \iff re^{2i\pi t} = r'e^{2i\pi t'}.$$

Let $f : X \longrightarrow Y = B^2(0,1)$ be defined by $f(r,t) = re^{2i\pi t}$. The map f is compatible with the equivalence relation. Since X is compact and Y is Hausdorff, by Corollary 2.3.7, f induces a homeomorphism $\bar{f} : [0,1]^2/_\sim \longrightarrow B^2(0,1)$. Since $[(r,t)] = \{(r,t)\}$, for all $(r,t) \in [0,1]^2$, then $[0,1]^2/_\sim$ is homeomorphic to $[0,1]^2$. Therefore, $B^2(0,1)$ and $[0,1]^2$ are homeomorphic, providing another proof of the equivalence in Example 1.5.14 when $n = 2$.

Example 2.3.10. Let $X = [0,1] \times [0,1]$ be the square where the bottom side points are identified, i. e., $(x,0) \sim (x',0)$ for all $x, x' \in [0,1]$. As intuitively seen, we prove that the resulting quotient space is homeomorphic to a triangle (see [61, Exercise 1.7.4]). Let T_0 be the triangle with vertices $(0,1); (1/2,0); (1,1)$ and $T_1 : \{(0,0); (1/2,0); (0,1)\}$, $T_2 : \{(1/2,0); (1,0); (1,1)\}$ the two triangles whose union is the complement of T_0 in X (Figure 2.1). Define the map $f : X \longrightarrow T_0$ by

$$f(x,y) = \begin{cases} (\frac{1-y}{2}, y), & (x,y) \in T_1, \\ (x,y), & (x,y) \in T_0, \\ (\frac{1+y}{2}, y), & (x,y) \in T_2. \end{cases}$$

Then f is continuous, surjective and agrees with the equivalence relation (f is said to be relation-preserving). Since X is compact and T_0 is Hausdorff, Corollary 2.3.7 provides a homeomorphism between $X/_\sim$ and T_0. The intersection of the triangles is the vertex $A = (1/2,0) = f(x,0)$, $0 \le x \le 1$.

Figure 2.1: Triangle as a quotient space of a square.

Example 2.3.11. On the unit circle $X = S^1$, let us identify the lower half of X with a point by setting $(x,y) \sim (x',y')$ whenever $y \le 0$ and $y' \le 0$ (see [61, Exercise 1.7.9]). Intuitively, we must bring together the diametral points $(1,0)$ and $(-1,0)$ by stretching the lower half of X (i. e., gluing together all its points). As a result, we obtain a circle once again. This is justified by taking the map $f : X = S^1 \longrightarrow Y = S^1$ given by $f(e^{i\theta}) = e^{i\phi(\theta)}$, where $0 \le \theta < 2\pi$ and

$$\phi(\theta) = \begin{cases} 2\theta, & 0 \le \theta \le \pi, \\ 2\pi, & \pi \le \theta < 2\pi. \end{cases}$$

Since ϕ is continuous, so is f. Clearly, $f(S^1) \subseteq S^1$ and the map $\phi : [0,2\pi) \longrightarrow [0,2\pi)$ is surjective. Then $S^1 \subseteq f(S^1)$, i. e., f is surjective. Corollary 2.3.7 implies that $S^1/_\sim$ and S^1 are homeomorphic.

Example 2.3.12. In Euclidean space $X = \mathbb{R}^n$, define the equivalence relation $x \sim y \iff x - y \in \mathbb{Z}^n$. The quotient space $X/_\sim$, denoted $\mathbb{T}^n$, is called the torus. In Example 2.2.5, we have explicitly shown that for $n = 1$, $X/_\sim$ is homeomorphic to S^1. For the general case, the map $f : \mathbb{R}^n \longrightarrow (S^1)^n$ defined by $f(t_1, t_2, \ldots, t_n) = (e^{2\pi i t_1}, e^{2\pi i t_2}, \ldots, e^{2\pi i t_n})$ is continuous and surjective. f induces a continuous bijection $\overline{f} : X/_\sim \longrightarrow (S^1)^n$. Hence, $f = \overline{f} \circ \pi$. The restriction of f to $[0,1]^n$ is still continuous, surjective. By Corollary 2.3.7, $\mathbb{T}^n$ is homeomorphic to $(S^1)^n$ (see Example 3.2.3 for an equivalent definition of the 2-dimensional torus $\mathbb{T}^2$).

Example 2.3.13. The circle S^1 can be regarded as a subspace of the set of complex numbers $\mathbb{C}$. Let $\mathbb{T}^n$ be the n-dimensional torus identified with the n-product $S^1 \times S^1 \times \cdots \times S^1$ in Example 2.3.12. The square map $f : \mathbb{T}^n \longrightarrow \mathbb{T}^n$ given by $f(z) = z^2 = (z_1^2, z_2^2, \ldots, z_n^2)$ is continuous and surjective, where $z = (z_1, z_2, \ldots, z_n)$. Furthermore, f agrees with the relation given by $z \sim z'$ if and only if $z_k = \pm z'_k$ for all $k = 1, 2, \ldots, n$. Since $\mathbb{T}^n$ is compact, by Corollary 2.3.7, the quotient space $\mathbb{T}^n/_\sim$ is homeomorphic to $\mathbb{T}^n$ (compare with Example 2.3.12). Then the torus $\mathbb{T}^n$ is homeomorphic to $(S^1)^n$ with the antipodal points in each circle S^1 identified.

Example 2.3.14. Let $X = \mathbb{R}$ and $[a, b] \subset X$. In X, define an equivalence relation $\sim$ by

$$x_1 \sim x_2 \quad \Longleftrightarrow \quad \begin{cases} \text{either } x_1 = x_2 \\ \text{or } x_1, x_2 \in [a, b]. \end{cases}$$

The quotient space $\mathbb{R}/_{\sim}$ looks like $\mathbb{R}$ with the interval $[a, b]$ shrunk to a point. Even though $\mathbb{R}$ is not compact, we can construct a surjective continuous function $f : X \longrightarrow Y = \mathbb{R}$, which agrees with the relation $\sim$. For example, let f be constant on $[a, b]$, increasing on the half-lines $(-\infty, a)$ and (b, ∞) with $\lim_{x\to\pm\infty} f(x) = \pm\infty$. Then f is continuous, surjective and closed. By Corollary 2.3.6, $\mathbb{R}/_{\sim}$ and $\mathbb{R}$ are homeomorphic. The situation is quite different if we replace $[a, b]$ by (a, b) even though $\mathbb{R}/\sim$ still resembles $\mathbb{R}$. In this case, the process of construction of a continuous map $f : X \longrightarrow Y = \mathbb{R}$, which is constant on (a, b) fails.

2.4 Further properties of the quotient space

Remark 2.4.1. Let $(X, \sim)$ and $(X', \sim')$ be two spaces with two equivalence relations. Let $f : X \longrightarrow X'$ be a continuous map such that for $x_1, x_2 \in X$,

$$x_1 \sim x_2 \quad \Longleftrightarrow \quad f(x_1) \sim' f(x_2).$$

We say that f preserves the equivalence relations. Let $\overline{f} : X/_{\sim} \longrightarrow X'/_{\sim'}$ be defined for $x \in X$ by $\overline{f}([x]) = [f(x)]$, i. e., the following diagram commutes:

$$\begin{array}{ccc} X & \xrightarrow{\quad f \quad} & X' \\ {\scriptstyle \pi_X}\downarrow & & \downarrow{\scriptstyle \pi_{X'}} \\ X/_{\sim} & \xrightarrow{\quad \overline{f} \quad} & X'/_{\sim'} \end{array}$$

Since π_X and $\pi_{X'}$ are surjective, $\overline{f}$ is well-defined. Since f and $\pi_{X'}$ are continuous and $\pi_{X'} \circ f = \overline{f} \circ \pi_X$, Theorem 2.2.6 implies that $\overline{f}$ is continuous. If f is further a homeomorphism, then $X/_{\sim}$ and $X'/_{\sim'}$ are homeomorphic.

Example 2.4.2. Let $X = (0, 1) \times (0, \infty)$ be a vertical infinite strip of Euclidean plane and $X' = (S^1 \setminus \{(-1, 0)\}) \times (0, \infty)$ a slit (cracked) infinite cylinder. Consider the equivalence relations $(s, t) \sim (1 - s, t)$, for $(s, t) \in X$ and $(z, y) \sim (\overline{z}, y)$, for $(z, y) \in X'$, where $\overline{z}$ is the conjugate of the complex number z. By Example 1.5.32, the punctured circle is homeomorphic to the open intervals $(-1, 1)$ and $(0, 1)$. The homeomorphism $h : (0, 1) \longrightarrow S^1 \setminus \{(-1, 0)\}$ given by $h(s) = e^{2i\pi s}$ satisfies $h(1 - s) = \overline{h(s)}$. Let $f : X \longrightarrow X'$ be defined by $f(s, t) = (h(s), t)$. Then f is a homeomorphism and

$$(s, t) \sim (1 - s, t) \quad \Longleftrightarrow \quad f(s, t) \sim f(1 - s, t).$$

Thus, f is compatible with the two equivalence relations. By Remark 2.4.1, the quotient spaces $X/\sim$ and $X'/\sim'$ are homeomorphic. Since $X/\sim$ is homeomorphic to X and the space X is homeomorphic to $(0,1)\times(0,1)$, we obtain that $X/\sim$ is homeomorphic to $(0,1)\times(0,1)$. Therefore, $X'/\sim'$ is homeomorphic to $(0,1)\times(0,1)$, and thus homeomorphic to the open disk of the plane.

Proposition 2.4.3. *The quotient space of a quotient space of X is homeomorphic to a quotient space of X.*

Proof. Let $\sim$ and $\sim_1$ be two equivalence relations in the spaces X and $X_1 = X/_\sim$, respectively, and let $X_2 = X_1/_{\sim_1}$. For $x \in X$, denote by $[x]$ the equivalence class of x in X_1 and $[[x]]_1$ the equivalence class of $[x]$ in X_2. Define the equivalence relation $\sim'$ in X by

$$x \sim' y \quad \Longleftrightarrow \quad [x] \sim_1 [y].$$

For $X' = X/_{\sim'}$, define the map $h : X' \longrightarrow X_2$ by $h([x]') = [[x]]_1$, where $[x]'$ refers to the equivalence class in X'. By definition, the following diagram commutes:

$$\begin{array}{ccc} X & \xrightarrow{\ \pi_1\ } & X_1 \\ {\scriptstyle \pi'}\downarrow & & \downarrow{\scriptstyle \pi_2} \\ X' & \xrightarrow{\ h\ } & X_2 \end{array}$$

where π_1, π_2, π' are the projection maps. Hence, h can be viewed as the induced map of the composition $\pi_2 \circ \pi_1$:

$$X \xrightarrow{\ \pi'\ } X' \xrightarrow{\ h\ } X_2, \qquad h \circ \pi' = \pi_2 \circ \pi_1$$

Since $\pi_2 \circ \pi_1$ is an identification map, by Theorem 2.3.3, h is a homeomorphism. □

Proposition 2.4.4. *If X is locally connected (locally path-connected, respectively), then the quotient space $X/_\sim$ is locally connected (locally path-connected, respectively).*

Proof. We prove the proposition in the case of locally connected spaces. Let $U \subset X/_\sim$ be an open set and $C \subset U$ a component. Then $\pi^{-1}(C) \subset \pi^{-1}(U)$, where π is the projection map and $\pi^{-1}(U)$ is open in X. Let $x \in \pi^{-1}(C)$ and C_x the component of x in $\pi^{-1}(U)$. Then

$$x \in C_x \subset \pi^{-1}(U) \quad \Longrightarrow \quad \pi(x) \in \pi(C_x) \subset \pi(\pi^{-1}(U)) \subset U.$$

Since C is a component, which contains $\pi(x)$ in U and $\pi(C_x)$ is connected, we have $\pi(C_x) \subset C \subseteq U$. Hence, $x \in C_x \subset \pi^{-1}(C)$. In addition, C_x is open because X is locally compact. Therefore, $\pi^{-1}(C)$ is open in X, i. e., C is open in $X/_\sim$. By Theorem 1.4.46, we conclude that $X/_\sim$ is locally connected. □

2.5 Equivalence relation defined by a set

Given a subset $A \subset X$, define an equivalence relation by $x_1 \sim x_2 \Longleftrightarrow x_1 = x_2$ or $x_1, x_2 \in A$. The quotient space $X/_\sim$ is obtained by identifying (or shrinking, collapsing, reducing) the subset A into a single element. This is why the quotient space is called an identifying space in this case. $X/_\sim$ is denoted X/A and is also called the quotient space of X modulo A. It consists of the partition whose members are the one-element sets $\{\{x\}, x \in X \setminus A\}$ together with the one-set element $\{A\}$.

As noticed in Remark 2.4.1, if X and $X' = h(X)$ are homeomorphic, then so are X/A and X'/A' where $A' = h(A)$.

$X \setminus A$ will stand for the complement of A in X, denoted also $X - A$ in some books. We hope this will cause no confusion throughout the book.

Proposition 2.5.1. *Let A be either open or closed in X. Then the difference set $X \setminus A$ and $(X/A) \setminus \pi(A)$ are homeomorphic.*

Proof. Let $\pi_c = \pi_{|X\setminus A}$, where $\pi : X \longrightarrow X/A$ is the canonical projection map. Then $\pi_c : X \setminus A \longrightarrow (X/A) \setminus \pi(A)$ is a continuous bijection. Moreover, if A is closed in X, then π_c is an open map. Indeed, every open subset $U \subset X \setminus A$ is open in X. Then $\pi_c(U)$ is open in X/A because $\pi^{-1}(\pi_c(U)) = U$. A similar reasoning holds if A is open in X to show that π_c is closed. □

The quotient space X/A is described in the following examples.

Example 2.5.2. By taking $f(t) = e^{2i\pi t}$, $t \in [0, 1]$ and appealing to Corollary 2.3.7, we obtain that $[0, 1]/_{\{0,1\}} \cong S^1$.

Example 2.5.3. Let $X = \mathbb{R}$ and $A = \mathbb{Z}$. Then the quotient space is $\mathbb{R}/\mathbb{Z} = \{\{\{x\}, n < x < n+1, n \in \mathbb{Z}\}, [0]\}$ where $[0] = [1] = \cdots = [n] = \cdots$. Thus, $\mathbb{R}/\mathbb{Z}$ is a collection of countably many circles (a bunch of circles or a bouquet of circles) with one common point (see Figure 2.2 and compare with Example 2.2.5, where the equivalence relation is different). The general definition of the wedge sum is given in Section 3.3 (see Definition 3.3.1(4)).

Example 2.5.4. Consider the n-sphere $X = S^n \subset \mathbb{R}^{n+1}$ and let $\mathrm{Eq} = S^n \cap \mathbb{R}^n$ be the equator (projection of the sphere onto the space $\mathbb{R}^n$). Define the relation $x \sim x' \Longleftrightarrow (x = x')$ or $x, x' \in \mathrm{Eq}$. Let Y be the union of two spheres having the origin as unique common point. Let $f : X \longrightarrow Y$ be the map defined by $f(x) = 0 \in \mathbb{R}^{n+1}$, if $x \in \mathrm{Eq}$ and $f(x) = x$ if else. Clearly, f is surjective and by the pasting Lemma A.2.1, f is continuous. Corollary 2.3.7 implies that the quotient space $S^n/(S^n \cap \mathbb{R}^n)$ obtained by sewing the equator to a point is homeomorphic to the union of two spheres (see Figure 2.3). Since $S^n \cap \mathbb{R}^n \cong S^{n-1}$, we

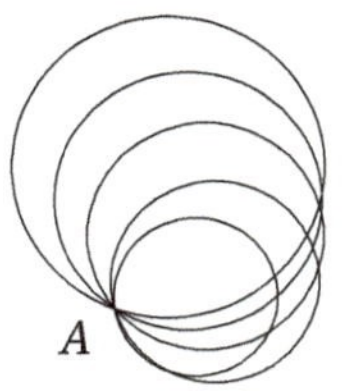

Figure 2.2: Bouquet of circles.

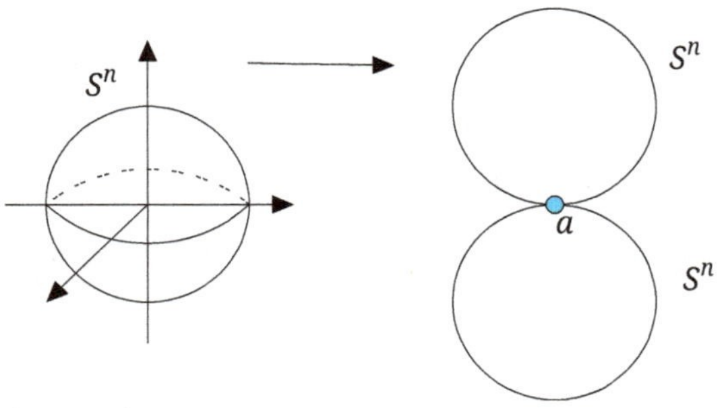

Figure 2.3: Sphere with equator identified with a point.

can write $S^n/S^{n-1} \cong S^n \vee S^n$, where we have anticipated by using the symbol $\vee$ for the wedge sum. The sphere is split into two spheres of same dimension.

Example 2.5.5. Let $B_+^2 = \{(x,y) \in B^2 : y \geq 0\}$ be the closed upper disk with the equivalence relation $(x,0) \sim (-x,0)$, $-1 \leq x \leq 1$ (see [90, Exercise 1.3.11(ii)]). Using complex notation for $z = (x,y)$, the map $f : X = B_+^2 \longrightarrow Y = B^2$ given by $f(z) = z^2$ is continuous, surjective and relation-preserving. Indeed,

$$f(z) = f(z') \iff z' = z \text{ or } z' = -z \iff z' = z \text{ or } z' = (x',0) = -z = -(x,0),$$

for all $-1 \leq x \leq 1$. By Corollary 2.3.7, $B_+^2/_\sim \cong B^2$.

Example 2.5.6. Let $S_+^2 = \{(x,y,z) \in S^2 : z \geq 0\}$ be the closed northern hemisphere of the 2-dimensional sphere with the equivalence relation $(x,y,0) \sim (-x,y,0)$ and $x^2 + y^2 = 1$ (see [90, Exercise 1.3.11(iii)]). Define the map $f : X = S_+^2 \longrightarrow Y = S^2$ by

$$f(x, r\cos\theta, r\sin\theta) = (x, r\cos 2\theta, r\sin 2\theta),$$

where $0 \leq \theta \leq \pi$, $0 \leq r \leq 1$ and $r^2 = 1 - x^2$. Clearly, f is continuous, surjective and relation-preserving. Indeed,

$$f(x,y,z) = f(x',y',z') \iff x = x', \quad \theta' = \theta + k\pi, \quad r' = r \quad (k = 0 \text{ or } k = \pm 1).$$

Hence,

$$f(x,y,z) = f(x',y',z') \iff x = x', \quad \theta = 0, \quad \theta' = \pi \quad (\text{or } \theta' = 0, \theta = \pi) \quad r' = r.$$

Finally,

$$f(x,y,z) = f(x',y',z') \iff (x,y,0) \sim (x,-y,0).$$

By Corollary 2.3.7, $S^2_+/_\sim \cong S^2$. Note that one can replace the northern hemisphere S^2_+ by the closed upper ball B^3_+ with $x^2 = 1 - r^2$ replaced by $x^2 \leq 1 - r^2$ and consider the equivalence relation $(x,y,0) \sim (-x,y,0)$ and $x^2 + y^2 \leq 1$. The resulting quotient space is $B^3_+/_\sim \cong B^3$.

A quotient space need not be Hausdorff. However, we have the following.

Proposition 2.5.7. *Let X be* a regular space *and $A \subset X$ a closed subset. Then the quotient space X/A is Hausdorff.*

Proof. Let $z, z' \in X/A$ be two distinct elements. If $z = \{x\}$ and $z' = \{x'\}$ with $x, x' \in X \setminus A$, then z and z' are separated by open neighborhoods because $X \setminus A$ is Hausdorff. If $z' = \{A\}$ and $z = \{x\}$, for some $x \in X \setminus A$, then $\pi^{-1}(z) = \{x\}$. Since X is regular (Definition 1.1.16), there exist disjoint open neighborhoods U and V of $\{x\}$ and A, respectively. Since A is closed in X, then by the proof of Proposition 2.5.1, π is an open map. Therefore, the images $\pi(U)$ and $\pi(V)$ are disjoint open neighborhoods of z and z', respectively. □

Proposition 2.5.8. *Let X be Hausdorff and $A \subset X$ a compact subset. Then the quotient space X/A is Hausdorff.*

Proof. Let $z, z' \in X/A$ be two distinct elements. Then either $z = \{x\}$ and $z' = \{x'\}$ with $x, x' \in X \setminus A$ or $z' = \{A\}$ and $z = \{x\}$, for some $x \in X \setminus A$. In the first case since X is Hausdorff, there exist open neighborhoods $U \in \mathcal{N}_x$ and $U' \in \mathcal{N}_{x'}$ such that $U \cap U' = \emptyset$. Let $V = U \cap (X \setminus A)$ and $V' = U' \cap (X \setminus A)$. Then U' and V' are disjoint open sets in $X \setminus A$. Hence, $\pi(U') = U'$ and $\pi(V') = V'$ are disjoint open neighborhoods of z, z', respectively, where π is the canonical projection map. In the second case, since X is Hausdorff, for all $a \in A$, there exist $U^x_a \in \mathcal{N}_x$ and $V^x_a \in \mathcal{N}_a$ such that $U^x_a \cap V^x_a = \emptyset$. Then $A \subseteq \bigcup V^x_a$. Since A is compact, $A \subseteq \bigcup_{i=1}^N V^x_{a_i} = V^x_0$. Let the open set $U^x_0 = \bigcap_{i=1}^N U^x_{a_i}$. Then $U^v_0 \cap V^x_0 = \emptyset$, where $U^x_0 \in \mathcal{N}_x$ and $V^x_0 \in \mathcal{N}_A$, proving the claim. □

Assume that X is compact. Then X/A is compact as the continuous image of a compact set. From either Proposition 2.5.7 (since a compact set is a regular space) or Proposition 2.5.8, we obtain the following.

Corollary 2.5.9. *Let X be compact Hausdorff and $A \subset X$ a closed subset. Then the quotient space X/A is compact Hausdorff.*

The following result is a consequence of [76, Lemma 73.3], which states that if X is normal and π is a closed quotient map, then $\pi(X)$ is normal.

Proposition 2.5.10. *Let X be a normal space (Definition 1.1.18) and $A \subset X$ a closed subset. Then the quotient space X/A is a normal space.*

Notice that A is also normal for the subspace topology.

Consider the closed unit ball $X = B^n$ and its boundary $A = S^{n-1}$ equipped with Euclidean norm

$$\|x\| = \left(\sum_{i=1}^{n} x_i^2\right)^{1/2}.$$

Denote by B^n/S^{n-1} the quotient space

$$\{[x],\ x \in B^n\} = \{\{S^{n-1}\}, \{\{x\},\ x \notin S^{n-1}\}\}.$$

Theorem 2.5.11. *For $n \geq 1$, B^n/S^{n-1} and S^n are homeomorphic.*

Proof. (1) Consider the homeomorphisms $h_1 : B^n \setminus S^{n-1} \longrightarrow \mathbb{R}^n$ and $h_2 : \mathbb{R}^n \longrightarrow S^n \setminus \{N_p\}$, where $N_p = (0, 0, \ldots, 0, 1) \in \mathbb{R}^{n+1}$ is the northern pole and h_1, h_2 are defined by $h_1(x) = \frac{x}{1-\|x\|}$ and

$$h_2(x) = \left(\frac{2x_1}{1+\|x\|^2}, \ldots, \frac{2x_n}{1+\|x\|^2}, \frac{\|x\|^2-1}{\|x\|^2+1}\right).$$

Then define the map

$$h(x) = \begin{cases} (h_2 \circ h_1)(x), & x \in B^n \setminus S^{n-1}, \\ N_p, & x \in S^{n-1}. \end{cases}$$

We have

$$h(x) = \left(\frac{2x_1(1-\|x\|)}{\|x\|^2 + (1-\|x\|)^2}, \ldots, \frac{2x_n(1-\|x\|)}{\|x\|^2 + (1-\|x\|)^2}, \frac{\|x\|^2 - (1-\|x\|^2)^2}{\|x\|^2 + (1-\|x\|^2)^2}\right).$$

Then $h(S^{n-1}) = N_p$, $\lim_{\|x\| \to 1} h(x) = p$, h is onto and

$$h : B^n \longrightarrow S^n \text{ is continuous,}$$
$$h : B^n \setminus S^{n-1} \longrightarrow S^n \setminus \{Np\} \text{ is one-to-one.}$$

(2) Note that

$$\{h^{-1}(y),\ y \in S^n\} = \{h^{-1}(y),\ y \in S^n \setminus \{p\}\} \cup \{S^{n-1}\}.$$

Since $h_{|B^n/S^{n-1}}$ is one-to-one, we get

$$\{h^{-1}(y),\ y \in S^n\} = \{\{x\},\ x \in B^n \setminus S^{n-1}\} \cup \{S^{n-1}\} = B^n/S^{n-1}.$$

Define the map $\Phi : B^n/S^{n-1} \longrightarrow S^n$ by $\Phi(\{h^{-1}(y)\}) = y$.
(a) Since h is surjective ($h^{-1}(y) \neq \emptyset$), so is Φ.
(b) h is one-to-one because $\{h^{-1}(y),\ y \in S^n\}$ is a partition of B^n/S^{n-1}.

Hence, h is bijective. Moreover, $\Phi \circ \pi = h$ and $\Phi^{-1} \circ h = \pi$, where $\pi : B^n \longrightarrow B^n/S^{n-1}$ is the quotient map. Using Proposition 2.1.9, we conclude that

$$\begin{aligned} h = \Phi \circ \pi \text{ continuous} &\implies \Phi \text{ continuous} \\ \pi = \Phi^{-1} \circ h \text{ continuous} &\implies \Phi^{-1} \text{ continuous.} \end{aligned}$$

Therefore, Φ is a homeomorphism. □

Exercises (Chapter 2)

1. Let n be a positive integer. Define on S^1 the relation $z \sim z' \iff z = z'$ or $z^n = (z')^n$. Determine the quotient space $S^1/_\sim$.

2. In Euclidean plane $\mathbb{R}^2$, define an equivalence relation by $(x,y) \sim (x',y')$ if and only if $xy = x'y'$. Prove that the quotient space $\mathbb{R}^2/_\sim$ is homeomorphic to $\mathbb{R}$.

3. In $\mathbb{R}$, define the relation $x \sim y \iff x - y \in \mathbb{Q}$. Prove that the topology of the quotient space $\mathbb{R}/_\sim$ is the trivial topology.

4. Let $U \subset X/_\sim$ be a nonempty subset. Prove that U is open if and only if there exists a saturated open subset $A \subset X$ such that $U = \pi(A)$.

5. Let $f : X \longrightarrow Y$ be a surjective continuous map and X, Y two spaces. Prove that f is an identification map if and only if

$$f^{-1}(C) \text{ closed in } X \iff C \text{ closed in } Y.$$

6. Prove that the quotient space $X/_\sim$ is Hausdorff if and only if for all $x,y \in X$, if $x \nsim y$, then there exist saturated open neighborhoods $U \in \mathcal{N}_x$ and $V \in \mathcal{N}_y$ such that $U \cap Y = \emptyset$.

7. Let $X/_\sim$ be a quotient space and $A = \{(x,y) \in X \times X : x \sim y\}$.
(1) Prove that if $X/_\sim$ is Hausdorff, then A is closed in $X \times X$.
(2) Assume that A is closed in $X \times X$ and X is compact. Prove that $X/_\sim$ is Hausdorff. (*Hint:* See [100, Proposition 2.1].)

8. On the real numbers $\mathbb{R}$, defined the equivalence relation $x \sim y$ if and only if $x = y$ or $x \neq 0$ and $y = 1/x$.

(1) Determine the quotient space $A/_{\sim}$.

(2) Show that the set $A = [1, +\infty)$ is not saturated.

(3) Find a saturated set, if any.

9. Let $f : X \longrightarrow Y$ be an identification map with X compact Hausdorff. Prove that the following statements are equivalent:

(1) Y is Hausdorff;

(2) f is closed;

(3) the set

$$\{(x, y) \in X \times X : f(x) = f(y)\}$$

is closed in $X \times X$ (see Exercise 3, Section 1.3).

10. Prove that the set of connected components endowed with the quotient topology is totally disconnected.

3 Topological constructions

By making use of some of the quotient spaces investigated in Chapter 2, our purpose in this chapter is to build up some new topological spaces. Some constructed surfaces will occupy major parts of this book, where they are used to illustrate theorems and examples. We quote the projective space (Section 3.1), the torus, the Möbius band and the Klein bottle (Section 3.2). Further spaces are introduced using quotient spaces and mappings between spaces. In Section 3.3, we will define the cone, the suspension space, the wedge sum and the smash product of spaces. Spaces attached by maps are particularly involved in cell homology. So, special attention is paid to the adjunction space defined in Section 3.4. In Section 3.5, we consider the special cases of the mapping cylinder and the mapping cone. Manifolds and their connected sums are discussed in Section 3.6 and Section 3.7. The polygonal presentations of manifolds are helpful in determining the fundamental groups of the connected sum of these manifolds. They are presented in Section 3.8.

3.1 The projective space

Definition 3.1.1. Let $X = S^n$ be the n-sphere with the equivalence relation defined by

$$x_1 \sim_1 x_2 \quad \Longleftrightarrow \quad \begin{cases} \text{either } x_1 = x_2 \\ \text{or } x_1 = -x_2, \end{cases}$$

where the equivalence classes are $[x]_1 = \{x, -x\}$ for $x \in S^n$. Let $f : X \longrightarrow Y$ be the function defined by $f(x) = (0, x)$, where $(0, x)$ is the line joining x and the origin of $\mathbb{R}^{n+1}$ and Y is the set of all lines passing through the origin of $\mathbb{R}^{n+1}$. Notice that f is continuous, surjective and agrees with $\sim_1$ so by Corollary 2.3.7 it induces a homeomorphism between $X/_{\sim_1}$ and Y. Moreover, X is compact and Y is Hausdorff. The quotient space $X/_{\sim_1}$, which is homeomorphic to Y is called *the real projective space* and denoted $\mathbb{RP}^n$ or sometimes $\mathrm{P}_n(\mathbb{R})$. Thus, $\mathbb{RP}^{n-1}$ can be identified with a subspace of $\mathbb{RP}^n$. The quotient map is denoted π_1. The subspace $\mathbb{RP}^2 \subset \mathbb{R}^3$ is called the real projective plane.

Remark 3.1.2. Owing to Remark 2.4.1, one can replace the sphere S^2 by the boundary of the unit 3-cube $[0, 1]^3$ to obtain the projective plane $\mathbb{RP}^2$ by suitable equivalence of relation on the opposite faces. See also Example 3.1.15.

We describe in detail the first four projective spaces for $n = 0, 1, 2, 3$.

Example 3.1.3. Since $S^0 = \{-1, +1\}$, $\mathbb{RP}^0$ is a single-element set.

Recall the following.

https://doi.org/10.1515/9783111517384-003

Definition 3.1.4. (1) An orthogonal matrix is a $n \times n$ matrix I_n such that $I_n^{-1} = I_n^t$, where I_n^t is the transpose of I_n.
(2) The group $\mathrm{SO}(n)$ is the group of special $n \times n$ orthogonal matrices of determinant 1.

The group $\mathrm{SO}(n)$ is identified with the group of rotations in $\mathbb{R}^n$.

Proposition 3.1.5. *The space* $\mathbb{RP}^1$, *the circle* S^1 *and the space* $\mathrm{SO}(2)$ *rotations in Euclidean plane are homeomorphic.*

Proof. (1) The space $\mathbb{RP}^1$ is homeomorphic to the set of lines passing through the origin of the plane $\mathbb{R}^2$. The square map $f : S^1 \longrightarrow S^1$ given by $f(z) = z^2$ is continuous, surjective and agrees with the equivalence relation $\sim_1$ that defines $\mathbb{RP}^1$. Since S^1 is compact Hausdorff, by Corollary 2.3.7, $\mathbb{RP}^1$ is homeomorphic to S^1.

(2) It is easy to see that the map $h : S^1 \longrightarrow \mathrm{SO}(2)$ defined by

$$h(e^{i\theta}) = \begin{pmatrix} \cos\theta & -\sin\theta \\ \sin\theta & \cos\theta \end{pmatrix}$$

is a homeomorphism. □

Corollary 3.1.6. *The group* $\mathrm{SO}(2)$ *is compact Hausdorff and path-connected.*

Proposition 3.1.7. *The space* $\mathbb{RP}^3$ *is homeomorphic to the group* $\mathrm{SO}(3)$.

Proof. Let $\mathbb{RP}^3$ be defined by Proposition 3.1.13 and $f : B^3 \longrightarrow \mathrm{SO}(3)$ the map given by $f(0) = \mathrm{I}_3$, the 3×3 matrix identity and $f(x)$ the rotation about the vector x through angle $\theta = \pi\|x\|$, for $x \neq 0$. Recall that a rotation about a vector $v = (a, b, c)$ through angle θ can be written as (see, e. g., [77]):

$$R_\theta(a,b,c) = \begin{pmatrix} a^2(1-\cos\theta)+\cos\theta & ab(1-\cos\theta)-c\sin\theta & ac(1-\cos\theta)+b\sin\theta \\ ab(1-\cos\theta)+c\sin\theta & b^2(1-\cos\theta)+\cos\theta & bc(1-\cos\theta)-a\sin\theta \\ ac(1-\cos\theta)-b\sin\theta & bc(1-\cos\theta)+a\sin\theta & c^2(1-\cos\theta)+\cos\theta \end{pmatrix}.$$

Hence, f is continuous. If $x = (x_1, x_2, x_3)$ lies on the sphere $S^2 = \partial B^3$, then the angle of rotation is $\theta = \pi$, which yields

$$f(x) = \begin{pmatrix} 2x_1^2 - 1 & 2x_1x_2 & 2x_1x_3 \\ 2x_1x_2 & 2x_2^2 - 1 & 2x_2x_3 \\ 2x_1x_3 & 2x_2x_3 & 2x_3^2 - 1 \end{pmatrix}.$$

Hence, $f(x) = f(y)$ if and only if $x = \pm y$ for all $x, y \in S^2$, i. e., f maps antipodal points to the same rotation. Consider the diagram

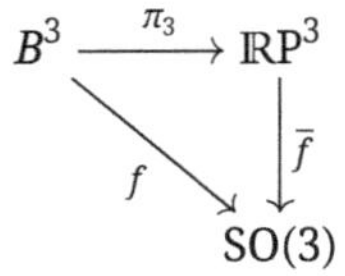

To check that f is surjective, let $R(v,\theta)$ be a rotation about a vector v of Euclidean space through angle θ. If $0 \le \theta \le \pi$, then $f(x) = R(v,\theta)$ for $x = \frac{\theta}{\pi}\frac{v}{\|v\|} \in B^3$. If $\pi \le \theta \le 2\pi$, then $0 \le 2\pi - \theta \le \pi$ and again we have $f(x) = R(v,\theta)$ for $x = \frac{2\pi-\theta}{\pi}\frac{v}{\|v\|} \in B^3$. Since B^3 is compact and SO(3) is Hausdorff, by Corollary 2.3.7, $\bar{f}$ is a homeomorphism. □

Some topological properties of the projective space are given by the following.

Proposition 3.1.8. *The projective space* $\mathbb{R}P^n$ *is:*
(1) *compact and path-connected;*
(2) *Hausdorff.*

Proof. (1) By Example 1.4.23 and Remark 2.2.4, $\mathbb{R}P^n$ is compact and path-connected in the quotient topology.

(2) By Exercise 1, Section 1.3, it is sufficient to prove that the diagonal

$$D = \{(x,y) \in \mathbb{R}^{n+1} \setminus \{0\} \times \mathbb{R}^{n+1} \setminus \{0\} : x \sim_1 y\}$$

is closed. The equivalence relation $\sim_1$ is given by Definition 3.1.1. Consider the continuous function $f(x,y) = \sum_{i\neq j}(x_i y_j - x_j y_i)^2$, for $x, y \in \mathbb{R}^{n+1} \setminus \{0\}$ and note that $D = f^{-1}(0)$. Indeed, since $x \sim_1 y$ means the existence of a real $t \neq 0$ such that $x = ty$, clearly $D \subset f^{-1}(0)$. Conversely, if $f(x,y) = 0$, then $x_i y_j = x_j y_i$ for all $i \neq j$. Then, for some $x_{i_0} \neq 0$, $y_j = \frac{x_j y_{i_0}}{x_{i_0}}$ for every $j \neq i_0$. Hence, $y = tx$ for $t = \frac{y_{i_0}}{x_{i_0}}$, i. e., $x \sim_1 y$ and $f^{-1}(0) \subset D$. □

From Proposition 3.1.7 and Proposition 3.1.8, we have the following.

Corollary 3.1.9. *The group* SO(3) *is a compact Hausdorff path-connected space.*

Proof. The path-connectedness of SO(3) can also be proved directly by considering the path $\varphi : [0,1] \longrightarrow \mathrm{SO}(3)$ defined by $\varphi(t) = R_{t\theta}(a,b,c)$, where the rotation $R_\theta(a,b,c)$ is introduced in the proof of Proposition 3.1.8. □

A further description of the projective space is provided by the following.

Proposition 3.1.10. *In the punctured space* $X = \mathbb{R}^{n+1} \setminus \{0\}$, *define an equivalence relation by*

$$x \sim_2 y \iff \exists \lambda \in \mathbb{R} \setminus \{0\}, \quad x = \lambda y.$$

Then $\mathbb{R}P^n$ *is homeomorphic to* $X/\sim_2$.

Proposition 3.1.10 provides a more direct description of the real projective space: $\mathbb{RP}^n$ is homeomorphic to the set of lines passing through the origin.

Proof. (1) Let π_2 be the projection map on X and $Y = S^n/_{\sim_1}$. Define the function $f = \pi_1 \circ r : X \longrightarrow Y$, where $r(x) = x/\|x\|$ is the radial retraction on S^n and $\pi_1(x) = [x]_1$ (see Definition 3.1.1). We first check that f agrees on $\sim_2$. We have

$$f(x) = f(y) \iff [x/\|x\|]_1 = [y/\|y\|]_1 \iff x/\|x\| \sim_1 y/\|y\|.$$

Then $x = -\frac{\|x\|}{\|y\|}y$, and thus $x \sim_2 y$. Conversely,

$$x \sim_2 y \iff \exists \lambda \in \mathbb{R}, \quad x = \lambda y.$$

Therefore, there exists $\lambda \neq 0$ such that $\|x\| = |\lambda|\|y\|$. Hence, $x/\|x\| = \frac{\lambda}{|\lambda|}y/\|y\|$, i. e.,

$$x/\|x\| = \pm y/\|y\| \iff x/\|x\| \sim_1 y/\|y\|.$$

Hence, $[x/\|x\|]_1 = [y/\|y\|]_1$ and so $f(x) = f(y)$.

(2) The map f is surjective and continuous as the composite of two continuous maps. We claim that f is an open map. Since r is an open map (see Exercise 1, Section 1.5), it suffices to prove that π_1 is open. To this end, let $V \subset S^n$ be an open set. Then $\pi_1^{-1}(\pi_1(V)) = V \cup (-V)$. Indeed, $x \in V \Longrightarrow \pi_1(x) \in \pi_1(V)$ and $x \in (-V) \Longrightarrow (-x) \in V \Longrightarrow \pi_1(-x) = \pi_1(x) \in \pi_1(V)$. Conversely, if $x \in \pi_1^{-1}(\pi_1(V))$, then $\pi_1(x) \in \pi_1(V)$, i. e., there exists $x' \in V$ such that $\pi_1(x) = \pi_1(x')$, i. e., $x = \pm x'$, and thus $x \in V \cup (-V)$. It follows that $\pi_1(V)$ is open for the quotient topology. By Corollary 2.3.6, we conclude that $\mathbb{R}^{n+1} \setminus \{0\}/_{\sim_2}$ and $S^n/_{\sim_1}$ are homeomorphic. Notice here that unlike Example 2.3.10, the space X need not be compact. □

Remark 3.1.11. Let $i : \mathbb{R}^n \hookrightarrow \mathbb{R}^{n+1}$ be the inclusion map defined by $i(x_1, x_2, \dots, x_n) = (x_1, x_2, \dots, x_n, 0)$. Then $i(S^n) \subseteq S^n$ and i induces an injective map $j : \mathbb{R}^n \setminus \{0\}/_{\sim_2} \hookrightarrow \mathbb{R}^{n+1} \setminus \{0\}/_{\sim_2}$ given by $j([x]) = [i(x)]$. Since $\mathbb{RP}^{n-1}$ is compact, $\mathbb{RP}^{n-1}$ can be thought of as a closed subset of $\mathbb{RP}^n$.

Definition 3.1.12. The points x and $-x$ on the sphere S^n are called *antipodal points* (they lie on the same line through the origin).

Proposition 3.1.13. *In the closed ball $X = B^n$, define an equivalence relation $\sim_3$ with the antipodal points of the boundary S^{n-1} identified and the interior points unchanged. Then $X/\sim_3$ is homeomorphic to $\mathbb{RP}^n$.*

Proof. The closed ball $X = B^n$ is homeomorphic to the closed northern hemisphere S^n_+ (see Example 1.5.22). Thus, consider B^n as a subset of S^n (the case $n = 2$ is discussed in [10, Example 13.6]). In B^n, define the equivalence relation $\sim_3$ as in Definition 3.1.1. By the way, notice that for $n = 1$, this process explains why $\mathbb{RP}^1$ is homeomorphic to S^1. Consider the diagram

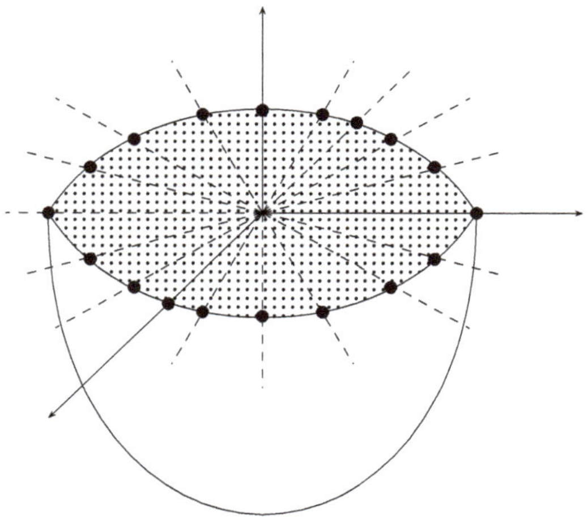

Figure 3.1: Projective plane.

$$\begin{array}{ccc} B^n & \xrightarrow{\quad i \quad} & S^n \\ \pi_3 \downarrow & & \downarrow \pi_1 \\ B^n/_{\sim_3} & \xrightarrow{\quad h \quad} & S^n/_{\sim_1} \end{array}$$

where i is the inclusion map and h is uniquely determined by $h \circ \pi_3 = \pi_1 \circ i$, i. e., the diagram commutes. Since i is continuous, by Theorem 2.2.6 h is continuous. The map h is a bijection. Since $B^n/_{\sim_3}$ is compact because B^n is and $S^n/_{\sim_1}$ is Hausdorff, by Theorem 1.5.36, $B^n/_{\sim_3}$ and $S^n/_{\sim_1}$ are homeomorphic. □

Remark 3.1.14. Proposition 3.1.13 provides a third way to introduce the nth projective space $\mathbb{R}P^n$. In order to define $\mathbb{R}P^n$, we point out that the identification $x \sim_3 -x$ is given on S^{n-1} rather than on S^n ($\mathbb{R}P^2$ is depicted in Figure 3.1).

Example 3.1.15. According to Proposition 3.1.13 and Remark 3.1.2, the projective plane $\mathbb{R}P^2$ can be also obtained from the square $I^2 = [0,1]^2$ via the equivalence relation $(0,y) \sim (1, 1-y)$ and $(x,0) \sim (1-x,1)$ for $(x,y) \in I^2$.

The complex projective space is defined in a similar way.

Definition 3.1.16. In the punctured complex space $X = \mathbb{C}^{n+1} \setminus \{0\}$, define the equivalence relation $x \sim y$ if and only if there exists $\lambda \in \mathbb{C}$ such that $y = \lambda x'$.

The quotient space $X/_\sim$, denoted $\mathbb{C}P^n$, is called *the complex projective space.*

Remark 3.1.17. The sphere $X = S^{2n+1}$ can be viewed as a subspace of $\mathbb{R}^{2n+2} \cong \mathbb{C}^{n+1}$ (in fact in $\mathbb{C}^{n+1} \setminus \{0\}$). $\mathbb{C}P^n$ can also be defined by the equivalence relation $z \sim -z$ or

$z_k \sim z'_k$ if and only if there exists $\lambda \in \mathbb{C}$ such that $z'_k = \lambda z_k$ for all $z = (z_1, \ldots, z_n, z_{n+1})$, $z' = (z'_1, \ldots, z'_n, z'_{n+1}) \in S^{2n+1}$.

Note that for all z_k, $k = 1, \ldots, n+1$, $\sum_{k=1}^{n+1} |z_k|^2 = 1$ so that $\lambda \in S^1$.

Remark 3.1.18. (1) Arguing as in Proposition 3.1.5, we can show that $\mathbb{C}\mathrm{P}^1$ is homeomorphic to S^2 (see Exercise 21, Sections 3.3–3.5).

(2) More properties of complex projective spaces are proposed in Exercises 15 and 16 of Sections 3.1–3.2 and Exercises 21 and 22 of Sections 3.3–3.5.

3.2 Torus, Möbius band and Klein bottle

Quotient and product spaces may be combined by considering the quotient space X/A, where X and A are product spaces. Geometrically, in order to define an equivalence relation one can construct quotient topologies by gluing some points of the subspace A together, but not necessarily all together. Next, some spaces are constructed through examples.

Example 3.2.1. Let $X = [0,1]^n$, $A = \partial X$ and identify all points of the boundary of X. Note that X is homeomorphic to the closed unit ball B^n_∞ in $\mathbb{R}^n$ for the max-norm $\|x\|_\infty = \max(|x_i|,\ 1 \le i \le n)$. The latter ball is homeomorphic to the closed unit ball B^n_2 for Euclidean norm $\|x\|_2 = \sqrt{\sum_{i=1}^n x_i^2}$, where a homeomorphism is given by $h(x) = \frac{\|x\|_2}{\|x\|_\infty} x$ if $x \neq 0$ and $h(0) = 0$. In addition, $h^{-1}(x) = \frac{\|x\|_\infty}{\|x\|_2} x$ if $x \neq 0$, $h^{-1}(0) = 0$ and $h(S^{n-1}) = A$. By Remark 2.4.1 and Theorem 2.5.11, X/A is homeomorphic to S^n (see Figure 3.2 for $n = 2$).

Example 3.2.2. Let $X = [0,1]^2$ and

$$A = \{(x,0), (x,1) : 0 \le x \le 1\}$$

be the set of opposite points on the horizontal edges of the square. Then X/A is just $X/_\sim$, where $(x,y) \sim (x',y') \iff x = x'$ and $y, y' \in \{0,1\}$, the quotient space obtained by gluing

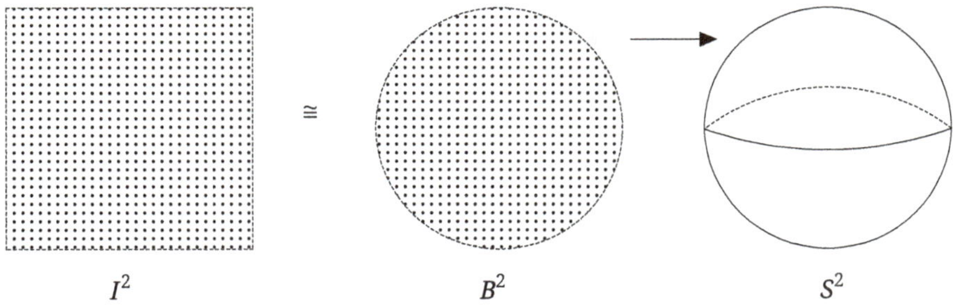

Figure 3.2: Sphere obtained from a square by identifying boundary points.

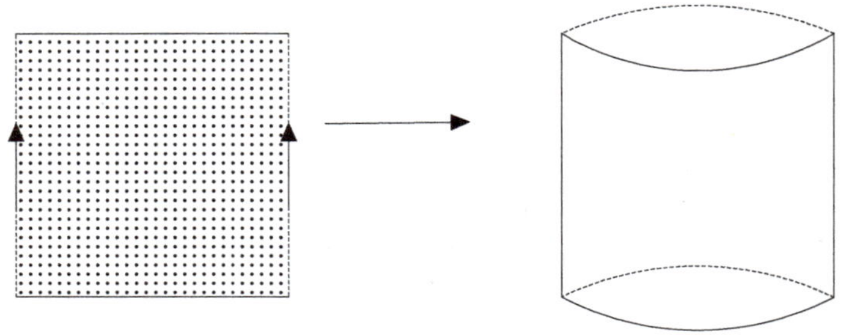

Figure 3.3: Cylinder obtained from a rectangle by gluing the vertical sides.

together the horizontal edges of the unit square. To determine the quotient space X/A, consider the function $f(x,y) = (\cos(2\pi x), \sin(2\pi x), y)$, for (x,y) in the compact space X. Since f is continuous and surjective onto the Hausdorff space $S^1 \times [0,1]$, by Corollary 2.3.7 X/A is homeomorphic to $S^1 \times [0,1]$, i. e., *the bounded cylinder* (see Figure 3.3). We say that the square has been stretched into a cylinder.

Example 3.2.3. In the space $X = [0,1]^2$, let us glue together both pairs of opposite edges (horizontal and vertical) and consider the subset $A = A_1 \cup A_2$, where

$$A_1 = \{(x,0),(x,1) : 0 \le x \le 1\} \quad \text{and} \quad A_2 = \{(0,y),(1,y) : 0 \le y \le 1\}.$$

The process used in Example 3.2.2 is continued by bending the vertical edges. To put it concisely, X/A is defined by $(x,y) \sim (x',y')$ if and only if $(x,y),(x',y') \in A_1$ or $(x,y),(x',y') \in A_2$. Corollary 2.3.7 can be again applied with the map $f(x,y) = (e^{2i\pi x}, e^{2i\pi y})$, which is continuous and surjective onto $S^1 \times S^1$. The resulting quotient space is *the torus* $\mathbb{T}^2$, which is homeomorphic to $S^1 \times S^1$ (see Figure 3.4 and Example 2.3.12).

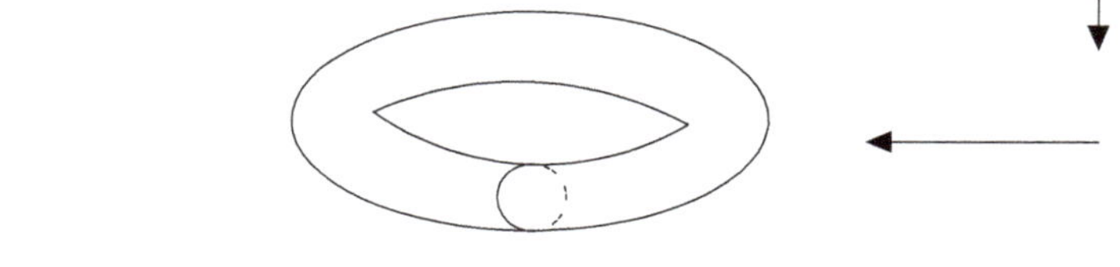

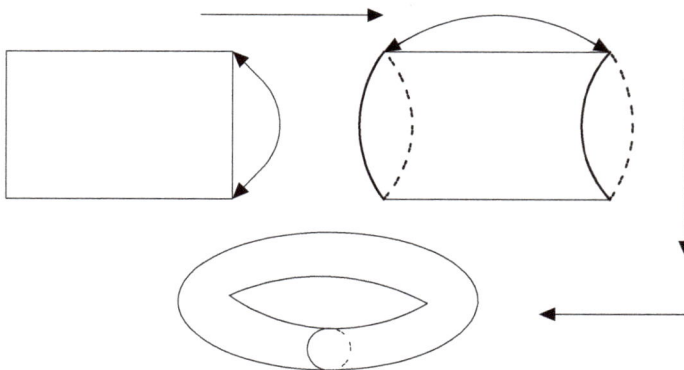

Figure 3.4: Torus obtained from a rectangle by gluing the opposite sides.

Analytically, the function f is the composition map of two "cylinder functions" $f = f_2 \circ f_1$ with $f_1(x,y) = (x, e^{2i\pi y})$ and $f_2(x,y) = (e^{2i\pi x}, y)$, each one representing the gluing of opposite edges.

According to the definitions in Examples 3.2.2 and 3.2.3, the following result is immediate.

Proposition 3.2.4. *On a cylinder* $X = S^1 \times [0,1]$, *define the equivalence relation* $(z,0) \sim (z,1)$. *Then* $X/_\sim$ *is homeomorphic to a torus.*

Example 3.2.5. (1) Let $X = [0,1]^2$ and the equivalence relation $(x,y) \sim (x',y')$ if and only if either $(x,y) = (x',y')$ or $(x' = 1 - x$ and $|y' - y| = 1)$. Equivalently, two different points (x,y), (x',y') are equivalent only if one point is on one horizontal edge, and the other point is on the opposite half of the other horizontal edge, while the vertical edges are not involved in this half-twist process. Then we can take as the set of identification

$$A_M = \{[0,1/2] \times \{0\}, [1/2,1] \times \{1\}\} \cup \{[0,1/2] \times \{1\}, [1/2,1] \times \{0\}\}.$$

The resulting quotient space is called *the Möbius band* (or Möbius strip) and is denoted $\mathbb{M}^2$ (see Figures 3.5 and 3.6).

(2) The *boundary of the Möbius band* is defined as $\partial \mathbb{M}^2 = Y/_\sim$, where $Y = \{0,1\} \times [0,1] \subset X$.

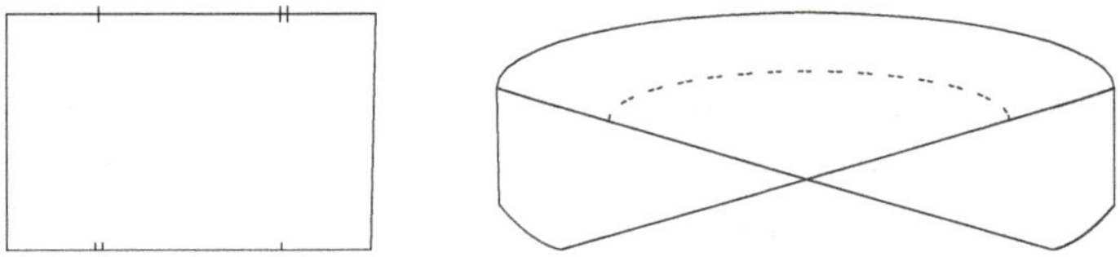

Figure 3.5: Möbius band: Polygonal representation.

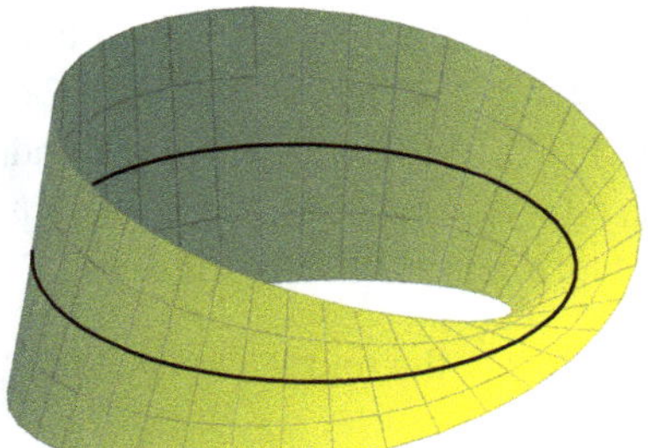

Figure 3.6: Möbius band: 3-D representation.

(3) The *open Möbius band* is obtained by identifying $(x, 0)$ and $(1 - x, 1)$, for $0 < x < 1$, i. e., without the end-points, i. e., the Möbius band without its boundary. It will be denoted $\mathring{\mathbb{M}}^2$.

(4) The range of $\{1/2\} \times [0, 1]$ under the quotient map π is called *the core circle of the Möbius strip* and is denoted $C(\mathbb{M}^2)$.
Since $\{1/2\} \times [0, 1] \cong [0, 1]$, by Example 2.3.8, $C(\mathbb{M}^2)$ is homeomorphic to a circle.
In sum,

$$\begin{aligned} \mathbb{M}^2 &= \pi([0,1] \times [0,1]), \quad & \mathring{\mathbb{M}}^2 &= \pi((0,1) \times [0,1]), \\ \partial\,\mathbb{M}^2 &= \pi(\{0,1\} \times [0,1]), \quad & C(\mathbb{M}^2) &= \pi(\{1/2\} \times [0,1]), \\ \mathbb{M}^2 &= \mathring{\mathbb{M}}^2 \cup \partial\,\mathbb{M}^2. \end{aligned}$$

The following proposition shows that the Möbius band can be defined from the cylinder $S^1 \times [0, 1]$.

Proposition 3.2.6. *On the cylinder $S^1 \times [0, 1]$, define the relation of equivalence $(z, 0) \sim_2 (-z, 1)$. Then $S^1 \times [0, 1]/_{\sim_2} \cong \mathbb{M}^2$.*

Proof. For $(x,y) \in I^2 = [0, 1]^2$, let $\theta(x,y) = y(1 - x) + x(1 - y) + y$ and $\phi(x,y) = e^{i\pi\theta(x,y)}$. Consider the commutative diagram

$$\begin{array}{ccc} I^2 & \xrightarrow{\quad\phi\quad} & S^1 \times I \\ \downarrow{\scriptstyle \pi_1} & & \downarrow{\scriptstyle \pi_2} \\ \mathbb{M}^2 = I^2/_{\sim_1} & \xrightarrow{\quad h\quad} & (S^1 \times I)/_{\sim_2} \end{array}$$

where π_1 (π_2, respectively) is the projection map for the equivalence relation $(x, 0) \sim_1 (1 - x, 1)$ on I^2 (and $(z, 0) \sim_2 (-z, 1)$ on $(S^1 \times I)$, respectively).

Since for all $x \in I$, $0 \le x \le \theta(x,y) \le 2 - x \le 2$, ϕ is surjective. Furthermore, $h \circ \pi_1 = \pi_2 \circ (\phi \times \mathrm{Id})$ and involved maps π_1, π_2, ϕ are continuous and surjective. Therefore, the induced map h is surjective and continuous by Theorem 2.1.9. So, it remains to prove that h is injective. Assume that $h(\pi_1(x,y)) = h(\pi_1(x',y'))$ for some (x,y) and (x',y') in I^2. Then $\pi_2((\phi(x,y)) = \pi_2(\phi((x',y'))$, i. e., $e^{i\pi\phi(x,y)} = -e^{i\pi\theta(x',y')} = e^{i\pi(\theta(x',y')+1+2k)}$, $y = 1$ and $y' = 0$. Hence, $\theta(x, 1) = \theta(x', 0) + 1 + 2k$ for some integer $k \in \mathbb{Z}$, i. e., $x' = 1 - x + 2k$. Since $-1 \le x' - 1 + x \le 1$, $k = 0$. Hence, $\theta(x, 1) = \theta(x', 0) + 1$. This yields, $x' = 1 - x$ and $(x, 1) \sim_1 (1-x, 0)$. Equivalently, $(x', 0) \sim_1 (1-x', 1)$, i. e., $\pi_1(x,y) = \pi_1(x',y')$, proving that h is injective. Since $\mathbb{M}^2$ is compact and $(S^1 \times I)/_{\sim_2}$ is Hausdorff, h is a homeomorphism. □

Similarly, the open Möbius band $\mathring{\mathbb{M}}^2$ can be defined from the unbounded cylinder $S^1 \times \mathbb{R}$.

Further to the core circle $C(\mathbb{M}^2)$ of the Möbius strip, the following result indicates a second interesting circle for the Möbius band.

Proposition 3.2.7. *The boundary of the Möbius band* $\mathbb{M}^2$ *is homeomorphic to the unit circle* S^1.

Proof. Let $X = \{0,1\} \times [0,1]$, $Y = S^1$ and consider the map $f : X \longrightarrow Y$ given by $f(s,t) = e^{i\pi(s+t)}$. Then f is continuous, onto and so is the induced map $\overline{f}$:

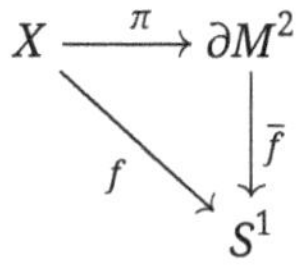

The map $\overline{f}$ is injective by definition of f and the commutativity of the diagram. Since X is compact and Y is Hausdorff, $\overline{f}$ is a closed map; hence, a homeomorphism between the boundary of the Möbius band and the unit circle. □

The next result shows that the Möbius band can be sewed along its boundary circle to the boundary of a closed disk to yield a projective plane.

Theorem 3.2.8. *The projective plane* $\mathbb{RP}^2$ *is homeomorphic to a Möbius band* $\mathbb{M}^2$ *attached to a closed disk* B^2 *along their boundaries.*

Proof. Consider the partition of the sphere $S^2 = S^2_N \cup \mathrm{Eq} \cup S^2_S$, where

$$S^2_N = \{(x_1,x_2,x_3) \in \mathbb{R}^3 : x_3 \geq 1/2\}, \quad S^2_S = \{(x_1,x_2,x_3) \in \mathbb{R}^3 : x_3 \leq -1/2\},$$

and

$$\mathrm{Eq} = \{(x_1,x_2,x_3) \in \mathbb{R}^3 : -1/2 \leq x_3 \leq 1/2\}$$

is the equatorial cylinder (see Figure 3.7). By Example 1.5.22, S^2_N and S^2_S are homeomorphic to the closed unit disk B^2. Clearly, S^2_N and S^2_S have same images via the projection map $\pi : S^2 \longrightarrow \mathbb{RP}^2$ and $\pi(S^2_N) \cong \pi(S^2_S) \cong \pi(B^2) \cong B^2$. Cutting the cylinder Eq and rolling it out, we get the rectangle R with half-opposite sides identified. By definition of the Möbius band, $\pi(\mathrm{Eq}) \cong \pi(R) \cong \mathbb{M}^2$. Since π is surjective, we get

$$\mathbb{RP}^2 = \pi(S^2) = \pi(S^2_N) \cup \pi(\mathrm{Eq}) \cup \pi(S^2_S) \cong B^2 \cup_\partial \mathbb{M}^2,$$

where $\cup_\partial$ means that the common intersection is the boundary circle of $\mathbb{M}^2$. Indeed,

$$\pi(S^2_N) \cap \pi(\mathrm{Eq}) \cong \pi(C) = C,$$

where C is the circle intersection of the sphere S^2 and the plane $x_3 = 1/2$. □

Remark 3.2.9. Consider the projective plane $\mathbb{RP}^2$ defined from the closed disk B^2 by identifying antipodal points on the boundary S^1 (see Proposition 3.1.13). In Figures 3.8

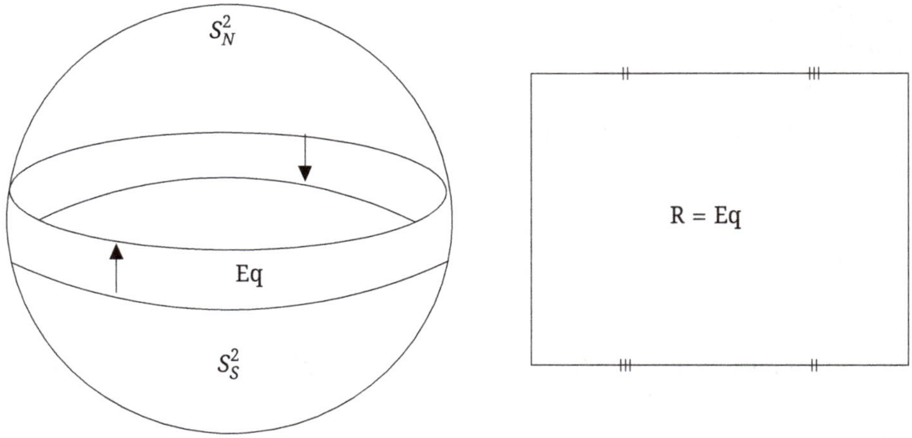

Figure 3.7: Sphere partition.

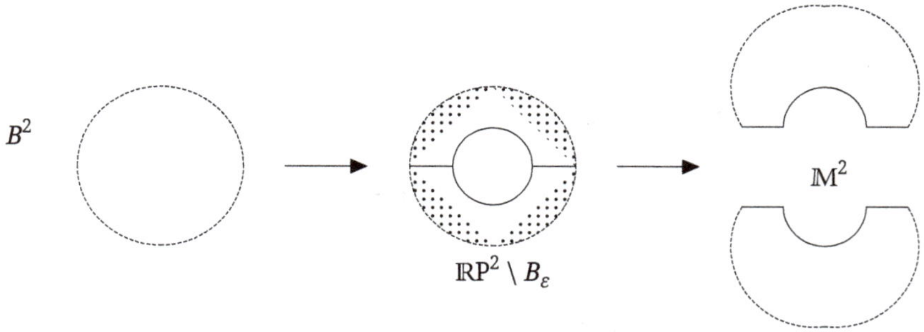

Figure 3.8: Möbius band as the detachment of a closed disk from a projective plane. (1) Ball representation.

and 3.9, the polygonal construction indicates how the Möbius band is obtained as the detachment of the small disk B_ε from $\mathbb{RP}^2$.

In Proposition 4.2.17, Proposition 7.2.33 and Proposition 7.2.37, we will discuss more topological structures of the Möbius band.

In the above examples, we have only indicated the points involved in the identification. No need to recall that all points of the given space are included in the partition of the quotient for the equivalence relation be reflexive.

Example 3.2.10. We can once again glue together both pairs of opposite edges of the unit square, but with the orientation reversed for one pair, i. e., we define

$$(x,y) \sim (x',y') \iff x' - x \in \{-1,0,1\} \text{ and } y' = y \text{ or } x' = 1 - x \text{ and } |y' - y| = 1.$$

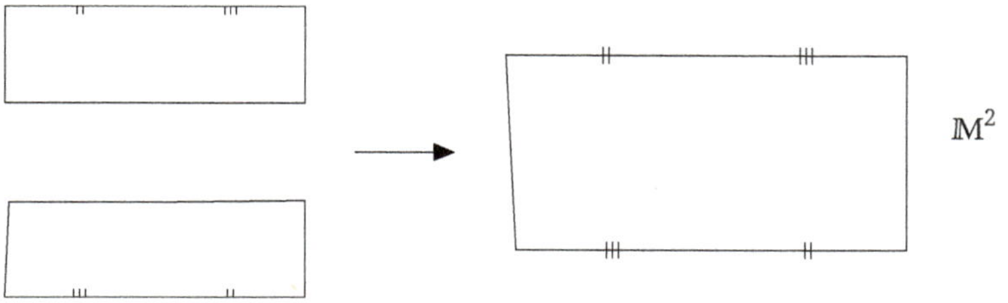

Figure 3.9: Möbius band as the detachment of a closed disk from a projective plane. (2) Polygonal representation.

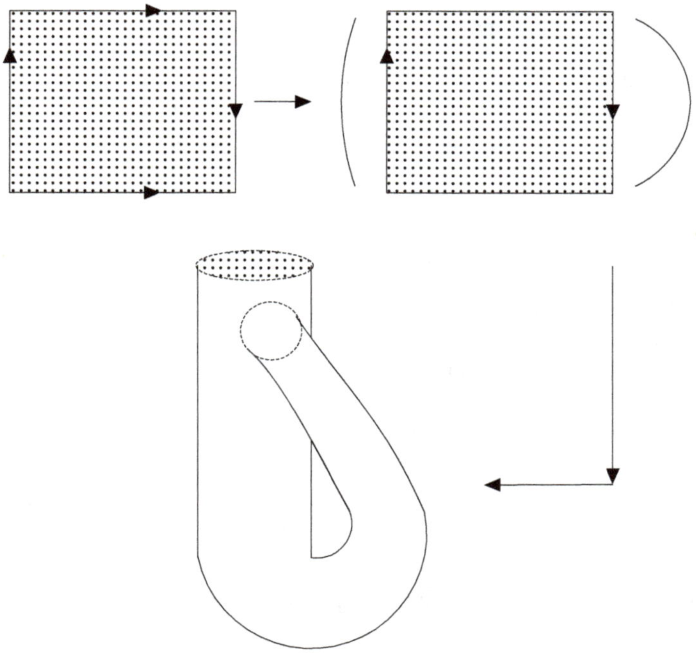

Figure 3.10: Klein bottle.

The set of identification points is

$$A_K = A_M \cup \{\{0\} \times [0,1], \{1\} \times [0,1]\}.$$

The constructed quotient space, called *Klein bottle*, is denoted $\mathbb{K}^2$ (see Figure 3.10).

Exercises (Sections 3.1–3.2)

1. Let $f : B^n \longrightarrow S^n$ be given by

$$f(x_1, x_2, \dots, x_n) = (2x_1\sqrt{1-\|x\|^2}, \dots, 2x_n\sqrt{1-\|x\|^2}, 2\|x\|^2 - 1).$$

(1) Check that f is continuous and surjective.
(2) Prove that B^n/S^{n-1} is a Hausdorff space.
(3) Deduce a second proof of Proposition 2.5.11.

2. Prove that $\mathbb{R}^n/B^n$ and $\mathbb{R}^n$ are homeomorphic.

3. (1) Prove that the function $f : \mathbb{R}^n \longrightarrow [0, \infty)$ defined by $f(x) = \|x\|$ is an open map. (*Hint:* For some $0 < y_0 \in V = f(U)$, where $U \subset \mathbb{R}^n$ is open, let $B(x_0, \alpha) \subset U$ and $y_0 = f(x_0)$. Show that for $0 < \beta < \frac{\alpha y_0}{\alpha+2y_0}$, we have $(y_0 - \beta, y_0 + \beta) \subset V$).
(2) In Euclidean plane $\mathbb{R}^2$, define the equivalence relation: $x \sim y$ if and only if x and y lie on the same circle centered at the origin (see [61, Exercise 1.9.76]). Prove that $\mathbb{R}^2/_\sim$ and $[0, \infty)$ are homeomorphic. (*Hint:* Use Corollary 2.3.6 and part (1).)

4. Identify all points on the southern hemisphere of the sphere S^2. Show that the resulting quotient space $S^2/_\sim$ is homeomorphic to S^2 (see [61, Exercise 1.9.77]). Notice that for the circle S^1 the question was already solved in Example 2.3.11.

5. In the disk $X = B^2$, define the equivalence relations: $(x,y) \sim_1 (-x,-y)$, $(x,y) \sim_2 (x,-y)$ and $(x,y) \sim_3 (-y,-x)$. In each case, determine the quotient space.

6. On the torus $\mathbb{T}^2 \cong S^1 \times S^1$, define the equivalence relation $(z, z') \sim (-z, \overline{z'})$. Show that $X/_\sim \cong \mathbb{K}^2$.

7. Let X be a locally connected space. Prove that the quotient space $X/_\sim$ is locally connected.

8. Let $h : S^1 \times S^1 \longrightarrow \mathbb{R}^3$ be given by $h(e^{i\theta}, e^{i\theta'}) = (x, y, z)$ (see [79]), where

$$x = (2 + \cos\theta')\cos\theta, \quad y = (2 + \cos\theta')\sin\theta, \quad z = \sin\theta'.$$

(1) Prove that h is an embedding map. Determine the inverse map h^{-1}.
(2) Deduce that the torus is homeomorphic to a surface in $\mathbb{R}^3$.
(3) Compare with the map $h(x_1, x_2, x_3, x_4) = (x_1(2+x_3), x_2(2+x_3), x_4)$ and find its inverse map h^{-1}.

9. (1) On the cylinder $C = S^1 \times [0,1]$, define the equivalence relation $(z, 0) \sim (-z, 0)$. Prove that $C/_\sim$ is homeomorphic to the Möbius band $\mathbb{M}^2$. (*Hint:* See [45, Example 1.35, p. 65].)

(2) Define the equivalence relation $(z,0) \sim (-z,0)$ and $(z,1) \sim (-z,1)$. Prove that $C/_\sim$ is homeomorphic to the Klein bottle $\mathbb{K}^2$.

(3) Deduce that two Möbius bands can be glued together along their boundaries to form a Klein bottle. This will be proved in Corollary 3.8.12 by another method (see Remark 3.8.13, also).

10. (1) Let $X = S^1 \times [0,1]$ be a cylinder and A, Y be two copies of the circle S^1. Let $f : S^1 \times \{0\} \cong S^1 \longrightarrow S^1$ be given by $f(z) = z^2$. Prove that the adjunction space $X \cup_{z^2} S^1$ is homeomorphic to a Möbius band $\mathbb{M}^2$. (*Hint:* Recall that $X \cup_{z^2} S^1$ is homeomorphic to $\pi(X)$, where $\pi : X \cup S^1 \longrightarrow X \cup_{z^2} S^1$ is the projection map (see Remark 3.4.3) and then use Question 9(1).)

(2) Deduce that $S^1 \times I \cup_{x \sim \pi(x)} \mathbb{RP}^1 \cong \mathbb{M}^2$. (*Hint:* Consider the composition

$$S^1 \xrightarrow{\ \pi\ } \mathbb{RP}^1 \xrightarrow{\ h\ } S^1, \qquad h \circ \pi : S^1 \longrightarrow S^1$$

where π is the projection map and h is the homeomorphism $h : S^1 \longrightarrow \mathbb{RP}^1$ is given by $h(z) = \{\omega, -\omega\}$ with $\omega^2 = z$ (see Proposition 3.1.5), then use Question (1). (The following definition of an $(n+1)$-dimensional Möbius band $\mathbb{M}^n$ is provided in [47, Example 2.3(b)]:

$$\mathbb{M}^{n+1} = C_\pi = S^n \times I \cup_\pi \mathbb{RP}^n,$$

where $\pi : S^n \longrightarrow \mathbb{RP}^n$ is the projection map and C_π the corresponding mapping cylinder. For $n = 1$, this is consistent with Question 2.)

(3) Prove that the Möbius band $\mathbb{M}^2$ and the circle S^1 are homotopically equivalent (see Section 4.2), providing a second proof of Proposition 4.2.17. (*Hint:* Use Question (1) and Theorem 4.2.18.)

11. Consider the cylinder $C = S^1 \times [0,1]$.

(1) Define the equivalence relation $(z,0) \sim (z^{-1},1)$. Prove that $C/_\sim$ is homeomorphic to the Klein bottle $\mathbb{K}^2$. (*Hint:* Start by rolling out the circle S^1.)

(2) In Euclidean plane $\mathbb{R}^2$, define the equivalence relation $(x,y) \sim (x,y+1)$ and then the relation $(x,y) \sim (x+1,-y)$. Show that the resulting quotient space is homeomorphic to the Klein bottle.

12. (1) Let $f : S^2 \longrightarrow \mathbb{R}^4$ be defined by $f(x_1,x_2,x_3) = (x_1^2 - x_2^2, x_1x_2, x_1x_3, x_2x_3)$. Show that f embeds the real projective plane $\mathbb{RP}^2$ onto $\mathbb{R}^4$.

(2) Discuss the case of the function $f : S^2 \longrightarrow \mathbb{R}^6$ is given by

$$f(x_1,x_2,x_3) = (x_1^2, x_2^2, x_3^2, x_1x_2, x_1x_3, x_2x_3).$$

13. Let $\pi_2 : \mathbb{R}^{n+1} \setminus \{0\} \longrightarrow \mathbb{RP}^n$ be the projection map, where $\mathbb{RP}^n$ is the real projective space defined by the following equivalence relation (see Proposition 3.1.10):

$$x \sim_2 y \iff \exists \lambda \in \mathbb{R} \setminus \{0\}, \quad x = \lambda y.$$

Prove that:
(1) π_2 is open but not closed;
(2) $\mathbb{RP}^n$ is homeomorphic to the quotient space obtained by restricting $\sim_2$ on the sphere S^n.

14. Prove that the projective space $\mathbb{RP}^n$ can be defined by the same equivalence relation $x \sim_1 -x$, for $x \in S^{n-1}$ but only on the closed northern hemisphere S^n_+.

15. Consider the complex projective space as a quotient space of the sphere S^{2n+1} (see Remark 3.1.17) and let $\pi : S^{2n+1} \longrightarrow \mathbb{CP}^n$ be the projection map (see Definition 3.1.16). Prove that:
(1) π is a closed map;
(2) $\mathbb{CP}^n$ is compact.

16. Let $f_n : B^{2n} \longrightarrow \mathbb{CP}^n$ be given by (see [28, p. 168])

$$f_n(x_1, \ldots, x_n, y_1, \ldots, y_n) = (z_1, \ldots, z_n, z_{n+1}),$$

where $z_k = x_k + iy_k$ $(1 \le k \le n)$ and $z_{n+1} = \sqrt{1 - \sum_{k=1}^{n}(x_k^2 + y_k^2)}$. Prove that:
(1) f_n is onto;
(2) f_n is one-to-one on the open ball $\mathring{B}^{2n}$;
(3) $\mathbb{CP}^n \setminus \mathbb{CP}^{n-1}$ and $\mathring{B}^{2n}$ are homeomorphic (compare with Example 3.4.19).

17. (1) Identify the points on the boundary circles of an annulus $A(0; 1, 2)$, which lie on the same ray from the origin. Show that the resulting quotient space $A/_\sim$ is homeomorphic to the torus $\mathbb{T}^2$ (see [61, Exercise 1.9.78]).
(2) Identify the points on the outer circle of a closed annulus A to one point and the points on the inner circle to a (different) point. Show that $A/_\sim$ is homeomorphic to S^2 (see [61, Exercise 1.9.79]).

18. Prove that the quotient space $\mathbb{M}^2/\partial\,\mathbb{M}^2$ is homeomorphic to the projective plane $\mathbb{RP}^2$.

19. In the diamond-shaped unit ball D^2, define the equivalence relation $(x, y) \sim (-x, y)$, for x, y on the boundary (see Figure 3.11).
(1) Show that $D^2/_\sim$ and S^2 are homeomorphic. (*Hint:* See [62, Proposition 6.1].)
(2) What happens if D^2 is the unit square $I^2 = I \times I$ with the equivalence relation $(0, t) \sim (t, 0)$ and $(t, 1) \sim (1, t)$, for $t \in [0, 1]$?

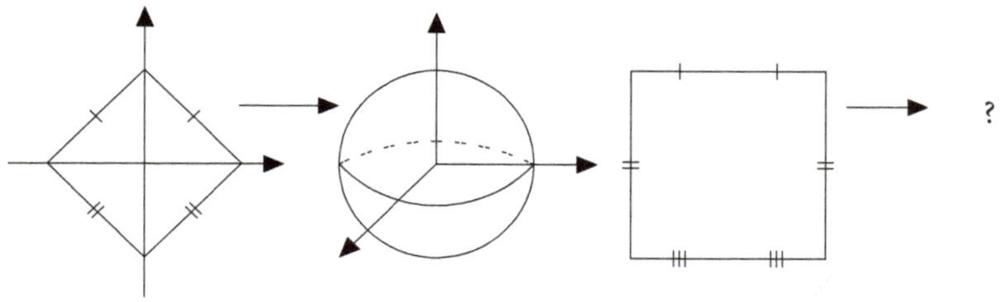

Figure 3.11: Ball identification.

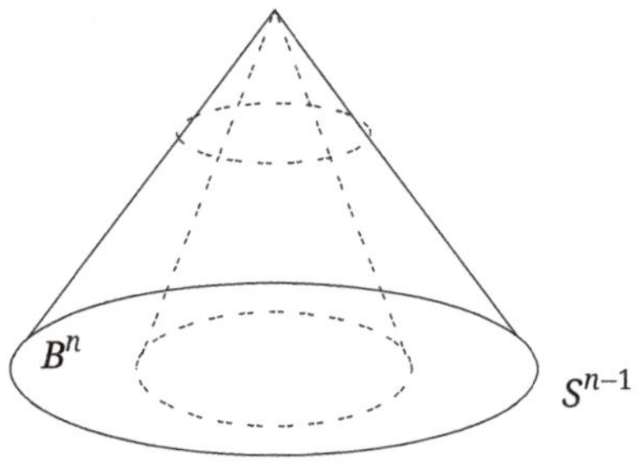

Figure 3.12: Cone over the unit sphere.

3.3 Cone, suspension, join, wedge and smash

Further to the spaces introduced in Sections 3.1–3.2, some general quotient spaces are defined. Recall that I denotes the unit closed interval $[0,1]$.

Definition 3.3.1. (1) *The cone* over a space X, denoted $C(X)$ is the quotient space defined on the product $X \times I$ by the equivalence relation:

$$(x,t) \sim (x',t') \iff \begin{cases} \text{either } (x,t) = (x',t') \\ \text{or } t = t' = 1, \end{cases}$$

that is to say $C(X) \cong (X \times I)/(X \times \{1\})$. The image $\pi(X \times \{1\})$ is called *the vertex or the apex* of X. The image $\pi(X \times \{0\}) \cong X$ is the base of the cone; see Figure 3.12.

(2) *The suspension* of a space X, denoted $\Sigma(X)$, is the quotient space defined by the equivalence relation:

$$(x,t) \sim (x',t') \iff \begin{cases} \text{either } (x,t) = (x',t') \\ \text{or } t = t' = 0 \\ \text{or } t = t' = 1, \end{cases}$$

that is to say

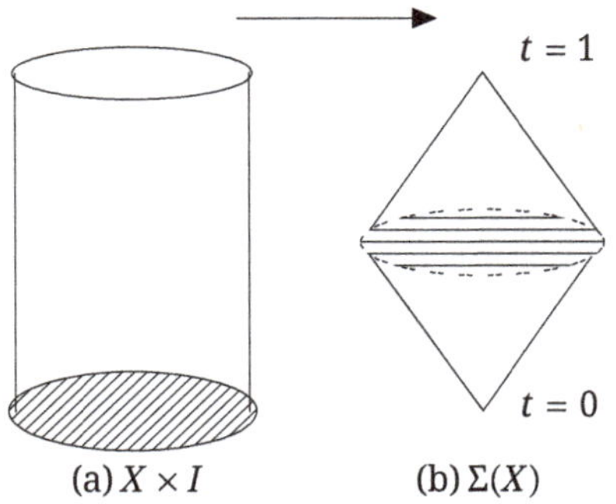

Figure 3.13: (a) Cylinder of base X; (b) Suspension of X.

$$\Sigma(X) \cong (X \times I/X \times \{0\}) \cup (X \times I/X \times \{1\}).$$

Notice that

$$\Sigma(X) \cong C(X)/X \times \{0\} \cong C(X)/X.$$

Figure 3.13 shows that a suspension is the result of two cones glued together at their base.

(3) *The join* of two spaces X and Y, denoted $X * Y$, is the quotient space of the product $X \times [0,1] \times Y$ with equivalence relation given by

$$(x,t,y) \sim (x',t',y') \iff \begin{cases} \text{either } (x,t,y) = (x',t',y') \\ \text{or } t = t' = 0 \text{ and } x = x' \\ \text{or } t = t' = 1 \text{ and } y = y'. \end{cases}$$

(4) Let X, Y be two disjoint spaces with $x_0 \in X$ and $y_0 \in Y$. *The wedge sum* of X and Y, denoted $X \vee Y$ (one-point union) is the quotient space of the union $X \cup Y$ by the equivalence relation given by

$$z \sim z' \iff \begin{cases} \text{either } z = z' \\ \text{or } z = x_0 \text{ and } z' = y_0. \end{cases}$$

(5) Let X, Y be two disjoint spaces with $x_0 \in X$ and $y_0 \in Y$. *The smash product* of X and Y, denoted $X \wedge Y$ is the quotient space $(X \times Y)/(X \vee Y)$.

Remark 3.3.2. A cone is a partition. which consists of:
(1) the slice $X \times \{1\}$, that is the top of $X \times I$ identified with a single point;
(2) the sets of points (x,t) with $x \in X$ and $0 \le t < 1$.

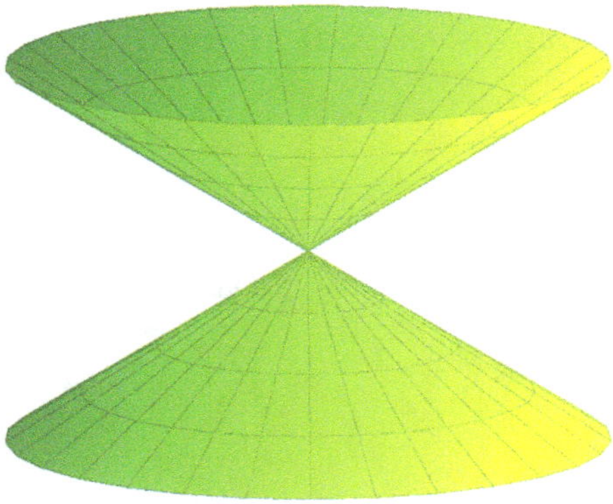

Figure 3.14: Double-sheet cone ($z^2 = x^2 + y^2$).

Remark 3.3.3. (1) Since the map $(x,t) \longmapsto (x, 1-t)$ is a homeomorphism, we can define a cone as the quotient $(X \times I)/(X \times \{0\})$.
(2) The embedding $h : X \longrightarrow C(X)$ such that $h(x) = [(x,0)] = \{(x,0)\}$ (see Exercise 1) identifies the space X with a subspace of its cone and its suspension, respectively.
(3) The cone $C(X)$ is convex whenever X is.

Example 3.3.4. The quotient space $C(S^1) = (S^1 \times I)/(S^1 \times \{1\})$ is the cone over the unit circle S^1 with vertex (apex) $(0,0,1)$. In Theorem 3.3.6, we will show that this cone is homeomorphic to the closed disk B^2.

Example 3.3.5. (1) The quotient space $\Sigma(S^1) = C(S^1)/S^1$ is a double cone over the unit circle S^1 with vertices $(0,0,1)$ and $(0,0,-1)$.
(2) Such a double cone is also homeomorphic to the double-sheet cone of equation $z^2 = x^2 + y^2$ pictured in Figure 3.14.

As for the sphere, we have the following result, where $\cong$ refers to some homeomorphism.

Theorem 3.3.6. *We have*
(1) $C(S^{n-1}) \cong C(B^{n-1}) \cong B^n$;
(2) (a) $\Sigma(S^{n-1}) \cong S^n$;
(b) $\Sigma(B^{n-1}) \cong B^n \vee B^n$.

Proof. (1) The map $f : S^{n-1} \times I \longrightarrow B^n$ defined by $f(x,t) = tx$ is continuous and surjective. f satisfies $f(x,t) = f(x',t')$ if and only if $(t = t' = 0)$ or $(t = t' \neq 0$ and $x = x')$. Thus, define $(x,t) \sim (x',t')$ if and only if $(x,t) = (x',t')$ and $t, t' \neq 0$ or (x,t) and (x',t') lie in $S^{n-1} \times \{0\}$. The space $S^{n-1} \times I$ is compact and B^n is Hausdorff. By Corollary 2.3.7, $C(S^{n-1}) = (S^{n-1} \times I)/(S^{n-1} \times \{0\})$ is homeomorphic to B^n. As for the cone over the ball B^{n-1}, just define the map $f : B^{n-1} \times I \longrightarrow B^n$ by $f(x,t) = (tx, t)$.

(2) (a) Since $\Sigma(X) \cong C(X)/X \cong (C(X)/X) \times \{0\}$, by Theorem 2.5.11 and part (1),

$$\Sigma(S^{n-1}) \cong (C(S^{n-1})/S^{n-1}) \times \{0\} \cong B^n/S^{n-1} \cong S^n.$$

(b) $\Sigma(B^{n-1}) \cong (C(B^{n-1})/B^{n-1}) \times \{0\} \cong B^n/B^{n-1} \cong B^n \vee B^n$. □

Corollary 3.3.7. *Let C^k and Σ^k denote k iterated cones and suspensions, respectively. Then*

$$B^n \cong C^{n-1}(S^1) \quad \textit{and} \quad S^n \cong \Sigma^{n-1}(S^1).$$

Example 3.3.8. In the cylinder $X = S^{n-1} \times [0,1] \subset \mathbb{R}^{n+1}$, define an equivalence relation by gluing together points of the top, i. e., $(x,y,1) \sim (x',y',1)$ for all $x,y \in S^{n-1}$ and keeping the points unchanged on the cylinder lateral surface. The obtained quotient space is a cone over S^{n-1}. By Theorem 3.3.6(1), $X/_\sim$ is homeomorphic to B^n.

Example 3.3.9. In the cylinder $X = S^{n-1} \times [0,1] \subset \mathbb{R}^{n+1}$, define an equivalence relation by gluing together points of the top and points of the bottom, i. e., $(x,y,0) \sim (x',y',0)$ and $(x,y,1) \sim (x',y',1)$ for all $x,y \in S^{n-1}$. We do not affect the points on the cylinder lateral surface. Then the constructed quotient space is a suspension (double cone) over S^{n-1}. By Theorem 3.3.6(2), $X/_\sim$ is homeomorphic to S^n.

Remark 3.3.10. Consider the disjoint union $X \cup Y$ with $x_0 \in X$ and $y_0 \in Y$. The union $Z = (X \times \{y_0\}) \cup (\{x_0\} \times Y)$ identifies with the wedge sum $X \vee Y$ (one-point union) (Definition 3.3.1(4)). Thus, the quotient $(X \times Y)/(X \vee Y)$ makes sense.

Definition 3.3.11. Let $X \subset \mathbb{R}^n$ be compact and identify $\mathbb{R}^n$ with a subspace of $\mathbb{R}^{n+1}$. Define the *geometric cone* of X subtended from the northern pole as

$$G(X) = \{tx + (1-t)N_p : x \in X, t \in I\},$$

where $N_p = (0,0,\dots,0,1) \in \mathbb{R}^{n+1}$ (or any point not belonging to X).

Proposition 3.3.12. *Let $X \subset \mathbb{R}^n$ be compact. Then the topological cone $C(X)$ and the geometric cone $G(X)$ are homeomorphic spaces.*

Proof. Consider the diagram

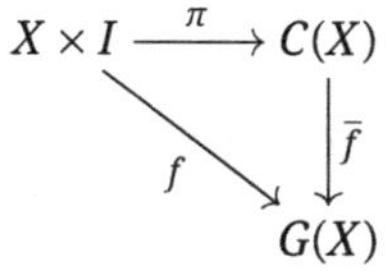

where $f(x,t) = tx + (1-t)N_p$. Then clearly f is continuous surjective. Moreover, by the identification $x = (x,0) \in \mathbb{R}^{n+1}$, we have

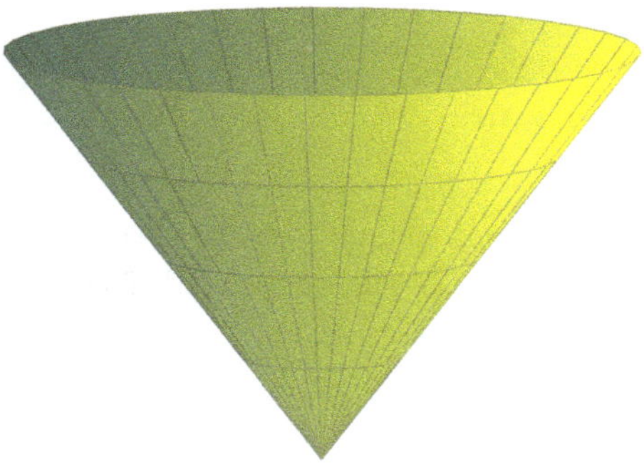

Figure 3.15: One-sheet cone.

$$\begin{aligned} f(x,t) = f(x',t') &\iff tx - t'x' = (t-t')N_p \\ &\iff (tx - t'x', 0) = (0, t - t'). \end{aligned}$$

This is equivalent to either ($t = t' = 0$ and $x, x' \in X$) or ($t = t' \neq 0$ and $x = x'$). Hence, f agrees with Definition 3.3.1(1) of the cone. By Corollary 2.3.7, f induces a homeomorphism $\bar{f}$. □

Example 3.3.13. For $X = S^1$, the geometric cone $G(X)$ is the conical surface $z = \sqrt{x^2 + y^2}$, $0 \leq z \leq 1$ with one sheet (see Figure 3.15).

After discussing some examples of cones and suspensions, we describe the wedge sum of n circles and the join of some spaces. The following definition shows that it coincides with the bouquet of circles.

Definition 3.3.14. *A bouquet of n circles* is a space $X = \bigvee_{k=n}^{n} S_k$, where S_k is homeomorphic to S^1, for each $k \in [1, n]$ and $S_k \cap S_{k'} = \{x_0\}$ for all $k \neq k'$. It is also called an n-fold rose (or an n-petaled rose).

Example 3.3.15. (1) By Example 2.5.4, the wedge sum of two spheres $S^n \vee S^n$ is homeomorphic to the quotient space S^n / S^{n-1}.
(2) In particular, the wedge sum $X \vee Y \cong S^1/_{\sim}$ is homeomorphic to the union of two circles meeting at some point, i. e., the figure eight (see Figure 1.9 in Example 1.5.17). A second proof will be given in Remark 6.2.8.

Example 3.3.16. (1) When the space X is compact and Y reduces to a single point $\{y_0\}$, the join $X * \{y_0\}$ is the cone over $X \times \{y_0\}$, which is homeomorphic to the cone over X (see [12, 5.7.4 (Corollary 1)]).
(2) Along similar lines, one can check that $\Sigma(X) \cong X * \{0, 1\}$.

Example 3.3.17. We have
(1) $S^{n-1} * \{0\} \cong B^n$;
(2) $B^n * \{0\} \cong B^{n+1}$.

The join of two spaces has simple properties.

Proposition 3.3.18. (1) *Commutativity:* $X * Y = Y * X$.
(2) *Associativity:* $(X * Y) * Z = X * (Y * Z)$.

Proof. (1) Let $p : X \times Y \longrightarrow Y \times X$ be the permutation $p(x,y) = (y,x)$. Clearly, p is continuous. Let $h : X * Y \longrightarrow Y * X$ be given by $h([(x,y,t)]) = [(y,x,t)]$. Then the following diagram commutes:

$$\begin{array}{ccc} X \times Y & \xrightarrow{\quad p \quad} & Y \times X \\ {\scriptstyle \pi_1}\downarrow & & \downarrow{\scriptstyle \pi_2} \\ X * Y & \xrightarrow{\quad h \quad} & Y * X \end{array}$$

where π_1 (π_2, respectively) is the projection map on the join $X * Y$ ($Y * X$, respectively). Thus, h is a homeomorphism.

(2) One can check that the mapping $h : (X * Y) * Z \longrightarrow X * (Y * Z)$ given by

$$h([([(x,y,t)],z,s)]) = \left[\left(x, \left[\left(\frac{1-t}{1-st}y, \frac{1}{1-st}z, s\right)\right], st\right)\right]$$

if $st \neq 1$ and x if $st = 1$ is a homeomorphism. □

Proposition 3.3.19. *The join $X * Y$ of two spaces X and Y is a path-connected space.*

Proof. Let $\pi(x,y,t)$ and $\pi(x',y',t')$ belong to $X * Y$. Then the map $\phi : I \longrightarrow X * Y$ given by

$$\phi(s) = \begin{cases} \pi(x,y,3s(1-t)+t), & 0 \le s \le 1/3, \\ \pi(x',y,2-3s), & 1/3 \le s \le 2/3, \\ \pi(x',y',t'(3s-2)), & 2/3 \le s \le 1 \end{cases}$$

is a path joining the two points. □

3.4 Adjunction space

Let X and Y be two nonempty disjoint spaces and $A \subset X$. We define a topology in $X \cup Y$ by declaring that U is open in $X \cup Y$ if $U \cap X$ and $U \cap Y$ are open sets.

Given a continuous map $f : A \longrightarrow Y$, define an equivalence relation in $X \cup Y$ by $a \sim f(a)$ for all $a \in A$.

Definition 3.4.1. The quotient space denoted $Z_f = X \cup_f Y = (X \cup Y)/_\sim$ is called *the adjunction space* (or attaching space). The function f is called *the attaching map* (or adjoining map).

We say that X is adjoined or attached (glued) to Y along f.

Remark 3.4.2. The equivalence classes of the adjunction space Z_f are:
(1) the one-element sets $\{\{x\}, x \in X \setminus A\}$;
(2) the one-element sets $\{\{y\}, y \in Y \setminus f(A)\}$;
(3) the sets $\{\{y, x \in f^{-1}(y)\}, y \in f(A)\}$.

Remark 3.4.3. (1) Clearly, Z_f coincides with $\pi(Y)$ whenever $A = X$, where $\pi : X \cup Y \longrightarrow Z_f$ is the canonical projection map.
(2) Also, $Z_f = \pi(X)$ if f is surjective.
(3) More generally, Z_f can be thought of as the union $\pi(X) \cup \pi(Y)$ with $\pi(X) \cap \pi(Y) = \pi(A) = \pi(f(A))$.

Before we present some examples of adjunction spaces, we recall the definition of the pushout of a diagram.

Definition 3.4.4. Consider the commutative diagram

$$\begin{array}{ccc} A & \xrightarrow{\quad f \quad} & B \\ \big\downarrow{\scriptstyle i} & & \big\downarrow{\scriptstyle j} \\ X & \xrightarrow{\quad g \quad} & Y \end{array}$$

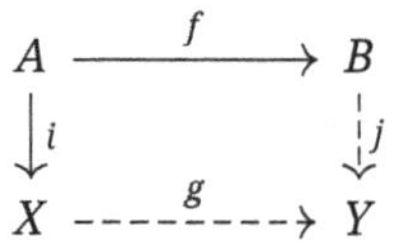

i. e., $g \circ i = j \circ f$. We call Y *a pushout* of the diagram if for every space Z and every morphisms $h : X \longrightarrow Z$, $k : B \longrightarrow Z$, which commute ($h \circ i = k \circ f$), there exists a unique morphism $\psi : Y \longrightarrow Z$ such that the following diagram commutes: $\psi \circ g = h$ and $\psi \circ j = k$:

$$\begin{array}{ccccc} A & \xrightarrow{f} & B & \xrightarrow{\quad k \quad} & \\ \big\downarrow{\scriptstyle i} & & \big\downarrow{\scriptstyle j} & & \searrow \\ X & \xrightarrow[g]{} & Y & \xrightarrow{\quad \psi \quad} & Z \\ & & \xrightarrow[\qquad h \qquad]{} & & \nearrow \end{array}$$

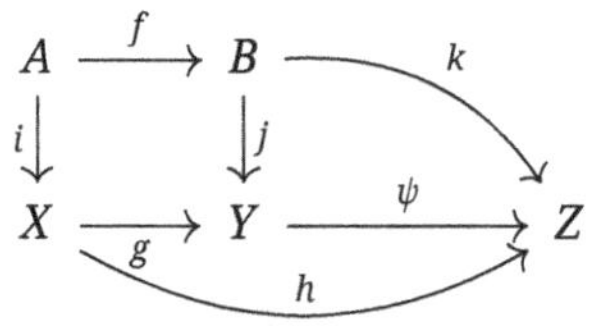

Example 3.4.5. The pasting lemma (Lemma A.2.1) may be formulated through the following commutative diagram, where $X = A \cup B$ is the pushout:

$$\begin{array}{ccccc} A \cap B & \xhookrightarrow{\quad i_1 \quad} & A & \xrightarrow{\quad g \quad} & \\ \big\downarrow{\scriptstyle i_2} & & \big\downarrow{\scriptstyle i_3} & & \searrow \\ B & \xhookrightarrow{\quad i_4 \quad} & X = A \cup B & \xrightarrow{\quad h \quad} & Y \\ & & \xrightarrow[\qquad f \qquad]{} & & \nearrow \end{array}$$

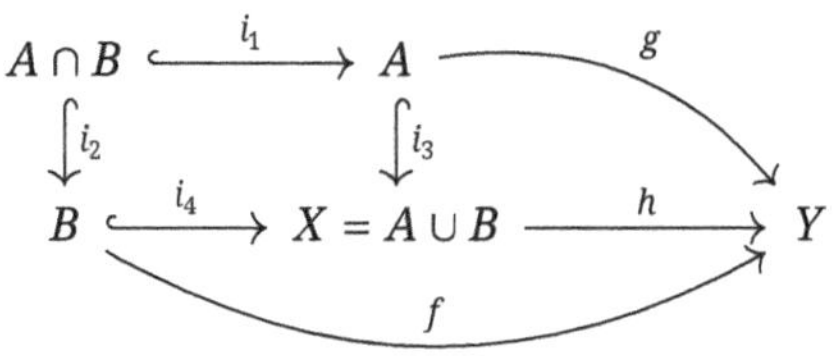

For $1 \leq k \leq 4$, the inclusion maps i_k trivially commute. The equality $g \circ i_1 = f \circ i_2$ means that f and g agree on the intersection of the closed (open) sets A and B. Finally, one of the equalities $h \circ i_4 = f$ or $h \circ i_3 = g$ both with a factorization result (Theorem 2.1.9) imply the continuity of the map h.

The spaces defined in Section 3.2 are immediate examples of adjunction spaces.

Example 3.4.6. Given the closed interval $I = [0,1]$, let

$$A_1 = I \times \{0\}; \quad A_2 = \{0\} \times I; \quad A_3 = I \times \{1\}; \quad A_4 = \{1\} \times I.$$

Then the adjunction space $Z_f = I^2 \cup_f I^2$ is:

(1) A cylinder (Example 3.2.2) when $f : A_1 \longrightarrow A_3$ is given by $f(x,0) = (x,1)$.
(2) A torus (Example 3.2.3) when $f : A_1 \cup A_2 \longrightarrow A_3 \cup A_4$ is defined by $f(x,0) = (x,1)$ and $f(0,y) = (1,y)$.
(3) A Möbius band (Example 3.2.5) when $f : A_1 \longrightarrow A_3$ is given by $f(x,0) = (1-x,1)$.
(4) A Klein bottle (Example 3.2.10) when $f : A_1 \cup A_2 \longrightarrow A_3 \cup A_4$ is given by $f(x,0) = (1-x,1)$ and $f(0,y) = (1,y)$.

In all cases, Y is the image set and $X = I^2 \setminus Y$, which guarantees $X \cap Y = \emptyset$.

The next two examples of adjunction spaces are examined in detail.

Example 3.4.7. Let $X = \{(x,y,z) \in \mathbb{R}^3 : x^2 + y^2 + z^2 \leq 1,\ z \leq -3/4\}$, $Y = \{(x,y,z) \in \mathbb{R}^3 : x^2 + y^2 + z^2 \leq 1,\ z \geq 3/4\}$, $A = \{(x,y,z) \in X : z = -3/4\}$ and $f : A \longrightarrow Y$ be given by $f(x,y,-3/4) = (x,y,3/4)$. Then $X \cup_f Y \cong B^3$. Indeed, let $h : X \cup Y \longrightarrow B^3$ be defined by

$$h(x,y,z) = \begin{cases} (x,y,3(z+1)), & \text{if } -1 \leq z \leq -3/4, \\ (x,y,2z-3/4), & \text{if } 3/4 \leq z \leq 7/8, \\ (x,y,-16z+15), & \text{if } 7/8 \leq z \leq 1. \end{cases}$$

Then h is continuous and surjective. By Corollary 2.3.7, Z_f and B^3 are homeomorphic (see Figure 3.16).

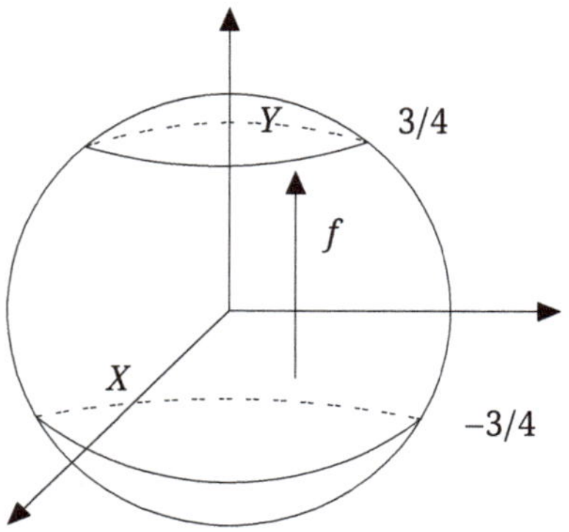

Figure 3.16: Attaching two ball parts (1).

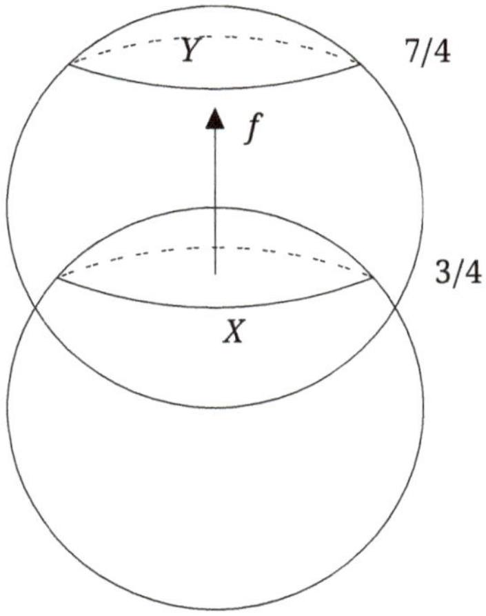

Figure 3.17: Attaching two ball parts (2).

Example 3.4.8. Let

$$X = \{(x,y,z) \in \mathbb{R}^3 : x^2 + y^2 + z^2 \le 1,\ z \le 3/4\},$$
$$Y = \{(x,y,z) \in \mathbb{R}^3 : x^2 + y^2 + (z-1)^2 \le 1,\ z \ge 7/4\},$$

and $A = \{(x,y,z) \in X : z = 3/4\}$. Let $f : A \longrightarrow Y$ be given by $f(x,y,z) = (x,y,z+1)$ for all $(x,y,z) \in A$, i. e., $f(x,y,z) \in \{(x,y,z) \in Y : z = 7/4\} \subset Y$ (see a gentle illustration in [53, Example 4.3.4]). Here again, $X \cup_f Y \cong B^3$. To check this, let $h : X \cup Y \longrightarrow B^3$ be given by

$$h(x,y,z) = \begin{cases} (x,y,-15z/7 + 1/7), & \text{if } -1 \le z \le 3/4, \\ (x,y,-6z + 49/4), & \text{if } 7/4 \le z \le 15/8, \\ (x,y,-24z + 46), & \text{if } 15/8 \le z \le 2. \end{cases}$$

Then h is continuous and surjective. By Corollary 2.3.7, Z_f and B^3 are homeomorphic. Unlike Example 3.4.11, the attached parts come from two different balls. The resulting adjunction space is homeomorphic to the ball $B^3(0,2)$ (see Figure 3.17).

Proposition 3.4.9. *Let X be a compact space, $A \subset X$ a closed subset and $Y = \{y_0\}$. Then the adjunction space $Z_{\{y_0\}} = X \cup_{\{y_0\}} \{y_0\}$ is homeomorphic to the quotient space X/A.*

Proof. Let $f(x) = y_0$ for all $x \in A$. Let $q : X \cup Y \longrightarrow Z_{\{y_0\}}$ be the projection map and q_X, q_Y denote the restrictions of q on X and Y, respectively. The map p will refer to the canonical projection on the quotient space X/A. For the map g given by $g(y_0) = \{A\}$, we have:

(1) $g \circ f = p \circ i$ for $g(f(a)) = g(y_0) = \{A\} = p(a) = (p \circ i)(a)$ for all $a \in A$.

(2) We have $q_Y \circ f = q_X \circ i$ because $q_Y(f(a)) = q_Y(y_0) = \{\{y_0, a\}, a \in A\} = q_X(a) = (q_X \circ i)(a)$ for all $a \in A$.

(3) Define the map $h : X/A \longrightarrow Z_f$ by $h \circ p = q_X$. Then $h \circ g = q_Y$ for

$$(h \circ g)(y_0) = h(\{A\}) = h(p(a)) = q_X(a) = q_Y(y_0), \quad \text{for all } a \in A.$$

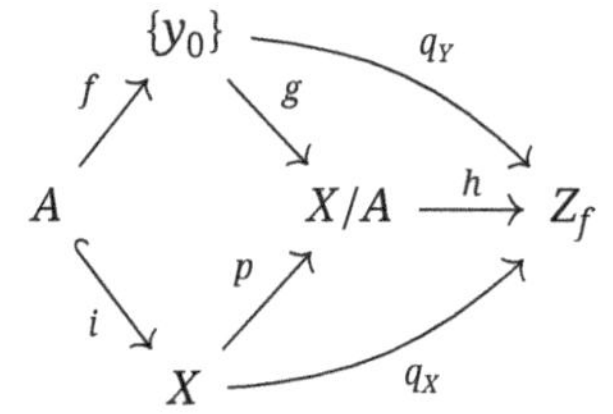

(4) The map h is clearly surjective and continuous by the factorization theorem because q_X is continuous and p is an identification map (Theorem 2.1.9).

(5) Furthermore, h is injective. Indeed, let x, x' be in X. By definition of h, $h(p(x)) = h(p(x'))$ is equivalent to $q_X(x) = q_X(x')$. Then the case when $x \in A$ and $x' \in X \setminus A$ is ruled out, so we only consider two cases:

(a) $x, x' \in X \setminus A$: since $q_X(x) = \{x\}$ for all $x \in X \setminus A$, we get $x = x'$;
(b) $x, x' \in A$: since $x \neq y_0$, $\{x, y_0\} = \{x', y_0\}$ implies $x = x'$.

(6) By Corollary 2.5.9, the quotient space X/A is compact Hausdorff. Appealing to Theorem 1.5.36, we conclude that $h : X/A \longrightarrow Z_f$ is a homeomorphism. □

The diagram in the proof of Proposition 3.4.9 shows that the quotient space X/A is a pushout according to Definition 3.4.4.

Remark 3.4.10. Proposition 3.4.9 shows that a quotient space X/A is a particular case of adjunction space.

Example 3.4.11. For $A = S^{n-1}$, $B^n \cup_{\{y_0\}} \{y_0\} \cong B^n/S^{n-1} \cong S^n$.

Example 3.4.12. In Definition 3.4.1, let $A = \{x_0\} \subset X$, $y_0 \in Y$, and $f(x_0) = y_0$. Then the adjunction space $Z_f = X \cup_f Y$ of X and Y is homeomorphic to the wedge sum $X \vee Y$. Thus, the wedge sum is a particular case of the adjunction space.

Example 3.4.13. Under the hypotheses of Proposition 3.4.9, let $A \subset X$, $y_0 \in Y$, and $f(x) = y_0$ for all $x \in A$ a constant map f. By Proposition 3.4.9 combined with Example 3.4.12, we have

$$\begin{aligned} X \cup_{\{y_0\}} Y &\cong \pi(X) \cup \pi(Y) \\ &\cong X/A \cup_{\{y_0\}} Y \\ &\cong X/A \vee Y. \end{aligned}$$

Example 3.4.14. We have $B^n \cup_{\{y_0\}} Y \cong S^n \vee Y$.

When the function f is just the identity on A, two simple examples are discussed.

Example 3.4.15. Take $Y = A \subset X$ and $f = \mathrm{Id}_{|A}$. Then the adjunction space $X \cup_{\mathrm{Id}_{|A}} A$ is homeomorphic to X. For instance,

$$B^n \cup_{\mathrm{Id}_{|S^{n-1}}} S^{n-1} \cong B^n.$$

Example 3.4.16. (1) Let X and Y be two copies of the ball B^n, $A = S^{n-1}$, and $f = \mathrm{Id}_{|S^{n-1}}$. Notice that the equivalence classes of Z_f are the same as those of B^n/S^{n-1}. Hence, $B^n \cup_{\mathrm{Id}_{|S^{n-1}}} B^n$ is homeomorphic to the sphere S^n, as expected. The proof can be conducted as in the proof of Proposition 3.4.9 with the adjunction space as a pushout, where $g(A) = \{A\}$.

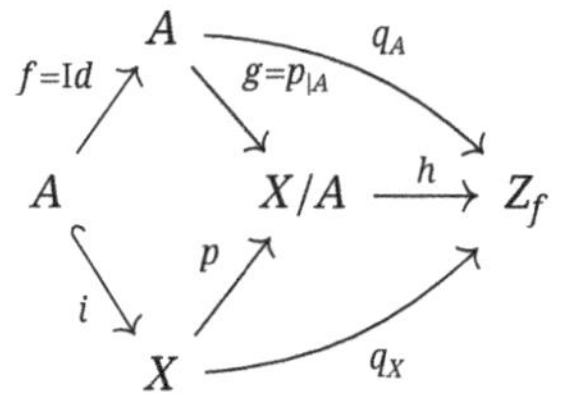

Experiment: Glue two identical disks along their boundaries. Then blow inside (or inflate) so as to separate the interior points (which are not involved in the gluing process, by the way). You get a balloon (a sphere).

(2) More generally, if $h : S^{n-1} \subset B^n \longrightarrow S^{n-1}$ is a homeomorphism, then using Alexander's trick, which provides a homeomorphism $\widetilde{h} : B^n \longrightarrow B^n$ (see Proposition 12.5.14), one can show that $B^n \cup_h B^n \cong S^n$ (see [42, Proposition (21.22)]). For example, $B^n \cup_{z\sim -z} B^n \cong S^n$.

Definition 3.4.17. Let $X = B^2$ and A, Y be two copies of the circle S^1. Let $f : S^1 \longrightarrow S^1$ be the nth power map given by $f(z) = z^n$, for some positive integer n. The adjunction space $B^2 \cup_{z^n} S^1$ denoted M_n is called *the Moore space.*

For $n = 1$, $M_1 = B^2 \cup_{\mathrm{Id}|S^1} S^1 \cong B^2$ (see Example 3.4.15). For $n = 2$, we have the following.

Proposition 3.4.18. *M_2 is homeomorphic to the projective plane $\mathbb{RP}^2$.*

Proof. Let $\pi : B^2 \longrightarrow \mathbb{RP}^2$ be the canonical projection map (Definition 3.1.1) given by

$$\pi(z) = \begin{cases} z, & z \in \mathring{B}^2, \\ \{z, -z\}, & z \in S^1. \end{cases}$$

Let $h : S^1 \longrightarrow \mathbb{RP}^1$ be the homeomorphism defined by $h(z) = \{\omega, -\omega\}$, $\omega^2 = z$ (see Proposition 3.1.5) and $g : B^2 \cup S^1 \longrightarrow \mathbb{RP}^2$ given by

$$g(z) = \begin{cases} \pi(z), & z \in B^2, \\ h(z), & z \in S^1, \end{cases}$$

where $B^2 \cup S^1$ is a disjoint union and $\mathbb{RP}^1$ is considered as a subset of $\mathbb{RP}^2$. Since π and h are continuous surjective, so is the map g. In addition, g agrees with the equivalence relation in $B^2 \cup S^1$. Indeed,

$$\begin{aligned} g(z) = g(z') \quad &\Longleftrightarrow \quad \begin{cases} \pi(z) = \pi(z'), & z, z' \in B^2, \\ h(z) = h(z'), & z, z' \in S^1, \\ \pi(z) = h(z'), & z \in B^2, z' \in S^1 \end{cases} \\ &\Longleftrightarrow \quad \begin{cases} z = z', & z, z' \in \mathring{B}^2, \\ z = \pm z', & z, z' \in \partial B^2, \\ z = z', & z,\ z' \in S^1, \\ \{z, -z\} = \{\omega, -\omega\}, & z \in \partial B^2,\ \omega^2 = z' \in S^1 \end{cases} \\ &\Longleftrightarrow \quad z \sim z', \end{aligned}$$

where $\sim$ is the equivalence relation defining the adjunction space. By Corollary 2.3.7, g induces a homeomorphism between $\mathbb{RP}^2$ and $M_2 = B^2 \cup_{z^2} S^1 = (B^2 \cup S^1)/\sim$. □

Example 3.4.19. Let $X = B^n$, $A = S^{n-1}$, $Y = \mathbb{RP}^{n-1}$ the $(n-1)$-projective space and $f = \pi$ the projection map ($\pi(x) = [x] = \{x, -x\}$). By the first definition of the projective space (Definition 3.1.1), the third one (Proposition 3.1.13), and since π is surjective, we have

$$\begin{aligned} B^n \cup_\pi \mathbb{RP}^{n-1} &= \{\{x\},\ x \in \mathring{B}^n\} \cup \{\{x, -x\},\ x \in S^{n-1}\} \\ &\cong B^n / \sim_3 \\ &\cong \mathbb{RP}^n. \end{aligned}$$

That is, the adjunction space $B^n \cup_f \mathbb{RP}^{n-1}$ is homeomorphic to the n-projective space $\mathbb{RP}^n$ for all n. Continuing the process of adjoining $\mathbb{RP}^{k-1}$ to the closed ball B^k, $k = n, n-1, \dots, 1$, we obtain the finite sequence

$$\begin{aligned} \mathbb{RP}^n &\cong B^n \cup_\pi B^{n-1} \cup_\pi \dots \cup_\pi B^2 \cup_\pi \mathbb{RP}^1 \\ &\cong B^n \cup_\pi B^{n-1} \cup_\pi \dots \cup_\pi B^2 \cup_\pi S^1 \\ &\cong B^n \cup_\pi B^{n-1} \cup_\pi \dots \cup_\pi [0, 1] \cup_\pi \{\mathrm{pt}\}, \end{aligned}$$

where $\{\mathrm{pt}\}$ is a point and the homeomorphism $\mathbb{RP}^1 \cong S^1$ follows from Proposition 3.1.5.

Remark 3.4.20. (1) The particular case $n = 2$ in Example 3.4.19 reads

$$\mathbb{RP}^2 \cong B^2 \cup_\pi \mathbb{RP}^1. \tag{3.1}$$

(2) By Proposition 3.1.5,

$$\mathbb{RP}^1 \cong S^1. \tag{3.2}$$

(3) Consider the composition diagram

$$\begin{array}{ccccc} A \subset X & \xrightarrow{\quad f \quad} & Y & \xrightarrow{\quad h \quad} & Y' \\ \big| & & & & \uparrow \\ \big\lfloor & & \xrightarrow{\qquad h \circ f \qquad} & & \big\rfloor \end{array}$$

where h is a homeomorphism. Then we can see that $X \cup_f Y \cong X \cup_{h \circ f} Y'$. Apply this to $X = B^2$, $Y = \mathbb{RP}^1$, $Y' = S^1$ and $h : \mathbb{RP}^1 \longrightarrow S^1$ a homeomorphism. Using (3.1) and (3.2), we get

$$\mathbb{RP}^2 \cong B^2 \cup_{h \circ \pi} S^1. \tag{3.3}$$

(4) By Proposition 3.4.18, $\mathbb{RP}^2 \cong B^2 \cup_{z^2} S^1$. In the proof of Proposition 3.4.18, the map $z \mapsto z^2$ is precisely the composite $h \circ \pi$, where $h : \mathbb{RP}^1 \longrightarrow S^1$ is some homeomorphism. Hence, $\mathbb{RP}^2 \cong B^2 \cup_{h \circ \pi} S^1$ and we recapture (3.3).

(5) Finally, Theorem 3.2.8 states that $\mathbb{RP}^2 \cong B^2 \cup_\partial \mathbb{M}^2$. To recap, we have

$$\mathbb{RP}^2 \cong B^2 \cup_{z^2} S^1 \cong B^2 \cup_\partial \mathbb{M}^2. \tag{3.4}$$

Example 3.4.19 motivates the definition of an n-skeleton.

Definition 3.4.21. (1) A *n-cell* is a space homeomorphic to the ball B^n.
(2) A cell is an n-cell for some $n = 1, \ldots$ called the dimension of the cell.
(3) A 0-cell consists of one point.

Remark 3.4.22. In [41], an n-cell is a pair homeomorphic to (B^n, S^{n-1}).

Definition 3.4.23. (1) A space X_n obtained by attaching an n-cell to another space X_{n-1} is called an *n-skeleton*:

$$X_n \equiv X_{n-1} \cup_f B^n, \quad \text{where } f : X_{n-1} \longrightarrow S^{n-1} \text{ is continuous.}$$

(2) The function f is the attaching map.
(3) The union $X = \bigcup_n X_n$ of n-skeletons is called a cell complex, or a *CW-complex*.

The terminology CW-complex goes back to John Henry Constantine Whitehead who introduced CW-complexes as a generalization of simplicial complexes (see Definition 10.1.19).

The letter "C" stands for "closure finiteness": a compact subset of a CW-complex intersects the interior of only finitely many cells. The letter "W" refers to "weak topology": a subset of a CW-complex is open if its restriction to each cell is open.

The cellular homology, not discussed in detail this book, investigates the topological structures of cell complexes. For interested readers, we recommend [10, Chapter IV, Sections 8–12], [24, Chapter 5], [33, Chapter 12, Lecture 13], [97, Chapter 6], among others.

A CW-complex space is locally contractible, according to the definition in Exercise 16, Chapter 7 (see [45, Proposition A.4, p. 522]).

A CW-complex space is paracompact (see [64, Theorem 4.2]).

Example 3.4.24. (1) Example 3.4.14 shows that the n-dimensional sphere is obtained by gluing one n-cell to a 0-cell.
(2) Theorem 2.5.11 may be expressed as $S^n \cong B^n \cup_{\mathrm{Id}} S^{n-1}$, showing that the sphere is an n-skeleton.

Example 3.4.25. By Example 3.4.19, the space $\mathbb{R}\mathrm{P}^n$ is an n-skeleton obtained from $\mathbb{R}\mathrm{P}^{n-1}$ by attaching the closed ball B^n.

Before ending this section, we investigate some general properties of the adjunction spaces. Let $\pi : X \cup Y \longrightarrow Z_f = X \cup_f Y$ be the continuous canonical projection map, where X and Y are arbitrary spaces. Assume that A *is a closed subset of* X.

Proposition 3.4.26. *If X and Y are compact (connected, path-connected, respectively), then Z_f is compact (connected, path-connected, respectively).*

Proof. (1) If X and Y are compact, then so are their images under the projection map π (Theorem 1.4.7). Therefore, the union $\pi(X) \cup \pi(Y) = Z_f$ is compact.

(2) Let X and Y be connected (path-connected, respectively). For $a \in A \subset X$, $\pi(a) = \pi(f(a))$. Then $\pi(X) \cap \pi(Y) \neq \emptyset$. Then the connectedness (path-connectedness, respectively) of $Z_f = \pi(X) \cup \pi(Y)$ follows from Theorem 1.4.12. □

Example 3.4.27. If X is compact (connected, path-connected), then so is the adjunction space $B^n \cup_f X$ obtained by attaching an n-cell to X.

Proposition 3.4.28. *The spaces Y and $\pi(Y)$ are homeomorphic.*

Proof. (1) Clearly, $\pi_{|Y}$ is injective. Indeed, for two different elements y_1 and y_2 in Y, we have $[y_1] \neq [y_2]$. Otherwise there exist x_1, x_2 in A such that $y_1 = f(x_1) \sim y_2 = f(x_2)$. Hence, $x_1 \sim x_2$ and $x_1, x_2 \in A$, which is not possible unless $x_1 = x_2$.

(2) To prove that $\pi_{|Y}$ is a closed map, let $V \subset Y$ be a closed set. Then $f^{-1}(V)$ is closed in A, i. e., $f^{-1}(V) = A \cap U$, where U is closed in X. Let $W = \pi(U \cup V)$, where $U \cup V$ is closed in $X \cup Y$. We aim to prove that $\pi(V) = \pi(Y) \cap W$ and W is closed in $X \cup_f Y$.

(a) For the first claim, notice that $V \subset Y$ and $V \subset U \cup V$ imply $\pi(V) \subset \pi(Y) \cap \pi(U \cup V) = \pi(Y) \cap W$. As for the reverse inclusion, it is a consequence of the identity

$$\pi(Y) \cap W = (\pi(Y) \cap \pi(U)) \cup (\pi(Y) \cap \pi(V))$$

and the inclusions

$$\pi(Y) \cap \pi(U) \subset \pi(A \cap U) = \pi(f^{-1}(V)) \subset \pi(V),$$

which result from the definition of the projection map π.

(b) Concerning the second claim, since $(U \cup V) \cap X = U$, $(U \cup V) \cap Y = V$ and $(U \cup V) \cap A = A \cap U$, by [25, Chapter XI, 6.2], $W = \pi(U \cup V)$ is closed in $X \cup_f Y$ if and only if $V \cup f(A \cap U)$ is closed in Y. The latter condition is immediately satisfied since $f(f^{-1}(V)) \subset V$ implies $V \cup f(A \cap U) = V \cup f(f^{-1}(V)) = V$, which is closed by assumption.

Hence, $\pi(V)$ is closed in $\pi(Y)$, i. e., $\pi_{|Y}$ is a closed map. Therefore, $\pi_{|Y}$ is an embedding onto $\pi(Y)$. □

Corollary 3.4.29. *The image $\pi(Y)$ is closed in Z_f.*

Proposition 3.4.30. *The spaces $X \setminus A$ and $\pi(X \setminus A)$ are homeomorphic.*

Proof. (1) The restriction map $\pi_{|X\setminus A}$ is injective because if $x, x' \in X \setminus A$ and $x \nsim x'$, then $[x] \neq [x']$.

(2) Let $U \subset X \setminus A$ be an open set. Since A is closed in X and $X \setminus A$ is open in X, U is open in X. Since $\pi(U) \subset X \setminus A$, we have $\pi(U) = \{\{x\}, x \in U\}$. Hence, $\pi(U)$ is open in $X \cup_f Y$. Finally, $\pi(U)$ is open in $(X \cup_f Y) \setminus \pi(Y) = \pi(X \setminus A)$. Indeed, by Corollary 3.4.29, $(X \cup_f Y) \setminus \pi(Y)$ is open in $X \cup_f Y$. Therefore, $\pi_{|X\setminus A}$ is an open map, hence an embedding onto $\pi(X \setminus A)$. □

Corollary 3.4.31. *The image $\pi(X \setminus A)$ is open in Z_f.*

The following remark discusses two particular and important cases.

Remark 3.4.32. (1) Suppose f is surjective. Then, by Remark 3.4.3 and Proposition 3.4.30, we have the disjoint union

$$Z_f = \pi(X) = \pi(X \setminus A) \cup \pi(A) \cong (X \setminus A) \cup \pi(A). \tag{3.5}$$

(2) Assume further that $A = X$. Again by Remark 3.4.3,

$$Z_f = \pi(X) = \pi(A) \cong \pi(f(A)) = \pi(Y). \tag{3.6}$$

Remark 3.4.33. (1) Propositions 3.4.28 and 3.4.30 state that the restriction of the projection map on either Y or $X \setminus A$ is an embedding.

(2) This means that Z_f can be thought of as the disjoint union of $X \setminus A$ and Y. Next, we present a second version of Proposition 3.4.30.

Proposition 3.4.34. *The spaces $X \setminus A$ and $Z_f \setminus \pi(Y)$ are homeomorphic.*

Proof. The restriction $h = \pi_{|X\setminus A}$ is the required homeomorphism. Indeed, $h(x) = \{x\}$ for all $x \in X \setminus A$. Hence, h is a continuous bijection. Furthermore, if $C \subset X \setminus A$ is a closed set, then due to the partition, we have $h(C) \cap \pi(Y) = \emptyset$ and $h^{-1}(h(C)) = C$. Hence, $h(C)$ is closed in $Z_f \setminus \pi(Y)$. □

As a consequence, we obtain the following result, confirming Remark 3.4.33.

Corollary 3.4.35. *The spaces $\pi(X \setminus A)$ and $\pi(Y)$ partition the space Z_f.*

Remark 3.4.36. By Proposition 3.4.28, we can define $j : Y \hookrightarrow Z_f$ as an inclusion map; hence, j is injective. Let $i : A \hookrightarrow X$ be a second inclusion map and $g = \pi_{|X}$ the restriction to X of the projection map. For all $a \in A$, we have

$$(j \circ f)(a) = f(a) \in Y \subset Z_f \quad \text{and} \quad (g \circ i)(a) = g(a) = \{a, f(a)\} \in Z_f.$$

By identification, we set $f(a) = g(a)$, i. e., $f = g_{|A}$ and $j \circ f = g \circ i$. Then the following diagram commutes

$$\begin{array}{ccc} A & \overset{i}{\hookrightarrow} & X \\ \downarrow f & & \downarrow g \\ Y & \underset{j}{\hookrightarrow} & Z_f \end{array}$$

A connection between the inclusion maps i and j is provided by the next result.

Proposition 3.4.37. *Let $i : A \hookrightarrow X$ the inclusion map. Then the inclusion map $j : Y \hookrightarrow Z_f$ is closed.*

Proof. Let $C \subseteq Y$ be a closed subset and $F = j(C)$. To prove that $\pi^{-1}(F)$ is closed in $X \cup Y$, note that

$$\begin{aligned} \pi^{-1}(F) &= (\pi^{-1}(F) \cap X) \cup (\pi^{-1}(F) \cap Y) \\ &= g^{-1}(F) \cup j^{-1}(F) \\ &= g^{-1}(j(C)) \cup j^{-1}(j(C)) \\ &= i(f^{-1}(C)) \cup C \\ &= f^{-1}(C) \cup C, \end{aligned}$$

where we have used that j is injective and $j \circ f = g \circ i \Longrightarrow g^{-1} \circ j = i \circ f^{-1}$. The continuity of f completes the proof. □

3.5 Mapping cylinder and mapping cone

Let $f : X \longrightarrow Y$ be a continuous map.

Definition 3.5.1. In the space $(X \times I) \cup Y$, define the equivalence relation $(x, 0) \sim f(x)$, for $x \in X$. The quotient space $(X \times I) \cup Y/_\sim$ is called *the mapping cylinder* and is denoted C_f.

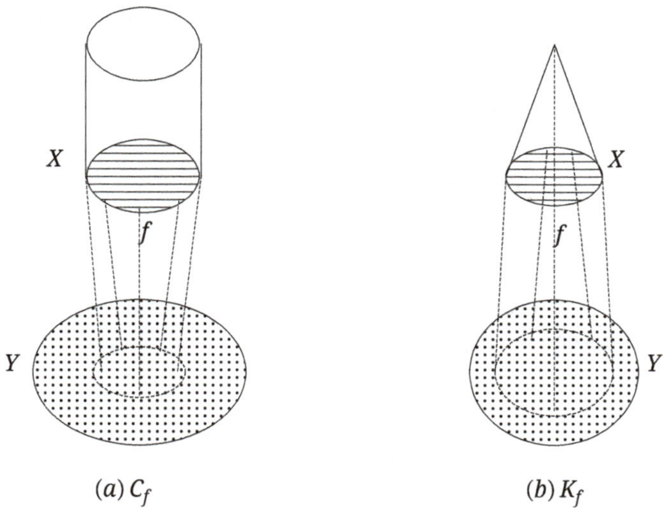

Figure 3.18: (a) Mapping cylinder; (b) Mapping cone.

Some properties of the mapping cylinder are explored in Exercise 21, Chapter 7. In Theorem 4.2.18, we will prove that C_f and Y are homotopically equivalent.

Definition 3.5.2. In the union $Z = (X \times I) \cup Y$, define an equivalence relation by

$$(x,1) \sim (x',1) \quad \text{and} \quad (x,0) \sim f(x), \quad \text{for all } x, x' \in X.$$

The quotient space $Z/_\sim$, denoted K_f, is called *the mapping cone* of f.

Remark 3.5.3. (1) The mapping cylinder C_f is the adjunction space of the cylinder $X \times I$ with basis X and the space Y along the map $f_0 : X \times \{0\} \subset X \times I \longrightarrow Y$, where $f_0(x,0) = f(x)$, i. e., $C_f \cong (X \times I) \cup_{f_0} Y$ (see Figure 3.18(a)).
(2) The mapping cone K_f is homeomorphic to $C_f/_{X\times\{1\}}$ (see Figure 3.18(b)).
(3) If $C(X)$ is the cone over the space X, then $K_f \cong C(X) \cup_f Y$.

Example 3.5.4. (1) The formulas $C_f \cong (X \times I) \cup_{f_0} Y$ and $K_f \cong C(X) \cup_f Y$ justify the designation of these spaces by mapping cylinder and mapping cone, respectively.
(2) If $X = Y$ and f is the identity map, then C_f is just the cylinder $X \times I$ and K_f is the cone over X.

Example 3.5.5. Let $X = Y = S^1$, and the map $f : S^1 \longrightarrow S^1$ be given by $f(z) = z^2$. By Definition 3.1.1, Theorem 3.3.6 and Proposition 3.4.18,

$$K_f \cong C(S^1) \cup_f S^1 \cong B^2 \cup_f S^1 \cong \mathbb{RP}^2.$$

Example 3.5.6. Let $X = S^n$, $Y = \mathbb{R}P^n$ and $f = \pi : S^n \longrightarrow \mathbb{R}P^n$ be the projection map given by $f(x) = \{x, -x\}$. By Theorem 3.3.6, the mapping cone of f is

$$K_f \cong C(S^n) \cup_f \mathbb{R}P^n \cong B^{n+1} \cup_f \mathbb{R}P^n \cong \mathbb{R}P^{n+1},$$

the latter isomorphism is proved in Example 3.4.19.

Example 3.5.7. Let $f : X \longrightarrow \{y_0\} \in Y$ be a constant map. Then:
(1) $C_f = C(X) \vee Y$ and $K_f = C(X) \cup_f Y \cong \Sigma(X) \vee Y$;
(2) in particular, if $Y = \{y_0\}$ is a singleton, then the mapping cylinder is just a cone $C_f = C(X)$ and the mapping cone a suspension $K_f = \Sigma(X)$.

Example 3.5.8. If $f = i : A \subset X \hookrightarrow X$ is the inclusion map, then $K_f \cong C(A) \cup_i X$.

Exercises (Sections 3.3–3.5)

1. (1) Prove that the map $h : X \longrightarrow C(X)$ given by $h(x) = [(x, 0)] = \{(x, 0)\}$ is an embedding.
(2) Deduce that $C(X)$ is Hausdorff if and only if X is.

2. Let $C(X)^- = \pi(X \times [0, 1/2]$ and $C(X)^+ = \pi(X \times [1/2, 1]$. Prove that $C(X)$, $C(X)^-$ and $C(X)^+$ are homeomorphic.

3. A continuous map $f : X \longrightarrow Y$ induces *a cone map* $C(f) : C(X) \longrightarrow C(Y)$ defined by

$$C(f)([(x, t)]) = [(f(x), t)].$$

By replacing $C(f)$ by $\Sigma(f)$, we obtain the definition of *a suspension map* $\Sigma(f) : \Sigma(X) \longrightarrow \Sigma(Y)$.

Prove that f is a homeomorphism if and only if $C(f)$ (resp., $\Sigma(f)$) is homeomorphism).

4. Show that $\Sigma(\mathrm{Id}_{|X}) \cong \mathrm{Id}_{|\Sigma(X)}$.

5. Describe C_f for $X = Y = S^1$ and $f(z) = z^2$.

6. Let $f : X \longrightarrow Y$ and $g : Y \longrightarrow Z$, where X, Y, Z are spaces. Prove that

$$\Sigma(g \circ f) = \Sigma(g) \circ \Sigma(f).$$

7. If X is compact (connected, respectively), show that so is $C(X)$.

8. If $C(X)$ is locally compact, show that X is compact.

9. (1) Prove that the cone $C(X)$ is path-connected.
(2) Deduce that every space can be embedded in a path-connected space.

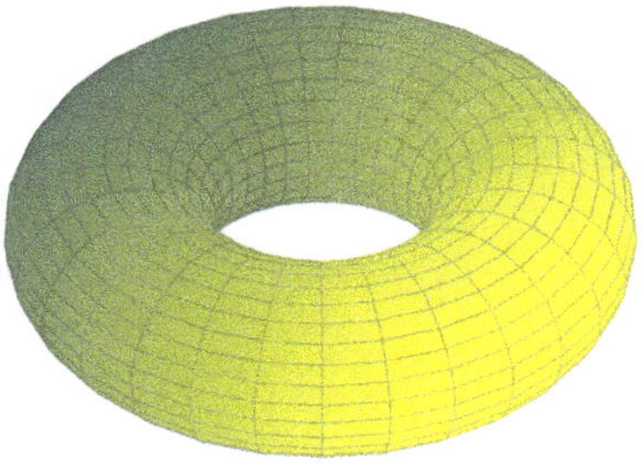

Figure 3.19: Torus embedded in Euclidean space.

10. Prove that the cone $C(X)$ is locally connected (resp., locally path-connected) if and only if X is locally connected (resp., locally path-connected).

11. Let $\mathbb{T}^2 \cong I^2/_\sim$ be the torus defined as the quotient space in Example 2.3.12 and let the map $f : I^2 \longrightarrow \mathbb{R}^3$ be given by

$$f(x,y) = ((2+\cos(2\pi x))\cos(2\pi y), (2+\cos(2\pi x))\sin(2\pi y), \sin(2\pi x)).$$

(1) Prove that f is continuous, surjective and agrees with the relation $\sim$.
(2) Deduce that the torus $\mathbb{T}^2$ is homeomorphic to the surface $S \subset \mathbb{R}^3$ whose equation is (see Figure 3.19)

$$(r-2)^2 + z^2 = 1 \quad (r = \sqrt{x^2+y^2}).$$

Compare with Exercise 8, Sections 3.1–3.2.

12. Prove that the wedge sum of two Hausdorff spaces is a Hausdorff space.

13. Prove that the wedge sum of two connected (path-connected, respectively) spaces is a connected (path-connected, respectively) space.

14. Let X and Y be two compact subspaces of $\mathbb{R}^n$.
(1) Prove that the join $X * Y$ is homeomorphic to the space Z of line segments $[x,y]$, where $x \in X, y \in Y$ and two segments can only intersect at an end-point.
(2) Deduce that $X * Y$ is homeomorphic to $(C(X) \times Y) \cup (X \times C(Y))$.

15. Prove that the join of two balls B^m and B^n is homeomorphic to the ball B^{m+n+1}.

16. (1) Check that $S^{n-1} * S^0 \cong S^n$ and $B^n * S^0 \cong B^{n+1} \vee B^{n+1}$.
(2) Prove that $S^m * S^n \cong B^{m+n+1}/S^{m+n}$.
(*Hint:* Use Exercise 14(2), Theorem 3.3.6 and Theorem 2.5.11.)
(3) Deduce that $S^m * S^n \cong S^{m+n+1}$.

17. Prove that the map $h : S^{m+n+1} \longrightarrow S^m * S^n$ given by $h(x,y) = rx/\|x\| + sy/\|y\|$, where

$$r\pi/2 = \sin^{-1}(\|x\|) \quad \text{and} \quad s\pi/2 = \sin^{-1}(\|y\|)$$

is a homeomorphism. (*Hint:* See [12, p. 173].)

18. On the space $X = B^n \times S^{n-1} \cup S^{n-1} \times B^n$, define the equivalence relation $(x,y) \sim (y,x)$. Show that $X/_\sim \cong S^{n-1} * \mathbb{R}P^{n-1}$.

19. Prove that if X and Y are Hausdorff spaces, then so are the mapping cylinder C_f and the mapping cone K_f.

20. Let $f : S^1 \subset B^2 \longrightarrow [-1,1]$ be given by $f(z) = \mathrm{Re}(z)$. Prove that the adjunction space $B^2 \cup_f [0,1]$ is homeomorphic to S^2.

21. Consider the identification

$$S^3 = \{(z_1, z_2) \in \mathbb{C}^2 : |z_1|^2 + |z_2|^2 = 1\} \subset B^4 \subset \mathbb{R}^4 \cong \mathbb{C} \times \mathbb{C}.$$

Prove that the map $h : S^3 \longrightarrow S^2$ given by

$$h(z_1, z_2) = \left(\frac{|z_1|^2 - |z_2|^2}{|z_1|^2 + |z_2|^2}, \frac{\overline{z_1}z_2 - z_1\overline{z_2}}{|z_1|^2 + |z_2|^2}, \frac{\overline{z_1}z_2 + z_1\overline{z_2}}{|z_1|^2 + |z_2|^2} \right)$$

induces a homeomorphism between $\mathbb{C}P^1$ and S^2. (*Hint:* See [66, Exercise 5.19].)

22. (1) For $n \geq 2$, let $f = \pi : S^{2n-1} \subset B^{2n} \longrightarrow S^{2n-1}/_\sim \cong \mathbb{C}P^{n-1}$ be the canonical projection map. Prove that the adjunction space $B^{2n} \cup_f \mathbb{C}P^{n-1}$ obtained by attaching an $2n$-cell to $\mathbb{C}P^{n-1}$ (see Definition 3.4.21) is homeomorphic to $\mathbb{C}P^n$. This is an equivalent version of Exercise 16, Sections 3.1–3.2. (*Hint:* See [42, Proposition (19.10)], actually the same proof as in [28, p. 168].)

(2) The case $n = 1$ is Exercise 22. For $n = 2$, check that one can consider as well the map $f : S^3 \longrightarrow S^2$ given by $f(z_1, z_2) = z_1 z_2^{-1}$ with the identifications $S^2 \cong \mathbb{C}P^1$ (Exercise 22) and $S^2 \cong \mathbb{C} \cup \{\infty\}$ (one-point compactification).

23. Let $Z_f = X \cup_f Y = (X \cup Y)/_\sim$ be the adjunction space of X and Y through a continuous map $f : A \longrightarrow Y$, where $A \subset X$.

(1) Prove that if X and Y are T_1 spaces (i. e., every one-set element is closed), then so is Z_f.
(*Hint:* See [50, Proposition 2.1].)

(2) Prove that this property no longer holds if T_1 is replaced by Hausdorff (T_2) or regular (T_3).
(*Hint:* See [50, Proposition 2.2].)

(3) Prove that if X and Y are Hausdorff and A is compact then Z_f is a Hausdorff space.
(*Hint:* See [92, Theorem 6.5.4].)

24. Let $Z_f = X \cup_f Y = (X \cup Y)/_\sim$ be the adjunction space of X and Y through a continuous map $f : A \longrightarrow Y$, where $A \subset X$. Prove that if X and Y are normal spaces (Definition 1.1.18) and A is closed, then Z_f is normal.

3.6 Manifolds

Definition 3.6.1. (1) *A topological manifold* X is a second countable locally Euclidean Hausdorff space. That is to say, X is a Hausdorff space, it admits a countable basis for its topology and there is a positive integer n such that every point in X has a neighborhood in X homeomorphic to an open subset of $\mathbb{R}^n$. n is called the dimension of X.

(2) An n-manifold M_n is a topological manifold of dimension n.

(3) *A surface* is a 2-manifold.

In the remainder of this book, a manifold means a topological manifold.

Remark 3.6.2. An n-manifold is locally compact and locally connected.

Remark 3.6.3. Since an n-manifold M_n is locally path-connected, by Corollary 1.4.54, M_n is connected if and only if it is path-connected.

Example 3.6.4. S^n is an n-manifold since for all $x \in S^n$ and $y \neq x$, $S^n \setminus \{y\}$ is an open neighborhood of x, which is homeomorphic to $\mathbb{R}^n$ by the stereographic projection.

Example 3.6.5. By Theorem 1.5.38, the cylinders $S^1 \times (0,1)$ and $S^1 \times \mathbb{R}$ are surfaces. However, $S^1 \times [0,1]$ is not a surface since points on $S^1 \times \{0\}$ and $S^1 \times \{1\}$ do not have neighborhoods homeomorphic to the unit open ball.

Example 3.6.6. If M_n is an n-manifold, then every open subset of M_n is an n-manifold.

Definition 3.6.7. (1) *An n-manifold with boundary* M_n is a second countable Hausdorff space such that every point has a neighborhood homeomorphic to an open subset of

$$H^n = \{(x_1, x_2, \dots, x_n) \in \mathbb{R}^n : x_n \geq 0\}.$$

(2) The subset of points, which have a neighborhood homeomorphic to an open subset of the closed n-dimensional upper half-space,

$$H^n = \{(x_1, x_2, \dots, x_n) \in \mathbb{R}^n : x_n > 0\},$$

is an n-manifold called *the interior* of M_n, and denoted $\operatorname{int}(M_n)$.

(3) The set $M_n \setminus \operatorname{int}(M_n)$ is called *the boundary of* M_n and denoted ∂M.

Remark 3.6.8. (1) The boundary of an n-manifold is the subset of points, which have a neighborhood homeomorphic to a

$$H^n = \{(x_1, x_2, \dots, x_n) \in \mathbb{R}^n : x_n = 0\}.$$

(2) The boundary of an n-manifold, if nonempty, is an $(n-1)$-manifold.

Example 3.6.9. The closed ball B^n is an n-manifold whose boundary is S^{n-1}.

Example 3.6.10. The cylinder $S^1 \times [0,1]$ is a 2-manifold with boundary given by the union $S^1 \times \{0\}$ and $S^1 \times \{1\}$.

Example 3.6.11. The sphere S^2, the torus $S^1 \times S^1$ and the projective plane $\mathbb{RP}^2$ are closed 2-manifolds.

Example 3.6.12. The Möbius band $\mathbb{M}^2$ is a 2-manifold with boundary.

Definition 3.6.13. A closed manifold is a compact manifold without boundary.

More details on manifolds and their properties can be found in [62, 95].

3.7 Connected sum of manifolds

Definition 3.7.1. Let M and N be two nonempty disjoint n-manifolds with boundary. The connected sum, denoted $M\#N$ of M and N is the n-manifold obtained by deleting an open n-ball in each manifold and gluing the resulting punctured manifolds by some homeomorphism along the boundaries. That is to say, $M\#N$ is the adjunction space $X/\sim$, where

$$X = M^\circ \cup_h N^\circ = (M \setminus \mathring{B}_1) \cup_h (N \setminus \mathring{B}_2), \quad h : \partial(M \setminus \mathring{B}_1) \longrightarrow \partial(M \setminus \mathring{B}_2)$$

and

$$x, x' \in X, \quad x \sim x' \quad \Longleftrightarrow \quad x' = x \quad \text{or} \quad x \in \partial(M \setminus \mathring{B}_1) \quad \text{and} \quad x' = h(x).$$

Remark 3.7.2. One can consider open sets $U \subset M$, $V \subset N$ homeomorphic to open n-balls and $f, g : \mathbb{R}^n \longrightarrow U, V$ two homeomorphisms. Then the attaching map is $h = g \circ f$ (see Figure 3.20).

Remark 3.7.3. (1) Clearly, the connected sum is a commutative and an associative operation (first mentioned in [55, Lemma 2.1]).

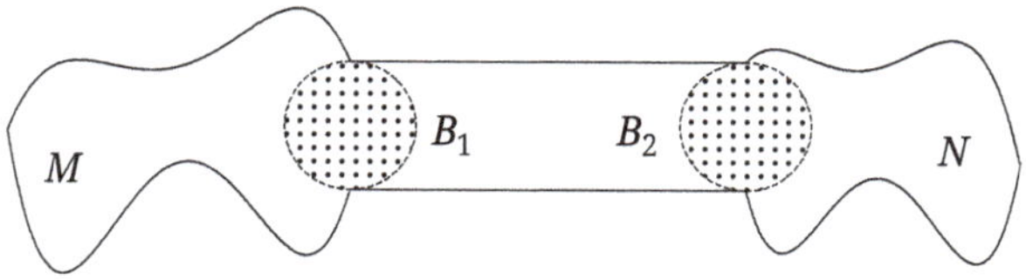

Figure 3.20: Connected sum, $M\#N$.

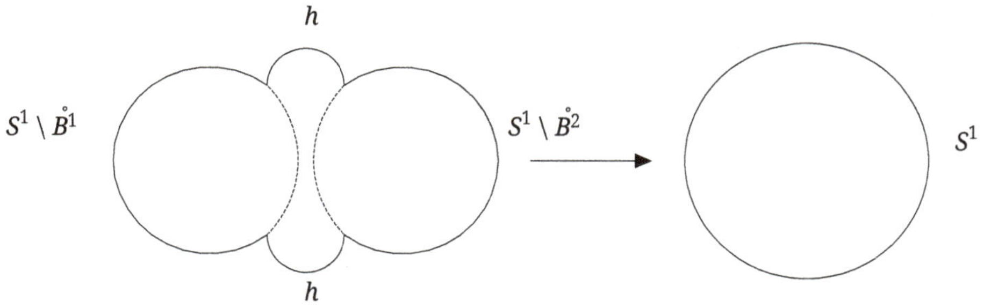

Figure 3.21: Connected sum $S^1 \# S^1 \cong S^1$.

(2) The connected sum does not depend on the choice of the balls B_i $(1 \leq i \leq 2)$ or the homeomorphism h (see [67, Theorem 5.1]).

The connected sum of manifolds occurs in the classification of compact surfaces, as summarized in the following theorem (see, e. g., [45, 67, 81]). The classification relies on the orientability of the surface (a surface is said to be orientable if it does not contain an open Möbius band within it). By Remark 3.8.13 and Theorem 3.2.8, the Klein bottle and the projective plane are not orientable.

Theorem 3.7.4. *Any compact surface S is either homeomorphic to a sphere, or to a connected sum of tori, or to a connected sum of projective planes.*

For the proof, we refer to [62, Theorem 6.15], [67, Proposition 5.1], [68, Theorem 5.1], [76, Theorem 77.5].

Example 3.7.5. If $M_1 = M_2 = S^1$, then $M_1 \# M_2 \cong S^1$ (see Figure 3.21).

The following result indicates that S^2 behaves as an identity element.

Proposition 3.7.6. *The connected sum of a surface M and the sphere S^2 is homeomorphic to M.*

Proof. Indeed, $M \# S^2 = (M \setminus \mathring{B}^2) \cup_h (S^2 \setminus \mathring{B}^2)$. Since the open disk $\mathring{B}^2$ is homeomorphic to the northern open hemisphere (Example 1.5.22), $S^2 \setminus \mathring{B}^2$ is homeomorphic to $S^2 \setminus \mathring{S}^2_-$. Therefore, $S^2 \setminus \mathring{B}^2$ is homeomorphic to the southern closed hemisphere, and consequently to the closed disk B^2. Finally, the result follows from the adjunction of the closed ball B^2 with $M \setminus \mathring{B}^2$ along the boundary S^1 via the identity map (we have deleted $\mathring{B}^2$ and then pasted back B^2). □

Example 3.7.7. In Figure 3.22, the process is described with $M = \mathbb{T}^2$. By the stereographic projection and Example 1.5.18, $S^2 \setminus \mathring{B}^2$ is homeomorphic to the closed ball B^2. The torus is ultimately obtained after filling in the hole of the perforated torus with the small disk by pasting the boundaries of the balls B^2 and $\mathring{B}^2$ using the homeomorphism h. The resulting space is $\mathbb{T}^2 \# S^2 \cong \mathbb{T}^2$.

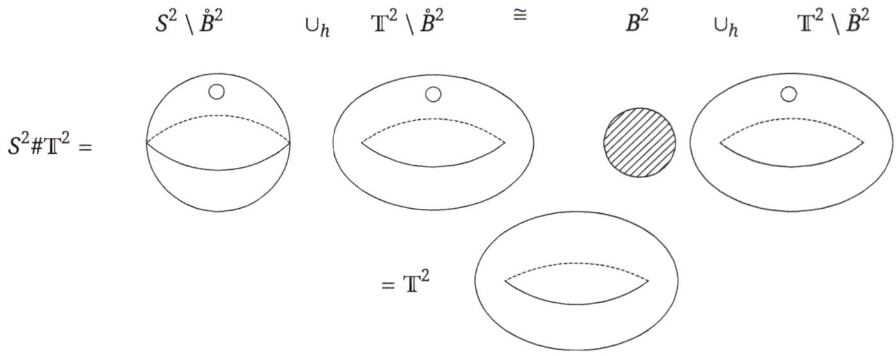

Figure 3.22: Connected sum $S^2\#\mathbb{T}^2 \cong \mathbb{T}^2$.

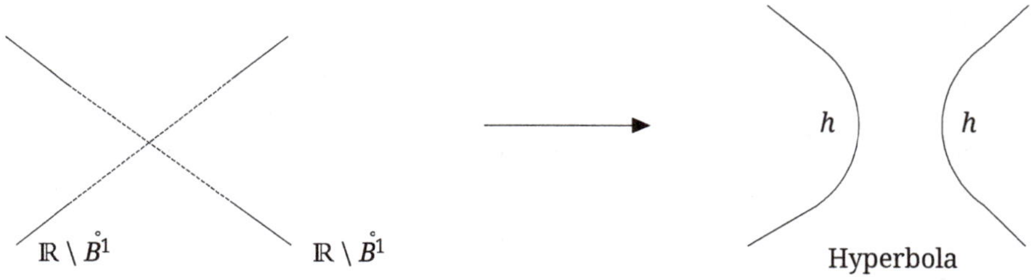

Figure 3.23: Connected sum $\mathbb{R}\#\mathbb{R}$ is a hyperbola.

Remark 3.7.8. By Example 1.5.22, the closed ball B^n is homeomorphic to the closed northern hemisphere S^n_+ with some homeomorphism h. Also, h maps S^{n-1} onto S^{n-1} and the open ball $\mathring{B}^n$ onto $S^n_+ \setminus S^{n-1}$. Then $S^n \setminus h(\mathring{B}^n)$ is homeomorphic to the closed southern hemisphere, hence homeomorphic to B^n. This explains how the small open disk was removed from S^n and then pasted back in both Example 3.7.7 and Figure 3.22.

Remark 3.7.9. Following Remark 3.7.8, we point out that the small open disk to be removed in the connected sum cannot be replaced by a point for otherwise the connected sum $M\#N$ becomes $M' \cup_h N'$, where $M' = M \setminus \{x_0\}$ and $N' = N \setminus \{y_0\}$ are the punctured manifolds. Thus, the map h is not defined and cannot be thought of as $h(x_0) = y_0$, which would give rise to the wedge sum $M' \vee N'$.

Example 3.7.10. The connected sum $\mathbb{R}\#\mathbb{R}$ gives rise to a hyperbola (see Figure 3.23).

3.8 Polygonal representation

It is possible to describe schematically the surfaces, which are already defined topologically as quotient spaces through Examples 3.2.2–3.2.10. A labeling scheme is a word

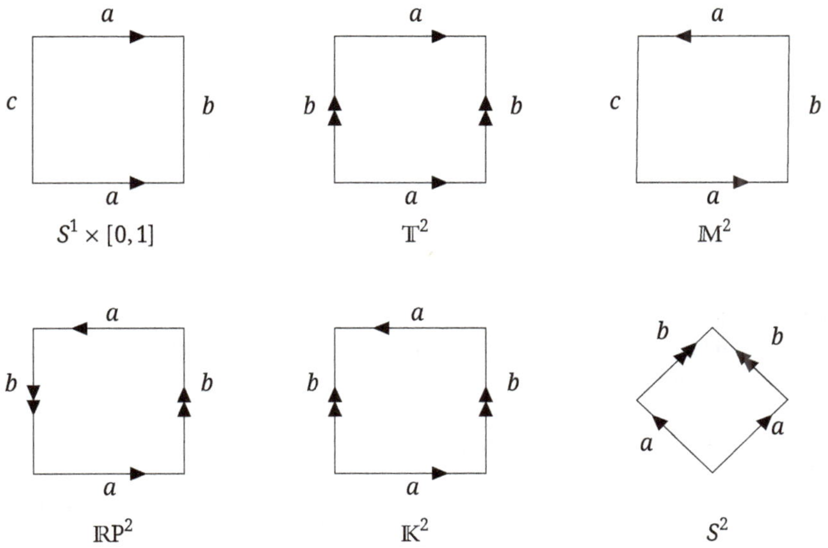

Figure 3.24: Polygonal representation of surfaces.

composed of the juxtaposition of a finite number of alphabetic letters (or symbols) that represent the edges of the surface arranged according to the order of the given orientation. For instance, when moving counterclockwise around the sides of the square, the letter a means that the arrow points clockwise and a^{-1} means that it points counterclockwise. Thus, a^{-1} refers to the edge with the reverse direction. Each surface is thus defined by *a word*. A surface is represented symbolically by $\langle S|\ a_1a_2\dots a_m\rangle$.

Example 3.8.1. The representation of the familiar surfaces are (see Figure 3.24).

Cylinder: $\langle a,b,c|\ aba^{-1}c\rangle$	Torus: $\langle a,b|\ aba^{-1}b^{-1}\rangle$
Möbius band: $\langle a,b,c|\ abac\rangle$	Real projective plane: $\langle a,b|\ abab\rangle$
Klein bottle: $\langle a,b|\ abab^{-1}\rangle$	Sphere: $\langle a,b|\ abb^{-1}a^{-1}\rangle$

The process allows us to make some geometrical transformations leaving unchanged the topological structure of the surface. We quote some of these elementary operations (see, e. g., [62, p. 133] and [76, Chapter 12]).

(1) *Cutting* a surface along an edge, denoted b, is the operation

$$\langle S|\ a_1a_2\dots a_m\rangle \longrightarrow \langle S|\ a_1a_2b, b^{-1}, a_3\dots a_m\rangle,$$

the comma meaning that the word is split into two words.

(2) *Pasting* (or gluing) is the inverse operation of cutting:

$$\langle S|\ a_1a_2b, b^{-1}, a_3 \dots a_m\rangle \longrightarrow \langle S|\ a_1a_2 \dots a_m\rangle.$$

(3) *Rotating* a surface is making a rotation (permutation) with edges orientation preserving:

$$\langle S|\ a_1a_2 \dots a_m\rangle \longrightarrow \langle S|\ a_2a_3 \dots a_ma_1\rangle,$$

where the edge a_1 has been rotated clockwise.

(4) *Reflecting* a surface S is to flip it over (to look at S in a mirror). A reflection is a homeomorphism because the orientations and orderings of the edges do not change:

$$\langle S|\ a_1a_2 \dots a_m\rangle \longrightarrow \langle S|\ (a_m)^{-1}(a_{m-1})^{-1} \dots (a_2)^{-1}(a_1)^{-1}\rangle.$$

(5) *Relabeling* a surface changes the occurrences of a symbol (or a group of symbols) to another symbol not already used, for instance,

$$\langle S|\ a_1a_2 \dots a_mb_1b_2 \dots b_m\rangle \longrightarrow \langle S|\ ab\rangle.$$

Example 3.8.2. Making a rotation shows that $aba^{-1}b \cong abab^{-1}$. This explains the representation of the Klein bottle in Figure 3.24.

Example 3.8.3. Different equivalent definitions of the projective space have been given so far. In Definition 3.1.1, $\mathbb{RP}^2$ is obtained as a quotient space from the sphere S^2. In Example 3.1.15, it is given from the square I^2. This explains the equivalence $abab \cong aa$ and justifies the relabeling operation (see Figure 3.25).

Example 3.8.4. The representation of the sphere comes from the equivalence relations considered in Exercise 19, Sections 3.1–3.2. The relation requires considering a diamond-shaped square or up to a rotation, a square. The equivalence $abb^{-1}a^{-1} \cong aa^{-1}$ justifies the relabeling operation (see Figure 3.26).

The following result allows us to derive the polygonal representation of the connected sum of two manifolds. For the proof, we refer the reader to [62, Proposition 6.12].

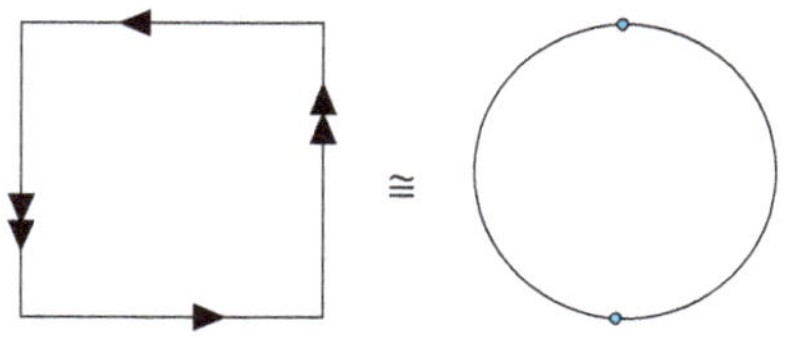

Figure 3.25: The projective plane $\mathbb{RP}^2$.

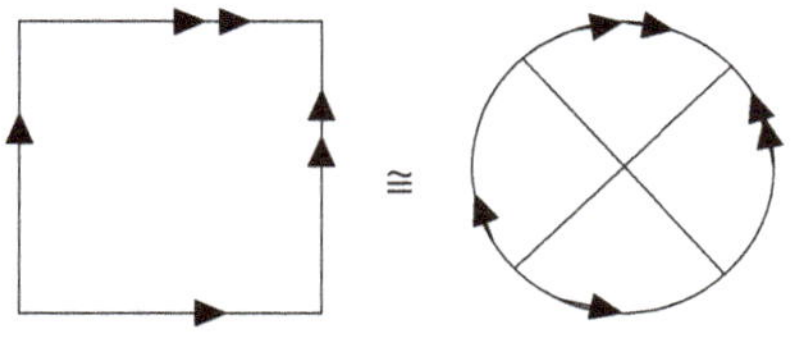

Figure 3.26: Equivalent form of the sphere S^2.

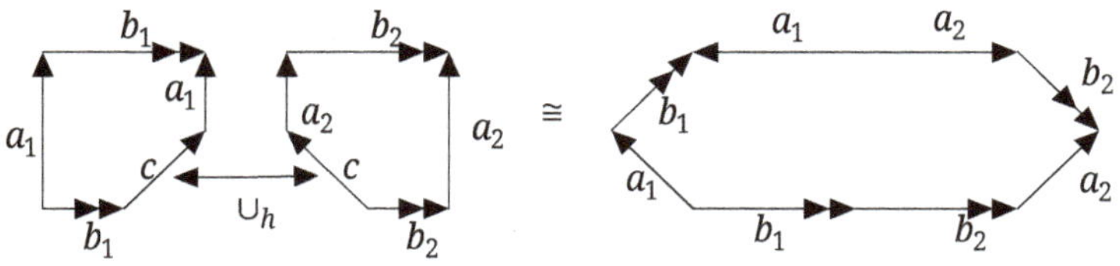

Figure 3.27: Polygonal representation of the connected sum of two tori.

Theorem 3.8.5. *Let S_1 and S_2 be two surfaces whose polygonal representations are $\langle S_1 | W_1 \rangle$ and $\langle S_2 | W_2 \rangle$, in which S_1 and S_2 are disjoint words. Then the connected sum $S_1 \# S_2$ has the representation $\langle S_1 \# S_2 | W_1 W_2 \rangle$, where $W_1 W_2$ denotes the concatenation (juxtaposition) of the words W_1 and W_2.*

Example 3.8.6. We have the connected sums
(1) $\mathbb{RP}^2 \# \mathbb{RP}^2 \cong \langle a_1, b_1, a_2, b_2 | a_1 b_1 a_1 b_1 a_2 b_2 a_2 b_2 \rangle \cong \langle a, b | aabb \rangle \cong \langle a, b | a^2 b^2 \rangle$;
(2) $\mathbb{M}^2 \# \mathbb{T}^2 \cong \langle a_1, b_1, c_1, a_2, b_2 | a_1 b_1 a_1 c_1 a_2 b_2 a_2^{-1} b_2^{-1} \rangle$;
(3) $\mathbb{RP}^2 \# \mathbb{K}^2 \cong \langle a_1, b_1, a_2, b_2 | a_1^2 b_1^2 a_2 b_2 a_2^{-1} b_2^{-1} \rangle$.

Example 3.8.7. By Theorem 3.8.5, the connected sum of two tori $X = \mathbb{T}^2 \# \mathbb{T}^2$ has the following representation:

$$\mathbb{T}^2 \# \mathbb{T}^2 \cong \langle a_1, b_1, a_2, b_2 | a_1 b_1 a_1^{-1} b_1^{-1} a_2 b_2 a_2^{-1} b_2^{-1} \rangle. \tag{3.7}$$

Schematically, the connected sum can be obtained by gluing along a common side of the polygons representing the tori with one small ball removed in each (perforated tori). Then the resulting surface is an octagon (in fact, each horizontal line in Figure 3.27 represents two sides) with the following coding:

$$\langle a_1, b_1, a_2, b_2 | b_1^{-1} a_1 b_1 a_1^{-1} a_2 b_2 a_2^{-1} b_2^{-1} \rangle,$$

which is equivalent to the representation (3.7) (see Figure 3.27).

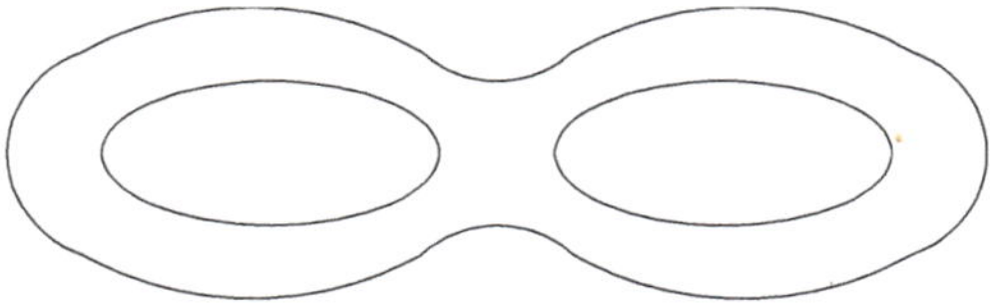

Figure 3.28: Connected sum of two tori ($\mathbb{T}^2 \# \mathbb{T}^2$).

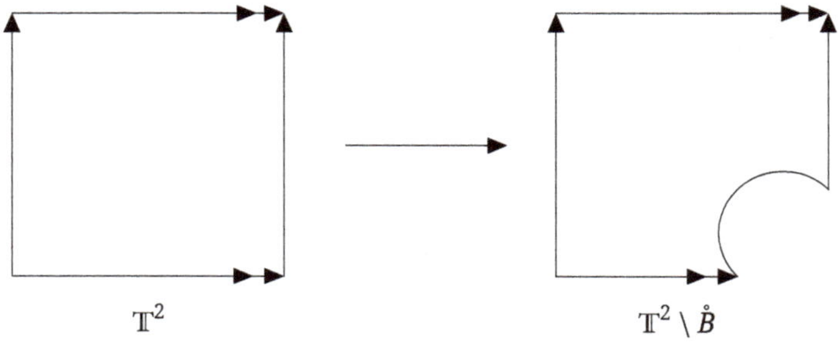

Figure 3.29: Torus and perforated torus.

Example 3.8.8. The connected sum of two tori

$$X = \mathbb{T}^2 \# \mathbb{T}^2 \cong \mathbb{T}^2 \setminus \mathring{B}_1 \cup_h \mathbb{T}^2 \setminus \mathring{B}_2$$

is a torus with two holes drilled through it (see Figure 3.27 and Figure 3.28). The space X can be thought of as a double torus with a pretzel shape. Figure 3.29 represents the perforated torus used to construct the connected sum.

Figure 3.27 shows that the perforated torus has the representation

$$\langle a, b, c | \ aba^{-1}c^{-1}b^{-1} \rangle.$$

In Figure 8.1, we will show that a perforated torus can be seen as the union of two cylinders. Its fundamental group will be derived (see Proposition 8.2.2).

Remark 3.8.9. The connected sum of two tori should not be confused with the wedge sum of two tori $Y = \mathbb{T}^2 \vee \mathbb{T}^2$ (see Figure 3.30).

The following result illustrates Theorem 3.7.4.

Proposition 3.8.10. $\mathbb{K}^2 \cong \mathbb{RP}^2 \# \mathbb{RP}^2$.

Proof. (1) *First proof.* Starting from the representation of the Klein bottle, we obtain the equivalent polygonal representations:

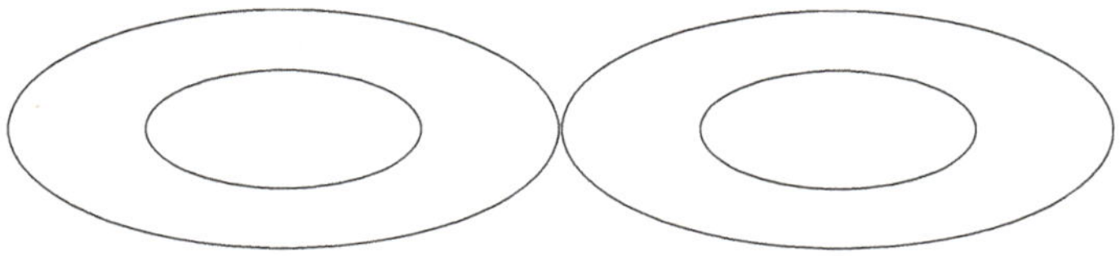

Figure 3.30: Wedge sum of two tori ($\mathbb{T}^2 \vee \mathbb{T}^2$).

$$\begin{aligned}\mathbb{K}^2 &\cong \langle a, b \mid abab^{-1} \rangle \quad \text{(using Example 3.8.1)}\\ &\cong \langle a, b, c \mid abc, c^{-1}ab^{-1} \rangle \quad \text{(cutting along an edge } c\text{)}\\ &\cong \langle a, b, c \mid bca, a^{-1}cb \rangle\\ &\quad \text{(rotating the first word, rotating and reflecting the second one)}\\ &\cong \langle b, c \mid bbcc \rangle \quad \text{(pasting along the edge } a \text{ and then rotating)}\\ &\cong \mathbb{RP}^2 \# \mathbb{RP}^2 \quad \text{(using Example 3.8.1, Theorem 3.8.5 and Example 3.8.6).}\end{aligned}$$

(2) *Second proof.* By Theorem 3.2.8, the complement $X = \mathbb{RP}^2 \setminus B^2$ is homeomorphic to the Möbius band $\mathbb{M}^2$. By Definition 3.7.1 of the connected sum, the union $X \cup_\partial X$ glued along the boundary of the open balls $\mathring{B}^2$ is exactly $\mathbb{RP}^2 \# \mathbb{RP}^2$. By Remark 3.8.13, two Möbius bands can be glued together along their boundaries to form a Klein bottle. Therefore, $\mathbb{RP}^2 \# \mathbb{RP}^2 \cong \mathbb{K}^2$. □

Remark 3.8.11. Proof of Proposition 3.8.10 shows that both diagrams $aabb$ and $aba^{-1}b$ lead to the same surface.

The following results shows that a Klein bottle can be decomposed to yield two Möbius bands. See Figures 3.31 and 3.32.

Corollary 3.8.12. *The Klein bottle is the adjunction space of two Möbius bands.*

Proof. By Theorem 3.2.8 and Proposition 3.8.10,

$$\begin{aligned}\mathbb{K}^2 &\cong \mathbb{RP}^2 \# \mathbb{RP}^2\\ &= (\mathbb{RP}^2 \setminus B^2) \cup_h (\mathbb{RP}^2 \setminus B^2)\\ &\cong \mathbb{M}^2 \cup_h \mathbb{M}^2,\end{aligned}$$

where $h : \partial B^2 \longrightarrow \partial B^2$ is some homeomorphism. □

Remark 3.8.13. Two Möbius bands glued together along their boundaries through a homeomorphism h form a Klein bottle in Figure 3.31. This is equivalent to saying that a Klein bottle may be decomposed in two Möbius bands as depicted in Figure 3.32.

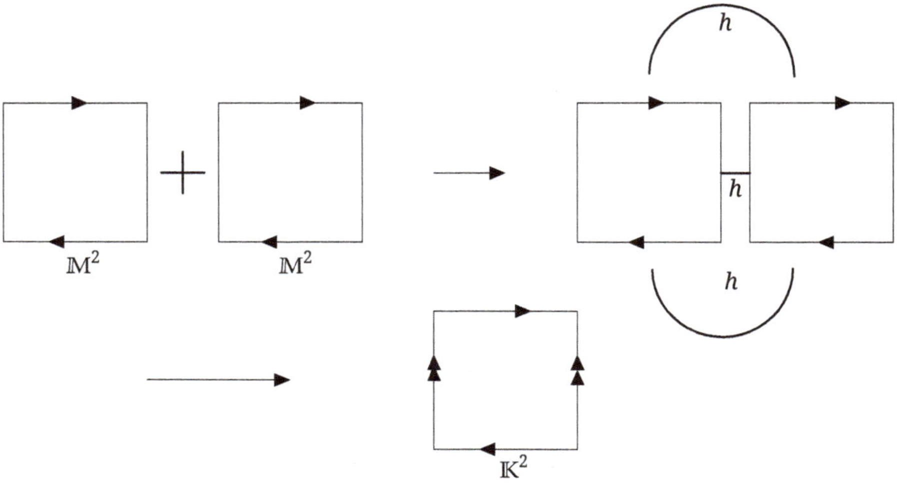

Figure 3.31: Adjunction space of two Möbius band: $\mathbb{M}^2 \cup \mathbb{M}^2 \cong \mathbb{K}^2$.

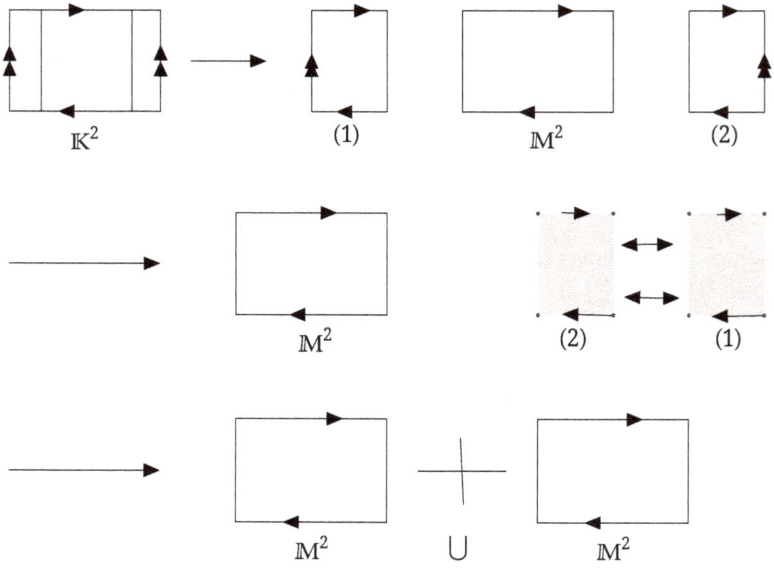

Figure 3.32: Decomposition of the Klein bottle: $\mathbb{K}^2 \cong \mathbb{M}^2 \cup \mathbb{M}^2$.

Exercises (Section 3.6–3.8)

1. Prove that

$$\mathbb{M}^2 \# \mathbb{T}^2 \cong \mathbb{M}^2 \# \mathbb{K}^2.$$

2. (a) Prove that

$$\mathbb{RP}^2 \# \mathbb{T}^2 \cong \mathbb{RP}^2 \# \mathbb{K}^2.$$

(b) Deduce that

$$\mathbb{RP}^2 \# \mathbb{RP}^2 \# \mathbb{RP}^2 \cong \mathbb{RP}^2 \# \mathbb{K}^2 \cong \mathbb{RP}^2 \# \mathbb{T}^2.$$

3. Prove that

$$\mathbb{T}^2 \# \mathbb{K}^2 \cong \mathbb{K}^2 \# \mathbb{K}^2.$$

4 The fundamental group

The fundamental group of a topological space was introduced for the first time by Henri Poincaré in 1895 to bridge the topological structures of spaces with the algebraic properties of some groups. The basic definition of the fundamental group employs the homotopy theory, which is developed in Section 4.1. The homotopy equivalence of spaces discussed in Section 4.2 is a concept weaker than the homeomorphism defined in Section 1.5 even though the homotopy equivalence implies the isomorphism of the associated fundamental groups (Section 4.5). Before introducing the fundamental group of a space in Section 4.4, the product of paths and the product of loops are introduced in Section 4.3. The remaining sections describe the fundamental group of the product of spaces (Section 4.6) and briefly discuss the topological groups (Section 4.7). Simply connected spaces and contractible spaces are investigated in Section 4.8 and Section 4.9, respectively. Both of spaces include many known spaces and surfaces, as described through numerous examples.

4.1 Homotopy theory

Let X, Y be two spaces and $f, g : X \longrightarrow Y$ two continuous functions. Recall that $I = [0,1]$ stands for the closed unit interval of the real line.

Definition 4.1.1. (1) We say that f and g are *homotopic* if there exists a continuous function

$$H : X \times I \longrightarrow Y$$

such that for all $x \in X$

$$H(x,0) = f(x) \quad \text{and} \quad H(x,1) = g(x).$$

We write $f \simeq g$ and call H a *homotopy*.

(2) If g is a constant function, i. e., $g(x) = y_0 \in Y$ for all $x \in X$, we say that f is *null-homotopic*.

Remark 4.1.2. A homotopy represents a continuous deformation of the map f to the map g. A homotopy is schematically represented by the commutative diagram

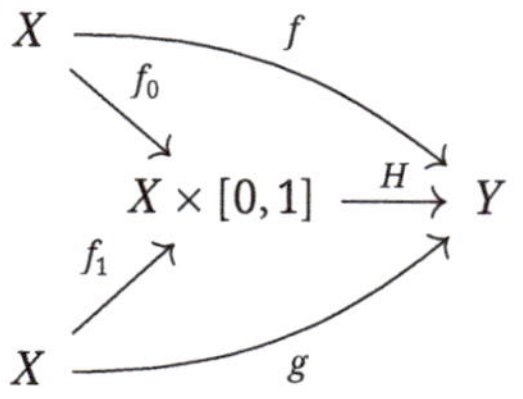

where f_0, f_1 are the embeddings $f_0(x) = (x, 0)$ and $f_1(x) = (x, 1)$.

https://doi.org/10.1515/9783111517384-004

Example 4.1.3. If X is connected and Y is discrete, then by Definition 1.4.1(4), every continuous function $f : X \longrightarrow Y$ is constant $f(x) = y_0 \in Y$ for all $x \in X$. Hence, every continuous function is null-homotopic with constant homotopy $H(t,x) = y_0$ for all $(t,x) \in I \times X$.

Example 4.1.4. (1) If the target space Y is a convex subset of a topological vector space, then for every space X, any two continuous functions $f, g : X \longrightarrow Y$ are homotopic via the line homotopy

$$H(x,t) = tf(x) + (1-t)g(x), \quad x \in X,\ t \in I.$$

(2) If Y is path-connected, then any two constant maps $f, g : X \longrightarrow Y, f(x) = y_0$ and $f(x) = y_1$ for all $x \in X$ are homotopic with the homotopy $H(x,t) = \phi(t)$, where ϕ is a path between y_0 and y_1.

Example 4.1.5. Let $X = [0,1]$ and $Y = [0,1] \cup [2,3]$. Then the continuous function $f(x) = x$ and $g(x) = x + 2$ are not homotopic for the graph of f to move continuously to the graph of g, it should leave the space Y.

Definition 4.1.6. (1) Let $f, g : I \rightarrow X$ be two paths with same end points: Same initial point x_0 and same final point x_1. We say that f and g are *path homotopic* if there exists a continuous mapping $H : I \times I \rightarrow X$ such that
  (a) $H(s,0) = f(s)$ and $H(s,1) = g(s)$ for all $s \in I$;
  (b) $H(0,t) = x_0$ and $H(1,t) = x_1$ for all $t \in I$.
  The map H is called *path homotopy* between f and g and we write $f \simeq_p g$.
(2) If $x_0 = x_1$, then H is called a *loop homotopy* (see Figure 4.1).

Example 4.1.7. Let $f, g : I \rightarrow X$, where $X = \mathbb{R}^2 \setminus (0,0)$ is the punctured plane, $f(s) = (\cos \pi s, \sin \pi s)$, and $g(s) = (\cos \pi s, 2 \sin \pi s)$. Then f and g are path homotopic with $x_0 = (1,0)$ and $x_1 = (-1,0)$ through the line homotopy

$$H(s,t) = (\cos \pi s, 2\sin \pi s - t \sin \pi s) = tf(s) + (1-t)g(s).$$

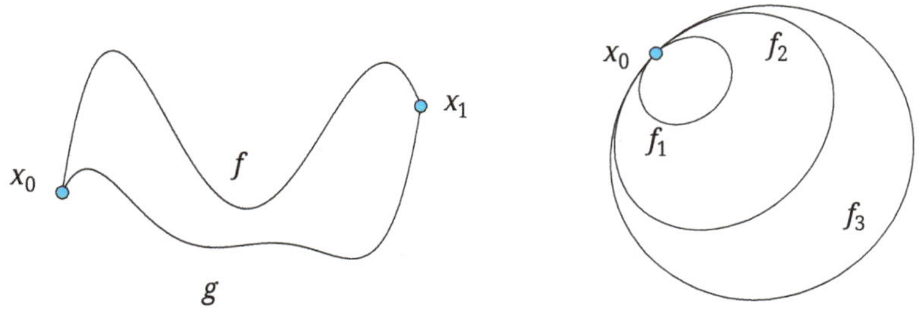

Figure 4.1: Path homotopy and loop homotopy.

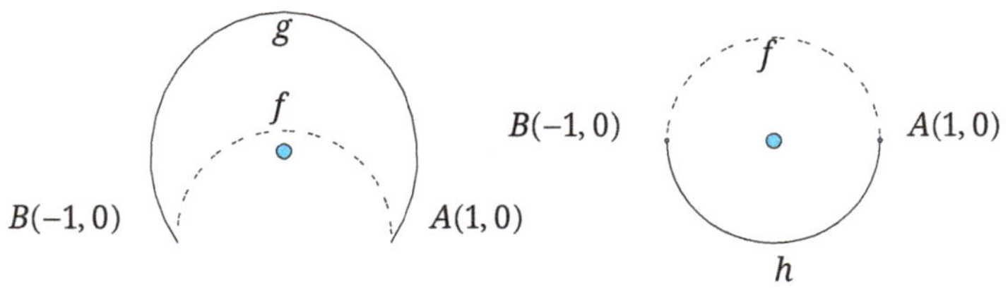

Figure 4.2: The maps f and g are homotopic (left) but f and h are not (right).

Indeed, $H(t,s) \neq 0$ for all $s,t \in [0,1]$ since

$$H(s,t)=0 \iff \begin{cases} \cos \pi s = 0, \\ 2\sin \pi s - t \sin \pi s = 0 \end{cases}$$
$$\iff \begin{cases} s = 1/2, \\ 2 - t = 0 \end{cases}$$
$$\iff \begin{cases} s = 1/2, \\ t = 2, \end{cases}$$

which is impossible.

However, if we consider $h(s) = (\cos \pi s, -\sin \pi s)$ then f and h are not line homotopic for $H(s,t) = (\cos \pi s, (2t-1)\sin \pi s) = 0$ for $s = t = 1/2$, and thus $H(1/2, 1/2) \notin X$. Even there is no homotopy between f and h because necessarily such a homotopy must meet the origin while moving from f to h (see Figure 4.2).

Proposition 4.1.8. *In the set of paths with same end points, $\simeq_p$ is an equivalence relation. The class $[f]$ will denote the class of f.*

Proof. (1) Reflexivity. $f \simeq_p f$, for each path f. The path homotopy is $H(s,t) = f(s)$, for $t,s \in I$. Clearly, H is continuous and satisfies
(a) $H(s,0) = H(s,1) = f(s)$;
(b) $H(0,t) = f(0) = x_0$;
(c) $H(1,t) = f(1) = x_1$.

(2) Symmetry. Let $f \simeq_p g$. Then there exists a continuous map $H : I \times I \longrightarrow X$ such that for all $s,t \in I$,

$$H(s,0) = f(s), \quad H(s,1) = g(s), \quad H(0,t) = x_0, \quad H(1,t) = x_1.$$

Let $G(s,t) = H(s, 1-t)$. Since H is continuous, G is continuous. Additionally,
(a) $G(s,0) = H(s,1) = g(s)$, $G(s,1) = H(s,0) = f(s)$,

(b) $G(0,t) = H(0,1-t) = x_0$, $G(1,t) = H(1,1-t) = x_1$.

Hence, G is a path homotopy between g and f, i. e., $g \simeq_p f$.

(3) Transitivity. Assume that $f \simeq_p g$ and $g \simeq_p h$. Then there exist path-homotopies H_1, H_2 between f, g and g, h, respectively. Define the map

$$H_3(s,t) = \begin{cases} H_1(s,2t), & t \in [0,1/2], \\ H_2(s,2t-1), & t \in [1/2,1]. \end{cases}$$

The maps H_1 and H_2 are continuous and satisfy $H_1(s,1) = g(s) = H_2(s,0)$, hence agree at $t = 1/2$. By Lemma A.2.1, H_3 is continuous. We further have

$$\begin{aligned} H_3(s,0) &= H_1(s,0) = f(s) \quad \text{and} \quad H_3(s,1) = H_2(s,1) = h(s), \\ H_3(0,t) &= \begin{cases} H_1(0,2t) = x_0, & t \in [0,1/2], \\ H_2(0,2t-1) = x_0, & t \in [1/2,1] \end{cases} \\ &= x_0, \\ H_3(1,t) &= \begin{cases} H_1(1,2t) = x_1, & t \in [0,1/2], \\ H_2(1,2t-1) = x_1, & t \in [1/2,1]. \end{cases} \\ &= x_1. \end{aligned}$$

□

Remark 4.1.9. The proof of the transitivity of the relation $\simeq$ in Proposition 4.1.8 goes through for any maps $f, g, h : X \longrightarrow Y$, showing that if $f \simeq g$ and $g \simeq h$, then $f \simeq h$.

Some immediate properties of homotopic maps are reported in the next two results.

Lemma 4.1.10. *Let $f \simeq g : X \longrightarrow Y$ and $h \simeq k : Y \longrightarrow Z$, where X, Y, Z are spaces. Then*

$$h \circ f \simeq k \circ g : X \longrightarrow Z.$$

Proof. Let F and G be homotopies joining f, g and h, k, respectively, and $H(x,t) = G(F(x,t),t)$. Then

$$\begin{aligned} H(x,0) &= G(F(x,0),0) = G(f(x),0) = h(f(x)), \\ H(x,1) &= G(F(x,1),1) = G(g(x),1) = k(g(x)). \end{aligned}$$

Hence, H is a homotopy between $h \circ f$ and $k \circ g$. □

Corollary 4.1.11. (1) *Let*

$$X \xrightarrow{\ f,g\ } Y \xrightarrow{\ h\ } Z$$

be continuous maps such that $f \simeq g$. Then $h \circ f \simeq h \circ g$.

(2) *Let*

$$X \xrightarrow{\quad h \quad} Y \xrightarrow{\quad f,g \quad} Z$$

be continuous maps such that $f \simeq g$. Then $f \circ h \simeq g \circ h$.

4.2 Homotopy equivalence

Definition 4.2.1. Two spaces X, Y are called *homotopically equivalent* (or homotopy equivalent) (or are of the same homotopy type) (or have the same homotopy type) if there exist continuous functions $f : X \longrightarrow Y$ and $g : Y \longrightarrow X$ such that $g \circ f \simeq \mathrm{Id}_X$ and $f \circ g \simeq \mathrm{Id}_Y$.

The functions f, g are called *homotopy equivalences.*

Example 4.2.2. Homeomorphic spaces are homotopically equivalent (but the converse is not true).

Example 4.2.3. By Example 4.1.4(1), convex spaces are homotopically equivalent.

Example 4.2.4. Euclidean space $X = \mathbb{R}^n$ and a one point set $Y = \{x_0\} \subset X$ are homotopically equivalent with homotopy equivalence maps given by the inclusion map $g = i : Y \hookrightarrow X$ and the constant map $f(x) = x_0$. In one hand, the composite $(g \circ f)(x) = g(x_0) = x_0$ for all $x \in X$ is homotopic to the identity Id_X with homotopy $H(t, x) = tx + (1 - t)x_0$ (because X is contractible; see Section 4.9). In the other hand, $(f \circ g)(x) = f(x) = x_0 = \mathrm{Id}_Y(x_0)$, for $x = x_0$. Notice however that $X = \mathbb{R}^n$ and $Y = \{x_0\}$ are not homeomorphic because there is no bijection between X and Y.

Example 4.2.5. The closed annulus $A(0; 1, 2)$ and the circle S^1 are homotopically equivalent with homotopy equivalence maps given by the inclusion map $f = i : S^1 \hookrightarrow A$ and the radial retraction $g = r : A \longrightarrow S^1$ (see Example 1.4.23) defined by $r(x) = x/\|x\|$ (or $g(r, \theta) = e^{i\theta}$ in polar coordinates). Indeed, $(g \circ f)(x) = g(x) = x/\|x\| = x$ for all $x \in S^1$ and $(f \circ g)(x) = g(x)$. We check that $g \simeq \mathrm{Id}_A$ with homotopy $H(t, x) = tx + (1 - t)g(x)$, $x \in A$, $t \in I$. We have

$$\begin{aligned}
\big\|tx + (1 - t)x/\|x\|\big\| &= \frac{\|t\|x\|x + (1 - t)x\|}{\|x\|} \\
&= \frac{\|(t\|x\| + 1 - t)x\|}{\|x\|} \\
&= \frac{(t\|x\| + 1 - t)\|x\|}{\|x\|} \\
&= t\|x\| + (1 - t)
\end{aligned}$$

and

$$1 = t + 1 - t \leq t\|x\| + (1 - t) \leq 2t + (1 - t) \leq 2, \quad \text{for all } x \in A, \text{ and } t \in I.$$

Example 4.2.6. Using Example 1.5.23 and Example 4.2.5, we obtain that the circle S^1 and the cylinder $S^1 \times [0,1]$ are homotopically equivalent. More generally, the sphere S^{n-1} and the cylinder $S^{n-1} \times [0,1]$ are homotopically equivalent. The homotopy equivalence maps are $f(x) = (x, 0)$ from S^{n-1} to $S^{n-1} \times [0,1]$ and $g(x,t) = f^{-1}(x,0) = x$ from $S^{n-1} \times [0,1]$ to S^{n-1}.

Example 4.2.7. More generally, for any space X, the spaces X and $X \times [0,1]$ are homotopically equivalent. The construction of homotopy equivalence maps involved is the same as in Example 4.2.6: the inclusion map $f = i : X \hookrightarrow X \times [0,1]$, where X is identified with the slice $X \times \{0\}$ and the projection map $g : X \times [0,1] \longrightarrow X$. Then $g \circ f = \mathrm{Id}_X$ and $g \circ f \simeq \mathrm{Id}_{X\times[0,1]}$ with the homotopy $H(x, s, t) = (x, st)$.

Example 4.2.8. The open annulus $A(0; 1, 2)$ and the circle $C(0; 3/2)$ are homotopically equivalent with homotopy $H(t, x) = tx + \frac{(1-t)}{2\|x\|}x$. By Theorem 1.5.37, we deduce that the punctured open ball $\mathring{B}^2 \setminus \{(0, 0)\}$ is homotopically equivalent to the circle $C(0; 3/2)$.

Example 4.2.9. The sphere $X = S^{n-1}$ and the punctured space $Y = \mathbb{R}^n \setminus \{0\}$ are homotopically equivalent with homotopy equivalence maps given by the inclusion map $f = i : X \hookrightarrow Y$ and the radial retraction $g : Y \longrightarrow X$ defined by $g(x) = x/\|x\|$.

Example 4.2.10. The spaces $S^1\times[0,1]$ and S^1 are homotopically equivalent but not homeomorphic for otherwise by Exercise 8, Section 1.5, $X = S^1 \times [0,1]$ would be homeomorphic to $Y = S^1 \times \{t_0\}$ for some $t_0 \in I$. Let h be such a homeomorphism. By removing a cut point $p = (x_0, t_0)$, we obtain that $X' = X \setminus \{p\}$ is homeomorphic to $Y' = Y \setminus \{h(p)\}$. Hence, $(0,1) \times ([0, t_0) \cup (t_0, 1))$ and $(0, 1)$ are homeomorphic, a contradiction.

Definition 4.2.11. A figure eight, denoted 8, is the wedge sum of two closed curves (see Example 1.5.33).

Definition 4.2.12. A figure theta, denoted Θ, is a surface homeomorphic to $S^1 \cup ([-1, 1] \times \{0\})$.

Figure eight is homeomorphic to number 8. Figure theta is homeomorphic to Greek letter Θ.

Proposition 4.2.13. *A closed disk with two holes $X = B^2 \setminus (B_1 \cup B_2)$ and figure eight are homotopically equivalent.*

Proof. The maps involved in the homotopy equivalence are the inclusion map $f = i : 8 \hookrightarrow X$ and a map $g : X \longrightarrow 8$ defined as follows. The value $g(x)$ is the intersection of the figure eight with the line (xx_1) when x lies in the right-half and the intersection of the figure eight with the line (xx_2) when x lies in the left-half of the circle C. When x lies on

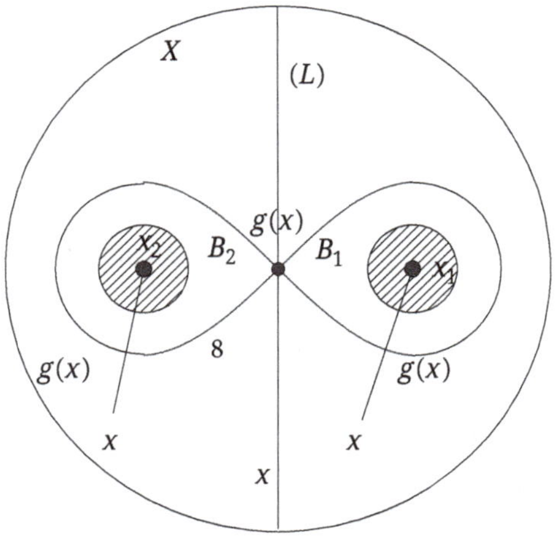

Figure 4.3: A doubly punctured closed disk and figure eight are homotopically equivalent.

the diameter (L), take the origin as $g(x)$ (se Figure 4.3). By definition of the projection, $g(f(x)) = g(x) = x$ for all $x \in 8$. Finally, $f(g(x)) = g(x)$ for all $x \in X$, where $g(x)$ and x are line homotopic since they belong to the line (xx_1). Hence, $g \simeq \mathrm{Id}_X$. □

Remark 4.2.14. (1) The proof of Proposition 4.2.13 still holds for an open disk with two holes $X' = \mathring{B}^2 \setminus (B_1 \cup B_2)$.

(2) Since the twice-punctured plane $\mathbb{R}^2 \setminus \{x_1, x_2\}$ is homeomorphic to X', it follows that $\mathbb{R}^2 \setminus \{x_1, x_2\}$, the sphere S^2 with three points removed and the figure eight are homotopically equivalent.

Proposition 4.2.15. *A closed disk with two holes $X = B^2 \setminus (\mathring{B}_1^2 \cup \mathring{B}_2^2)$ and Figure Θ are homotopically equivalent.*

Proof. A figure theta is homeomorphic to $S^1 \cup \ell$, where $\ell = (-a, a)$. In Figure 4.4, $x_1 = (0, 1/2)$, $x_2 = -x_1$ are centers of small balls, and $a = (1, 0)$. The maps involved are the inclusion map $f = i : \Theta \hookrightarrow X$ and the radial retraction $g = r : X \longrightarrow \Theta$ given by

$$r(x) = \begin{cases} (x_k x) \cap \ell, & x \in \mathring{D}, \\ (x_k x) \cap S_k^1, & x \notin D, \end{cases}$$

where S_1^1 (S_2^1, respectively) denotes the upper half-circle (lower half-circle, respectively) of S^1 and D refers to the rhombus (diamond-shaped figure) with vertices $a, x_1, -a, x_2$ and $k = 1, 2$. Then $g(f(x)) = g(x) = x$ for all $x \in \Theta$ and $f(g(x)) = g(x)$ for all $x \in X$, where $g(x)$ and x are straight-line homotopic because they belong to the line (xx_k). Hence, $g \simeq \mathrm{Id}_X$. □

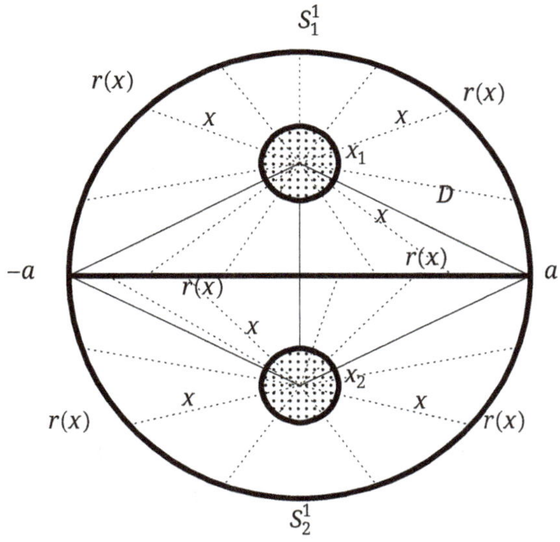

Figure 4.4: A twice-punctured closed disk and a figure theta are homotopically equivalent.

Corollary 4.2.16. *The figure eight and the figure theta are homotopically equivalent.*

Proposition 4.2.17. *The Möbius band is homotopically equivalent to a circle.*

Proof. By Theorem 2.5.11, we can identify S^1 with $[0,1]/\{0,1\}$. Let $\pi_1 : I \longrightarrow S^1$ be the corresponding projection map and $\pi_2 : I^2 \longrightarrow \mathbb{M}^2$ the canonical projection on the Möbius band defined by Example 3.2.5. The equatorial map $\alpha : I \longrightarrow \mathbb{M}^2$ is defined by $\alpha(x) = \pi_2(1/2, x) = [(1/2, x)]$. This is a loop since $\alpha(0) = [(1/2, 0)] = [(1/2, 1)] = \alpha(1)$. Define $f : S^1 \longrightarrow \mathbb{M}^2$ by $f([x]) = (f \circ \pi_1)(x) = \alpha(x)$, i. e.,

$$I \xrightarrow{\pi_1} S^1 \xrightarrow{f} \mathbb{M}^2.$$

Since α is continuous and π_1 is an identification map, the factorization Theorem 2.1.9 entails that f is continuous. The map $g : \mathbb{M}^2 \longrightarrow S^1$ defined by $g([(x,y)]) = [y]$ is well-defined since

$$[(x,0)] = [(1-x,1)] \quad \Longrightarrow \quad g([x,0]) = [0] = [1] = g([(1-x,1)]).$$

Consider the compositions

$$I^2 \xrightarrow{pr_2} I \xrightarrow{\pi_1} S^1 \quad \text{and} \quad I^2 \xrightarrow{\pi_2} \mathbb{M}^2 \xrightarrow{g} S^1,$$

where both π_1 and the projection onto the second variable pr_2 are continuous. Since $g \circ \pi_2 = \pi_1 \circ pr_2$, by Theorem 2.1.9, g is continuous. In addition,

$$(g \circ f)([x]) = g([1/2, x]) = [x] \quad \Longrightarrow \quad g \circ f = \mathrm{Id}_{|S^1}$$

and

$$(f \circ g)([(x,y)]) = f([y]) = [(1/2, y)] = \alpha(y).$$

Finally, $H([x,y],t) = [(\frac{1}{2} + (x - f \circ g)t, y)]$ is a homotopy between $f \circ g$ and $\mathrm{Id}_{|\mathbb{M}^2}$, which completes the proof. □

Let $f : X \longrightarrow Y$ be the attaching map and C_f the mapping cylinder (see Definition 3.5.1). The following result holds (see [25, Chapter XVIII, 4.2] or [51, Corollary 1.2]).

Theorem 4.2.18. *The spaces C_f and Y are homotopically equivalent.*

Proof. Since X identifies with the slice $X \times \{0\}$, X can be embedded in the mapping cylinder C_f. We denote by $i : X \hookrightarrow C_f$ the injection. Let $\pi : X \times I \longrightarrow C_f$ be the surjective canonical map. Let $j = \pi_{|Y}$ and $\rho : C_f \longrightarrow$ be given by $\rho([(x,t)]) = f(x)$ and $\rho([y]) = y$, respectively. Then j is a homeomorphism. The maps j and ρ are shown to be the homotopy equivalence maps. We have the diagrams of compositions:

$$X \xrightarrow{i} C_f \xrightarrow{\rho} Y \quad \text{and} \quad Y \xrightarrow{j} C_f \xrightarrow{\rho} Y,$$

where $(\rho \circ j)(y) = \rho([y]) = y$, i. e., $\rho \circ j = \mathrm{Id}_{|Y}$. So, we only need to prove that the composite $j \circ \rho : C_f \xrightarrow{\rho} Y \xrightarrow{j} C_f$ is homotopic to the identity map $\mathrm{Id}_{|C_f}$, where

$$(j \circ \rho)([(x,t)]) = j(f(x)) = [f(x)] \quad \text{and} \quad (j \circ \rho)([y]) = j(y) = [y].$$

Let

$$h_1 : (X \times I) \times I \longrightarrow C_f \quad \text{and} \quad h_2 : Y \times I \longrightarrow C_f$$

be given by $h_1((x,t),s) = [(x, t(1-s)]$ and $h_2(y,t) = [y]$. Then, for all $x \in X$, $s \in I$,

$$h_1((x,0),s) = [(x,0)] = [f(x)] = h_2(f(x),s).$$

Then the map $h = h_1 \vee h_2 : C_f \times I \longrightarrow C_f$ is continuous and satisfies

$$\begin{cases} h((x,t),0) = [(x,t)], & h(y,0) = [y], \\ h((x,t),1) = [(x,0)] = [f(x)], & h(y,1) = [y]. \end{cases}$$

Therefore, $h(\cdot,0) = \mathrm{Id}_{|C_f}$, $h(\cdot,1) = (j \circ \rho)(\cdot)$, and h is the required homotopy. □

We end this section by a useful result in connection with Exercise 11 of this chapter.

Proposition 4.2.19. *Let X, Y be two homotopically equivalent spaces. Then X is path-connected if and only if Y is path-connected.*

Proof. Suppose that X is path-connected and let y_1, y_2 in Y two arbitrary points. Since X and Y are homotopically equivalent, there exist continuous maps

$$X \xrightarrow{f} Y \xrightarrow{g} X \xrightarrow{f} Y$$

such that $f \circ g \simeq \mathrm{Id}_Y$. Let $x_1 = g(y_1)$ and $x_2 = g(y_2)$. Since $f(X)$ is path-connected (Theorem 1.4.20), there exists a path connecting $f(x_1) = (f \circ g)(y_1)$ and $f(x_2) = (f \circ g)(y_2)$. Since $f \circ g \simeq \mathrm{Id}_Y$, there is a path between $f(x_1)$ and y_1 and a path between $f(x_2)$ and y_2. By transitivity of the homotopy (Proposition 4.1.8), there exists a path between y_1 and y_2, proving that Y is path-connected. □

4.3 Product of paths and product of classes

Definition 4.3.1. Let f be a path from x_0 to x_1 and g a path from x_1 to x_2. Define *the product* $h = f * g$ of paths f and g to be the path from x_0 to x_2 given by

$$h(s) = \begin{cases} f(2s), & s \in [0, 1/2], \\ g(2s-1), & s \in [1/2, 1]. \end{cases}$$

When $x_0 = x_1 = x_2$, we talk about product of loops (or *loop product*).

Remark 4.3.2. (1) Since f and g are continuous and coincide at $s = 1/2 : f(1) = g(0) = x_1$, h is continuous by Lemma A.2.1 and satisfies $h(0) = f(0) = x_0$ and $h(1) = g(1) = x_2$. Hence, h is a path from x_0 to x_2.

(2) The product h may be thought of as a point moving along f at twice the normal rate arriving at x_1 when $s = 1/2$ and then moving along g at twice the normal rate, arriving at x_2 when $s = 1$.

The distributivity of the composition of the product of paths is immediately checked.

Proposition 4.3.3. *Let*

$$I \xrightarrow{\quad f, g \quad} X \xrightarrow{\quad h \quad} Y$$

be continuous maps. Then

$$h \circ (f * g) = (h \circ f) * (h \circ g).$$

The following result will useful in the sequel.

Lemma 4.3.4. *Let $f, g : I \longrightarrow X$ be two paths from x_0 to x_1 and $k, h : I \longrightarrow X$ two paths from x_1 to x_2. We have*

$$\begin{cases} f \simeq g \\ k \simeq h \end{cases} \Longrightarrow \quad f * k \simeq g * h.$$

Proof. Let $F(s,t)$ and $G(s,t)$ be homotopies between f, g and k, h, respectively. Define the map

$$H(s,t) = \begin{cases} F(2s,t), & 0 \le s \le 1/2,\ t \in I, \\ G(2s-1,t), & 1/2 \le s \le 1,\ t \in I. \end{cases}$$

Since $F(1,t) = G(0,t) = x_1$ agree at $s = 1/2$ and F, G are continuous, H is continuous. Furthermore, $H(0,t) = F(0,t) = x_0$, $H(1,t) = G(1,t) = x_2$,

$$\begin{aligned} H(s,0) &= \begin{cases} F(2s,0) = f(2s), & 0 \le s \le 1/2, \\ G(2s-1,0) = k(2s-1), & 1/2 \le s \le 1 \end{cases} \\ &= (f * k)(s), \end{aligned}$$

and

$$\begin{aligned} H(s,1) &= \begin{cases} F(2s,1) = g(2s), & 0 \le s \le 1/2, \\ G(2s-1,1) = h(2s-1), & 1/2 \le s \le 1 \end{cases} \\ &= (g * h)(s). \end{aligned}$$

Hence, H is a path homotopy between $f * k$ and $g * h$. □

Corollary 4.3.5. (1) *If $f \simeq g$, then whenever the product makes sense, we have*

$$f * h \simeq g * h \quad \text{and} \quad h * f \simeq h * g.$$

(2) *The maps f and g are homotopic if and only if $f * \overline{g}$ is null-homotopic, where $\overline{g}(s) = g(1-s)$.*

Definition 4.3.6. The path product $*$ induces *a product operation* on path-homotopy classes defined by

$$[f] * [g] = [f * g]. \tag{4.1}$$

Remark 4.3.7. This new operation is well-defined. Indeed, let $f' \simeq_p f$ and $g' \simeq_p g$. By Lemma 4.3.4, $f * g \simeq_p f' * g'$, i. e., $[f' * g'] = [f * g]$.

Remark 4.3.8. Part (2) of Corollary 4.3.5 can be directly checked

$$
\begin{aligned}
f * \overline{g} \simeq e_{x_0} \quad &\Longleftrightarrow \quad [f * \overline{g}] = [e_{x_0}] \\
&\Longleftrightarrow \quad [f * \overline{g}] * [g] = [e_{x_0}] * [g] \\
&\Longleftrightarrow \quad [f * (\overline{g} * g)] = [e_{x_0} * g] \\
&\Longleftrightarrow \quad [f * e_{x_0}] = [e_{x_0} * g] \\
&\Longleftrightarrow \quad [f] = [g] \\
&\Longleftrightarrow \quad f \simeq g.
\end{aligned}
$$

4.4 The group $\pi_1(X, x_0)$

Definition 4.4.1. The set of all equivalence classes (the quotient space) under homotopy of the loops based at x_0 is called *the fundamental group* of X relative to the base point x_0. It is denoted $\pi_1(X, x_0)$ (π for Пoincaré), i. e.,

$$\pi_1(X, x_0) = \{[f],\ f : I \longrightarrow X \text{ continuous and } f(0) = f(1) = x_0\}. \tag{4.2}$$

The group $\pi_1(X, x_0)$ is also called the Poincaré group or first homotopy group.

Example 4.4.2. Let $X \subseteq \mathbb{R}^n$ be convex and $x_0 \in X$. Then $\pi_1(X, x_0) = \{[e_{x_0}]\}$, where e_{x_0} is the constant loop defined by $e_{x_0}(s) = x_0$ for all $s \in I$. Indeed, every loop f based at x_0 is homotopic to e_{x_0} with homotopy

$$H(s, t) = tf(s) + (1 - t)e_{x_0}(s),$$

because $H(s, 0) = e_{x_0}(s)$ and $H(s, 1) = f(s)$. Furthermore,

$$H(0, t) = tf(0) + (1 - t)e_{x_0}(0) = t\,x_0 + (1 - t)x_0 = x_0$$

and

$$H(1, t) = tf(1) + (1 - t)e_{x_0}(1) = t\,x_0 + (1 - t)x_0 = x_0.$$

Hence,

$$f \simeq_p e_{x_0} \quad \Longleftrightarrow \quad [f] = [e_{x_0}],$$

for every loop f based at x_0.

Example 4.4.3. By Example 4.1.3, the fundamental group of a discrete space is the trivial group.

Here and hereafter, we will use for simplicity of notation $\simeq$ instead of $\simeq_p$ for loop homotopy.

Theorem 4.4.4 (Group properties). *The pair* $(\pi_1(X, x_0), *)$ *is a group. The* identity element *is the class of the constant loop*

$$e_{x_0} : I \longrightarrow X, \; e_{x_0}(s) = x_0, \quad \textit{for all } s \in I. \tag{4.3}$$

Proof. (1) The class $[e_{x_0}]$ *is the identity element in* $\pi_1(X, x_0)$. For every loop f,

$$\begin{aligned} & [e_{x_0}] * [f] = [f] * [e_{x_0}] = [f] \\ \Longleftrightarrow \quad & [e_{x_0} * f] = [f * e_{x_0}] = [f] \\ \Longleftrightarrow \quad & e_{x_0} * f \simeq f * e_{x_0} \simeq f. \end{aligned}$$

Consider the homotopy

$$H(s,t) = \begin{cases} f(\frac{2s}{1+t}), & 0 \le s \le \frac{1+t}{2}, \\ x_0, & \frac{1+t}{2} \le s \le 1. \end{cases}$$

Then

$$\begin{aligned} H(s,0) &= \begin{cases} f(2s), & 0 \le s \le 1/2, \\ x_0, & 1/2 \le s \le 1, \end{cases} \\ &= (f * e_{x_0})(s), \quad \text{for all } s \in [0,1], \end{aligned}$$

$H(s,1) = f(s)$ and $H(0,t) = f(0) = x_0 = H(1,t)$ for all $s, t \in [0,1]$. Hence, H is a loop homotopy between $f * e_{x_0}$ and f. Additionally, it can be checked that the map H given by

$$H(s,t) = \begin{cases} x_0, & 0 \le s \le \frac{1-t}{2}, \\ f(\frac{2s+t-1}{t+1}), & \frac{1-t}{2} \le s \le 1 \end{cases}$$

is a homotopy between $e_{x_0} * f$ and f.

(2) *The inverse (or reverse) of* $[f] \in \pi_1(X, x_0)$ *is* $[\bar{f}]$, where $\bar{f}(s) = f(1-s)$, i. e.,

$$\begin{aligned} [f] * [\bar{f}] = [e_{x_0}] \quad & \Longleftrightarrow \quad [f * \bar{f}] = [e_{x_0}] \\ & \Longleftrightarrow \quad f * \bar{f} \simeq e_{x_0} \end{aligned}$$

and $\bar{f} * f \simeq e_{x_0}$. To check this, let the homotopy

$$H(s,t) = \begin{cases} f(2st), & 0 \le s \le 1/2, \\ \bar{f}(2t(s-1)+1), & 1/2 \le s \le 1. \end{cases}$$

Then

$$
\begin{aligned}
H(s,0) &= \begin{cases} f(0), & 0 \le s \le 1/2, \\ \overline{f}(1), & 1/2 \le s \le 1 \end{cases} \\
&= x_0, \\
H(s,1) &= \begin{cases} f(2s), & 0 \le s \le 1/2, \\ \overline{f}(2s-1), & 1/2 \le s \le 1, \end{cases} \\
&= (f * \overline{f})(s), \quad \text{for all } s \in I, \\
H(0,t) &= f(0) = x_0, \quad \text{and} \quad H(1,t) = \overline{f}(1) = x_0, \quad \text{for all } t \in I.
\end{aligned}
$$

Therefore, H is a loop homotopy between e_{x_0} and $f * \overline{f}$, i. e., $f * \overline{f} \simeq e_{x_0}$. As a consequence, $\overline{f} * \overline{\overline{f}} \simeq e_{x_0}$. This means that $\overline{f} * f \simeq e_{x_0}$ and $[f * \overline{f}] = [\overline{f} * f] = [e_{x_0}]$, i. e.,

$$[f] * [\overline{f}] = [\overline{f}] * [f] = [e_{x_0}].$$

(3) *The operation* $(*)$ *is associative:*

$$([f] * [g]) * [h] = [f] * ([g] * [h]).$$

We have

$$
\begin{aligned}
(f * g) * h(s) &= \begin{cases} (f * g)(2s), & 0 \le s \le 1/2, \\ h(2s-1), & 1/2 \le s \le 1 \end{cases} \\
&= \begin{cases} f(4s), & 0 \le s \le 1/4, \\ g(4s-1), & 1/4 \le s \le 1/2, \\ h(2s-1), & 1/2 \le s \le 1, \end{cases} \\
f * (g * h)(s) &= \begin{cases} f(2s), & 0 \le s \le 1/2, \\ (g * h)(2s-1), & 1/2 \le s \le 1 \end{cases} \\
&= \begin{cases} f(2s), & 0 \le s \le 1/2, \\ g(4s-2), & 1/2 \le s \le 3/4, \\ h(4s-3), & 3/4 \le s \le 1. \end{cases}
\end{aligned}
\tag{4.4}
$$

Let

$$
\begin{aligned}
k(s) &= \begin{cases} f(s), & 0 \le s \le 1, \\ g(s-1), & 1 \le s \le 2, \\ h(s-2), & 2 \le s \le 3, \end{cases} \\
m_1(s) &= \begin{cases} 2s, & 0 \le s \le 1/2, \\ 4s-1, & 1/2 \le s \le 1 \end{cases}
\end{aligned}
$$

and

$$m_2(s) = \begin{cases} 4s, & 0 \le s \le 1/2, \\ 2s+1, & 1/2 \le s \le 1. \end{cases}$$

Then

$$f * (g * h) = k \circ m_1 \quad \text{and} \quad (f * g) * h = k \circ m_2.$$

Finally, a loop homotopy between $f * (g * h)$ and $(f * g) * h$ is given by

$$H(s,t) = k\big((1-t)m_1(s) + tm_2(s)\big).$$

Indeed,

$$\begin{cases} H(s,0) = k(m_1(s)) = k \circ m_1(s); \quad H(s,1) = k(m_2(s)) = k \circ m_2(s), \\ H(0,t) = k[(1-t)m_1(0) + tm_2(0)] = k(0) = x_0, \\ H(1,t) = k[(1-t)m_1(1) + tm_2(1)] = k[(1-t)3 + t(3)] = k(3) = h(1) = x_0. \end{cases}$$

Therefore,

$$f * (g * h) \simeq (f * g) * h,$$

proving associativity. □

Let X, Y be two spaces and $h : X \longrightarrow Y$ be a continuous map. For some $x_0 \in X$, let $y_0 = h(x_0)$. Notice that if the map f is a loop based at x_0, then the composite $h \circ f : I \longrightarrow Y$ is a loop based at the point y_0. Thus, we can define

$$h_* : \pi_1(X, x_0) \to \pi_1(Y, y_0)$$

by

$$[f] \longmapsto [h \circ f].$$

Definition 4.4.5. The map h_* is called the *homomorphism induced* by h.

Remark 4.4.6. (1) The map h_* is well-defined since if $[f] = [f']$, then $f \simeq f'$ which implies, by Lemma 4.1.11, $h \circ f \simeq h \circ f'$ and then $[h \circ f] = [h \circ f']$.
(2) The map h_* is a homomorphism, because by Proposition 4.3.3, we have

$$\begin{aligned} h_*([f] * [g]) &= h_*([f * g]) \\ &= [h \circ (f * g)] = [(h \circ f) * (h \circ g)] \\ &= [h \circ f] * [h \circ g] = h_*([f]) * h_*([g]). \end{aligned}$$

4.5 Isomorphic fundamental groups

Theorem 4.5.1 (Change of base point). *Let $x_0, x_1 \in X$ be two points connected by a path. Then the fundamental groups $\pi_1(X, x_0)$ and $\pi_1(X, x_1)$ are isomorphic. In particular, if X is path-connected, then for all $x_0, x_1 \in X$, $\pi_1(X, x_0)$ and $\pi_1(X, x_1)$ are isomorphic.*

Proof. (1) Step 1. Let γ be a path from x_0 to x_1 and let f be a loop based at x_0. Then $(\overline{\gamma} * f) * \gamma$ defines a loop based at x_1, where $\overline{\gamma}(s) = \gamma(1-s)$ is the inverse of γ. Indeed, using the formula (4.4), we have

$$((\overline{\gamma} * f) * \gamma)(s) = \begin{cases} \overline{\gamma}(4s), & 0 \le s \le 1/4, \\ f(4s-1), & 1/4 \le s \le 1/2, \\ \gamma(2s-1), & 1/2 \le s \le 1. \end{cases}$$

For $s = 0$, $\overline{\gamma}(0) = \gamma(1) = x_1$ and for $s = 1$, $\gamma(1) = x_1$. Then $(\overline{\gamma} * f) * \gamma$ is a loop based at the point x_1.

(2) Step 2. Define the change of base point map (see Figure 4.5)

$$\phi : \pi_1(X, x_0) \to \pi_1(X, x_1)$$

by

$$\phi([f]) = [(\overline{\gamma} * f) * \gamma]. \tag{4.5}$$

We prove that ϕ is an isomorphism. First, ϕ is well-defined since by Corollary 4.3.5,

$$f \simeq g \quad \Rightarrow \quad (\overline{\gamma} * f) * \gamma \simeq (\overline{\gamma} * g) * \gamma.$$

(a) The map ϕ is a homomorphism. We need to show that

$$\phi([f] * [g]) = \phi([f]) * \phi([g]), \quad \text{for all loops } f, g \text{ based at } x_0.$$

Equivalently,

$$\phi([f * g]) = \phi([f]) * \phi([g]),$$

or

$$[(\overline{\gamma} * (f * g)) * \gamma] = [(\overline{\gamma} * f) * \gamma] * [(\overline{\gamma} * g) * \gamma].$$

We have

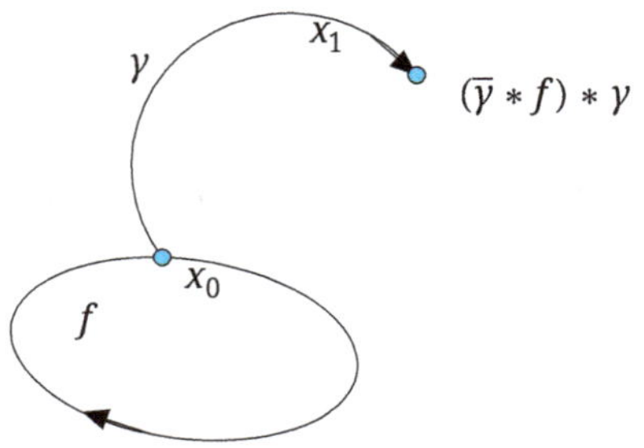

Figure 4.5: Change of base point x_0 to x_1 via a path γ.

$$\begin{aligned}[(\overline{\gamma} * f) * \gamma] * [(\overline{\gamma} * g) * \gamma] &= [((\overline{\gamma} * f) * \gamma) * ((\overline{\gamma} * g) * \gamma)] \\ &= [(\overline{\gamma} * f) * (\gamma * \overline{\gamma}) * (g * \gamma)] \\ &= [(\overline{\gamma} * f) * e_{x_0} * (g * \gamma)] \\ &= [(\overline{\gamma} * f) * (g * \gamma)] \\ &= [\overline{\gamma} * (f * g) * \gamma].\end{aligned}$$

(b) The map ϕ is a bijection. Let the map

$$\psi : \pi_1(X, x_1) \to \pi_1(X, x_0)$$

be defined by

$$\psi([f]) = [(\gamma * f) * \overline{\gamma}].$$

Then

$$\begin{aligned}(\psi \circ \phi)([f]) &= \psi([(\overline{\gamma} * f) * \gamma]) \\ &= [\gamma * ((\overline{\gamma} * f) * \gamma) * \overline{\gamma}] \\ &= [(\gamma * \overline{\gamma}) * f * (\gamma * \overline{\gamma})] \\ &= [f] \\ &= (\phi \circ \psi)([f]),\end{aligned}$$

i. e., $\psi \circ \phi$ and $\phi \circ \psi$ are identity maps in $\pi_1(X, x_0)$ and $\pi_1(X, x_1)$, respectively. Therefore, ϕ, ψ are isomorphisms. □

Let α be a path from x_0 to x_1 and β a path from x_1 to x_2 and $\gamma = \alpha * \beta$ their path product. Let $\phi = \phi_\gamma$ be the isomorphism given by (4.5). The following result shows that to the product of paths corresponds a composition of isomorphisms.

Proposition 4.5.2. *We have*

$$\phi_{\alpha*\beta} = \phi_\beta \circ \phi_\alpha.$$

Proof. It is clear that $\overline{\alpha * \beta} = \overline{\beta} * \overline{\alpha}$, where $\overline{\alpha}$ is the inverse path of α. Using the associativity of the path product, we get

$$\begin{aligned}
\phi_\gamma([f]) &= [\overline{\gamma} * f * \gamma] \\
&= [\overline{(\alpha * \beta)} * f * (\alpha * \beta)] \\
&= [\overline{\beta} * \overline{\alpha} * f * \alpha * \beta] \\
&= [\overline{\beta} * (\overline{\alpha} * f * \alpha) * \beta] \\
&= \phi_\beta([\overline{\alpha} * f * \alpha]) \\
&= \phi_\beta(\phi_\alpha([f])).
\end{aligned}$$

□

We now present a more general result than the one given by Theorem 4.5.1. First, we check the following.

Lemma 4.5.3. *We have*

$$(k \circ h)_* = k_* \circ h_*. \tag{4.6}$$

Proof. For every loop f based at x_0, we have

$$\begin{aligned}
(k \circ h)_*([f]) &= [(k \circ h) \circ f] \\
&= [k \circ (h \circ f)] = k_*([h \circ f]) \\
&= k_*(h_*([f])) = (k_* \circ h_*)([f]).
\end{aligned}$$

□

Theorem 4.5.4. *Let $p, q : X \longrightarrow Y$ be homotopic maps with some homotopy H. For some $x_0 \in X$, let $\gamma(t) = H(x_0, t)$ and define the change of base point map*

$$\phi : \pi_1(Y, p(x_0)) \longrightarrow \pi_1(Y, q(x_0))$$

by (4.5). *Then the following diagram commutes (i. e., $q_* = \phi \circ p_*$):*

$$\begin{array}{ccc}
\pi_1(X, x_0) & \xrightarrow{q_*} & \pi_1(Y, q(x_0)) \\
& {\scriptstyle p_*}\searrow & \uparrow{\scriptstyle \phi} \\
& & \pi_1(Y, p(x_0))
\end{array}$$

Proof. Let $f : I \longrightarrow X$ be a loop based at x_0. Define the map

$$G(s,t) = \begin{cases}
\overline{\gamma}(2s), & 0 \le s \le \frac{1-t}{2}, \\
H(f(\frac{4s+2t-2}{3t+1}), t), & \frac{1-t}{2} \le s \le \frac{t+3}{4}, \\
\gamma(4s-3), & \frac{t+3}{4} \le s \le 1.
\end{cases}$$

By Lemma A.2.1, G is continuous. Indeed,

(1) $\overline{\gamma}(2\frac{1-t}{2}) = \gamma(t)$,

(2) $H(f(\frac{4\frac{1-t}{2}+2t-2}{3t+1}), t) = H(f(0), t) = H(x_0, t) = \gamma(t)$,

(3) $H(f(\frac{4\frac{t+3}{4}+2t-2}{3t+1}), t) = H(f(1), t) = H(x_0, t) = \gamma(t)$,

(4) $\gamma(4\frac{t+3}{4} - 3) = \gamma(t)$.

(5) For $t = 0$,

$$\begin{aligned} G(s, 0) &= \begin{cases} \overline{\gamma}(2s), & 0 \le s \le 1/2, \\ H(f(4s-2), 0), & 1/2 \le s \le 3/4, \\ \gamma(4s-3), & 3/4 \le s \le 1 \end{cases} \\ &= \begin{cases} \overline{\gamma}(4s), & 0 \le s \le 1/4\ (s' = 2s), \\ H(f(4s-1), 0), & 1/4 \le s \le 1/2\ (s' = s + 1/4), \\ \gamma(2s-1), & 1/2 \le s \le 1\ (s' = \frac{s+1}{s}) \end{cases} \\ &= \overline{\gamma} * (p \circ f) * \gamma, \end{aligned}$$

for $H(f(4s-1), 0) = (p \circ f)(4s-1)$.

(6) For $t = 1$,

$$\begin{aligned} G(s, 1) &= \begin{cases} \overline{\gamma}(0), & s = 0, \\ H(f(s), 1), & 0 \le s \le 1, \\ \gamma(1), & s = 1 \end{cases} \\ &= \begin{cases} \overline{\gamma}(0), & s = 0, \\ (q \circ f)(s), & 0 \le s \le 1, \\ \gamma(1), & s = 1 \end{cases} \\ &= (q \circ f)(s). \end{aligned}$$

(7) $G(0, t) = \overline{\gamma}(0) = \gamma(1)$.

(8) $G(1, t) = \gamma(1)$, where $\gamma(1) = H(x_0, 1) = q(x_0)$.

Therefore, G is a loop homotopy between $\overline{\gamma} * (p \circ f) * \gamma$ and $q \circ f$ at the base point $q(x_0)$. Hence,

$$[q \circ f] = [\overline{\gamma} * (p \circ f) * \gamma].$$

Finally, for every loop f based at x_0, we have

$$\begin{aligned} q_*([f]) &= \phi([p \circ f]) \\ &= \phi(p_*([f])) \\ &= (\phi \circ p_*)([f]), \end{aligned}$$

and the claim of the theorem follows. □

The following result shows that the fundamental group is an invariant of the homotopy type (i. e., homotopically equivalent spaces have isomorphic fundamental groups).

Corollary 4.5.5. *If $f : X \longrightarrow Y$ is a homotopy equivalence, then $f_* : \pi_1(X, x_0) \longrightarrow \pi_1(Y, f(x_0))$ is an isomorphism for all $x_0 \in X$.*

Proof. Let $g : Y \longrightarrow X$ be the homotopy inverse of f, that is to say $f \circ g \simeq \mathrm{Id}_Y$ and $g \circ f \simeq \mathrm{Id}_X$. By Theorem 4.5.4, $(g \circ f)_* = \phi \circ (\mathrm{Id}_X)_*$, where the change of base point

$$\phi : \pi_1(X, x_0) \longrightarrow \pi_1(X, (g \circ f)(x_0))$$

is an isomorphism because x_0 and $(g \circ f)(x_0)$ are path-connected via the path $\gamma(t) = H(x_0, t)$, $t \in I$ (Theorem 4.5.1), where H is a homotopy joining Id_X and $g \circ f$. Hence, $\phi \circ \mathrm{Id}_{\pi_1(X,x_0)} = \phi = g_* \circ f_*$. In the same way, there exists an isomorphism ψ such that $\psi = f_* \circ g_*$. In conclusion, f_* and g_* are invertible. □

Corollary 4.5.6. *If two path-connected spaces have the same homotopy type, then their fundamental groups are isomorphic.*

Example 4.5.7. Using Example 1.5.14, Example 4.2.6, Example 4.2.9 and Corollary 4.5.5, we have

$$\pi_1(\mathbb{R}^n \setminus \{0\}) \cong \pi_1(S^{n-1} \times [0,1]) \cong \pi_1(\mathring{B}^n \setminus \{0\}).$$

Remark 4.5.8. Notice that $p \simeq q \not\Longrightarrow p_* = q_*$. However, we have the following.

Corollary 4.5.9. *Let $p, q : X \longrightarrow Y$ be homotopic maps with homotopy H satisfying*

$$H(x_0, t) = p(x_0) = q(x_0), \quad \text{for all } t \in I,$$

for some $x_0 \in X$. Then $p_* = q_*$.

Proof. Let $\gamma(s) * \overline{\gamma}(s) = p(x_0) = q(x_0) = e_{p(x_0)}$, for $s \in I$. Then the change of the base point given by the map (4.5) is $\phi = \mathrm{Id}_{\pi_1(X,x_0)}$. Hence, $[\overline{\gamma} * f * \gamma] = [f]$. □

4.6 Fundamental group of the product

A natural way to compute the fundamental group of the product of two spaces X and Y is now presented. Let $(x_0, y_0) \in X \times Y$ and $P_X : X \times Y \longrightarrow X$, $P_Y : X \times Y \longrightarrow Y$ be the projection maps. Define the map

$$\phi : \pi_1(X \times Y, (x_0, y_0)) \longrightarrow \pi_1(X, x_0) \times \pi_1(Y, y_0)$$

by

$$\phi([f]) = ([P_X \circ f], [P_Y \circ f])$$

for every loop f based at (x_0, y_0). To prove that ϕ is a group isomorphism, we will make use of Lemma 4.1.11.

Theorem 4.6.1. *The space $\pi_1(X \times Y)$ is isomorphic to $\pi_1(X) \times \pi_1(Y)$.*

Proof. (1) The map ϕ is well-defined. Let $f, g : I \longrightarrow X \times Y$ be two loops based at (x_0, y_0) and such that $f \simeq g$. By application of Lemma 4.1.11 with $h = P_X$ and then $h = P_Y$, we obtain $P_X \circ f \simeq P_X \circ g$ and $P_Y \circ f \simeq P_Y \circ g$. Then $[P_X \circ f] = [P_X \circ g]$ and $[P_Y \circ f] = [P_Y \circ g]$. Hence,

$$[f] = [g] \quad \Rightarrow \quad \phi([f]) = \phi([g]).$$

Hence, $P_X \circ f \simeq P_X \circ g$. Along similar lines, we can show that $P_Y \circ f \simeq P_Y \circ g$. Therefore, $[P_X \circ f] = [P_X \circ g]$ and $[P_Y \circ f] = [P_Y \circ g]$, i. e., $\phi([f]) = \phi([g])$.

(2) The map ϕ is a homomorphism. Let $f, g : I \longrightarrow X \times Y$ be two loops based at (x_0, y_0). Then $P_X \circ (f * g) = (P_X \circ f) * (P_X \circ g)$ and $P_Y \circ (f * g) = (P_Y \circ f) * (P_Y \circ g)$. Hence,

$$\begin{aligned}
\phi([f] * [g]) &= \phi([f * g]) \\
&= ([P_X \circ (f * g)], [P_Y \circ (f * g)]) \\
&= ([(P_X \circ f) * (P_X \circ g)], [(P_Y \circ f) * (P_Y \circ g)]) \\
&= ([P_X \circ f], [P_Y \circ f]) * ([P_X \circ g], [P_Y \circ g]) \\
&= \phi([f]) * \phi([g]),
\end{aligned}$$

as claimed. We have used the following definition of the operation $*$ in the product space:

$$([f_1], [f_2]) * ([g_1], [g_2]) = ([f_1] * [g_1], [f_2] * [g_2]). \tag{4.7}$$

(3) The map ϕ is surjective. Given $([f_1], [f_2]) \in \pi_1(X, x_0) \times \pi_1(Y, y_0)$, let $f : I \longrightarrow X \times Y$ be the loop based at (x_0, y_0) and defined by $\phi([f]) = ([f_1], [f_2])$, i. e., ϕ is surjective.

(4) The map ϕ is injective. Assume that $\phi([f]) = \phi([g])$. Then $[P_X \circ f] = [P_X \circ g]$ and $[P_Y \circ f] = [P_Y \circ g]$, that is to say $P_X \circ f \simeq P_X \circ g$ and $P_Y \circ f \simeq P_Y \circ g$. Hence, $f \simeq g$ (componentwise equivalence). Therefore, $[f] = [g]$. □

4.7 Fundamental group of a topological group

Definition 4.7.1. A group $(G, \cdot)$ with identity element e is *a topological group* if G is a topological space such that the product map $(x, y) \longmapsto x \cdot y : G \times G \longrightarrow G$ and the inverse map $x \longmapsto x^{-1} : G \longrightarrow G$ are continuous.

Example 4.7.2. (1) Additive Euclidean space $\mathbb{R}^n$ is a topological group.

(2) The circle S^1 with multiplication of complex numbers is a topological group.

Example 4.7.3. (1) The general linear group of invertible matrices, denoted GL_n, with the group structure of matrix multiplication is a topological group. Since the map $A \longrightarrow \det A$ is continuous, GL_n is an open subset of the space $\mathbb{M}_n$ of $n \times n$ matrices.
(2) The orthogonal group $\mathrm{O}(n)$ of $n \times n$ orthogonal matrices (whose transpose is equal to the inverse) is a topological subgroup of GL_n.

Remark 4.7.4. From Definition 4.7.1, we can see that the inverse map $x \longmapsto x^{-1} : G \longrightarrow G$ is a homeomorphism and the same holds for the maps $x \longmapsto xg : G \longrightarrow G$ and $x \longmapsto gx : G \longrightarrow G$ for any element $g \in G$. Therefore,
(1) left cosets and right cosets of an open (closed, respectively) subgroup are open (closed, respectively),
(2) the left cosets $(xH)_{x \in G}$ constitute a partition of the group G,
(3) if H and K are open, then $KH = \bigcup_{k \in K} kH$ and $HK = \bigcup_{h \in H} hK$ are open as the union of open sets.

Proposition 4.7.5. *Let $(G, \cdot)$ be a topological group and $H \subset G$ an open subgroup. Then H is also closed.*

Proof. Let K be the complement of the subgroup H. According to Remark 4.7.4, K can be expressed as the union $K = \bigcup_{k \notin H} kH$. Since H is open, K is open. Hence, H is closed. □

Proposition 4.7.6. *Let $(G, \cdot)$ be a topological group and $H \subset G$ a subgroup. Then the closure $\overline{H}$ of H is a subgroup of G.*

Proof. Let $x \in \overline{H}$ and $U \in \mathcal{N}_{x^{-1}}$ an open neighborhood of x^{-1}. By continuity of the inverse map $\varphi : x \longmapsto x^{-1}$, $\varphi^{-1}(U)$ is an open neighborhood of x, hence intersects H. Then $\varphi(\varphi^{-1}(U) \cap H) \neq \emptyset$, which in turn implies $U \cap \varphi(H) \neq \emptyset$. Since H is a subgroup, $\varphi(H) = H$, and thus $U \cap H \neq \emptyset$, proving that $x^{-1} \in \overline{H}$. Similarly, we can show that if $x, y \in \overline{H}$, then $xy \in \overline{H}$. □

Given two loops f and g based at an element e of a path-connected topological group $(G, \cdot)$, define the operation

$$(f \cdot g)(t) = f(t) \cdot g(t), \quad t \in I.$$

We have the following.

Proposition 4.7.7. (1) *The path $f \cdot g$ is a loop at e.*
(2) *The operation $\cdot$ induces a group operation on $\pi_1(G, e)$, denoted $\cdot$ as well.*
(3) *The fundamental group $\pi_1(G, e)$ is Abelian.*
(4) *In the topological group G, the operations $*$ and $\cdot$ coincide.*

Proof. (1) We have $(f \cdot g)(0) = (f \cdot g)(1) = e \cdot e = e$.
(2) We have $[f] \cdot [g] = [f \cdot g]$.

(3) Let f, g be two loops based at e. We claim that $f * g$ and $g * f$ are loop homotopic. Define the paths α, β from 0 to 1 on the interval I to be

$$\alpha(s) = \begin{cases} 2s, & s \in [0, 1/2], \\ 1, & s \in [1/2, 1] \end{cases}$$

and

$$\beta(s) = \begin{cases} 0, & s \in [0, 1/2], \\ 2s - 1, & s \in [1/2, 1]. \end{cases}$$

By definition of the product $f * g$ (see Section 4.3),

$$(f * g)(s) = f(\alpha(s)) \cdot g(\beta(s)) \quad \text{and} \quad (g * f)(s) = f(\beta(s)) \cdot g(\alpha(s)).$$

The paths α and β are further homotopic via the line homotopy $H_1(s,t) = t\alpha(s) + (1-t)\beta(s)$. Conversely, $H_2(s,t) = t\beta(s) + (1-t)\alpha(s)$ is a homotopy between β and α. Then clearly $H(s,t) = f(H_1(s,t)).g(H_2(s,t))$ is a loop homotopy from $g * f$ to $f * g$.

(4) Let H be the homotopy in part (c). Notice that

$$H(s, 1/2) = f\left(\frac{1}{2}\alpha(s) + \frac{1}{2}\beta(s)\right).g\left(\frac{1}{2}\beta(s) + \frac{1}{2}\alpha(s)\right) = f(s).g(s) = (f.g)(s),$$

whereas $H(s,1) = f(\alpha(s)) \cdot g(\beta(s)) = (f * g)(s)$. Therefore, $f \cdot g$ and $f * g$ are homotopic. □

4.8 Simply connected space

Definition 4.8.1. *A simply connected* space is a path-connected space whose fundamental group is the trivial group (consisting only of the identity element).

Remark 4.8.2. If X is simply connected, then $\pi_1(X, x_0)$ is trivial for some $x_0 \in X$. Let $x_1 \in X$ be such that $x_1 \neq x_2$. Since X is path-connected, by Theorem 4.5.1, $\pi_1(X, x_0)$ and $\pi_1(X, x_1)$ are isomorphic. As a consequence $\pi_1(X, x_1)$ is also trivial, which shows why the base point is not involved in the definition of a simply connected space.

Theorem 4.8.3. *Let X be a path-connected space. Then X is simply connected if and only if any two paths having same end points are path homotopic.*

Proof. (1) Let X be a simply connected space and f, g two paths from x_0 to x_1. Then $f * \overline{g}$ is a loop based at x_0, where $\overline{g}(t) = g(1-t)$ is the inverse of g. Since X is simply connected, $f * \overline{g}$ is homotopic to e_{x_0}. By Corollary 4.3.5(2), f and g are homotopic.

(2) Conversely, if any two paths with same end points are path homotopic, then for all loops f based at x_0, we have $f \simeq e_{x_0} \iff [f] = [e_{x_0}]$. Hence, $\pi_1(X, x_0) = \{[e_{x_0}]\}$ and the space X is simply connected. □

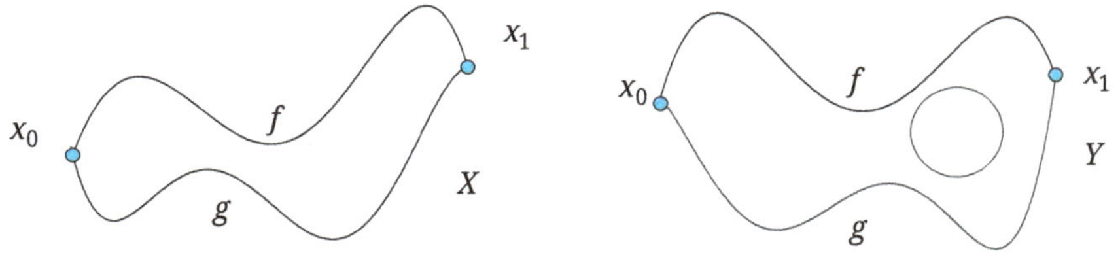

Figure 4.6: X is simply connected, Y is not.

Example 4.8.4. The torus $S^1 \times S^1$ is not simply connected. Actually, a loop that goes around the small circle cannot be shrunk to a point unless one cuts the torus.

Proposition 4.8.5. *Let X, Y be two homotopically equivalent spaces. Then X is simply connected if and only if Y is simply connected.*

Proof. (1) *First proof.* Since X and Y are homotopically equivalent, there exist continuous maps

$$X \xrightarrow{f} Y \xrightarrow{g} X \xrightarrow{f} Y$$

such that $f \circ g \simeq \mathrm{Id}_Y$ and $g \circ f \simeq \mathrm{Id}_X$. Suppose that X is simply connected and let $\phi, \psi : [0,1] \longrightarrow Y$ be two continuous paths such that $\phi(0) = \psi(0) = y_1 \in Y$ and $\phi(1) = \psi(1) = y_2 \in Y$. Let $\widetilde{\phi}, \widetilde{\psi} : [0,1] \longrightarrow X$ be the composite maps $\widetilde{\phi} = g \circ \phi$ and $\widetilde{\psi} = g \circ \psi$. Then $\widetilde{\phi}(0) = \widetilde{\psi}(0) = g(y_1) \in X$ and $\widetilde{\phi}(1) = \widetilde{\psi}(1) = g(y_2) \in X$. Since X is simply connected, by Theorem 4.8.3 $\widetilde{\phi}$ and $\widetilde{\psi}$ are path-homotopic. By composition with map f and appealing to Corollary 4.1.11, we obtain that $(f \circ g) \circ \phi \simeq (f \circ g) \circ \psi$. Since $f \circ g \simeq \mathrm{Id}_Y$, we deduce that $\phi \simeq \psi$, as desired.

(2) *A second* and shorter *proof* follows from Proposition 4.2.19 and Corollary 4.5.5. □

Figure 4.6 illustrates the concept of simple connectedness.

4.9 Contractible space

Definition 4.9.1. A space X is said to be *contractible* if there exist a point $x_0 \in X$ and a continuous map $H : X \times I \longrightarrow X$ such that $H(x,0) = x$ and $H(x,1) = x_0$ for all $x \in X$. This means that the identity Id_X is null-homotopic, i. e., $\mathrm{Id}_X \simeq e_{x_0}$.

Example 4.9.2. A convex subset $C \subseteq \mathbb{R}^n$, including open and closed balls, is contractible to each of its points. Indeed, for every $x_0 \in C$, $H(x,t) = tx_0 + (1-t)x$, $0 \le t \le 1$, $x \in C$ is a homotopy between Id_C and e_{x_0}.

Example 4.9.3. Consider the comb space (see Figure 1.6)

$$Cb = \left\{(x_1, x_2) \in \mathbb{R}^2 : \left(x_1 = 0, 1, \frac{1}{2}, \frac{1}{3}, \frac{1}{4}, \ldots; \ 0 \le x_2 \le 1\right)\right\} \cup (I \times \{0\}),$$

that is $Cb = \overline{\bigcup_{n=1}^{\infty}(\{\frac{1}{n}\} \times I)} \cup (I \times \{0\})$. Let $H : Cb \times I \longrightarrow Cb$ be defined by $H(x,t) = (x_1, (1-t)x_2)$, for $x = (x_1, x_2) \in Cb$. Then H is a homotopy between Cb and the closed interval $[0,1]$ of the x_1-axis. Since $[0,1]$ is contractible as a convex subset of $\mathbb{R}$, the comb space Cb is contractible.

Example 4.9.4. The infinite broom (see Figure 1.7)

$$\left\{(x_1, x_2) \in \mathbb{R}^2 : \ 0 \le x_1 \le 1, \ x_2 = \frac{x_1}{n}, \ n = 1, 2, \ldots\right\} \cup (I \times \{0\})$$

is contractible to the origin with homotopy $H(x,t) = tx$.

Proposition 4.9.5. *A cone $C(X)$ over a space X is contractible.*

Proof. The homotopy $H : C(X) \times I \longrightarrow C(X)$ given by $H([(x,t)], s) = [(x, t + (1-t)s)]$ maps the identity to the constant $[(x,1)] = X \times \{1\}$. Hence, $C(X)$ is contractible to its apex, namely the slice $X \times \{1\}$. □

Proposition 4.9.6. *Let $X = S_1^1 \vee S_2^1$ be the wedge of two circles, where $S_1^1 = C(-a; 1)$ and $S_2^1 = C(a; 1)$, where $a = (1,0)$. Then the subspace $Y = X \setminus \{(-2,0), (2,0)\}$ is contractible to the intersection point $(0,0)$.*

In Proposition 8.2.1, we will prove that the whole space $X = S_1^1 \vee S_2^1$ is not contractible.

Proof. (1) Using complex representation, let

$$X_1 = S_1^1 \setminus \{(-2,0)\} = \{z \in \mathbb{C} : z = -a + e^{i\theta_1}, \ -\pi < \theta_1 < \pi\},$$
$$X_2 = S_2^1 \setminus \{(2,0)\} = \{z \in \mathbb{C} : z = a + e^{i\theta_2}, \ 0 < \theta_2 < 2\pi\},$$

and $Y = X_1 \cup X_2$ (see Figure 4.7). A suitable homotopy between $\mathrm{Id}_{|Y}$ and $(0,0)$ is given by

$$H(z,t) = \begin{cases} -a + e^{it\theta_1}, & z \in X_1, \\ a + e^{i(1-t)\pi + it\theta_2}, & z \in X_2. \end{cases}$$

(2) In terms of rectangular coordinates, a homotopy is constructed in [60, Example 4.12]:

$$H(x_1, x_2, t) = \begin{cases} \frac{(1-t)x}{\|((1-t)x_1+1, (1-t)x_2)\|}, & (x_1, x_2) \in X_1, \\ \frac{(1-t)x}{\|((1-t)x_1-1, (1-t)x_2)\|}, & (x_1, x_2) \in X_2. \end{cases}$$

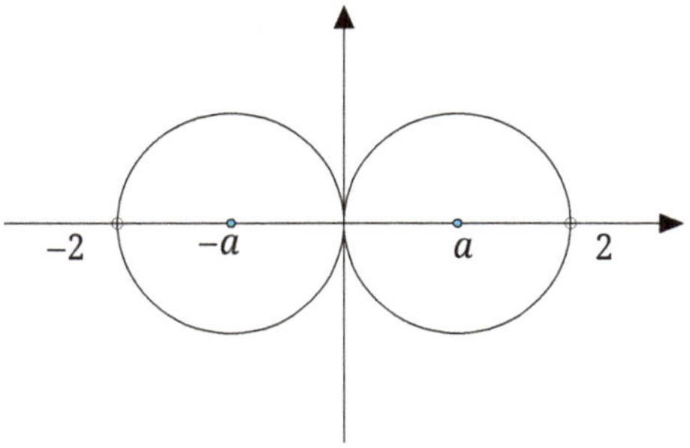

Figure 4.7: Punctured wedge of circles.

Clearly, H is well-defined, continuous, $H(x_1, x_2, 1) = 0$, and $H(x_1, x_2, 0) = (x_1, x_2)$ for all $(x_1, x_2) \in Y$. □

We presently return to some characteristic properties of the contractible spaces.

Theorem 4.9.7. *A space X is contractible if and only if for every space Y, any two continuous maps $f, g : Y \longrightarrow X$ are homotopic.*

Proof. (1) If X is contractible, then $\mathrm{Id}_X \simeq e_{x_0}$, for some $x_0 \in X$. By Lemma 4.1.11, $\mathrm{Id}_X \circ f \simeq e_{x_0} \circ f$, that is $f \simeq e_{x_0} \circ f$. Moreover, $g \simeq e_{x_0} \circ g$. But $(e_{x_0} \circ f)(y) = e_{x_0}(f(y)) = x_0$ for all $y \in Y$ and similarly $(e_{x_0} \circ g)(y) = e_{x_0}(g(y)) = x_0$ for all $y \in Y$. Hence, $e_{x_0} \circ f = e_{x_0} \circ g$, which implies by transitivity $f \simeq g$ (see Remark 4.1.9). Then the maps f and g are homotopic.

(2) Conversely, assume the condition of the theorem holds and take $Y = X, f = \mathrm{Id}_X$ and $g = e_{x_0}$, for some $x_0 \in X$. We obtain that Id_X and e_{x_0} are homotopic. Hence, X is contractible. □

Corollary 4.9.8. *A space X is contractible to x_0 if and only if X is contractible to every point $x \in X$.*

Proof. Clearly, if X is contractible to every point $x \in X$, then it is contractible to x_0. Conversely, if $e_{x_0} \simeq \mathrm{Id}_X$ for some $x_0 \in X$, then by Theorem 4.9.7, e_x and $e_{x_0} : X \longrightarrow X$ are homotopic for every $x \in X$. Hence, $e_x \simeq \mathrm{Id}_X$, i. e., X is contractible to x. □

Here are two further characterizations of contractibility.

Corollary 4.9.9. *A space X is contractible if and only if for every space Y, any continuous map $f : Y \longrightarrow X$ is null-homotopic.*

Proof. (1) If X is contractible, then by Theorem 4.9.7, the maps f and $g = e_{x_0}$ for some $(x_0 \in X)$ are homotopic. Hence, f is null-homotopic.

(2) Conversely, let $f, g : Y \longrightarrow X$ be two continuous maps. By assumption f and g are null-homotopic, and thus, by Remark 4.1.9, they are homotopic. By Theorem 4.9.7, X is contractible. □

Theorem 4.9.10. *A space X is contractible if and only if for every space Y, any continuous map $f : X \longrightarrow Y$ is null-homotopic.*

Proof. (1) If X is contractible, there exists $x_0 \in X$ such that $\mathrm{Id}_X \simeq e_{x_0}$. By Lemma 4.1.11, $f \circ e_{x_0} \simeq f \circ \mathrm{Id}_X = f$. But for all $x \in X$,

$$(f \circ e_{x_0})(x) = f(e_{x_0}(x)) = f(x_0) := y_0.$$

Hence, $f \simeq y_0$, proving that f is null-homotopic.

(2) Conversely, if we take $Y = X$ and $f = \mathrm{Id}_X$, then Id_X is null-homotopic, which means that X is contractible. □

Using the transitivity of the homotopy relation (see Remark 4.1.9), we get the following characterization.

Corollary 4.9.11. *The following statements are equivalent:*
(1) *A space X is contractible.*
(2) *For every space Y, any two continuous maps $f, g : X \longrightarrow Y$ are homotopic.*
(3) *For every space Y, any two continuous maps $f, g : Y \longrightarrow X$ are homotopic.*

Corollary 4.9.12. *Every contractible space is path-connected.*

Proof. Let $x_0, x_1 \in X$ and e_{x_0}, e_{x_1} the corresponding constant maps. Since X is contractible, by Theorem 4.9.7 applied with $Y = X$, $f = e_{x_0}$ and $g = e_{x_1}$, we get $e_{x_0} \simeq e_{x_1}$. Hence, there exists a homotopy $H : X \times I \longrightarrow X$ such that $H(x, 0) = e_{x_0}(x) = x_0$ and $H(x, 1) = e_{x_1}(x) = x_1$ for all $x \in X$. Then, for some $x \in X$, the map $f(t) = H(x, t)$ is a path from x_0 to x_1 because f is continuous, $f(0) = H(x, 0) = x_0$, and $f(1) = H(x, 1) = x_1$. □

The following example follows from Corollary 4.9.12 and Corollary 4.9.5.

Example 4.9.13. Over any space X, the cone $C(X)$ and the suspension $\Sigma(X)$ are path-connected.

Example 4.9.14. The topologist's sine function $\mathscr{S}$ (see Figure 1.3) is connected but, unlike the comb space, $\mathscr{S}$ is not path-connected (see Remark 1.4.24). Hence, $\mathscr{S}$ is not a contractible space.

Corollary 4.9.15. *Every contractible space is simply connected (hence has a trivial fundamental group).*

Proof. By Corollary 4.9.12, X is path-connected. To show that X is simply connected, we appeal to Theorem 4.8.3 and consider two paths $f, g : I \longrightarrow X$ having same endpoints. Since X is contractible, by Theorem 4.9.7 applied with $Y = I$, we find that the maps f and g are homotopic. Theorem 4.8.3 implies that X is simply connected. □

Remark 4.9.16. In Section 6.3, Corollary 6.3.4, we will show that the sphere S^n in $\mathbb{R}^{n+1}$ is simply connected for $n > 2$. In Section 11.1, Theorem 11.1.5, we will prove that S^n is not contractible for $n \geq 1$, showing that the converse of Corollary 4.9.12 fails to hold.

Before we present a concrete example of contractible spaces, we give another characterization of a contractible space.

Proposition 4.9.17. *A space X is contractible if and only if X homotopically equivalent to a one-point set.*

Proof. (1) Let X be contractible. Then there exists $x_0 \in X$ such that $\mathrm{Id}_X \simeq e_{x_0}$. The constant function $f(x) = x_0$, $x \in X$ and the trivial map $g : x_0 \longrightarrow X$ satisfy $g \circ f = \mathrm{Id}_X$ and $f \circ g = \mathrm{Id}_{|x_0}$.

(2) Conversely, let $x_0 \in X$ and $f : X \longrightarrow x_0$, $g : x_0 \longrightarrow X$ be such that $g \circ f \simeq \mathrm{Id}_X$ and $f \circ g \simeq \mathrm{Id}_{|x_0}$. Then, for $x_1 = g(x_0)$, $g \circ f = e_{x_1}$, and thus $\mathrm{Id}_X \simeq e_{x_1}$, i. e., X is contractible. □

Corollary 4.9.18. *Two contractible spaces have the same homotopy type.*

Corollary 4.9.19. *Let X and Y be homotopically equivalent spaces. Then X is contractible if and only if Y is contractible.*

Proposition 4.9.20. *If X is contractible, then for every space Y, the product $X \times Y$ is homotopically equivalent to Y.*

Proof. Let $H : X \times [0,1] \longrightarrow X$ be a homotopy between X and a point $x_0 \in X$. Then $H(x,0) = x$ and $H(x,1) = x_0$ for all $x \in X$. Let $f : X \times Y \longrightarrow Y$ be the projection along the second argument $f(x,y) = y$ and $g : Y \longrightarrow X \times Y$ the embedding $g(y) = (x_0,y)$. Then for all $y \in Y$,

$$(f \circ g)(y) = f(x_0,y) = y = \mathrm{Id}_Y(y) \quad \text{and} \quad (g \circ f)(x,y) = g(y) = (x_0,y).$$

Let $K : (X \times Y) \times I \longrightarrow X \times Y$ be given by $K(x,y,t) = (H(x,t),y)$. Then

$$K(x,y,0) = (H(x,0),y) = (x,y) = \mathrm{Id}_{X\times Y}(x,y),$$
$$K(x,y,1) = (H(x,1),y) = (x_0,y) = g(y) = (g \circ f)(x,y).$$

Therefore, $f \circ g$ is the identity on Y and $g \circ f$ is homotopic to the identity on $X \times Y$. □

Corollary 4.9.21. *If X and Y are contractible, then the product $X \times Y$ is contractible.*

The converse holds, too.

Proposition 4.9.22. *If $X \times Y$ is contractible, then both X and Y are contractible.*

Proof. Let H be a homotopy between $X \times Y$ and a point $(x_0,y_0) \in X \times Y$. Then $H_1(x,t) = \mathrm{pr}_1(H(x,y_0,t))$ and $H_2(y,t) = \mathrm{pr}_2(H(x_0,y,t))$ are homotopies between X and x_0, Y and y_0, respectively. Here, pr_1 and pr_2 stand for the projections onto the first and the second argument, respectively. □

Additional properties of contractible spaces are presented in Section 7.1.

Example 4.9.23. The space $B^2 \times S^1$ and S^1 are homotopically equivalent.

Definition 4.9.24. A nonempty subset A of a vector space X is *star-convex* (or starlike or star-shaped) if there exists some $x_0 \in A$ such that the line segment $[x_0, x]$ lies entirely in A for all $x \in A$. x_0 is called the center of A.

Recall that $[x_0, x] = \{tx_0+(1-t)x,\ 0 \le t \le 1\}$ denotes a closed segment with end-points x_0 and x in X. Clearly, a convex set is star-convex. Some star-convex and nonstar-convex sets are drawn in Figure 4.6. We have the following.

Proposition 4.9.25. *Every star-convex space is contractible (every loop can be contracted to the center).*

Proof. Let A be star-convex with center $x_0 \in A \subset X$ and let $x \in A$ be arbitrarily chosen. Then the map $H : X \times I \longrightarrow X$ defined by $H(x,t) = tx_0+(1-t)x$ is a homotopy between the identity Id_X and the constant map e_{x_0}. Indeed, H is continuous and satisfies $H(x,0) = x$ and $H(x,1) = x_0$ for all $x \in X$. Hence, X is contractible. □

Example 4.9.26. If a space X is contractible, then the product $X \times I$ is contractible. This result was directly proved in Example 4.2.7.

Remark 4.9.27. In summary, we have the chain of proper inclusions

$$\begin{aligned} \text{convex} &\subset \text{star-convex} \\ &\subset \text{contractible} \\ &\subset \text{simply connected} \\ &\subset \text{path-connected} \subset \text{connected.} \end{aligned}$$

Remark 4.9.28. In Figure 4.8, the first three figures are contractible, hence simply connected spaces. The annulus A is not star-convex because for each $x_0 \in A$, $0 \in [x_0, -x_0]$ but $0 \notin A$. By Example 4.2.5 and Example 4.2.8, annulus A and a circle have same homotopy equivalence. In Theorem 6.1.4, we will prove that a circle has a fundamental group homeomorphic to $\mathbb{Z}$. Hence, A is not simply connected.

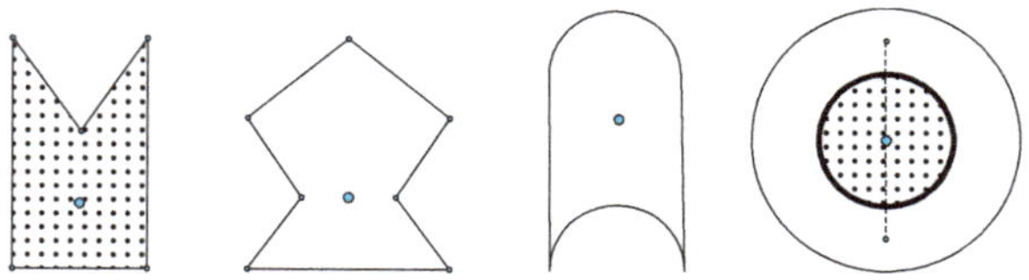

Figure 4.8: The first three sets are star-convex but not convex. The annulus is not star-convex.

Exercises (Chapter 4)

1. Prove that the composite $f * (g * h)$ is homotopic to the composite

$$(f * g * h)(s) = \begin{cases} f(3s), & 0 \le s \le 1/3, \\ g(3s-1), & 1/3 \le s \le 2/3, \\ h(3s-2), & 2/3 \le s \le 1, \end{cases}$$

providing a second proof of the associativity of operation $*$ in the proof of Theorem 4.4.4.

2. Prove that an alternative homotopy between $((f * g) * h)(s)$ and $(f * (g * h))(s)$ is given by

$$H(s,t) = \begin{cases} f(\frac{4s}{t+1}), & 0 \le s \le \frac{t+1}{4}, \\ g(4s-t-1), & \frac{t+1}{4} \le s \le \frac{t+2}{4}, \\ h(\frac{4s-t-2}{2-t}), & \frac{t+2}{4} \le s \le 1. \end{cases}$$

3. Prove or disprove the equality $f * g = g * f$ whenever the products make sense.

4. Let $f, g : I \longrightarrow X$ be two paths from x_0 to x_1 and $k, h : I \longrightarrow X$ two paths from x_1 to x_2. Prove that

$$\begin{cases} f * k \simeq g * h \\ \text{and} \\ f \simeq g \end{cases} \implies k \simeq h$$

(this cancellation property can be thought of as a converse to Lemma 4.3.4.)

5. For $x, y \in X$, let $e_x, e_y : X \longrightarrow X$ be the constant maps. Prove that e_x and e_y are homotopic if and only if x and y lie on the same component of X.

6. Show that if f has a left homotopy equivalence inverse and a right homotopy equivalence inverse, then f has a homotopy equivalence inverse. (*Hint:* See [25, Theorem 1.4, Chapter XVIII]).

7. Find out whether the geometric forms depicted in Figure 4.9 are homeomorphic or homotopically equivalent.

8. Find out whether the geometric forms depicted in Figure 4.10 are homeomorphic or homotopically equivalent.

9. Prove that S^1 and the set $X = S^1 \cup \ell$, where ℓ is the line joining $(1, 0)$ to $(2, 0)$ are homotopically equivalent. (*Hint:* See [2, Example 2.2.8].)

10. Let $A = \{1/n,\ n = 1, 2, \ldots\}$ and $B = A \cup \{0\}$. Show that A and B have not the same homotopy type.

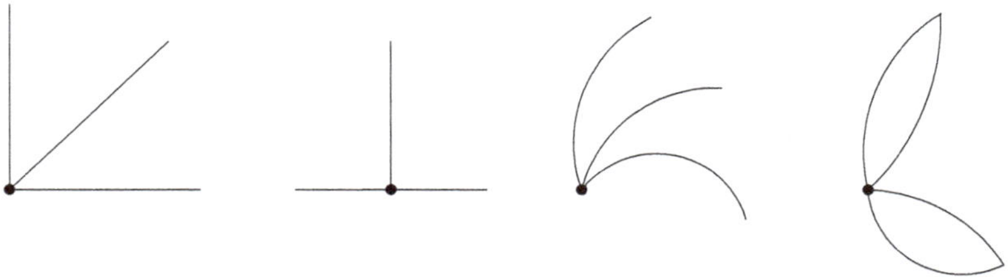

Figure 4.9: Homeomorphic? Homotopy equivalent?

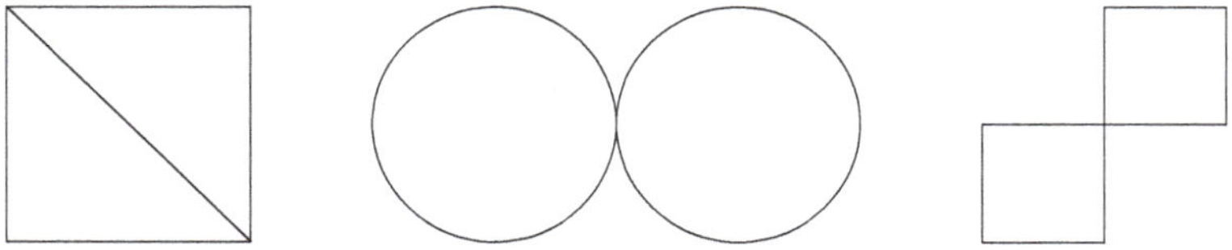

Figure 4.10: Homeomorphic? Homotopy equivalent?

11. Let X and Y be two homotopically equivalent spaces.
(1) Prove that there is a bijection between the set of connected (resp., path-connected) components of X and the set of connected (resp., path-connected) components of Y. (*Hint:* See [83, Corollary 1.17].)
(2) Deduce that X is connected if and only if Y is connected. (*Hint:* See [17, Proposition 6.20].)

12. Let $x_0 \in X$ and C be the path-component, which contains x_0. Prove that $\pi_1(X, x_0) \cong \pi_1(C, x_0)$. (*Hint:* Notice that every loop in X based at x_0 must stay in C or prove that the induced map of the inclusion is an isomorphism.)

13. Let $A \subset X$ be a path-connected subspace and $x_0 \in A$. Prove that the induced injection $i_* : \pi_1(A, x_0) \longrightarrow \pi_1(A, x_0)$ is surjective if and only if every path in X with endpoints in A is homotopic to a path in A.

14. Define $\pi_0(X)$ to be the set of all path-components of X (see Definition 1.4.34). By Theorem 1.4.35, their union is X. $\pi_0(X)$ is called the zeroth fundamental group (though this is not a group).
(1) (a) Assume that X is path-connected. Prove that $\pi_0(X)$ identifies with the space X.
(b) Assume that X is totally path-disconnected. Prove that $\pi_0(X)$ is the set $\{\{x\} : x \in X\}$.
(c) Assume that X is locally path-connected. Prove that the space $\pi_0(X)$ is discrete.
(2) Prove that $\pi_0(X)$ is homeomorphic to the space $[S^0, X]$ of homotopy classes of continuous maps $f : S^0 \longrightarrow X$ such that $f(-1) = x_0$, where $x_0 \in X$. (*Hint:* One may

consider the map $\psi : [S^0, X] \longrightarrow \pi_0(X)$, which assigns to $[f]$ the path-component of $f(1)$.)

15. Given two spaces X, Y and a continuous map $f : X \longrightarrow Y$, let $f_* : \pi_0(X) \longrightarrow \pi_0(Y)$ be the induced homomorphism, where $\pi_0(X)$ is introduced in Exercise 14.
(a) Show that f_* is well-defined.
(b) Let $g : Y \longrightarrow Z$. Prove that $(g \circ f)_* = g_* \circ f_*$.

16. A space X is called H-space (Hopf space) if there exist $x_0 \in X$ and a map $h : X \times X \longrightarrow X$ such that $h(x, x_0) = h(x_0, x) = x$ for all $x \in X$.
(1) Prove that every topological group is a H-space.
(2) Prove that if X is an H-space, then $\pi_1(X)$ is an Abelian group. (*Hint:* See [46, Theorems 4–18] or [83, Theorem 3.20].)

17. Consider the commutative diagram (see Remark 2.4.1), where $\overline{f}([x]) = [f(x)]$ is the induced homomorphism.

$$\begin{array}{ccc} S^2 & \xrightarrow{\quad f \quad} & S^1 \\ \downarrow{\scriptstyle \pi} & & \downarrow{\scriptstyle \pi} \\ \mathbb{RP}^2 & \xrightarrow{\quad \overline{f} \quad} & \mathbb{RP}^1 \end{array}$$

Prove that $\overline{f}$ induces a group homomorphism $\overline{f}_*$ between the fundamental groups.

18. Let $A \subset \mathbb{R}^2$ be bounded. Show that A is simply connected if and only if A and $\mathbb{R}^2 \setminus A$ are connected.

19. Prove that any two paths in $\mathbb{R}^3 \setminus \{0\}$ are homotopic.

20. Let $X = \{x, y\}$ be the Sierpiński space with the topology $\{\emptyset, X, \{x\}\}$. Prove that X is contractible.

21. Prove that the image of a simply connected set under a continuous function need not be simply connected.

22. Let $f : X \longrightarrow Y$ be a continuous map. Prove that the following statements are equivalent: (*Hint:* See [5, Proposition 1.4.9], [42, Proposition (21.14)].)
(1) f is null-homotopic;
(2) f extends continuously to $\tilde{f} : C(X) \longrightarrow Y$, where $C(X)$ is the cone over X.

23. Let K be a simple closed curve in $\mathbb{R}^3$ (Definition 1.5.5). Prove that $\pi_1(\mathbb{R}^3 \setminus K)$ and $\pi_1(S^3 \setminus K)$ are isomorphic.

24. Show that the subspace of $\mathbb{R}^3$, which is the union of the spheres of radius $1/n$ and center $(1, 0, 0)$ for $n = 1, 2, \ldots$ is simply connected.

5 Covering maps and lifting maps

Covering maps $p : \widehat{X} \longrightarrow X$ and covering spaces $\widehat{X}$ are first introduced in Section 5.1. They play a key role in determining some higher homotopy groups. Covering maps are particular identification maps. We study their topological properties. Section 5.2 is devoted to some important results on the existence and uniqueness of the lifting map by using covering maps. We consider separately the case of a loop, a path and then the general case. The lifting homotopies are also discussed together with the Monodromy theorem.

5.1 Covering maps

Recall that $\mathscr{N}_x$ denotes the set of all open neighborhoods of an element x of a space X.

Definition 5.1.1. Let $X, \widehat{X}$ be two spaces and $p : \widehat{X} \longrightarrow X$ a continuous map.

(1) The map p is called *a covering map* if for every $x \in X$, there exist $V \in \mathscr{N}_x$ and a family of disjoint open sets $(U_\alpha)_{\alpha\in J}$ in $\widehat{X}$ such that $p^{-1}(V) = \bigcup_{\alpha\in J} U_\alpha$ and the restriction $p_{|U_\alpha} : U_\alpha \longrightarrow V$ is a homeomorphism for all $\alpha \in J$.

(2) The neighborhood V is said to be *evenly covered* by p or *admissible*.

(3) The space $\widehat{X}$ is called *a covering space* (or a total space) and X *the base space* (often denoted in the literature E and B, respectively).

(4) Each element U_α of the partition is called *a sheet* (or a slice) for p over V.

(5) For each $x \in X$, the set $p^{-1}(x)$ is called *the fiber* of p over (at) x.

The simplest example of a covering map is the identity map on a space.

Proposition 5.1.2. *Let $p : \widehat{X} \longrightarrow X$ be a covering map. Then:*

(1) *p is surjective;*

(2) *p is open;*

(3) *For each $x \in X$, the fiber $p^{-1}(x)$ is discrete in the subspace topology.*

Proof. (1) Let $x \in X$, $V \in \mathscr{N}_x$ and $(U_\alpha)_{\alpha\in J}$ the open sets given by the definition of p. Since $p : U_\alpha \longrightarrow V$ is a homeomorphism for each $\alpha \in J$, there is some $\widehat{x} \in \widehat{X}$ such that $p(\widehat{x}) = x$.

(2) Let $\widehat{x} \in \widehat{X}$, $U \in \mathscr{N}_{\widehat{x}}$ and $V \in \mathscr{N}_{p(\widehat{x})}$. Then there exist $\alpha_0 \in J$ and $\widehat{x} \in U_{\alpha_0} \subset p^{-1}(V)$. Since $U \cap U_{\alpha_0}$ is open in U_{α_0} and $p_{|\{U_{\alpha_0}\}}$ is a homeomorphism, the set $p(U \cap U_{\alpha_0})$ is open in V. Hence, $p(U \cap U_{\alpha_0})$ is open in X, proving our claim (one can apply directly Proposition 1.5.34(1)).

(3) Let $x \in X$ and $\widehat{x} \in p^{-1}(x)$. Then there exists a neighborhood $U_{\widehat{x}} \in \mathscr{N}_{\widehat{x}}$ such that the restriction $p_{|U_{\widehat{x}}} : U_{\widehat{x}} \longrightarrow p(U_{\widehat{x}})$ is a homeomorphism. Since f is locally injective, $\{\widehat{x}\} = U_{\widehat{x}} \cap p^{-1}(x)$, proving the claim. □

Using Proposition 2.1.3, we deduce the following.

Corollary 5.1.3. *Every covering map is an identification map.*

https://doi.org/10.1515/9783111517384-005

Proposition 5.1.4. *A covering map $p : \widehat{X} \longrightarrow X$ is a local homeomorphism.*

Proof. Let $\hat{x} \in \widehat{X}$ and $x = p(\hat{x})$. Then there exists $V \in \mathcal{N}_x$ satisfying the definition of p. Let U_a be an open set such that $\hat{x} \in U_a$. Then $U_a \in \mathcal{N}_{\hat{x}}$ and $p_{|U_a} : U_a \longrightarrow V$ is a homeomorphism, as claimed. □

Remark 5.1.5. (1) Since a covering map p is a local homeomorphism, $\widehat{X}$ is locally compact (locally connected, locally path-connected, respectively) whenever X is, and conversely.
(2) If $p : \widehat{X} \longrightarrow X$ is a covering map, then X is an n-manifold if and only if $\widehat{X}$ is an n-manifold.

In general, the converse of Proposition 5.1.4 does not hold true as shows the map $p : (0, +\infty) \longrightarrow S^1$ given by $p(t) = e^{2i\pi t}$ (see [76, Example 2, Section 53]). However, we have a kind of converse for Proposition 5.1.4, allowing construction of covering maps.

Proposition 5.1.6. *Let $\widehat{X}$ and X be Hausdorff spaces such that $\widehat{X}$ is compact and X is connected. Then every local homeomorphism $p : \widehat{X} \longrightarrow X$ is a covering map.*

Proof. (1) Since p is a local homeomorphism, p is open by Corollary 1.5.35. Hence, $p(\widehat{X})$ is open in X. The space $p(\widehat{X})$ is further compact as the continuous image of the compact space $\widehat{X}$, hence $p(\widehat{X})$ is also closed in X. Since X is a connected space, $p(\widehat{X}) = X$, i. e., p is surjective.

(2) Since p is continuous by Proposition 1.5.34, $p^{-1}(x)$ is closed for all $x \in X$. Hence, $p^{-1}(x)$ is compact in the compact set $\widehat{X}$. The space $p^{-1}(x)$ is discrete because p is a local homeomorphism (see part (3), proof of Proposition 5.1.2). Hence, $p^{-1}(x)$ is finite.

(3) Let $x \in X$. Since p is a surjective local homeomorphism, for every $\hat{x} \in p^{-1}(x)$, there exists an open neighborhood $\mathcal{O}_{\hat{x}} \in \mathcal{N}_{\hat{x}}$ such that $p_{|O_{\hat{x}}} : \mathcal{O}_{\hat{x}} \longrightarrow p_{|\mathcal{O}_{\hat{x}}}(\mathcal{O}_{\hat{x}})$ is a homeomorphism. Consider the finite intersection of open sets $V = \bigcap_{\hat{x} \in p^{-1}(x)} p_{|\mathcal{O}_{\hat{x}}}(\mathcal{O}_{\hat{x}}) \in \mathcal{N}_x$ and the open sets $U_{\hat{x}} = p^{-1}(V) \cap \mathcal{O}_{\hat{x}}$, $\hat{x} \in p^{-1}(x)$. Then, by definition of V, we have $p^{-1}(V) = \bigcup_{\hat{x} \in p^{-1}(x)} U_{\hat{x}}$ and the restriction $p : U_{\hat{x}} \longrightarrow V$ is a homeomorphism for each $\hat{x} \in p^{-1}(x)$. □

Proposition 5.1.7. *For $n \geq 1$, the sphere S^n is a covering space of the real projective space $\mathbb{RP}^n$.*

Proof. The covering map is given by the canonical projection map $p = \pi$. Indeed, for each $x \in \mathbb{RP}^n$ and each $V \in \mathcal{N}_x$, there exist $\hat{x} \in p^{-1}(x)$ and a sufficiently small positive real number $\delta > 0$ such that $p^{-1}(V) = (U) \cup (-U)$ and $U = B(\hat{x}, \delta) \subset S^n_+$, where S^n_+ is the northern hemisphere (upper cap) $\{(x_1, x_2, \ldots, x_n, x_{n+1}) : x_{n+1} \geq 0\}$. If $(0, z) \in \mathbb{RP}^n$ is the x_{n+1}-axis, then $p^{-1}(0, z) = p^{-1}(p(N_p)) = \{N_p\} \cup \{S_p\}$, which consists of the two poles of the sphere S^n because $p^{-1}(p(U)) = (U) \cup (-U)$. □

Remark 5.1.8. From Example 3.6.4, Remark 5.1.5 and Proposition 5.1.7, the projective space $\mathbb{RP}^n$ is an n-manifold. It is further compact connected by Proposition 3.1.8.

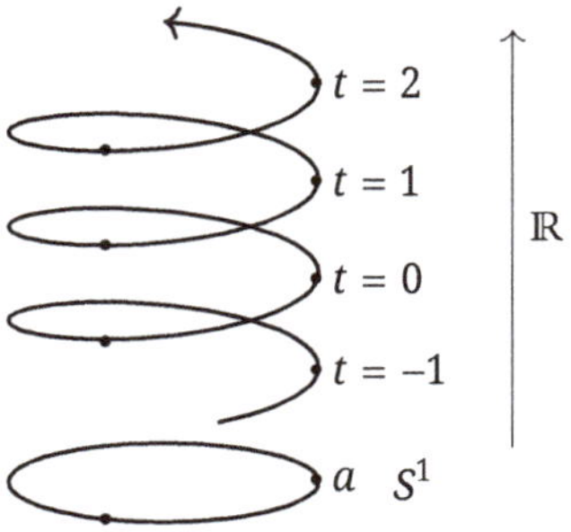

Figure 5.1: Standard covering map.

Definition 5.1.9. A covering map $p : \widehat{X} \longrightarrow X$ is said to be an *n-fold covering map* (or an n-sheet covering map) if the cardinality of each fiber $p^{-1}(x)$ equals n, for each $x \in X$.

Remark 5.1.10. (1) In the proof of Proposition 5.1.7, each inverse image of the projection map p consists of two points. Then p is a *double covering map.*
(2) If $\widehat{X}$ is compact, $p^{-1}(x)$ for each $x \in X$ (see the proof of Part (2), Proposition 5.1.6). Then every covering map $p : \widehat{X} \longrightarrow X$ is an n-fold covering map, where n is the cardinality of $p^{-1}(x)$.

Definition 5.1.11. When the covering space $\widehat{X}$ is simply connected (see Definition 4.8.1), the covering map p is said to be *universal.*

Example 5.1.12. Since by Corollary 9.2.2, the fundamental groups $\pi_1(S^n)$ are trivial for all $n \geq 2$, the canonical projection map $p = \pi$ of Proposition 5.1.7 is universal.

A standard example of a universal covering map is given by the next theorem (see Figure 5.1).

Theorem 5.1.13. *The map $p : \mathbb{R} \longrightarrow S^1$ given by $p(t) = e^{2\pi i t}$ is a universal covering map.*

Proof. We only check that p is a covering map. Let $x = (x_1, x_2) \in S^1$, $x_1 > 0$ and $U \in \mathcal{N}_x$ be on the right arc of S^1. Then

$$p^{-1}(V) = \bigcup_{n \in \mathbb{Z}} U_n,$$

where $U_n = (n - 1/4, n + 1/4)$ and $p : U_n \longrightarrow V$ is a homeomorphism. Indeed, the map $p : \widehat{U_n} \longrightarrow V$ is continuous and bijective, hence a homeomorphism. Notice that $p^{-1}(x_1, x_2) = n + \frac{1}{2\pi} \sin^{-1}(x_2)$ and $p(n \pm 1/4) = \pm i$, where $i^2 = -1$. □

Remark 5.1.14. (1) In Theorem 5.1.13, the map p is continuous, surjective and open because of Proposition 1.5.34 and Theorem 5.1.13. By Proposition 2.1.3, we already know that p is an identification map, which is consistent with Corollary 5.1.3.

(2) The universal covering map p and the identification map $p_0 : [0,1] \longrightarrow S^1$ given by $p_0(t) = e^{2\pi i t}$ discussed in Example 2.1.5 have distinct domains. p_0 was proved to be closed (using the compactness of $[0,1]$) but not open. Here, p is open but not closed. Indeed, for $n = 2, \ldots$, the set $C = \{n + \frac{1}{n},\ n = 2,3,\ldots\}$ is closed in $\mathbb{R}$ but $p(C) = \{e^{\frac{2\pi i}{n}},\ n = 2,3,\ldots\}$ is not closed since it does not contain the limit point 1.

Theorem 5.1.15. *The nth power map $p_n(z) = z^n : S^1 \longrightarrow S^1$ is an n-fold covering map.*

Proof. Let $Z = e^{i\theta_0} \in S^1$ and $V_\varepsilon = \{z = e^{i\theta} : \theta_0 - \varepsilon < \theta < \theta_0 + \varepsilon\}$ with some $0 < \varepsilon < \pi$. Let

$$U_k = \left\{z = e^{i\alpha_k} : \frac{\theta_0 - \varepsilon}{n} + \frac{2k\pi}{n} < \alpha_k < \frac{\theta_0 + \varepsilon}{n} + \frac{2k\pi}{n}\right\}.$$

Then V_ε and U_k $(0 \le k \le n-1)$ are open sets. Moreover, for all $k \in [0, n-1]$, $U_k \cap U_{k+1} = \emptyset$ for $0 < \varepsilon < \pi$. Therefore,

$$\frac{\theta_0 + \varepsilon}{n} + \frac{2k\pi}{n} < \frac{\theta_0 - \varepsilon}{n} + \frac{2(k+1)\pi}{n}.$$

Finally, $p_n^{-1}(V_\varepsilon) = \bigcup_{k=0}^{k=n-1} U_k$. Indeed, if $z = e^{i\alpha} \in p_n^{-1}(V_\varepsilon)$, then $p_n(z) = e^{ni\alpha} \in V_\varepsilon$. Hence, $\theta_0 - \varepsilon < n\alpha < \theta_0 + \varepsilon$, and thus $\frac{\theta_0-\varepsilon}{n} < \alpha < \frac{\theta_0+\varepsilon}{n}$, i. e., $z \in U_0$. Conversely, if $z = e^{i\alpha_k} \in U_k$ for some $k \in [0, n-1]$, then $p_n(z) = e^{ni\alpha_k} = e^{ni\alpha_k + 2ik\pi}$ with $\theta_0 - \varepsilon + 2k\pi < n\alpha_k + 2k\pi < \theta_0 - \varepsilon + 2k\pi$, i. e., $\theta_0 - \varepsilon < n\alpha_k < \theta_0 + \varepsilon$. Hence, $p_n(z) \in V_\varepsilon$. Since each U_k contains only one nth root, the map $p_n : U_k \longrightarrow V_\varepsilon$ is a homeomorphism, for each $k \in [0, n-1]$. □

Remark 5.1.16. Theorem 5.1.15 shows that the covering space is not unique because further to $\mathbb{R}$, S^1 is an n-fold covering of itself with covering map given by the nth power map $p(z) = z^n$.

We end this section with some useful results. For the first two theorems, we refer to [76, Theorem 53.2, Theorem 53.3].

Theorem 5.1.17. *Let $p : \widehat{X} \longrightarrow X$ be a covering map, $X_0 \subset X$ a subspace and $\widehat{Y} = p^{-1}(X_0)$. Then the restriction $p : \widehat{Y} \longrightarrow X_0$ is a covering map.*

Proof. Let $x_0 \in X_0$, $V \in \mathcal{N}_{x_0}$ and $(U_\alpha)_{\alpha \in J}$ be the open covering of $p^{-1}(V)$. Then $V \cap X_0 \in \mathcal{N}_x$ in X_0 and $(U_\alpha \cap X_0)_{\alpha \in J}$ is an open covering of $p^{-1}(V \cap X_0)$. Finally, for each $\alpha \in J$, $p_{|\{U_\alpha \cap X_0\}}$ is homeomorphism onto $V \cap X_0$. □

Theorem 5.1.18. *Let $p : \widehat{X} \longrightarrow X$ and $p' : \widehat{X'} \longrightarrow X'$ be two covering maps. Then the product map*

$$p \times p' : \widehat{X} \times \widehat{X'} \longrightarrow X \times X'$$

is a covering map.

Proof. Given $(x, x') \in X \times X'$, let $V \in \mathcal{N}_x$, $V' \in \mathcal{N}_{x'}$, $(U_\alpha)_{\alpha\in J}$ and $(U'_\alpha)_{\alpha\in J}$ be the open covers of $p^{-1}(V)$ and $p^{-1}(V')$, respectively. Then $p^{-1}(V \times V') = \bigcup_{\alpha\in J} U_\alpha \times U'_\alpha$ and for all $\alpha \in J$, $p : U_\alpha \times U'_\alpha \longrightarrow V \times V'$ is a homeomorphism. □

Example 5.1.19. If p is the standard covering map given by Theorem 5.1.13, then the product map $p \times \mathrm{Id}_{|\mathbb{R}^2} : \mathbb{R} \times \mathbb{R}^2 \longrightarrow S^1 \times \mathbb{R}^2$ given by $(e^{2\pi i t}, x)$ is a covering map.

Example 5.1.20. If $p(t) = e^{2\pi i t}$, then the product map $p\times p : \mathbb{R}\times\mathbb{R} \longrightarrow S^1\times S^1$ is a covering map of the torus $\mathbb{T}^2$.

Example 5.1.21. The torus $\mathbb{T}^2$ is also covered by the cylinder $S^1 \times \mathbb{R}$, which in turn is covered by $\mathbb{R}^2$.

The following result is useful for the so-called Monodromy theorem.

Theorem 5.1.22. *Let $p : \widehat{X} \longrightarrow X$ be a covering map and $f, g : Y \longrightarrow \widehat{X}$ two continuous maps such that $p \circ f = p \circ g$ and $f(y_0) = g(y_0) = \widehat{x_0} \in \widehat{X}$ for some $y_0 \in Y$. If Y is connected and $\widehat{X}$ is Hausdorff, then f and g are identical.*

Proof. Let $E = \{y \in Y : f(y) = g(y)\}$. Since $y_0 \in E$, $E \neq \emptyset$. Since $\widehat{X}$ is Hausdorff, E is closed. To prove that E is open, let $y_0 \in E$ and $x_0 = p(f(y_0)) = p(g(y_0)) \in X$. Then there exist $V \in \mathcal{N}_X$ and disjoint open sets $(U_\alpha)_{\alpha\in J}$ such that $p^{-1}(V) = \bigcup_{\alpha\in J} U_\alpha$ and the restriction $p_{|\{U_\alpha\}} : U_\alpha \longrightarrow V$ is a homeomorphism for all $\alpha \in J$. Let $\alpha_0 \in J$ be such that $f(y_0) = g(y_0) \in U_{\alpha_0}$. Then y_0 belongs to the intersection $F = f^{-1}(U_{\alpha_0}) \cap g^{-1}(U_{\alpha_0})$, which is open in Y. To prove that $F \subseteq E$, let $y \in F$. By supposition, $f(y)$, $g(y) \in U_{\alpha_0}$ and $p(f(y)) = p(g(y))$. Since $p : U_{\alpha_0} \longrightarrow V$ is a homeomorphism, $f(y) = g(y)$, i. e., $y \in E$, proving the claim. Since Y is connected, we conclude that $E = Y$, proving the theorem. □

5.2 Lifting maps

In this section, we discuss some lifting properties of a map $f : Y \longrightarrow X$, where X is a space and Y is path-connected and locally path-connected, i. e., the existence and uniqueness of a map $\widehat{f} : Y \longrightarrow \widehat{X}$ satisfying $p \circ \widehat{f} = f$. The case when $X = S^1$ and $Y = I$ is of particular interest with regards to the path lifting property. In Theorem 5.2.1, we investigate in detail the case when f is a loop in S^1. The cases of a loop and a path are discussed in Subsection 5.2.1 and Subsection 5.2.2, respectively. Some consequences are then proved including the so-called Monodromy theorem. Finally, the general lifting case is studied in Subsection 5.2.3. We will make use of the results of Subsection 5.2.2.

5.2.1 Loop lifting

Consider the unit circle of the complex plane

$$S^1 = \{(x_1, x_2) \in \mathbb{R}^2 : x_1^2 + x_2^2 = 1\}.$$

Let $f : I \longrightarrow S^1$ be a loop based at $a = (1,0)$, $p : \mathbb{R} \longrightarrow S^1$ the projection map (universal covering map) given by $p(t) = e^{2\pi i t}$, and $p_0 = p_{|I}$ its restriction to the interval $I = [0,1]$. By Theorem 5.1.17, p_0 is still a covering map. Note that the fibers are the integers $p^{-1}(a) = \mathbb{Z} \subset \mathbb{R}$.

We start with two fundamental results. For the first one, we refer for instance to [38, Chapitre VI, Lemme 2.2].

Theorem 5.2.1. *There exists a unique path $\widehat{f} : I \longrightarrow \mathbb{R}$ such that $\widehat{f}(0) = 0$ and $p \circ \widehat{f} = f$:*

$$\begin{array}{ccc} I & \xrightarrow{\ f\ } & S^1 \\ & \searrow^{\widehat{f}} & \uparrow p \\ & & \mathbb{R} \end{array}$$

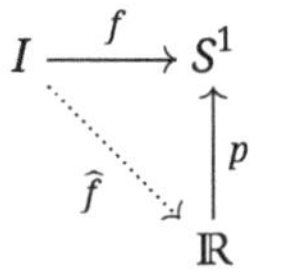

Proof. Note that $p : (-1/2, 1/2) \longrightarrow S^1 \setminus \{-a\}$ is a homeomorphism. Indeed, p is continuous, injective and it is a homomorphism, i. e., $p(t+s) = p(t)p(s)$. More generally, the product of a loop that winds around m times and a loop that winds around n times is a loop, which winds around $(m+n)$ times. Furthermore, $p : [-\alpha, \alpha] \longrightarrow S^1$ is a homeomorphism on the compact subinterval $[-\alpha, \alpha]$ for every $0 < \alpha < 1/2$, whence the claim. Note that $-a = -p(0) = p(1/2) = p(-1/2)$. Let $q : S^1 \setminus \{-a\} \longrightarrow (-1/2, 1/2)$ be the inverse homeomorphism ($q(a) = 0$).

(1) Existence of $\widehat{f}$. Since the loop f is uniformly continuous on the compact interval I (Definition 1.2.6), there exists $n \in \mathbb{N}$ such that for all $t, t' \in I$,

$$|t - t'| \le \frac{1}{n} \quad \Longrightarrow \quad \|f(t) - f(t')\| < 1.$$

Hence, $f(t) \ne -f(t')$, and thus $f(t)\overline{f(t')} \ne -1$, where $\overline{f(t')}$ is the conjugate of the complex number $f(t')$. Indeed, since $\overline{f(t')} = \frac{1}{f(t')}$,

$$f(t)\overline{f(t')} = \frac{f(t)}{f(t')} \ne -1 \quad \Longleftrightarrow \quad f(t) \ne -f(t').$$

Let $\{(\frac{k}{n}, \frac{k+1}{n}),\ 0 \le k \le n-1\}$ be a covering of the compact interval I. For $\frac{k}{n} \le t \le \frac{k+1}{n}$,

$$\left|t - \frac{k}{n}\right| \le \frac{1}{n} \quad \text{and} \quad \left|\frac{i}{n} - \frac{i-1}{n}\right| = \frac{1}{n}$$

so that

$$f(t)\overline{f\left(\frac{k}{n}\right)} \ne -a \quad \text{and} \quad f\left(\frac{i}{n}\right)\overline{f\left(\frac{i-1}{n}\right)} \ne -a.$$

As a consequence, both $f(t)\overline{f(\frac{k}{n})}$ and $f(\frac{i}{n})\overline{f(\frac{i-1}{n})}$ belong to $S^1 \setminus \{-a\}$. So, we can define

$$\widehat{f}(t) = q\Big(f(t)\overline{f\Big(\frac{k}{n}\Big)}\Big) + \sum_{i=1}^{k} q\Big(f\Big(\frac{i}{n}\Big)\overline{f\Big(\frac{i-1}{n}\Big)}\Big),$$

for $\frac{k}{n} \leq t \leq \frac{k+1}{n}$. The function $\widehat{f}$ is continuous and satisfies

(a) $\widehat{f}(0) = q(f(0)) = q(a) = 0$;

(b) $p \circ \widehat{f} = f$ because $p \circ q = \mathrm{Id}_{|S^1\setminus\{-a\}}$ and $t = 0 \Longrightarrow k = 0$.

(2) Uniqueness of $\widehat{f}$. Let $\widehat{f}_1, \widehat{f}_2 : I \longrightarrow \mathbb{R}$ be such that $\widehat{f}_1(0) = \widehat{f}_2(0) = 0$ and $p \circ \widehat{f}_1 = p \circ \widehat{f}_2 = f$. Since for each $\alpha \in \mathbb{R}$, $p : (\alpha - 1/2, \alpha + -1/2) \longrightarrow S^1 \setminus \{-p(\alpha)\}$ is a homeomorphism, $\widehat{f}_1 = \widehat{f}_2$. Since p is homeomorphism, one can replace $(-\frac{1}{2}, \frac{1}{2})$ by any interval (a, b) with $b - a \leq 1$. □

Definition 5.2.2. A map $\widehat{f}$ such that $p \circ \widehat{f} = f$ is called *a lifting* (or a lift) of f.

Definition 5.2.3. The terminal point $\widehat{f}(1) \in p^{-1}(a) = \mathbb{Z}$ is called *the winding number* (or *the degree*) of the loop f and is denoted $\omega(f)$.

Example 5.2.4. The constant loop e_1 based at a in S^1 has the trivial lifting $\widehat{e_1} \equiv 0$ because for all $t \in I$, $a = e_1(t) = e^{2\pi i \widehat{e_1}(t)}$. Hence, $\omega(e_1) = 0$.

Example 5.2.5. Let $\gamma : I \longrightarrow \mathbb{R} \setminus \{0\}$ be a simple closed curve (Definition 1.5.5), say represented by a continuous map $f : I \longrightarrow S^1$. By Theorem 5.2.1, there exists a unique path $\widehat{f} : I \longrightarrow \mathbb{R}$ such that $\widehat{f}(0) = 0$ and $p \circ \widehat{f} = f$, i. e., $e^{2\pi i \widehat{f}} = f(t)$. Assume $\widehat{f}$ to be continuously differentiable and compute the Cauchy integral

$$\frac{1}{2\pi i} \oint_{\gamma} \frac{dz}{z} = \frac{1}{2\pi i} \int_0^1 \frac{2\pi i \widehat{f}'(t) z dt}{z} = \int_0^1 \widehat{f}'(t) dt = \widehat{f}(1) = \omega(f).$$

If $f(t) = e^{2\pi i n t}$, then we obtain the Cauchy formula commonly used in complex analysis:

$$\oint_{\gamma} \frac{dz}{z} = 2\pi i n.$$

The following result concerns the homotopy lifting map and is proved exactly as Theorem 5.2.1 was. We omit the proof.

Theorem 5.2.6. *For each continuous map $H : I \times I \longrightarrow S^1$ such that $H(0,0) = a$, there exists a unique continuous map (lifting of H) $\widehat{H} : I \times I \longrightarrow \mathbb{R}$ such that $\widehat{H}(0,0) = 0$ and $p \circ \widehat{H} = H$.*

5.2.2 Path lifting

We state without proof the case of path lifting and homotopy lifting, where Y is the interval I and X is any space. The proofs of Theorem 5.2.7 and Theorem 5.2.8 can be found, e. g., in [16, Theorem 5.3, Theorem 5.4]. The first one concerns the path lifting while the second one shows the existence and uniqueness of homotopy lifting.

Theorem 5.2.7. *Let $p : \widehat{X} \longrightarrow X$ be a* covering map *and $f : I \longrightarrow X$ a path such that $f(0) = x_0$. If $\widehat{x_0} \in \widehat{X}$ is such that $p(\widehat{x_0}) = x_0$, there exists a unique* lifting map *$\widehat{f} : I \longrightarrow \widehat{X}$ such that $p \circ \widehat{f} = f$ and $\widehat{f}(0) = \widehat{x_0}$.*

Theorem 5.2.8. *Let $p : \widehat{X} \longrightarrow X$ be a* covering map *and $H : I \times I \longrightarrow X$ a homotopy such that $H(0,0) = x_0$. If $\widehat{x_0} \in \widehat{X}$ is such that $p(\widehat{x_0}) = x_0$, then there exists a* unique lifting homotopy *$\widehat{H} : I \times I \longrightarrow \widehat{X}$ such that $p \circ \widehat{H} = H$ and $\widehat{H}(0,0) = \widehat{x_0}$.*

Corollary 5.2.9. *The cardinality of* card $p^{-1}(x)$ *is constant on X.*

Proof. Let $x_0, x_1 \in X$ and γ be a path joining x_0 and x_1. Let $\widehat{x_0} \in p^{-1}(x_0)$, i. e., $p(\widehat{x_0}) = x_0$. By Theorem 5.2.7, there exists a unique lift path $\widehat{\gamma} : I \longrightarrow \widehat{X}$ such that $\widehat{\gamma}(0) = \widehat{x_0}$ and $p \circ \widehat{\gamma} = \gamma$. Let $f : p^{-1}(x_0) \longrightarrow p^{-1}(x_1)$ be defined by $f(\widehat{x}) = \widehat{\gamma}(1) = x_1$ so that $\widehat{\gamma}(1) \in p^{-1}(x_1)$. Then the map $g : p^{-1}(x_1) \longrightarrow p^{-1}(x_0)$ given by $f(\widehat{x}) = \overline{\widehat{\gamma}}(1)$ is the inverse of the map f. Here, $\overline{\gamma}$ is the inverse of the path γ. □

The lifting homotopy property can be used to show that the covering map has a special induced homomorphism. The following result is called the Monodromy theorem, allowing a certain kind of cancellation. We assume X to be path-connected and locally path-connected. To prove the result, we will appeal to Theorem 5.1.22.

Theorem 5.2.10. *Let $p : \widehat{X} \longrightarrow X$ be a covering map and $f, g : I \longrightarrow \widehat{X}$ two paths such that $f(0) = g(0) = \widehat{x_0} \in \widehat{X}$. Then f and g are path-homotopic if and only if $p \circ f \simeq p \circ g$.*

Proof. (1) The necessary condition is just Corollary 4.1.11.

(2) Let us prove the sufficiency of the condition. Set $(p \circ f)(0) = (p \circ g)(0) = x_0 = p(\widehat{x_0})$ and $(p \circ f)(1) = (p \circ g)(1) = x_1$ and let $H : I \times I \longrightarrow X$ be a homotopy between $p \circ f$ and $p \circ g$. By Theorem 5.2.8, there exists a lifting homotopy $\widehat{H}$ such that $\widehat{H}(0,0) = \widehat{x_0}$ and $p \circ \widehat{H} = H$. We have $(p \circ \widehat{H})(t,0) = H(t,0) = p \circ f(t)$ and $\widehat{H}(0,0) = f(0) = \widehat{x_0}$. Theorem 5.1.22 with $Y = I$ yields $\widehat{H}(t,0) = f(t)$ for all $t \in I$. Also, $\widehat{H}(t,1) = g(t)$ for all $t \in I$. Viewed as a function of the second argument, note that $(p \circ \widehat{H})(0,t) = H(0,t) = x_0$ and $p \circ \widehat{x_0} = x_0$. By the uniqueness of the lifting homotopy, we deduce that $\widehat{H}(0,t) = \widehat{x_0}$ for all $t \in I$. Similarly, $(p \circ \widehat{H})(1,t) = H(1,t) = x_1$ and $p \circ f(1) = p \circ g(1) = x_1$ imply $\widehat{H}(1,t) = f(1) = g(1)$ for all $t \in I$. In conclusion, we have shown that $\widehat{H}$ is a homotopy between f and g (and that f, g have same terminal points), as claimed. □

Corollary 5.2.11. *Let $p : \widehat{X} \longrightarrow X$ be a covering map, $\widehat{x} \in \widehat{X}$ and $x = p(\widehat{x})$. Then the induced homomorphism*

$$p_* : \pi_1(\widehat{X}, \widehat{x}) \to \pi_1(X, x)$$

is a monomorphism (Definition B.1.2*).*

Proof. Let $p_*([f]) = p_*([g])$, where f and g are loops in $\widehat{X}$ based as $\hat{x}$. Then

$$[p \circ f] = [p \circ g] \quad \Longleftrightarrow \quad p \circ f \simeq p \circ g,$$

which implies by Theorem 5.2.10, that $f \simeq g$, i. e., $[f] = [g]$, as claimed. □

Corollary 5.2.12. *The covering space of a simply connected, locally path-connected space is a simply connected, locally path-connected space.*

The following extension of Theorem 5.2.13 comes from [16, Chapter 5, Exercise 9].

Theorem 5.2.13. *Let $p : \widehat{X} \longrightarrow X$ be a covering map and $f : Y \longrightarrow X$ a continuous map with Y simply connected. Let $H : Y \times I \longrightarrow X$ be a homotopy such that $H(y, 0) = (p \circ f)(y)$ for all $y \in Y$. Then there exists a lifting homotopy $\widehat{H} : Y \times I \longrightarrow \widehat{X}$ such that $p \circ \widehat{H} = H$ and $\widehat{H}(y, 0) = f(y)$ for all $y \in Y$.*

Remark 5.2.14. The power map $f : S^1 \longrightarrow S^1$ given by $f(z) = z^2$ (see Theorem 5.1.15) shows that a covering map is not necessarily injective even though p_* is injective by Corollary 5.2.11. However, we have the following.

Corollary 5.2.15. *Let $p : \widehat{X} \longrightarrow X$ be a covering map such that the base space X is simply connected. Then p is an isomorphism.*

Proof. Given Proposition 5.1.2, it is only needed to prove that p is injective. Let $\widehat{x_1}, \widehat{x_2} \in \widehat{X}$ be such that $p(\widehat{x_1}) = p(\widehat{x_2}) = x_0 \in X$. Since $\widehat{X}$ is path connected, there exists a path $\gamma : I \longrightarrow \widehat{X}$ such that $\gamma(0) = \widehat{x_1}$ and $\gamma(1) = \widehat{x_2}$. Then $(p \circ \gamma)(0) = (p \circ \gamma)(1) = x_0$, i. e., $p \circ \gamma$ is a loop in X. Since X is simply connected, there exists a homotopy $H : I \times I \longrightarrow X$ such that $H(s, 0) = (p \circ \gamma)(s)$ and $H(s, 1) = x_0$ for all $s \in [0, 1]$. Theorem 5.2.13 applied for $Y = I$ yields some homotopy $\widehat{H} : I \times I \longrightarrow \widehat{X}$ such that $\widehat{H}(s, 0) = \gamma(s)$ for all $s \in I$. In particular, $\widehat{H}(0, 0) = \widehat{H}(1, 0)$ implies that $\widehat{x_1} = \widehat{x_2}$, proving that p is one-to-one. □

5.2.3 General lifting map

Consider the general case, where X and Y are arbitrary spaces. In this situation, a sufficient and necessary condition for existence of the lifting map is imposed.

Theorem 5.2.16. *Let $p : \widehat{X} \longrightarrow X$ be a* covering map, *$\widehat{x_0} \in \widehat{X}$ and $x_0 = p(\widehat{x_0})$. Let $f : Y \longrightarrow X$ be a continuous map such that $f(y_0) = x_0$ for some $y_0 \in Y$. Suppose that Y is path-connected and locally path-connected. Then there exists a unique* lifting map *$\hat{f} : Y \longrightarrow \widehat{X}$ such that $p \circ \hat{f} = f$ and $\hat{f}(y_0) = \widehat{x_0}$ if and only if $f_*(\pi_1(Y, y)) \subseteq p_*(\pi_1(\widehat{X}, \widehat{x_0}))$. Here, f_* and p_* are the induced homomorphisms of f and p, respectively.*

Proof. (1) To prove that the condition is necessary, let $\hat{f}$ be such a lifting map. By (4.6), $p_* \circ \widehat{f_*} = f_*$. For the composition to hold, we must have $\widehat{f_*}(\pi_1(Y, y_0)) \subseteq \pi_1(\widehat{X}, \widehat{x_0})$.

(2) Conversely, assume that the inclusion condition holds. For each $y \in Y$, let γ be a path from y to y_0 and set $g = f \circ \gamma$ (g depends on γ). Then $g(1) = f(y_0) = x_0$. By Theorem 5.2.7, there exists a unique lifting map $\widehat{g} : I \longrightarrow \widehat{X}$ such that $p \circ \widehat{g} = g$ and $\widehat{g}(1) = \widehat{x_0}$. Define $\widehat{f} : Y \longrightarrow \widehat{X}$ by $\widehat{f}(y) = \widehat{g}(0)$. Then, for each $y \in Y$,

$$(p \circ \widehat{f})(y) = p(\widehat{g}(0)) = g(0) = f(y).$$

To check that $\widehat{f}$ does not depend on the path γ, let γ' be another path from y to y_0, $g' = f \circ \gamma'$ and $h = f \circ (\gamma'^{-1} * \gamma) = (f \circ \gamma'^{-1}) * (f \circ \gamma)$. Then $\gamma'^{-1} * \gamma$ is a loop in Y based at y_0. Hence, h is a loop based at x_0, which lifts to

$$\widehat{h} = \widehat{f \circ \gamma'^{-1}} * \widehat{f \circ \gamma} = \widehat{g'}^{-1} * \widehat{g}.$$

Indeed, by Proposition 4.3.3 and the uniqueness of the lifting map, $\widehat{f * g} = \widehat{f} * \widehat{g}$, which also results from Proposition 4.3.3 together with $(\widehat{f \circ a})^{-1} = \widehat{f \circ a^{-1}}$ for every path a. Furthermore, $[h] \in f_*(\pi_1(Y, y_0))$ implies by the condition of the theorem that $[h] \in p_*(\pi_1(\widehat{X}, \widehat{x_0}))$. By [76, Theorem 54.6(c)], we deduce that $\widehat{h}$ is a loop based at x_0. Hence,

$$\widehat{h}(0) = \widehat{g'}^{-1}(0) = \widehat{g'}(1) = \widehat{h}(1) = \widehat{g}(1),$$

as claimed. The continuity of the map $\widehat{f}$ is proved as in [76, Lemma 79.1]. $\qquad\square$

Remark 5.2.17. In Theorem 5.2.1 and Theorem 5.2.7, the condition $f_*(\pi_1(Y, y)) \subseteq p_*(\pi_1(\widehat{X}, \widehat{x_0}))$ is automatically satisfied since $Y = I$ has a trivial fundamental group.

Example 5.2.18. Let $p(t) = e^{2i\pi t} : \widehat{X} = \mathbb{R} \longrightarrow X = S^1$ be the universal covering map of the circle (see Theorem 5.1.13).

(1) Let $f : Y = S^n \longrightarrow X = S^1$ be a continuous map, $n > 1$. Since $\pi_1(S^n)$ is the trivial group for all $n \geq 2$, by Theorem 5.2.16, f has a lifting $\widehat{f}$:

$$
\begin{array}{ccc}
Y = S^n & \xrightarrow{\ f\ } & X = S^1 \\
 & \searrow{\scriptstyle \widehat{f}} & \Big\uparrow{\scriptstyle p} \\
 & & \widehat{X} = \mathbb{R}
\end{array}
$$

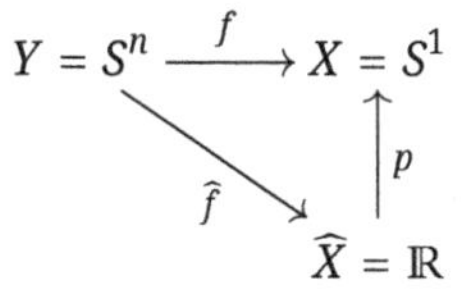

(2) If $f : Y = S^1 \longrightarrow X = S^1$ is a continuous map, then the condition of Theorem 5.2.16 holds only if f is null-homotopic, which is not the case when f is, e. g., the identity map. Thus, Theorem 5.2.16 does not provide a lifting map of f.

(3) By Corollary 6.1.11, the spaces $\pi_1(\mathbb{R}P^n)$ and $\mathbb{Z}/2$ are isomorphic for all $n \geq 2$. Then the same reasoning as in (2) shows that for a continuous map $f : Y = \mathbb{R}P^n \longrightarrow X = S^1$ have a lifting map, f must be null-homotopic.

(4) However, if $f : Y = S^1 \longrightarrow X = S^1$ is a continuous map, one can lift the composite $g = f \circ p : \mathbb{R} \longrightarrow S^1$. In fact, the condition of Theorem 5.2.16 holds since $\pi(\mathbb{R})$ is the trivial group. Thus, there exists a lifting map $\widehat{g}$ such that $p \circ \widehat{g} = f \circ p$:

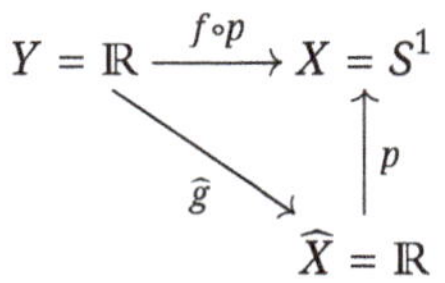

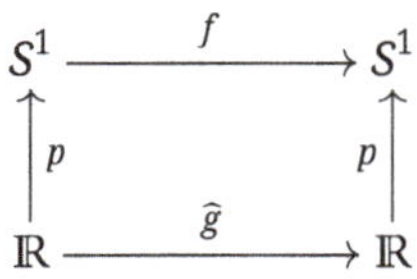

Consider three important cases:
(a) If $f(z) = z$ is the identity map, then $p \circ \widehat{g} = g$ implies $e^{2i\pi\widehat{g}(t)} = f(e^{2i\pi t}) = e^{2i\pi t}$. By uniqueness of the lifting map, $\widehat{g}(t) = t$ for all $t \in \mathbb{R}$.
(b) If $f(z) = z^n$ is the nth power map, then $e^{2i\pi\widehat{g}(t)} = e^{2in\pi t}$. By uniqueness of the lifting map, we get $\widehat{g}(t) = nt$, $t \in \mathbb{R}$.
(c) If $f(z) = R_\theta(z) = ze^{i\theta}$ is the rotation of angle θ, then $e^{2i\pi\widehat{g}(t)} = e^{2i\pi t}e^{i\theta}$. Hence, $\widehat{g}(t) = t + \theta/2\pi$, $t \in \mathbb{R}$.

Corollary 5.2.19. *Let $p : \widehat{X} \longrightarrow X$ be a* covering map *with $\widehat{x_0} \in \widehat{X}$ and $p(\widehat{x_0}) = x_0$. Let $f : Y \longrightarrow X$ be a continuous map such that $f(y_0) = x_0$. Suppose that Y is simply connected and locally path-connected. Then there exists a unique* lifting map *$\widehat{f} : Y \longrightarrow \widehat{X}$ such that $p \circ \widehat{f} = f$ and $\widehat{f}(y_0) = \widehat{x_0}$.*

Remark 5.2.20. Let $X = S^1, \widehat{X} = \mathbb{R}$, $x_0 = a$, $\widehat{x_0} = 0$ and $f : I \longrightarrow S^1$ a loop based at the point a (instead of a path). Then Corollary 5.2.19 is just Theorem 5.2.1.

Exercises (Chapter 5)

1. Consider the projection on the first coordinate $p : X \times Y \longrightarrow X$. If Y is supplied with the discrete topology, show that p is a covering map. (*Hint:* Consider $U_y = V \times \{y\}$, where $y \in Y$ and V is any neighborhood of a point $x \in X$.)

2. Let $p : \widehat{X} \longrightarrow X$ be a covering map with X Hausdorff. Prove that $\widehat{X}$ is Hausdorff.

3. Let $p : \widehat{X} \longrightarrow X$ be a continuous map and V an open connected subset of X, which is evenly covered by p.
(1) Prove that p maps each component of $p^{-1}(V)$ homeomorphically onto V.
(2) Consider the commutative diagram:

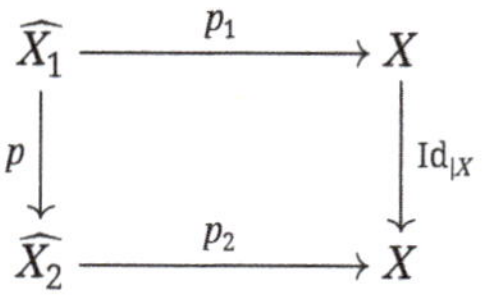

where X is locally connected, p is surjective and p_1, p_2 are two covering maps. Show that p is a covering map, too. (*Hint:* See [94, Corollary 13].)

4. Let $p : X \longrightarrow Y$ and $q : Y \longrightarrow Z$ be two covering maps. Prove that if every fiber of q is finite, then the composition map $q \circ p : X \longrightarrow Z$ is a covering map. (*Hint:* See [76, Exercise 4, Section 53].)

5. Let $p : \widehat{X} \longrightarrow X$ be a covering map such that every fiber is finite. Prove that $\widehat{X}$ is compact if and only if X is compact.

6. Prove that the map $p : \mathbb{R}^2 \longrightarrow S^1 \times \mathbb{R}$ given by $p(x, y) = (e^{2\pi i x}, y)$ is a covering map.

7. Prove that the map $p : \mathbb{T}^2 \longrightarrow \mathbb{T}^2$ given by $p(z, z') = (z, z'^2)$ is a covering map.

8. Let $\pi : \mathbb{C}^{n+1} \setminus \{0\} \longrightarrow \mathbb{C}\mathrm{P}^n$ be the projection map, where $\mathbb{C}\mathrm{P}^n$ is the complex projective space (see Definition 3.1.16). Prove that:
(1) π is surjective;
(2) $\mathbb{C}\mathrm{P}^1$ is homeomorphic to S^2;
(3) for all $x \in \mathbb{C}\mathrm{P}^n$ $(n \geq 2)$, $\pi^{-1}(x)$ is homeomorphic to S^1;
(4) for all $x \in \mathbb{C}\mathrm{P}^n$ $(n \geq 2)$, there exists $V \in \mathcal{N}_x$ such that $\pi^{-1}(V)$ is homeomorphic to $V \times S^1$;
(5) S^{2n+1} is a covering space over $\mathbb{C}\mathrm{P}^n$ with fiber S^1. (*Hint:* See [36, Exercise II.13.9].)

9. Let X be a locally path-connected space. Prove that a continuous map $p : \widetilde{X} \longrightarrow X$ is a covering map if and only if for each path-component C of X, $p_{|p^{-1}(C)} : p^{-1}(C) \longrightarrow C$ is a covering map. (*Hint:* See [92, Proposition 15.1.5].)

10. (1) Prove that the map $p : \mathbb{C} \longrightarrow \mathbb{C} \setminus \{0\}$ given by $p(z) = e^z$ is a covering map.
(2) Find $\pi_1(\mathbb{C} \setminus \{0\})$. (*Hint:* See [36, p. 138] and Example 6.2.11.)

11. Let $p : \widehat{G} \longrightarrow G$ be a homomorphism of topological groups, which is a covering map. Prove that if G is Abelian, then so is $\widehat{G}$.

12. Let X be path-connected and $p : \widehat{X} \longrightarrow X$ a compact continuous covering map. Show that $\pi_1(X)$ is finite.

6 Fundamental groups of the circle and the sphere

The fundamental group of the circle S^1 is first computed in Section 6.1, where the winding number of a loop is defined. We present three proofs that $\pi_1(S^1)$ is isomorphic to an infinite cyclic group. The process of the proof makes use of the definition and the properties of the fundamental group developed in Chapter 4. The general case of the sphere S^n ($n \geq 1$) is discussed in Section 6.3. More precisely, we show that for $n \geq 2$, the sphere S^n is simply connected. Furthermore, the fundamental groups computed in this chapter will help in investigating some classes of maps defined on circles (Section 6.2) and spheres (Section 6.4). Maps on circles are related to the degree of a map whereas maps on spheres are involved in the Borsuk–Ulam theorem; they will be revisited at the end of this book (Section 12.6).

6.1 Fundamental group of the circle

Let $a = (1, 0) \in S^1$. We start with a simple result.

Proposition 6.1.1. *Homotopic loops have same winding number.*

Proof. Let $p(t) = e^{2\pi it}$ be the universal covering map of the circle. Let f and g be homotopic loops based at a and $H : I \times I \longrightarrow S^1$ a homotopy joining f and g. By Theorem 5.2.6, let $\widehat{H}$ be the lifting of H. Then $p(\widehat{H}(s, 1)) = H(s, 1) = a$, i. e., $\widehat{H}(s, 1) \in p^{-1}(a) = \mathbb{Z}$ for all $s \in I$. Since $\widehat{H}$ is continuous, it is constant, i. e., $\widehat{H}(s, 1) = p^{-1}(a)$ for all $s \in I$. In addition, $(p \circ \widehat{H})(0, t) = f(t)$ and $(p \circ \widehat{H})(1, t) = g(t)$. Hence, $\widehat{f}(t) = \widehat{H}(0, t)$ and $\widehat{g}(t) = \widehat{H}(1, t)$ with $\widehat{f}(1) = \widehat{g}(1) = p^{-1}(a)$, as claimed. □

The converse of Proposition 6.1.1 holds true.

Proposition 6.1.2. *Two loops are homotopic if and only if they have the same winding number.*

Proof. Let f and g be two loops based at a and $\widehat{f}, \widehat{g}$ their lifting maps, respectively with same winding number $\omega(f) = \widehat{f}(1) = \widehat{g}(1) = \omega(g) = d$. Consider a homotopy $\widehat{H} : I \times I \longrightarrow \mathbb{R}$ joining $\widehat{f}$ and $\widehat{g}$. Then, for $s, t \in I$, $\widehat{H}(s, 0) = \widehat{g}(s)$, $\widehat{H}(s, 1) = \widehat{f}(s)$, $\widehat{H}(0, t) = 0$ and $\widehat{H}(1, t) = d$. Finally, the composite $H(s, t) = (p \circ \widehat{H})(s, t)$ is a loop homotopy between f and g. □

We will return back to Proposition 6.1.2 to discuss the higher dimension in Proposition 12.5.2 (Hopf's classification theorem).

Example 6.1.3. The clockwise path $f_+(t) = e^{2\pi it}$ and the counterclockwise path $f_-(t) = e^{-2\pi it}$ are not path homotopic. It straightforward to check that $\omega(f_+) = \widehat{f_+}(1) = 1$ and $\omega(f_-) = \widehat{f_-}(1) = -1$. Then the claim follows from Proposition 6.1.1.

https://doi.org/10.1515/9783111517384-006

We are now in position to compute the fundamental group of the circle S^1. For sake of completeness, we present three ways to prove that the fundamental group $\pi_1(S^1)$ is isomorphic to the additive group of integers $\mathbb{Z}$, i. e., $\pi_1(S^1)$ is an infinite cyclic group.

Theorem 6.1.4. *The map*

$$\Psi : \pi_1(S^1) \longrightarrow \mathbb{Z},$$

which assigns to a class $[f]$ the winding number $\omega(f)$ of the loop f is an isomorphism.

Proof. (1) The map Ψ is a group homomorphism, i. e., for every loops f, g based at a,

$$\Psi([f] * [g]) = \omega(f) + w(g).$$

(a) *First proof.* By Theorem 5.2.1, let $\widehat{f}$ and $\widehat{g}$ be the lifting maps of the loops f and g, respectively, and let $\widehat{h} = \widehat{g} + \omega(f)$. We have $\widehat{h}(0) = 0 + \omega(f) = \widehat{f}(1)$. Thus, we can define the product $\widehat{f} * \widehat{h}$, which satisfies $(\widehat{f} * \widehat{h})(0) = \widehat{f}(0) = 0$ and

$$p \circ (\widehat{f} * \widehat{h}) = (p \circ \widehat{f}) * (p \circ \widehat{h}) = f * (p \circ \widehat{g}) = f * g$$

because

$$(p \circ \widehat{h})(t) = e^{2\pi i(\widehat{g}(t)+\omega(f))} = e^{2\pi i\widehat{g}(t)} = (p \circ \widehat{g})(t).$$

Hence, $\widehat{f} * \widehat{h}$ is a lifting of $f * h$. As a consequence,

$$\begin{aligned} \Psi([f] * [g]) &= \Psi([f * g]) = \omega(f * g) \\ &= (\widehat{f} * \widehat{h})(1) = \widehat{h}(1) = \widehat{g}(1) + \widehat{f}(1) \\ &= w(g) + \omega(f). \end{aligned}$$

(b) *Second* (and equivalent) *proof.* Given the lifting maps $\widehat{f}$ and $\widehat{g}$ of the loops f and g, respectively, define the function:

$$\widehat{h} = \begin{cases} \widehat{f}(2t), & 0 \leq t \leq 1/2, \\ \widehat{g}(2t-1) + \widehat{f}(1), & 1/2 \leq t \leq 1. \end{cases}$$

Then $\widehat{h}$ is continuous, $\widehat{h}(0) = \widehat{f}(0) = 0$ and $p \circ \widehat{h} = f * g$. By uniqueness of the lifting map, $\widehat{h} = \widehat{f * g}$. Using the definition of $\widehat{h}$, we obtain

$$\Psi([f] * [g]) = \Psi([f * g]) = \omega(f * g) = \widehat{h}(1) = \widehat{f}(1) + \widehat{g}(1) = \omega(f) + \omega(g).$$

The claim of part (1) is complete.

(2) The map Ψ is surjective. Let $n \in \mathbb{Z}$ and $\widehat{f}(t) = tn$. Then $f(t) = (p \circ \widehat{f})(t) = e^{2\pi int}$ is a loop in S^1 such that $\Psi([f]) = \widehat{f}(1) = n$.

(3) Ψ is injective. Let $\Psi([f]) = \widehat{f}(1) = 0$. Then $\widehat{f} : I \longrightarrow \mathbb{R}$ is a loop in $\mathbb{R}$. Since the real line is simply connected, we find $\pi_1(\mathbb{R}) = \{0\}$ and then $[f] = 0$. □

By Example 2.2.5 and Corollary 4.5.5, $\pi_1(\mathbb{R}/\mathbb{Z}) \cong \mathbb{Z}$.

A second proof consists in defining an isomorphism from $\mathbb{Z}$ to $\pi_1(S^1)$ (see, e. g., [45, Theorem 1.7, p. 29]). We have the following.

Theorem 6.1.5. *For $n = 1, 2, \ldots$, let $w_n(s) = (\cos 2\pi ns, \sin 2\pi ns)$, $s \in I$. Then the map $\Phi : \mathbb{Z} \to \pi_1(S^1)$ defined by $\Phi(n) = [w_n]$ is an isomorphism.*

Remark 6.1.6. For $n = 1, 2, \ldots$, w_n is a loop based at the point $(1, 0) \in S^1$. Since S^1 is path-connected, by Theorem 4.5.1, $\pi_1(S^1)$ does not depend on the base point.

Proof of Theorem 6.1.5. (1) The map Φ is well-defined. Let $\widehat{\omega}_n(t) = nt : I \longrightarrow \mathbb{R}$ and $\omega_n = p \circ \widehat{\omega}_n$. Let $[\omega_n] = \Phi(n)$ and $[\gamma_n] = \Phi(n)$, where $\omega_n, \gamma_n : I \longrightarrow S^1$ are two loops. Consider the line homotopy $H : I \times I \longrightarrow \mathbb{R}$ joining $\widehat{\omega}_n$ and $\widehat{\gamma}_n$ and let $\gamma_n = p \circ \widehat{\gamma}_n$ and $G = p \circ H$. Then

$$\begin{aligned} G(s,0) &= p(H(s,0)) = p(\widehat{\omega}_n(s)) = \omega_n(s), \\ G(s,1) &= p(H(s,1)) = p(\widehat{\gamma}_n(s)) = \gamma_n(s), \\ G(0,t) &= p(H(0,t)) = 0, \\ G(1,t) &= p(H(1,t)) = n. \end{aligned}$$

This shows that G is a homotopy between ω_n and γ_n, hence $[\omega_n] = [\gamma_n]$.

(2) The map Φ is a group homomorphism. Define the shifting map $\tau_n : \mathbb{R} \longrightarrow \mathbb{R}$ by $\tau_n(x) = x + n$. Then $\tau_m \circ \widehat{\omega_n}(t) = nt + m$ and

$$\widehat{\omega_m} * (\tau_m \circ \widehat{\omega_n})(t) = \begin{cases} 2mt, & 0 \le t \le 1/2, \\ n(2t-1) + m, & 1/2 \le t \le 1 \end{cases}$$

whereas $\widehat{\omega_{m+n}}(t) = (m+n)t$. Hence, $\widehat{\omega_{m+n}}$ and $\widehat{\omega_m} * (\tau_m \circ \widehat{\omega_n})$ are line homotopic; hence, $\omega_{m+n} \simeq p \circ (\widehat{\omega_m} * (\tau_m \circ \widehat{\omega_n}))$. But $\widehat{\omega_n}$ is homotopic to $\tau_m \circ \widehat{\omega_n}$ as a shifting. Therefore,

$$\begin{aligned} \Phi(m) * \Phi(n) &= [\omega_m] * [\omega_n] \\ &= [p \circ \widehat{\omega_m}] * [p \circ \widehat{\omega_n}] \\ &= [(p \circ \widehat{\omega_m}) * (p \circ \widehat{\omega_n})] \\ &= [p \circ (\widehat{\omega_m} * (\tau_m \circ \widehat{\omega_n}))] \\ &= [\omega_{m+n}]. \end{aligned}$$

(3) The map Φ is surjective. Let $[f] \in \pi_1(S^1)$, where $f(0) = f(1) = a$. Since $0 \in p^{-1}(a) = \mathbb{Z}$, by Theorem 5.2.1, there exists a unique lifting $\widehat{f} : I \longrightarrow \mathbb{R}$ such that $\widehat{f}(0) = 0$ and $p \circ \widehat{f} = f$. Hence, $p(\widehat{f}(1)) = f(1) = a$ and so $\widehat{f}(1) \in p^{-1}(a) = \mathbb{Z}$, which means that $\widehat{f}$ is a

path from 0 to $\widehat{f}(1)$. Since $\mathbb{R}$ is convex, we get $\widehat{f} \simeq \widehat{\omega}_{\widehat{f}(1)}$ and $p \circ \widehat{f} = p \circ \widehat{\omega}_{\widehat{f}(1)} = \omega_{\widehat{f}(1)}$. Then the equality of classes follows:

$$[p \circ \widehat{f}] = [p \circ \omega_{\widehat{f}(1)}] = [\omega_{\widehat{f}(1)}] = \Phi(\widehat{f}(1)) = [f].$$

(4) The map Φ is injective. Let $\Phi(m) = \Phi(n)$, for some $m, n \in \mathbb{Z}$, i. e., $\omega_n \simeq \omega_m$. Let $H : I \times I \longrightarrow S^1$ be a loop homotopy between ω_m and ω_n, i. e., for all $s, t \in I$,

$$\begin{aligned} H(s,0) &= \omega_m(s), \\ H(s,1) &= \omega_n(s), \\ H(0,t) &= H(1,t) = a. \end{aligned}$$

By Theorem 5.2.1, there exists a unique homotopy lifting $\widehat{H} : I \times I \longrightarrow \mathbb{R}$ such that $\widehat{H}(0,0) = 0$ and $p \circ \widehat{H} = H$. Hence, $(p \circ \widehat{H})(s,0) = H(s,0) = \omega_m(s)$ and $(p \circ \widehat{\omega}_m)(s) = \omega_m(s)$. Furthermore, $(p \circ \widehat{H})(s,1) = H(s,1) = \omega_n(s)$ and $(p \circ \widehat{\omega}_n)(s) = \omega_n(s)$. By uniqueness of the lifting maps $\widehat{\omega}_m$ and $\widehat{\omega}_n$, we get $\widehat{\omega}_n(s) = \widehat{H}(s,1)$ and $\widehat{\omega}_m(s) = \widehat{H}(s,0)$. Finally, since $n = \widehat{\omega}_n(1) = \widehat{H}(1,1)$ and $m = \widehat{\omega}_m(1) = \widehat{H}(1,0)$, we conclude that $m = n$ for $(p \circ \widehat{H})(1,t) = H(1,t) = a$ for all $t \in I$ and $\widehat{H}(1,t) = \widehat{H}(1,0) + k_t$, where $k_t \in \mathbb{Z}$ and $\lim_{t \to 0} k_t = 0$ implies $k_t \equiv 0$. □

Example 6.1.7. The spaces B^2 and S^1 are not homotopically equivalent since their fundamental groups are not isomorphic.

Example 6.1.8. Let $\mathbb{T}^2 \cong S^1 \times S^1$ be the torus (see Example 2.3.12). Combining Theorem 4.6.1 and Theorem 6.1.5, we find that

$$\pi_1(\mathbb{T}^2) \cong \pi_1(S^1 \times S^1) \cong \pi_1(S^1) \times \pi_1(S^1) \cong \mathbb{Z} \times \mathbb{Z}.$$

More generally,

$$\pi_1(\mathbb{T}^n) \cong \pi_1(\overbrace{S^1 \times S^1 \times \cdots \times S^1}^{n}) \cong \overbrace{\mathbb{Z} \times \mathbb{Z} \times \cdots \times \mathbb{Z}}^{n}.$$

Example 6.1.9. Since B^2 is convex, the solid torus $B^2 \times S^1$ has the homotopy type of the circle, hence $\pi_1(B^2 \times S^1) \cong \mathbb{Z}$. The same holds for the open solid torus $\mathring{B}^2 \times S^1$, which is homeomorphic to $\mathbb{R}^2 \times S^1$.

As for the third way to see that the fundamental group of S^1 is isomorphic to $\mathbb{Z}$, we consider a general framework with some covering map $p : \widehat{X} \longrightarrow X$, $x_0 \in X$ and $p(\widehat{x_0}) = x_0$, where $\widehat{x_0} \in \widehat{X}$. Then define

$$\Psi : \pi_1(X, x_0) \longrightarrow p^{-1}(x_0)$$

by $\Psi([f]) = \omega(f) = \widehat{f}(1)$, where $\widehat{f}$ is the lifting map of f starting at $\widehat{x_0}$ (see Theorem 5.2.7).

Theorem 6.1.10. (1) *If $\widehat{X}$ is path-connected, then Ψ is surjective.*
(2) *If $\widehat{X}$ is simply connected, then Ψ is bijective.*
(3) *The map Ψ is a group homomorphism.*

By setting $\widehat{X} = \mathbb{R}$, $X = S^1$, $p(t) = e^{2\pi i t}$, $p^{-1}(x_0) = \mathbb{Z}$, $x_0 = a = (1,0)$ and $\widehat{x_0} = 0$ in Theorem 6.1.10, we recover Theorem 6.1.4.

Prof of Theorem 6.1.10. (1) Let $\widehat{x_1} \in p^{-1}(x_0)$ and $\widehat{f}$ a path in $\widehat{X}$ from $\widehat{x_0}$ to $\widehat{x_1}$. Then $f = p \circ \widehat{f}$ is a loop in X based at x_0, which satisfies $\Psi([f]) = \widehat{x_1}$. Hence, Ψ is surjective.

(2) Let $[f], [g] \in \pi_1(X, x_0)$ be such that $\Psi([f]) = \Psi([g])$. Then $\widehat{f}(1) = \widehat{g}(1)$, where $\widehat{f}$ and $\widehat{g}$ are the lifting maps of f and g, respectively. Since $\widehat{X}$ is simply connected, there is a path homotopy $\widehat{H}$ in X between $\widehat{f}$ and $\widehat{g}$. Then $H = p \circ \widehat{H}$ is a homotopy between f and g, i. e., $[f] = [g]$, proving that Ψ is one-to-one, hence bijective.

(3) The map Ψ is a homomorphism as in Theorem 6.1.4. □

Theorem 6.1.10 enables us to determine the fundamental group of the real projective space $\mathbb{RP}^n = S^{n-1}/_\sim$.

Corollary 6.1.11. *For $n \geq 2$, the spaces $\pi_1(\mathbb{RP}^n)$ and $\mathbb{Z}/2$ are isomorphic, where $\mathbb{Z}/2 = \mathbb{Z}/2\mathbb{Z}$ is the finite cyclic group of order 2 of integers modulo 2.*

The order of a group is its cardinality.

Proof. Just apply Theorem 6.1.10(2) with $\widehat{X} = S^{n-1}$, $X = S^{n-1}/_\sim \cong \mathbb{RP}^{n-1}$ and $p = \pi : S^{n-1} \longrightarrow S^{n-1}/_\sim$ the projection map. Since for all $x \in X$, $p^{-1}([x]) = \{x, -x\}$, the cardinality of $\pi_1(\mathbb{RP}^n)$ equals 2. Then $\pi_1(\mathbb{RP}^n)$ is isomorphic to $\mathbb{Z}/2$. □

6.2 Maps on the circle and degree of a map

We first extend the notion of the degree given in Definition 5.2.3 to any continuous map $f : S^1 \longrightarrow S^1$. Let $A = [0,1]/\{0,1\}$ and consider a homeomorphism $h : A \longrightarrow S^1$ (see Theorem 2.5.11). The canonical projection $\pi : [0,1] \longrightarrow A$ is defined by $\pi(t) = [t]$, $t \in I$. Consider the composition

$$[0,1] \xrightarrow{\ \pi\ } A \cong S^1 \xrightarrow{\ f\ } S^1, \qquad g = f \circ h \circ \pi : [0,1] \longrightarrow S^1$$

Then the composite $g : I \longrightarrow S^1$ is continuous and satisfies

$$g(0) = f(h([0])) = f(h(\{0,1\})) = f(h([1])) = g(1),$$

i. e., g is a loop whose winding number is given by Definition 5.2.3. By uniqueness of the winding number, we define the degree of f as the winding number of the loop g.

Definition 6.2.1. We have

$$\deg f = \omega(g).$$

By Proposition 6.1.1, homotopic loops have same degree.

Example 6.2.2. Let $f : S^1 \longrightarrow S^1$ be a continuous map.
(1) If f is constant, then $\deg f = \widehat{g}(1) = \widehat{g}(0) = 0$ (see Example 5.2.4).
(2) If f is the identity, then $\widehat{g} = t$ is the lifting map of g. Hence, $\deg f = \widehat{g}(1) = 1$.
(3) Since $z\overline{z} = 1$, the conjugation (inverse) map $f : S^1 \longrightarrow S^1$ given by $f(z) = \overline{z}$ has degree -1 for $\widehat{g}(t) = -t$.

More generally, we have the following.

Proposition 6.2.3. *Let $f : S^1 \longrightarrow S^1$ be the nth power map $f(z) = z^n$. Then $\deg f = n$.*

Proof. First proof. By identification of π and the covering map p given by $p(t) = e^{2\pi i t}$, the composition $g = f \circ p$ in the above diagram yields $g(t) = e^{2\pi n i t}$. Also, $\deg f = \omega(g)$. In the proof of Theorem 6.1.4, we have checked that $\widehat{g} = nt$ is the lifting map of g. Hence, $\deg f = \widetilde{g}(1) = n$, as claimed. As z turns once around S^1, $f(z)$ turns around S^1 n times (see also Example 5.2.18(4)(b)).

Second proof. We argue by induction. For $n = 1$, the statement is Example 6.2.2(1). Assume that $f^n(z) = z^n$ has degree n and consider the composition

$$S^1 \overset{\mathrm{Id}\times f^n}{\longrightarrow} S^1 \times S^1 \overset{m}{\longrightarrow} S^1,$$

where $m(z, z') = z.z'$ is the multiplication of complex numbers and let $f^{n+1} = m\circ(\mathrm{Id}\times f^n)$. First, we compute the induced map of m. By the identification provided by the isomorphism $\Phi(n) = [e^{2\pi n i t}, t \in I] : \mathbb{Z} \to \pi_1(S^1)$ of Theorem 6.1.5, for all integers k, k', we have

$$\begin{aligned} m_*(k, k') &= m_*([e^{2\pi k i t}], [e^{2\pi k' i t}]) \\ &= [e^{2\pi k i t}.e^{2\pi k i t}]) \\ &= [e^{2\pi(k+k') i t}] \\ &= k + k'. \end{aligned}$$

Then the induced map of the composition is

$$f_*^{n+1}(k) = m_*(\mathrm{Id} \times f_*^n)(k) = m_*(k, nk) = k + nk = (1 + n)k,$$

which completes the proof.

A third proof will be given in Proposition 12.5.8. □

Remark 6.2.4. Proposition 6.2.3 shows that the winding number $\omega(f)$ measures how many times the loop f wraps around the circle.

Remark 6.2.5. (1) Let $\gamma : I \longrightarrow X$ be a loop based at a point $x_0 \in X$. Since $I/_{\{0,1\}}$ and S^1 are homeomorphic, let $h : I/_{\{0,1\}} \longrightarrow S^1$ be a homeomorphism and $\pi : I \longrightarrow I/_{\{0,1\}}$ the projection map. For each $x \in S^1$, let $[t] = h^{-1}(x) \in I/_{\{0,1\}}$. If $t \in (0,1)$, then $\pi^{-1}([t]) = t$. Otherwise $\gamma(t) = x_0$. Then we may define $f : S^1 \longrightarrow X$ in a unique way by $f(x) = \gamma(t)$. Of course, f is continuous.

(2) Conversely, given a continuous map $f : S^1 \longrightarrow X$ such that $f(a) = x_0$, the composite function $\gamma = f \circ h \circ \pi$ is a loop at x_0.

Remark 6.2.6. Remark 6.2.5 allows us to identify the loop γ and the function f. Let $[S^1, X]$ refer to the set of homotopy classes of continuous maps $f : S^1 \longrightarrow X$ such that $f(a) = x_0$, where $a = (1,0) \in S^1$ and $x_0 \in X$. Thus, $[S^1, X]$ identifies with the fundamental group $\pi_1(X, x_0)$.

Let X be a path-connected space. From Proposition 6.1.2, we have the following.

Corollary 6.2.7. *The map* $\Phi : [S^1, X] \longrightarrow \pi_1(X)$, *which assigns to* $[f]$ *the class* $[g]$ *is bijective.*

Here, the class $[f]$ is relative to the homotopy of continuous maps from S^1 to X with $f(a) = x_0$, whereas the class $[\gamma]$ is relative the homotopy of loops in X and γ is as given by Remark 6.2.5.

Remark 6.2.8. In $[S^1, X]$, we can define an operation as follows.

Let the equivalence relation in S^1 be such that $(1,0) \sim (-1,0)$. Then the quotient space $S^1/_\sim$ is homeomorphic to the union of two circles C_1 and C_2, where $C_1 = C((0,1);1)$ and $C_2 = C((0,-1);1)$ are joined at the origin. Indeed, the figure eight formed by the union $C_1 \cup C_2$ is obtained by squeezing the middle of the circle S^1 while joining the opposite points $(-1,0)$ and $(1,0)$. Let h be the map defined by

$$h(x) = \begin{cases} x, & x \in S^1_+ \cup S^1_-, \\ 0, & x = (1,0) \text{ or } x = (-1,0), \end{cases}$$

where S^1_+, and S^1_- refer to the upper and lower half-circle, respectively. Then $\overline{h}$ is a homeomorphism, as shown in the following diagram, where $h = \overline{h} \circ \pi$:

$$S^1 \xrightarrow{\quad \pi \quad} S^1/_\sim \xrightarrow{\quad \overline{h} \quad} C_1 \cup C_2$$

Let $f, g : S^1 \longrightarrow X$ be such that $f(a) = g(a) = x_0$ and define $f \overline{*} g : S^1 \longrightarrow X$ by

$$(f \overline{*} g)(x) = \begin{cases} f(x), & h(x) \in C_1, \\ g(x), & h(x) \in C_2. \end{cases} \tag{6.1}$$

Note that $x \in S^1_+$ ($x \in S^1_-$, respectively) means $h(x) \in C_1$ ($h(x) \in C_2$, respectively). Furthermore, S^1_+ and S^1_- are homeomorphic to C_1 and C_2 with the origin removed in each circle, respectively. Finally, since f and g agree on $(1, 0)$, $f \bar{*} g$ is continuous by Lemma A.2.1. For $f \bar{*} g$ to map the point a to x_0, it is sufficient to shift the circles C_1, C_2 horizontally so that the tangent point is $(1, 0)$ instead of $(0, 0)$. Therefore, the operation $\bar{*}$ is well-defined.

Next, we present an extension result for maps on the circle taking values in a space X. The general case of the sphere will be discussed in Theorem 6.4.1.

Theorem 6.2.9. *Let X be a space, $a = (1, 0) \in S^1$, $x_0 \in X$, and $f : S^1 \longrightarrow X$ a continuous map. The following statements are equivalent:*
(1) *f is null-homotopic,*
(2) *f extends continuously to $\widetilde{f} : B^2 \longrightarrow X$;*
(3) *The induced homomorphism $f_* : \pi_1(S^1, a) \longrightarrow \pi_1(X, x_0)$ is the trivial homomorphism.*

Proof. (1) $\Longrightarrow$ (2): (a) *First proof.* Let $H : S^1 \times I \longrightarrow X$ be continuous and satisfies $H(x, 0) = x_0$, $H(x, 1) = f(x)$ for all $x \in S^1$. The following diagram commutes

$$
\begin{array}{ccc}
S^1 \times I & \xrightarrow{\ H\ } & X \\
& {\scriptstyle \pi} \searrow & \uparrow {\scriptstyle \widetilde{f}} \\
& & B^2
\end{array}
$$

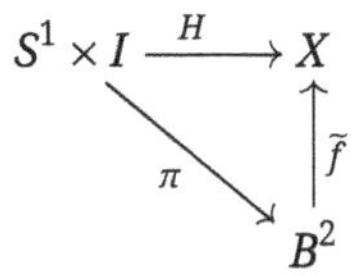

where $\pi(x, t) = tx$.
(a) The map π is well-defined for $\|tx\| = t \in [0, 1]$.
(b) The map π is continuous because $\lim_{(t_n, x_n) \to (t, x)} (t_n x_n) = tx$.
(c) π is injective since if $(x_1, t_1), (x_2, t_2) \in S^1 \times I$ are such that $t_1 x_1 = t_2 x_2 \neq 0$, then

$$
\|t_1 x_1\| = \|t_2 x_2\| \quad \Longleftrightarrow \quad t_1 = t_2,
$$

which implies $x_1 = x_2$.
(d) The map π is surjective. Given some $y \in B^2$, let $t = \|y\| \in I$ and consider two cases.
   (i) If $y = 0$, then $\pi(x, 0) = 0 = y$ for all $x \in S^1$.
   (ii) If $y \neq 0$, then $\pi(\frac{1}{\|y\|} y, \|y\|) = y$, proving the claim.
(e) The map π is a closed map. Let $C = C_1 \times C_2$ be closed in $S^1 \times I$ and $D = \pi(C)$. Let $(y_n)_n \subset D$ be a sequence, which converges to some limit $y \in B^2$. Then there exist sequences $x_n \in C_1$ and $t_n \in C_2$ such that $y_n = t_n x_n$. Hence, $\|y_n\| = t_n$ converges to $\|y\|$ in $\mathbb{R}$. Since $(x_n)_n \subset C_1$, there exists a subsequence $(x_n)_k$, which converges to some limit $x \in C_1$. Therefore, $\lim_{k \to \infty} y_{n_k} = \|y\| x$ in B^2, where $y_{n_k} = t_{n_k} x_{n_k}$. By uniqueness of the limit, $y = \|y\| x$, with $x \in C_1$ and $\|y\| \in C_2$. Finally, $y \in D$, showing the claim.
(f) We have $\pi^{-1}(\{0\}) = \{(x, t) \in S^1 \times I : tx = 0\} = \{(x, t) \in S^1 \times I : t = 0\} = S^1 \times \{0\}$.

(g) We have $\pi : S^1 \times I \longrightarrow B^2 \setminus \{(0,0)\}$ is a homeomorphism with inverse

$$\pi^{-1}(y) = (1/\|y\|, \|y\|), \quad \text{for all } y \in B^2 \setminus \{(0,0)\}.$$

Finally, define the map

$$\tilde{f}(x) = \begin{cases} (H \circ \pi^{-1})(x), & x \neq 0, \\ x_0, & x = 0. \end{cases}$$

As $x \to 0$, $\tilde{f}(x)$ converges to $H(S^1 \times I) = x_0 = \tilde{f}(0)$. In conclusion, $\tilde{f}$ is the required extension.

(b) *Second* (and shorter) *proof.* Identify S^1 with the set of complex numbers $x = e^{i\theta}$ and B^2 with the set of complex numbers $x = re^{i\theta}, 0 < r \leq 1$. Then just take $\tilde{f}(x) = H(e^{i\theta}, r)$.

(c) *Third proof.* (Using Chapter 7). Since by Proposition 7.3.15, the pair (B^2, S^1) has the HEP, just apply Corollary 7.3.9.

(2) $\Longrightarrow$ (3): Let $i : S^1 \hookrightarrow B^2$ be the inclusion map and $\tilde{f} : B^2 \longrightarrow X$ a continuous extension. Then $f = \tilde{f} \circ i : S^1 \longrightarrow X$ and so $f_* = (\tilde{f} \circ i)_* = \tilde{f}_* \circ i - *$, where $i_* : \pi_1(S^1, a) \longrightarrow \pi_1(B^2, a) = \{0\}$. Then i_* is trivial and so is $f_* : \pi_1(S^1, a) \longrightarrow \pi_1(X, x_0)$. Hence,

$$f_*([f]) = \tilde{f}_*(i_*([\phi])) = \tilde{f}_*(0)$$

for every loop ϕ in S^1, showing that $\tilde{f}_*$ is constant.

(3) $\Longrightarrow$ (1): Since f_* is trivial, $f \circ f \simeq e_{x_0}$ for every loop ϕ in S^1 based at a. In particular, for the loop p_0, there exists a homotopy $H : I \times I \longrightarrow X$ such that $H(s, 0) = f(p_0(s))$ and $H(s, 1) = H(0, t) = H(1, t) = x_0$ for all $s, t \in I$. Then H is compatible with the projection of the square $I \times I$ onto the cylinder $S^1 \times I$. Hence, there exists a homotopy $\widehat{H} : S^1 \times I \longrightarrow X$ such that $\widehat{H}(x, 0) = f(x)$ and $\widehat{H}(x, 1) = \widehat{H}(a, t) = x_0$ for all $t \in I$, $x \in S^1$. Therefore, $f \simeq e_{x_0}$. □

Example 6.2.10. The identity on S^1 is not null-homotopic because the induced homomorphism Id_* is the identity on $\mathbb{Z}$.

Example 6.2.11. Consider the inclusion map $i : S^1 \hookrightarrow \mathbb{R}^2 \setminus \{0\}$. Since $\pi_1(\mathbb{R}^2 \setminus \{0\}) \cong \pi_1(S^1) \cong \mathbb{Z}$, the induced homomorphism i_* is not trivial. Hence, i is not null-homotopic. Another way to see this result is to compose i with the radial retraction $r : \mathbb{R}^2 \setminus \{0\} \longrightarrow S^1$ given by $r(x) = x/\|x\|$. In fact since $r \circ i = i$, then $(r \circ i)_* = r_* \circ i_* = i_*$. This means that i_* is left invertible, hence injective but not the trivial homomorphism.

Corollary 6.2.12. *Let X be a path-connected space. Then X is simply connected if and only if every continuous map $f : S^1 \longrightarrow X$ extends continuously to $\tilde{f} : B^2 \longrightarrow X$. Equivalently, f is null-homotopic, i. e., for every such a map, the induced homomorphism $f_* : \pi_1(S^1, a) \longrightarrow \pi_1(X, x_0)$ is trivial.*

6.3 Fundamental group of the sphere

We start with a factorization theorem. This is a first version of the so-called Seifert–van Kampen theorem (Section 8.1). The proof relies on a known result in topology, namely Lemma A.3.1.

Theorem 6.3.1. *Let $X = U \cup V$, where U and V are open subsets. Suppose that U, V and $X_0 = U \cap V$ are nonempty and path-connected. Then, for every $x_0 \in X_0$ and every loop f in X based at x_0, there exists a finite number of loops $g_1, g_2, \dots, g_n$ based at x_0 such that*

$$[f] = [g_1] * [g_2] * \cdots * [g_n], \tag{6.2}$$

and each loop g_i $(1 \le i \le n)$ lies either in U or V.

Proof. Let $f : I \longrightarrow X$ be a loop based at x_0. By Lemma A.3.1, there exists a subdivision $0 = t_0 < t_1 < t_2 < \cdots < t_n < t_{n+1} = 1$ of $[0,1]$ such that for all $k \in [0, n+1]$, $f(t_k) \in X_0$. Moreover, either $f_k(t) \in X_1$ for all $t \in I$ or $f_k(t) \in X_2$ for all $t \in I$, where $f_k = f \circ \alpha_k$ and $\alpha_k(t) = t_{k+1}t + (1-t)t_k$ is the line homotopy joining the endpoints t_k and t_{k+1} of the interval $[t_k, t_{k+1}]$ $(k = 0, \dots, n)$. We have

$$\begin{aligned} f_k(0) &= f(\alpha_k(0)) = f(t_k) \in X_0, \\ f_k(1) &= f(\alpha_k(1)) = f(t_{k+1}) \in X_0, \\ f_{k+1}(0) &= f(\alpha_{k+1}(0)) = f(t_{k+1}) = f_k(1) \end{aligned}$$

and the product $f_k * f_{k+1}$ makes sense. We have

$$\begin{aligned} f_0 * f_1 * \cdots * f_n &= (f \circ \alpha_0) * (f \circ \alpha_1) * \cdots * (f \circ \alpha_n) \\ &= f \circ (\alpha_0 * \alpha_1 * \dots \alpha_n) \end{aligned}$$

because $f \circ (g * h) = (f \circ g) * (f \circ h)$ and recurrently for the product of n terms. The product $\alpha_0 * \alpha_1 * \dots \alpha_n$ is the concatenation of continuous line segments defined on the subdivision $(0, \frac{1}{2}, \frac{3}{4}, \frac{7}{8}, \dots, \frac{2^n-1}{2^n}, 1)$. By convexity of $\mathbb{R}$, this path is homotopic to the identity $\mathrm{Id}_{|[0,1]}$. Hence, $f_0 * f_1 * \cdots * f_n \simeq f$, i. e.,

$$[f] = [f_0 * f_1 * \cdots * f_n] = [f_0] * [f_1] * \cdots * [f_n].$$

For each $k \in [0, n-1]$, let γ_k be the path joining x_0 and $f(t_k)$ in the path-connected subspace X_0. Then

$$\begin{aligned} [f] &= [f_0] * [\overline{\gamma_1} * \gamma_1] * [f_1] * [\overline{\gamma_2} * \gamma_2] * \cdots * [\overline{\gamma_n} * \gamma_n][f_n] \\ &= [f_0 * \overline{\gamma_1}] * [\gamma_1 * f_1 * \overline{\gamma_2}] * \cdots * [\gamma_n * f_n]. \end{aligned}$$

Since for each k, f_k lies either in U or in V, the paths $f_0 * \overline{\gamma_1}$, $\gamma_1 * f_1 * \overline{\gamma_2}$, $\dots$, $\gamma_n * f_n$ lie entirely either in U or in V, as claimed. □

Remark 6.3.2. Let

$$\begin{aligned} i_* &: \pi_1(U, x_0) \hookrightarrow \pi_1(X, x_0), \\ j_* &: \pi_1(V, x_0) \hookrightarrow \pi_1(X, x_0) \end{aligned}$$

be the homomorphisms induced by the inclusion maps i and j, respectively. Theorem 6.3.1 states that $\pi_1(X, x_0)$ is generated by $i_*(\pi_1(U, x_0))$ and $j_*(\pi_1(V, x_0))$. Recall that a group G is generated by a family of elements $\{a_\alpha\}$ if every element of G can be written as a product of powers of the elements $\{a_\alpha\}$ (see Definition B.3.8).

Corollary 6.3.3. *Let $X = U \cup V$, where U and V are nonempty open sets, which are simply connected and $U \cap V$ is path-connected. Then X is simply connected.*

Proof. Since U and V are simply connected, the factor $[g_k]$ in the factorization (6.2) is trivial for each $k \in [0, n]$. Then so is $[f]$, which shows that $\pi_1(X, x_0)$ is a trivial group. The space X is path-connected as union of two path-connected spaces with nonempty intersection. Therefore, $\pi_1(X) = \pi_1(X, x)$ is the trivial group for every $x \in X$. □

Corollary 6.3.4. *For all $n \geq 2$, the sphere S^n is simply connected.*

Proof. First proof (using Corollary 6.3.3).

Let $N_p = (0, 0, \dots, 0, 1) \in \mathbb{R}^{n+1}$ and $S_p = (0, 0, \dots, 0, -1) \in \mathbb{R}^{n+1}$ be the northern pole and the southern pole, respectively. By Example 1.5.26, $U = S^n \setminus \{N_p\}$ and $\mathbb{R}^n$ are homeomorphic under the stereographic projection. Moreover, $S^n \setminus \{N_p\}$ and $S^n \setminus \{S_p\}$ are homeomorphic under the reflection map

$$h(x_1, x_2, \dots, x_{n+1}) = (x_1, x_2, \dots, x_n, -x_{n+1}).$$

Therefore, U and $V = S^n \setminus \{S_p\}$ are contractible, hence simply connected open sets. Again by the stereographic projection, the intersection $U \cap V = S^n \setminus \{N_p, S_p\}$ is homeomorphic to $\mathbb{R}^n \setminus \{0\}$, hence path-connected. By Corollary 6.3.3, we conclude that the sphere $S^n = U \cup V$ is simply connected for all $n \geq 2$.

Second proof (using Theorem 6.3.1). Let f be a loop based at $x_0 \in S^n$ and $x \notin \{x_0, N_p, S_p\}$. Define the open subsets $U = S^n \setminus \{x, N_p\}$ and $V = S^n \setminus \{x, S_p\}$. Then U and V are homeomorphic to $\mathbb{R}^n \setminus \{0\}$, thus path-connected. Likewise $U \cap V \cong S^n \setminus \{x, N_p, S_p\}$, which is homeomorphic to $\mathbb{R}^n \setminus \{0, 1\}$, is path-connected. By Theorem 6.3.1, f is homotopic to the product of loops $g = g_1 * g_2 * \cdots * g_n$, where each loop g_i $(1 \leq i \leq n)$ lies either in U or V. Hence, all of the g_i $(1 \leq i \leq n)$ and g miss the point x. This means that the loop g lies in $S^n \setminus \{x\}$, which is homeomorphic to the contractible Euclidean space $\mathbb{R}^n$. Therefore, f and g are null-homotopic. □

Remark 6.3.5. In the proof of Corollary 6.3.4, we have also proved that one can homotope any loop on the sphere S^n $(n \geq 2)$ to obtain a nonsurjective loop.

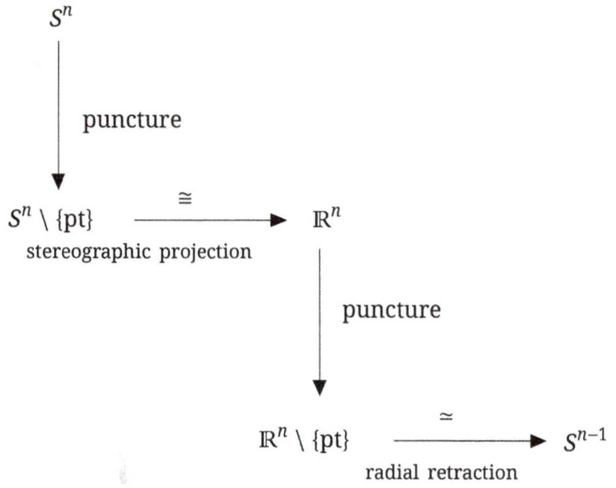

Figure 6.1: Reduction of dimension of the sphere after two punctures.

Remark 6.3.6. The proof of Corollary 6.3.4 shows that for $n \geq 2$, the double-punctured sphere $S^n \setminus \{N_p, S_p\}$ is homeomorphic to the punctured space $\mathbb{R}^n \setminus \{0\}$, which can directly be deduced from the stereographic projection. By Theorem 1.5.38, these spaces are also homeomorphic to the cylinders $S^{n-1} \times (0, \infty)$ and $S^{n-1} \times (0, 1)$, providing useful information for the subsequent computations of the fundamental groups (see also Remark 1.5.39).

Remark 6.3.7. Example 4.2.9 both with Remark 6.3.6 can be summarized in Figure 6.1. The diagram shows the effects of the stereographic projection and the radial retraction on the reduction of dimension of S^n.

Example 6.3.8. By Example 1.5.13, Example 4.2.6, Example 4.2.9, Corollary 4.5.5, Theorem 6.1.4 and Corollary 6.3.4, we have

$$\begin{aligned}\pi_1(\mathbb{R}^n \setminus \{0\}) &\cong \pi_1(S^{n-1} \times [0,1]) \\ &\cong \pi_1(\mathring{B}^n \setminus \{0\}) \\ &\cong \begin{cases} \mathbb{Z}, & n = 2, \\ 0, & n = 1 \text{ or } n > 2. \end{cases}\end{aligned}$$

The following application comes from [66, Corollary 11.28]. The case $n = 2$ will be discussed in Proposition 8.2.18.

Corollary 6.3.9. *The complement in $\mathbb{R}^n$ ($n \geq 3$) of a finite number of points is simply connected.*

Proof. Let $X = \mathbb{R}^n \setminus \{x_1, x_2, \ldots, x_k\}$, $k \geq 1$. We proceed by induction on k. For $k = 1$, X is homeomorphic to $S^n \setminus \{N_p, S_p\}$ by Remark 6.3.6. Example 6.3.8 and Corollary 6.3.4 imply that X is simply connected. Suppose the result is true for all $m < k$ and let $f : \mathbb{R}^n \longrightarrow \mathbb{R}$ be a linear map such that $f(x_1) \neq f(x_2)$, say $f(x_1) < f(x_2)$. Then $X = U \cup V$, with the open sets

$$U = \{x \in X : f(x) < f(x_2)\} \quad \text{and} \quad V = \{x \in X : f(x) > f(x_1)\}.$$

Clearly, $U = f^{-1} = \{-\infty, f(x_2)\}$ are $V = f^{-1} = \{f(x_1), \infty\}$ are nonempty open subsets of X. By the inductive assumption, U and V are simply connected. For some $m \leq k$, the intersection $U \cap V \cong \mathbb{R}^n \setminus \{x_1, x_2, \ldots, x_m\}$ is path-connected by supposition. By Corollary 6.3.3, $X = U \cup V$ is s imply connected. □

Applying stereographic projections, we get the following.

Corollary 6.3.10. *For $n \geq 2$ and $k \geq 1$, the sphere with k points removed $S^n \setminus \{x_1, x_2, \ldots, x_k\}$ is simply connected.*

Using the covering projection map of Exercise 8, Chapter 5 and Theorem 6.1.10, we derive the fundamental groups of the complex projective space.

Corollary 6.3.11.

$$\pi_1(\mathbb{C}\mathrm{P}^n) \cong \begin{cases} \pi_1(S^2), & \text{if } n = 1, \\ \pi_1(S^{2n+1}), & \text{if } n \geq 2. \end{cases}$$

This with Corollary 6.3.4 yield

Corollary 6.3.12. *For all $n \geq 1$, $\mathbb{C}\mathrm{P}^n$ is simply connected.*

6.4 Maps on the sphere and the Borsuk–Ulam theorem

In order to prove the Borsuk–Ulam theorem for the sphere in Euclidean space, we start with the following extension theorem, which generalizes Theorem 6.2.9 for higher dimensions. The result will be used in the definition of the higher homotopy groups (Section 9.1). The general definition of the degree of maps from S^n to S^n will be investigated in Section 12.5.

Theorem 6.4.1. *Let X be a space and $n \geq 1$. The following statements are equivalent:*
(1) *A continuous map $f : S^n \longrightarrow X$ is null-homotopic.*
(2) *The map f extends continuously to $\tilde{f} : B^{n+1} \longrightarrow X$.*

Proof. (1) $\Longrightarrow$ (2): *First proof.* Let $H : S^n \times I \longrightarrow X$ be continuous and satisfies $H(x, 0) = f(x)$ and $H(x, 1) = x_0$ for all $x \in S^n$. Define

$$\widetilde{f}(x) = \begin{cases} x_0, & 0 \le \|x\| \le 1/2, \\ H(x/\|x\|, 2 - 2\|x\|), & 1/2 \le \|x\| \le 1. \end{cases}$$

Since for $\|x\| = 1/2, H(2x, 1) = x_0$, by the pasting lemma A.2.1, $\widetilde{f}$ is continuous. For $\|x\| = 1$, we have $\widetilde{f}(x) = H(x,) = f(x)$. Hence, $\widetilde{f}$ is the required extension.

Second proof. By anticipation, Proposition 7.3.15 states that the pair (B^{n+1}, S^n) has the HEP. So, just apply Corollary 7.3.9.

(2) $\Longrightarrow$ (1): Let $\widetilde{f}$ be a continuous extension of f and let $x_0 \in S^n$. Then the map $H : S^n \times I \longrightarrow X$ defined by $H(x, t) = \widetilde{f}((1-t)x + tx_0)$ makes sense because $(1-t)x + tx_0 \in B^{n+1}$ for all $x \in S^n$. H is further a homotopy between f and the constant map $\widetilde{f}(x_0)$. □

Remark 6.4.2. Theorem 6.4.1 also derives from Theorem 3.3.6 both with Exercise 21, Chapter 4.

Before we discuss some properties of antipode-preserving maps, we recall an extension result for null-homotopic maps (Definition 4.1.1) due to K. Borsuk, when the target space Y is the sphere S^n. For the proof, we refer to [46, Theorems 4–5].

Proposition 6.4.3. *Let E be a separable metric space, $A \subset E$ a closed subset, and $f : A \longrightarrow S^n$ a null-homotopic map. Then f has a null-homotopic extension $\widetilde{f} : E \longrightarrow S^n$.*

Definition 6.4.4. A map $f : S^n \longrightarrow \mathbb{R}^n$ is said to be *antipode-preserving* if $f(x) = -f(-x)$ for all $x \in S^n$.

Example 6.4.5. In S^1, a rotation of angle θ is antipode-preserving.

The following result is known as the Borsuk–Ulam antipode theorem (see, e. g., [36, Theorem 6.4], [76, Theorem 57.1]).

Theorem 6.4.6. *For every continuous map $f : S^2 \longrightarrow \mathbb{R}^2$, there exists $x \in S^2$ such that $f(x) = f(-x)$.*

Proof. The map $g(x) = f(x) - f(-x)$ is antipode-preserving and continuous. If the conclusion of the theorem does not hold, then $g(x) \neq 0$ for all $x \in S^2$. Since $\frac{g(1,0,0)}{\|g(1,0,0)\|} \in S^1$, let θ be the angle of the rotation, which maps $\frac{g(1,0,0)}{\|g(1,0,0)\|} \in S^1$ to a, i. e., $\frac{g(1,0,0)}{\|g(1,0,0)\|} e^{i\theta} = a$. Define a function $h : B^2 \longrightarrow S^1$ by

$$h(x_1, x_2) = \frac{g(x_1, x_2, \sqrt{1 - \|x\|^2})}{\|g(x_1, x_2, \sqrt{1 - \|x\|^2})\|} e^{i\theta}.$$

Then h is continuous, odd on S^1 and satisfies $h(a) = a$. Let $h_0 = h \circ p_0$, where $p_0(t) = e^{2\pi i t}$, $t \in [0, 1]$ is the covering map given by Theorem 5.1.13. Then $h_0(0) = h_0(1) = a$, i. e., h_0 is a loop based at a. We claim that the winding number $\omega(h_0)$ of h_0 (Definition 5.2.3) is both zero and nonzero, leading to a contradiction.

Claim 1: $\omega(h_0) = 0$. Let e_1 be the constant loop based at a in S^1. Since $e_1 \simeq p_0$ and $(h \circ e_1)(t) = h(a) = a$ for all $t \in I$, by Corollary 4.1.11 $e_1 = h \circ e_1 \simeq h \circ p_0 = h_0$. By

Proposition 6.1.1, h_0 and e_1 have same winding number. Using Example 5.2.4, we conclude that $\omega(h_0) = \omega(e_1) = \widehat{e_1}(1) = 0$.

Claim 2: $\omega(h_0) \neq 0$. Let $\widehat{h_0}$ be the lifting map of h_0, i. e., $\widehat{h_0}(0) = 0$ and $p_0 \circ \widehat{h_0} = h_0 = h \circ p_0$. Since h is odd on S^1, we have for all $t \in (0, 1/2)$,

$$\begin{aligned} e^{2\pi i \widehat{h_0}(t+1/2)} &= (h \circ p_0)(t + 1/2) = h(e^{2\pi i(t+1/2)}) \\ &= h(-p_0(t)) = -h(p_0(t)) = -h_0(t) \\ &= -e^{2\pi i \widehat{h_0}(t)} = e^{\pi i(2\widehat{h_0}(t)+1)}. \end{aligned}$$

Hence for every $t \in (0, 1/2)$, there exists an integer $n_t \in \mathbb{Z}$ such that

$$2\widehat{h_0}(t + 1/2) = 2\widehat{h_0}(t) + 1 + 2n_t,$$

where the function $n_t(t) = \widehat{h_0}(t+1/2) - \widehat{h_0}(t) - 1/2$ is continuous because $\widehat{h_0}$ is continuous. Additionally, n_t takes values in the set of integers $\mathbb{Z}$. Thus, n_t is constant, i. e., $n_t = n \in \mathbb{Z}$ for all $t \in (0, 1/2)$. In particular, for $t = 0$ and $1/2$, $\widehat{h_0}(1/2) - 1/2 = n$ and $\widehat{h_0}(1) - \widehat{h_0}(1/2) - 1/2 = n$, respectively. Therefore, $\widehat{h_0}(1) = 2n + 1$ and then $\omega(h_0) = \widehat{h_0}(1) = 2n + 1 \neq 0$. □

Remark 6.4.7. In the proof of Theorem 6.4.6, there is a possibility to get rid of the rotation of angle θ in the definition of the function h. This consists in appealing to Exercise 6 of this chapter. Then the function h is homotopic to a continuous map $\widetilde{h}$ satisfying $\widetilde{h}(a) = a$. Therefore, a contradiction is reached with the function $\widetilde{h}$. However, $\widetilde{h_0} = \widetilde{h} \circ p_0$ is homotopic to $h_0 = h \circ p_0$. Thus, h_0 and $\widetilde{h_0}$ have same winding number by Proposition 6.1.2.

In Claim 2 of the proof of Theorem 6.4.6, we proved that given an odd function h on S^1, its degree (the winding number of the loop h_0), is an odd number. Thus, we state the Borsuk theorem.

Theorem 6.4.8. *Let $f : S^1 \longrightarrow S^1$ be a continuous antipode-preserving map. Then* $\deg f$ *is odd.*

Example 6.4.9. In S^1, a rotation R_θ of angle θ has odd degree. Actually, we will prove in Proposition 12.5.5 that $\deg R_\theta = 1$.

Remark 6.4.10. Assuming a spherical geometry of the earth, one can take as function $f(x) = (T, p)$, where T, p are the temperature and the pressure at position x. Theorem 6.4.6 says that there are at least two antipodal points on the earth where the temperature and the pressure are the same.

Corollary 6.4.11. *For $1 \leq i \leq 3$, let C_i be three closed subsets such that $\mathbb{R}^2 = \bigcup_{i=1}^{i=3} C_i$. Then at least of one of these subsets contains a pair of antipodal points.*

Proof. Define the map $f : S^2 \longrightarrow \mathbb{R}^2$ by $f(x) = (\mathrm{d}(x, C_1), \mathrm{d}(x, C_2))$. By continuity of the distance and Theorem 6.4.6, there exists at least one point $x \in S^2$ such that $\mathrm{d}(x, C_i) =$

$d(-x, C_i)$, for $i = 1, 2$. Hence, $x \in C_i$ if and only if $-x \in C_1$ for all $i = 1, 2$. If x and $-x$ are neither in C_1 nor in C_2, then the set of antipodal points $\{x, -x\}$ lies in C_3, proving the claim. □

There are several equivalent formulations of the Borsuk–Ulam theorem. Some versions are presented here.

Theorem 6.4.12. *Let $f : S^2 \longrightarrow \mathbb{R}^2$ be continuous and antipode-preserving. Then there exists $x \in S^2, f(x) = 0$.*

Proof. By Theorem 6.4.6 and the hypothesis, there exists $x \in S^2$ such that $f(x) = f(-x)$ and $f(x) = -f(-x)$. Then $f(x) = 0$. □

Theorem 6.4.13. *Theorem* 6.4.12 *is equivalent to Theorem* 6.4.6.

Proof. (1) Theorem 6.4.12 is a consequence of Theorem 6.4.6.

(2) Conversely, assume that Theorem 6.4.12 hold and let $g(x) = f(x) - f(-x)$. Then $g(x) = -g(-x)$ for all $x \in S^2$ and there exists $x \in S^2$ such that $g(x) = 0$, i. e., $f(-x) = f(-x)$, proving Theorem 6.4.6. □

Theorem 6.4.14. *There is no continuous antipode-preserving map $f : S^2 \longrightarrow S^1$.*

Proof. If such a continuous antipode-preserving map exists, then Theorem 6.4.12 implies that $f(x) = 0$ for some $x \in S^2$ but this is impossible since $0 \notin S^1$. □

Theorem 6.4.15. *Theorem* 6.4.14 *is equivalent to Theorem* 6.4.6.

Proof. (1) Theorem 6.4.14 follows from Theorem 6.4.12, which is equivalent to Theorem 6.4.6.

(2) Conversely, let Theorem 6.4.14 hold and assume by contradiction that $f(x) \neq -f(-x)$ for all $x \in S^2$. The function $g(x) = f(x) - f(-x)$ is antipode-preserving and satisfies $g(x) \neq 0$ for all $x \in S^2$. Thus we may define the map $h : S^2 \longrightarrow S^1$ by $h(x) = \frac{g(x)}{\|g(x)\|}$, which is odd and antipode-preserving, contradicting the hypothesis of Theorem 6.4.14. □

Theorem 6.4.16. *There is no continuous map $f : B^2 \longrightarrow S^1$, which is antipode-preserving on S^1.*

Proof. Define the map $g : S^2 \longrightarrow S^1$ by $g(x_1, x_2, x_3) = f(x_1, x_2)$. If f is antipode-preserving on S^1, then so would be g, contradicting Theorem 6.4.14. □

Theorem 6.4.17. *Theorem* 6.4.16 *is equivalent to Theorem* 6.4.12.

Proof. (1) Theorem 6.4.16 follows from Theorem 6.4.14, which is equivalent to Theorem 6.4.12.

(2) Conversely, let Theorem 6.4.16 hold and, by contradiction, let $f : S^2 \longrightarrow \mathbb{R}^2$ be continuous and $f(x) \neq 0$ for all $x \in S^2$. Define the function $g : S^2 \longrightarrow S^1$ by $g(x) = \frac{f(x)}{\|f(x)\|}$. Then g is continuous and antipode-preserving, for f is. This contradicts Theorem 6.4.16. □

Theorem 6.4.18. *Let $f : S^1 \longrightarrow S^1$ be a continuous antipode-preserving map. Then f is not null-homotopic.*

Proof. If f is null-homotopic, then by Theorem 6.2.9 f extends continuously to $\widetilde{f} : B^2 \longrightarrow X$. Since $\widetilde{f}$ is antipode-preserving on S^1, Theorem 6.4.16 is contradicted. □

Theorem 6.4.19. *Theorem* 6.4.18 *is equivalent to Theorem* 6.4.14.

Proof. (1) Theorem 6.4.18 follows from Theorem 6.4.16, which is equivalent to Theorem 6.4.14.

(2) Conversely, let Theorem 6.4.18 hold and, by contradiction, let $f : S^2 \longrightarrow S^1$ be a continuous antipode-preserving map. Then $g = f_{|S^1} : S^1 \longrightarrow S^1$ is a continuous antipode-preserving map. By hypothesis, g is not null-homotopic. Let $h = f_{|S^2_+}$, where S^2_+ is the closed northern hemisphere. The latter hemisphere is homeomorphic to the closed ball B^2 by Example 1.5.22. Then h is a continuous antipode-preserving map and $h = \widetilde{g} : B^2 \longrightarrow S^1$. By Theorem 6.2.9, g is null-homotopic, a contradiction. □

Theorems 6.4.6–6.4.18 provide five equivalent versions of the Borsuk theorem for some classes of continuous maps.

Corollary 6.4.20. *The following statements, which are equivalent, hold true:*
(1) *If $f : S^2 \longrightarrow \mathbb{R}^2$ is antipode-preserving, then there exists $x \in S^2$ such that $f(x) = f(-x)$.*
(2) *If $f : S^2 \longrightarrow \mathbb{R}^2$ is antipode-preserving, then there exists $x \in S^2$ such that $f(x) = 0$.*
(3) *If $f : S^2 \longrightarrow S^1$, then f is not antipode-preserving.*
(4) *If $f : B^2 \longrightarrow S^1$, then $f_{|S^1}$ is not antipode-preserving.*
(5) *If $f : S^1 \longrightarrow S^1$ is antipode-preserving, then f is not null-homotopic.*

An intuitively predicted result is the following.

Corollary 6.4.21. *The sphere S^2 cannot be embedded in the Euclidian plane.*

The one-dimensional case (see Exercise 6, Section 1.5) follows from connectedness arguments only.

Borsuk–Ulam (Theorem 6.4.8) is a slightly more general than Theorem 6.4.18, which follows as a direct consequence.

Exercises (Chapter 6)

1. Let $f : S^1 \longrightarrow S^1$ be a continuous map such that $f \circ f$ is null-homotopic. Show that the induced homomorphism $f_* : \mathbb{Z} \longrightarrow \mathbb{Z}$ is trivial.

2. Let $a : S^1 \longrightarrow S^1$ (or $a : B^2 \longrightarrow B^2$) be given by $a(x) = -x$. Show that a and the identity map are homotopic (a is called the antipodal map; see Definition 12.5.20). (The result will be proved for higher dimensions in Corollary 12.5.26).

3. Let $f : S^1 \longrightarrow S^1$ be an embedding. Prove that f is a homeomorphism.

4. Let X be a path-connected space.
(1) Prove that a space X is simply connected if and only if all continuous maps $f : S^1 \longrightarrow X$ are homotopic. (*Hint:* Use Theorem 6.2.9 both with the identification $S^1 \cong [0,1]/_{\{0,1\}}$.)
(2) Deduce that a continuous map $f : S^1 \longrightarrow \mathbb{R}^2$ such that $f(x) \neq 0$ for all $x \in S^1$ is not null-homotopic.

5. (1) Let X be a path-connected, locally path-connected and simply connected space.
(a) Prove that every continuous map $f : X \longrightarrow S^1$ is null-homotopic.
(b) Deduce that for all $n \geq 2$, any continuous map $f : S^n \longrightarrow S^1$ is null-homotopic.
(2) Consider the more general case of a path-connected and locally path-connected space X such that $\pi_1(X)$ is finite.

6. Let R_θ be some rotation and ϕ a loop in S^1. Prove that ϕ and $R_\theta \circ \phi$ have the same winding number.

7. Prove that every continuous $f : S^1 \longrightarrow S^1$ (not necessarily surjective) is homotopic to a continuous map $g : S^1 \longrightarrow S^1$ with $g(a) = a$, where $a = (1,0)$ (*Hint:* Just make the rotation $R(z) = \frac{a}{f(a)} z$ and show that R has degree 1, a result to be proved in Proposition 12.5.5.)

8. Let $f : S^1 \longrightarrow S^1$ be a null-homotopic map. Prove the existence of points $x, y \in S^1$ such that $f(x) = x$ and $f(y) = -y$. (The result is presented for higher dimensions in Exercise 7, Chapter 12.)

9. Let $f : S^1 \longrightarrow S^1$ be a continuous map. Prove that f has at least $|\deg f - 1|$ fixed points.

10. Let $f : S^1 \longrightarrow S^1$ be a reflection map (see Definition 12.5.20). Prove that f has at least 2 fixed points. (*Hint:* Use Proposition 12.5.23.)

11. (1) Let X be a space and $f : X \longrightarrow S^1$ be a continuous map, which is not surjective. Prove that f is null-homotopic.
(2) Let $f : S^1 \longrightarrow S^1$ be a continuous map such that $\deg f \neq 0$. Deduce that f is surjective. (The result will be proved for higher dimensions in Exercise 8, Chapter 12.)

12. Let $\widehat{f}(t) = 8\pi t(1-t) : [0,1] \longrightarrow \mathbb{R}$ and $f(t) = e^{2i\pi \widehat{f}(t)} : [0,1] \longrightarrow S^1$.
(1) Show that f induces a continuous surjective map $\overline{f} : [0,1]/_{\{0,1\}} \longrightarrow S^1$.
(2) Prove that $\widehat{f}$ is the lifting map of f with $\widehat{f}(0) = 0$.
(3) Deduce that $\deg \overline{f} = 0$. Conclude.

13. Let $f : S^1 \longrightarrow S^1$ be given by

$$f(z) = \begin{cases} z^{2k}, & \text{if } z \in S^1_+, \\ z^{-2k}, & \text{if } z \in S^1_-, \end{cases}$$

where $k \in \mathbb{Z}$ and S^1_+, S^1_- refer to the northern hemisphere and the southern hemisphere, respectively. Prove that f is continuous, surjective and null-homotopic. Conclude. (*Hint:* See [36, Exercise 1, Section 1, Chapter IV)].)

14. (1) Let $f : S^1 \longrightarrow S^1$ be a continuous map such that $\deg f \neq 1$. Prove that f has a fixed point $x \in S^1$. (*Hint:* Use Exercise 9, this section.)
(2) Deduce that every null-homotopic continuous map $f : S^1 \longrightarrow S^1$ has at least one fixed point. (This result will be extended to any dimension in Corollary 12.5.29.)
(3) Deduce that the algebraic equation $z^n = 1$ has n roots in $\mathbb{C}$. (*Hint:* Use Proposition 12.5.8.)

15. Let $f : \mathbb{T}^2 \longrightarrow S^1$ be a continuous map, where $\mathbb{T}^2$ is the 2-dimensional torus. Prove that $f(x) \neq f(-x)$ for all $x \in \mathbb{T}^2$.

16. Let $f : S^2 \longrightarrow S^1$ be a continuous antipode-preserving map. By identifying $\mathbb{RP}^n$ and $S^n/\sim$ (Definition 3.1.1), consider the diagram

$$\begin{array}{ccc} S^2 & \xrightarrow{\quad f \quad} & S^1 \\ {\scriptstyle \pi_2}\downarrow & & \downarrow{\scriptstyle \pi_1} \\ \mathbb{RP}^2 & \xrightarrow{\quad \overline{f} \quad} & \mathbb{RP}^1 \end{array}$$

where $\pi_k : S^k \longrightarrow \mathbb{RP}^k$ $(k = 1, 2)$ are covering maps (see Proposition 5.1.7).
(1) Check that the diagram is commutative and that $\overline{f}$ is continuous (see Exercise 6, Chapter 4).
(2) Write the diagram for the induced homomorphisms.
(3) Let $\gamma_2 \in S^2$ be a path such that $\gamma_2(0) = (0, 0, 1)$ and $\gamma_2(1) = (0, 0, -1)$. Prove that $\gamma_1 = \overline{f} \circ \pi_2 \circ \gamma_2$ is a loop.
(4) Prove that $\widehat{\gamma_1} = f \circ \gamma_2$, where $\widehat{\gamma_1}$ is the lifting map of γ_1.
(5) Prove that γ_1 is not null-homotopic.
(6) Determine the induced homomorphism $(\overline{f})_*$ and prove that $\overline{f}$ is null-homotopic.

(Exercise 12 provides an alternative proof to Theorem 6.4.14.)

17. Let $f : S^1 \longrightarrow S^1$ be a continuous antipode-preserving map and $p : S^1 \longrightarrow S^1$ the second power covering map $p(z) = z^2$.
(1) Prove that:
  (a) $p \circ f$ is constant on the fibers $p^{-1}(z)$, $z \in S^1$;
  (b) there exists a continuous map $h : S^1 \longrightarrow S^1$ such that $h \circ p = p \circ f$. (*Hint:* Use Theorem 2.1.11.) Let h_* be the induced homomorphism.
(2) Let $\widetilde{g}$ be a path such that $\widetilde{g}(0) = -\widetilde{g}(1)$ and $g = p \circ \widetilde{g}$. Prove that:
  (a) $h_*([g]) = [(f \circ \widetilde{g})^2]$;
  (b) $f \circ \widetilde{g}$ is not trivial;

(c) h_* is injective.

(3) Deduce that

(a) f_* is injective; (*Hint:* Use Corollary 5.2.11.)

(b) f is not null-homotopic.

(Note: This exercise provides an alternative proof to Theorem 6.4.18.)

18. Let $f : \Omega \subseteq \mathbb{R}^n \longrightarrow \mathbb{R}^2$ be a continuous map such that Ω is open and $n \geq 3$. Prove that f is not one-to-one.

19. Let $f : \mathbb{RP}^n \longrightarrow \mathbb{R}^2$ be a continuous map with some $n \geq 2$. Prove that f is null-homotopic.

20. Let $f : \mathbb{RP}^n \longrightarrow \mathbb{T}^n$ be a continuous map with some $n \geq 2$.

(1) Prove that there exists a lifting map $\widehat{f} : \mathbb{RP}^n \longrightarrow S^n$ such that $p \circ \widehat{f} = f$, where p is a universal covering map.

(2) Prove that $\widehat{f}$ and then f are null-homotopic. (*Hint:* Use Theorem 5.2.16.)

21. Prove that the inclusion map $S^1 \hookrightarrow \mathbb{R}^2 \setminus \{0\}$ is not null-homotopic.

22. Let $f : S^n \longrightarrow X$ be continuous map and X a contractible space. Prove that f has a continuous extension $\widetilde{f} : B^{n+1} \longrightarrow X$.

23. (1) Let X be a space and $f, g : X \longrightarrow S^n$ two continuous maps such that $f(x)+\lambda g(x) \neq 0$ for all $x \in X$ and all $\lambda > 0$. Show that f and g are homotopic.

(2) (a) Deduce a second proof that for $n \geq 1$, S^n is path-connected (Example 1.4.23 provides a direct proof.)

(b) Let $x_0 \in S^n$ and $X = S^n \setminus \{x_0\}$. Show that $\mathbb{R}^{n+1} \setminus X$ is path-connected.

24. For some space X and $n \geq 1$, let $f : X \longrightarrow S^n$ be continuous and not surjective map (i. e., f does not fill the entire sphere). Prove that f is null-homotopic. (Note: For $X = S^1$ and $n = 1$, this is Exercise 11 of this chapter.) (*Hint:* Use either Example 1.5.26 and Corollary 4.9.9 or question (1) of Exercise 23 above with $g(x) = -x_0$ for some $x_0 \in S^n$.)

25. Let $f : S^k \longrightarrow S^n$ be a continuous map and $k < n$.

(1) Prove that f is null-homotopic. (*Hint:* Show that f is homotopic to a nonsurjective map and use Exercise 24 above. Also see [25, Theorem 4.2, Chapter XVI], [46, Corollaries 6–41], [83, Theorem 7.5]. In Proposition 9.2.4, a direct proof using higher homotopy groups is provided.)

(2) Deduce that f extends continuously to $\widetilde{f} : B^{k+1} \longrightarrow S^n$. (*Hint:* use Theorem 6.4.1).

26. Let $f : I \longrightarrow S^n$ be a path with end-points x_0, x_1. For every $x \neq x_0, x_1$, prove the existence of a path $g : I \longrightarrow S^n \setminus \{x\}$ homotopic to f. This is called the free point lemma. (*Hint:* See [91, Lemma 2.6].)

27. Let X be connected and locally path-connected and $f : X \longrightarrow S^1$ a continuous map. Prove that there exists a lifting $\widehat{f} : X \longrightarrow \mathbb{R}$ if and only if the induced homomorphism f_* is trivial.

28. Let $f : \mathbb{RP}^n \longrightarrow \mathbb{RP}^n$ be a continuous map for some $n \geq 2$ and $\pi : S^n \longrightarrow \mathbb{RP}^n$ the canonical surjective map. Prove that there exists a unique lifting map $\widehat{g} : S^n \longrightarrow S^n$ such that $\pi \circ \widehat{g} = g = f \circ \pi$. (*Hint:* Use Proposition 5.1.7 and Theorem 5.2.16.)

29. (1) Let X be path-connected and $x_0, x_1 \in X$. Prove that the following statements are equivalent:
(a) $\pi_1(X)$ is Abelian.
(b) Any two paths f, g from x_0 to x_1 have same lifting maps. (*Hint:* See [76, Exercise 3, Section 52].)
(2) Prove that $\pi_1(S^1)$ and $\pi_1(\mathbb{T}^2)$ are Abelian.

30. (1) Prove that $\mathbb{T}^m$ and S^n are homeomorphic if and only if $m = n = 1$ (*Hint:* Consider the fundamental groups.)
(2) Deduce that for all $n \geq 2$, S^n is not homeomorphic to $(S^1)^n$ (see Example 2.3.12).

7 Borsuk's theory of retracts

Apart from the circle and the sphere, the computations of the fundamental groups of some spaces and manifolds require the knowledge of some additional topological properties. The retraction theory was introduced by Karol Borsuk in the 1930s in his thesis and plays a central role in the computation of the fundamental groups. In this chapter, the primary definitions of the retraction map and the retract set are reported in Section 7.1. Several properties of deformation retracts are investigated in 7.2. Section 7.3 is devoted to the homotopy extension property of pairs, also due to Borsuk. The retracts of the adjunctions spaces are described in Section 7.4. The concept of absolute retract is presented in Section 7.5. However, a substantial application of Chapter 4 and Chapter 6 is the nonretraction of the disk, equivalent to Brouwer's fixed-point theorem for the disc. This is discussed in Section 7.6.

7.1 Retract set and retraction map

7.1.1 Main properties

Definition 7.1.1. Let X be a space and $A \subset X$ a nonempty subset.

(1) The set A is called a *retract* of X if there exists a continuous map $r : X \longrightarrow A$ such that $r(x) = x$, for all $x \in A$, that is to say $r = \mathrm{Id}_{|A}$, where $\mathrm{Id}_{|A}$ is the identity map on A.

(2) The map r is said to be a *retraction*.

Remark 7.1.2. (1) Let $i : A \hookrightarrow X$ be the inclusion map (or the canonical injection) on A defined by $i(x) = x$ for all $x \in A$ and A a retract of X. Then i is continuous and $r \circ i = \mathrm{Id}_{|A}$. Hence, r is right invertible.

(2) Thus, a retraction is a continuous map $r : X \longrightarrow A$ such that $r \circ i = \mathrm{Id}_{|A}$.

Example 7.1.3. Every space retracts onto any of its points $A = \{x_0\}$ with constant retraction $r(x) = x_0$.

Example 7.1.4. Let X be a normed space and $A = B(x_0, R)$ a closed ball centered at $x_0 \in X$ with radius R. Then A is a retract of X with retraction given by

$$\widetilde{r}(x) = \begin{cases} x, & \text{if } \|x - x_0\| \leq R, \\ x_0 + R\frac{x-x_0}{\|x-x_0\|}, & \text{if } \|x - x_0\| \geq R. \end{cases}$$

By the pasting lemma, $\widetilde{r}$ is continuous because x and $x_0 + R\frac{x-x_0}{\|x-x_0\|}$ agree for $\|x - x_0\| = R$. Moreover, $\|\widetilde{r}(x) - x_0\| \leq R$ for all $x \in X$ shows that $\widetilde{r} : X \longrightarrow A$. Hence, $\widetilde{r}$ is a retraction, called *the ball retraction*.

https://doi.org/10.1515/9783111517384-007

Example 7.1.5. In Euclidean space $X = \mathbb{R}^n$, the unit sphere

$$S^{n-1} = \{x \in \mathbb{R}^n : \|x\| = 1\}$$

is a retract of the punctured space $\mathbb{R}^n \setminus \{0\}$ with radial retraction $r(x) = x/\|x\|$.

Example 7.1.6. The unit circle S^1 is a retract of the wedge sum of two circles $S_1^1 \vee S_2^1$ with radial retraction $r(x) = x/\|x\|$, where $S_1^1 = C(0;1)$ and $S_2^1 = C(2;1)$ are copies of S^1. More generally, S^1 is a retract of any subset A such that $S^1 \subset A \subset \mathbb{R} \setminus \{0\}$ (see Subsection 7.1.3).

Example 7.1.7. Let X, Y be two spaces and $y_0 \in Y$. The projection along the first coordinate $X \times Y \longrightarrow X \times \{y_0\}$ is a retraction; hence, up to some homeomorphism every product of spaces, retracts to each one of the spaces.

Let X and Y be two disjoint spaces. The following result is immediate.

Proposition 7.1.8. *If A and B are retracts of X and Y, respectively, then*
(1) *$A \times B$ is a retract of $X \times Y$;*
(2) *$A \vee B$ is a retract of $X \vee Y$.*

Example 7.1.9. (1) $S^1 \cong S^1 \times \{(1,0)\}$ is a retract of the torus $\mathbb{T}^2 \cong S^1 \times S^1$. Actually, a retraction is merely the projection $r(x,y) = x$.
(2) The torus $\mathbb{T}^2$ is a retract of $\mathbb{T}^2 \times \mathbb{T}^2$.
(3) The wedge sum of two tori $\mathbb{T}^2 \vee \mathbb{T}^2$ retracts onto a figure eight.

Example 7.1.10. Let $(H, \langle\cdot\rangle)$ be a Hilbert space and $C \subset H$ a nonempty closed convex subset. As a consequence of Lemma A.6.1, the projection map P in H is a retraction from H to C. Indeed, P is continuous, $P^2 = P$, and P is surjective since if $y \in C$, $P(y) = y$ implies that $P(P(y)) = y$, i. e., $P(x) = y$, with $x = P(y) \in C$ (elements of C are obviously self-projected).

Example 7.1.11. A two-element set $A = \{x_1, x_2\}$ in $X = \mathbb{R}^2$ is not a retract of Euclidean plane. Otherwise, let $r : \mathbb{R}^2 \longrightarrow A$ be a retraction and $X_1 = r^{-1}(\{x_1\})$, $X_2 = r^{-1}(\{x_2\})$ the inverse image sets. Since r is continuous, X_1, and X_2 are closed sets. They are disjoint by definition and nonempty because $x_1 \in X_1$ and $x_2 \in X_2$. Since r is surjective, $X = X_1 \cup X_2$, infringing the connectedness of the plane.

Example 7.1.12. As in Example 7.1.11, $\{a, b\}$ is not a retract of the closed interval $[a, b]$. Indeed, by connectedness of the interval $[a, b]$, a retraction $r : [a, b] \longrightarrow \{a, b\}$ should be constant, contradicting $r(a) = a$ and $r(b) = b$ (see Definition 1.4.1(4)).

Example 7.1.13. The closed interval $[a, b]$ is a retract of the real line $\mathbb{R}$ under the retraction

$$r(x) = \begin{cases} x, & x \in [a, b], \\ a, & x \leq a, \\ b, & x \geq b. \end{cases}$$

Example 7.1.14. The open interval (a, b) is not a retract of the real line $\mathbb{R}$ because by continuity we must have $r(a) = a$ and $r(b) = b$.

The proof of the following result is left as an easy exercise (Exercise 10, this chapter).

Proposition 7.1.15. *A retraction is an identification map.*

Remark 7.1.16. We can see that identification maps encompass continuous surjective maps, which are open (or closed) (Proposition 2.1.3), continuous maps with right-inverse continuous maps (Corollary 2.1.6), quotient maps (Remark 2.2.3), covering maps (Corollary 5.1.3) and retraction maps (Proposition 7.1.15).

The retract property of sets is preserved under homeomorphism.

Theorem 7.1.17. *Let $h : X \longrightarrow X'$ be a homeomorphism and $A \subset X$ a retract of X. Then $A' = h(A)$ is a retract of X'.*

Proof. Let $r : X \longrightarrow A$ be a retraction and $r' = h \circ r \circ h^{-1}$:

$$X' \xrightarrow{\quad h^{-1} \quad} X \xrightarrow{\quad r \quad} A \xrightarrow{\quad h \quad} A', \qquad r' = h \circ r \circ h^{-1} : X' \longrightarrow A'$$

Then, for all $y \in A' = h(A)$, we have $h^{-1}(y) \in A$ and $r(h^{-1}(y)) = h^{-1}(y)$. As a consequence,

$$r'(y) = (h \circ r \circ h^{-1})(y) = h(r(h^{-1}(y))) = h(h^{-1}(y)) = y,$$

proving that A' is a retract of X'. □

Example 7.1.18. As in Example 7.1.5, the circle S^1 is a retract of the punctured closed disk $B^2 \setminus \{(x_0, y_0)\}$ under the radial retraction, where (x_0, y_0) is any interior point to the ball B^2. By Example 1.5.14, the disk B^2 is homeomorphic to the square I^2 with the boundary ∂B^2 mapped homeomorphically to ∂I^2. Hence, the boundary ∂I^2 is a retract of the punctured square $I^2 \setminus \{(x_0, y_0)\}$, providing an alternative proof of [90, Example 1.4.4.4], where a retraction $r : I^2 \setminus \{(0, 0)\} \longrightarrow \partial I^2$ is written down explicitly.

A further topological property of retract sets is given by the following result.

Theorem 7.1.19. *Let X be a Hausdorff space and $A \subset X$ a retract subset. Then A is a closed set.*

Proof. Let $A \subset X$ be a retract and $B = X \setminus A$ its complement. To show that B is open, let $x \in B$ and $y = r(x) \in A$. Since X is Hausdorff, there exist disjoint open sets $U \in N(x)$ and $V \in N(y)$ such that $U \cap V = \phi$. Let $W = U \cap r^{-1}(V)$. Since r is continuous, W is open. Since $x \in W$, it suffices to prove that $W \subset B$. Let $z \in W$. Then $z \in U$ and $r(z) \in V$. Since $U \cap V = \phi$, $z \notin V$ and $z \neq r(z)$. Hence, $z \notin A$, i. e., $z \in B$, as claimed. □

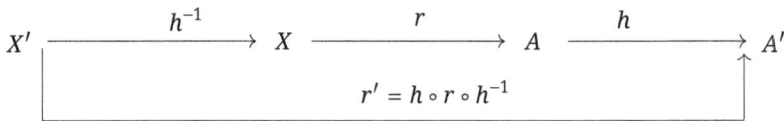

Remark 7.1.20. In general, the converse of Theorem 7.1.19 is not true (Corollary 7.1.38 provides a counterexample). Consider Example 7.1.3 with a space, which is not a T_1 space.

However, the following result generalizes Example 7.1.10.

Theorem 7.1.21. *If C is a closed convex subset of a normed space X, then C is a retract of X.*

Proof. (1) *First proof.* Let $\mathrm{Id}_{|C} : C \longrightarrow C$ be the restriction of the continuous identity map on C. By Dugundji's lemma A.5.2, $\mathrm{Id}_{|C}$ admits a continuous extension $\widetilde{\mathrm{Id}_{|C}} : X \longrightarrow X$ such that $\widetilde{\mathrm{Id}_{|C}}(X) \subset \overline{\mathrm{Co}}(\mathrm{Id}_{|C}(C)) = \overline{\mathrm{Co}}(C) = C$, for C is closed and convex. Let $r = \widetilde{\mathrm{Id}_{|C}}$. Then r is continuous and $r(x) = \mathrm{Id}_{|C}(x) = x$ for all $x \in C$. Hence, $r : X \longrightarrow C$ is a retraction.

(2) *Second proof.* Using Minkowski's functional $j : X \longrightarrow [0, \infty)$ (see Lemma A.8.1), we can see that the retraction r can be given explicitly by $r(x) = \frac{x}{\max(1, j(x))}$. We check the proof for some closed convex subset $C \subset \mathbb{R}^n$ and suppose, without loss of generality, that it contains the origin. Indeed, by Lemma A.8.1(3), $r(x) = x$ for every $x \in C$. In addition, $j(x) \geq 1$ for every $x \notin C$. Otherwise, by definition of the infimum, for all $\varepsilon > 0$, there exists $\lambda_\varepsilon > 0$ such that $x \in \lambda_\varepsilon C$ and $j(x) \leq \lambda_\varepsilon < j(x) + \varepsilon < 1$. As a consequence, $x = \lambda_\varepsilon y + (1 - \lambda_\varepsilon)0$ for some $y \in C$. Since C is convex, $x \in C$, a contradiction. Hence, $j(r(x)) = j(\frac{x}{j(x)}) = 1$ showing that $r(x) \in \partial C$, i. e., $r(X \setminus C) \subset \partial C \subset C$. Therefore, r is a retraction. □

Remark 7.1.22. (1) The sphere $A = S^{n-1}$ is not convex but it is a retract of the subspace $X = \mathbb{R}^n \setminus \{0\}$.

(2) Let $A = \partial B$ be the unit sphere in an infinite-dimensional normed space X. Then A is a retract of the closed ball B (Proposition 11.4.3), which is in turn a retract of X (Example 7.1.4). By Proposition 7.1.26, it follows that A is a retract of X. However, A is not convex since $\mathrm{Conv}(A) = B$ (Proposition A.1.7).

Parts (1) and (2) show that the converse of Theorem 7.1.21 does not hold true.

7.1.2 Characterizations

Theorem 7.1.23. *A continuous map $r : X \longrightarrow A$ is a retraction if and only if r is surjective and $r \circ r = r$.*

Proof. (1) Let r be a retraction. Then, for all $x \in A$, there exists $x' = x \in A$ such that $r(x') = x$. Thus, r is surjective and $r(x) \in A$ for all $x \in X$. Hence, $r(r(x)) = r(x)$ if and only if $r^2(x) = r(x)$ for all $x \in X$. Thus, r is a projection.

(2) Conversely, assume that $r : X \longrightarrow A$ is surjective and satisfies $r^2 = r$. Let $x \in A$. Since r is surjective, there exists $x' \in X$ such that $r(x') = x$. Hence, $r^2(x') = r(x)$ and since r is a projection ($r^2 = r$), $r(x') = r(x)$, i. e., $x = r(x)$. Hence, $r(x) = x$ for all $x \in A$, i. e., r is a retraction. □

The following result shows that the concept of retraction is intimately related to the extendability problem and provides a different characterization of a retract set.

Theorem 7.1.24. *Let $A \subset X$ be a subset of a space. Then*

$$A \text{ retract of } X \quad \Longleftrightarrow \quad \begin{cases} \text{for every space } Y, \text{ every continuous map } f : A \longrightarrow Y \\ \text{has a continuous extension } \widetilde{f} : X \longrightarrow Y, \end{cases}$$

meaning that $\widetilde{f} \circ i = f$, where $i : A \hookrightarrow X$ is the inclusion map:

$$\begin{array}{ccc} A & \xrightarrow{f} & Y \\ \downarrow{\scriptstyle i} & \nearrow{\scriptstyle \widetilde{f}} & \\ X & & \end{array}$$

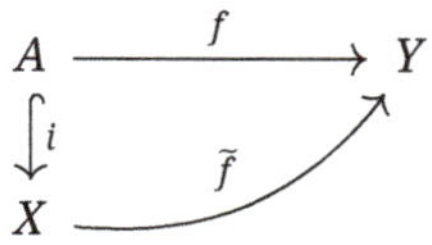

Proof. (1) If A is a retract of X, let $r : X \longrightarrow A$ be a retraction. Then $\widetilde{f} = f \circ r : X \longrightarrow A \longrightarrow Y$ is continuous and satisfies $\widetilde{f}(x) = f(r(x)) = f(x)$ for all $x \in A$.

(2) Conversely, let $\mathrm{Id}_{|A}$ be the identity map on A. By hypothesis, $\mathrm{Id}_{|A}$ has a continuous extension $\widetilde{\mathrm{Id}_{|A}} : X \longrightarrow A$. Moreover, $\widetilde{\mathrm{Id}_{|A}}(x) = \mathrm{Id}_{|A}(x) = x$, for all $x \in A$. Hence, the map $r = \widetilde{\mathrm{Id}_{|A}}$ is a retraction of X on A according to Definition 7.1.1. □

Theorem 7.1.25. *A space X is contractible if and only if X is a retract of the cone $C(X)$ over X.*

Proof. (1) Assume X to be contractible and let $H : X \times I \longrightarrow X$ be a homotopy between the identity on X and a point $x_0 \in X$. Then the map $r : C(X) \longrightarrow X$ given by $r([(x,t)]) = H(x,t)$ is the required retraction.

(2) Conversely, if the condition of the theorem holds, then the map $H : X \times I \longrightarrow X$ given by $H(x,t) = r([(x,t)])$ is the required homotopy.

An alternative way to prove the converse part is to appeal to Proposition 4.9.5, which states that the cone $C(X)$ of a space X is a contractible space. If X is a retract of the cone $C(X)$, then X is contractible by Theorem 7.1.29. □

7.1.3 Hereditary properties

It is clear from the definition that if $A \subset B \subset X$ and A is a retract of X, then A is a retract of B. The question is which properties can a retract set $A \subset X$ inherit from the whole space X? First, consider a "transitivity" property of retract sets.

Proposition 7.1.26. *If $A \subset B \subset X$, where A is a retract B and B is a retract of X, then A is a retract of X.*

Proof. The resulting retraction is just the composite of the two retractions. □

Remark 7.1.27. In general, if $A \subset B \subset X$ and B is a retract X, A is not necessarily a retract of X. For instance, the closed unit ball B^2 is a retract of Euclidean plane by Example 7.1.4. However, the sphere $S^1 \subset B^2$ is not a retract of Euclidean plane (see Example 7.1.39).

By Theorem 1.4.20, we have the following.

Theorem 7.1.28. *Let X be path-connected and $A \subset X$ a retract. Then A is path-connected.*

We suggest a direct proof.

Proof. Let $a_0, a_1 \in A$ and $f : [0,1] \longrightarrow X$ be a path joining a_0 and a_1 in X, i. e., $f(0) = a_0$ and $f(1) = a_1$. Let $r : X \longrightarrow A$ be a retraction and $g = r \circ f$:

$$
\begin{array}{ccc}
[0,1] & \xrightarrow{\ f\ } & X \\
& {\scriptstyle g}\searrow & \downarrow {\scriptstyle r} \\
& & A
\end{array}
$$

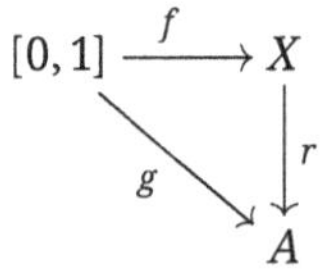

Then $g : [0,1] \longrightarrow A$ is continuous and satisfies $g(0) = r(f(0)) = r(a_0) = a_0$ and $g(1) = r(f(1)) = r(a_1) = a_1$, since $a_0, a_1 \in A$. Hence, g is a path joining a_0 and a_1 in A. □

Theorem 7.1.29. *Let X be contractible and $A \subseteq X$ a retract. Then A is contractible.*

Proof. (1) *First proof.* Since X is contractible, there exist $x_0 \in X$ and a homotopy H between Id_X and x_0. Let $G : A \times I \longrightarrow A$ be given by $G(x,t) = r(H_{|A \times I}(x,t))$, where $H_{|A \times I}$ is the restriction of H on $A \times I$. Then G is continuous, $G(x,0) = r(H_{|A \times I}(x,0)) = r(x) = x$ and for all $x \in A$, $G(x,1) = r(H_{|A \times I}(x,1)) = r(x_0)$, by definition of the homotopy H. Hence, G is a homotopy between $\mathrm{Id}_{|A}$ and the element $r(x_0) \in A$.

(2) *Second proof.* Since X is contractible, by Theorem 7.1.25, X is a retract of the cone $C(X)$ over X. Let $r : C(X) \longrightarrow X$ be such a retraction. Clearly, the restriction map $r_{|C(A)}$ is a retraction of $C(A)$ onto the subset A. Again Theorem 7.1.25 implies that A is contractible. □

Example 7.1.30. The comb space $X = Cb$ is not a retract of the square I^2 (see [92, Example 14.1.9]). By contradiction, assume that $r : I^2 \longrightarrow X$ is such a retraction and let $x_0 = (0, 1/2) \in X$. Then $r(x_0) = x_0$. Since r is continuous, there exists some $\delta > 0$ such that $r(I^2 \cap B(x_0, \delta)) \subset X \cap B(x_0, 1/4)$. Since $U = I^2 \cap B(x_0, \delta)$ is connected, the image by r is connected, too. Since $x_0 \in r(U)$, $r(U)$ is contained in the component of x_0, i. e., the set $E = \{(0, x_2) : 1/4 < x_2 < 3/4\}$. However, for $n = 1, 2, \dots (1/n, 1/2) \in U \cap X$ but $r(1/n, 1/2) = (1/n, 1/2) \notin E$, a contradiction.

Remark 7.1.31. The converse of Theorem 7.1.29 does not hold true as shows the comb space, which is contractible by Example 4.9.3. Also, the square I^2 is contractible. However, by Example 7.1.30, the comb space is not a retract of the square.

Theorem 7.1.32. *Let A be a retract of X and $a \in A$. Then the fundamental group $\pi_1(A, a)$ is isomorphic to a subgroup of the group $\pi_1(X, a)$.*

Proof. By Remark 7.1.2, we know that $r \circ i = \mathrm{Id}_{|A}$, where $r : X \longrightarrow A$ is a retraction, $i : A \hookrightarrow X$ the inclusion map, and $\mathrm{Id}_{|A}$ the identity map on A. Then the induced homomorphism

$$(r \circ i)_* = (\mathrm{Id}_{|A})_* \tag{7.1}$$

is an isomorphism on $\pi_1(A, a)$. By Lemma 4.5.3, $(r \circ i)_* = r_* \circ i_*$. Then $r_* \circ i_* = (\mathrm{Id}_{|A})_* = \mathrm{Id}_{|\pi_1(A,a)}$, showing that i_* is left invertible. As a consequence, $i_* : \pi_1(A, a) \longrightarrow \pi_1(X, a)$ is a monomorphism (and $r_* : \pi_1(X, a) \longrightarrow \pi_1(A, a)$ is an epimorphism). Therefore, $\pi_1(A, a)$ is isomorphic to the range $i_*(\pi_1(A, a))$, which is a subgroup of $\pi_1(X, a)$. □

Remark 7.1.33. The proof of Theorem 7.1.32 also shows that if A is a retract of a space X, then the induced homomorphism i_* of the inclusion map i is injective.

Example 7.1.34. Since $\pi_1(\mathbb{RP}^1) \cong \mathbb{Z}$ and $\pi_1(\mathbb{RP}^n) \cong \mathbb{Z}/2\mathbb{Z}$, $\mathbb{RP}^1$ is not a retract of $\mathbb{RP}^n$ for all $n \geq 2$.

Corollary 7.1.35. *Let X be a simply connected space and $A \subset X$ a retract. Then A is simply connected.*

Proof. Since X is simply connected, X is path-connected and the fundamental group $\pi_1(X)$ is trivial. By Theorem 7.1.28, the retract A is path-connected, too. By Theorem 7.1.32, $\pi_1(A)$ is isomorphic to a subgroup of $\pi_1(X)$, which contains only the identity element. Therefore, $\pi_1(A)$ is a trivial subgroup and A is simply connected. □

Remark 7.1.36. If A is not a retract of X, Corollary 7.1.35 fails as shows the subspace S^1 of B^2.

Example 7.1.37. Since S^1 is a retract of $\mathbb{R}^2 \setminus \{0\}$ under the radial retraction and S^1 is not simply connected, we deduce by Corollary 7.1.35 that $\mathbb{R}^2 \setminus \{0\}$ is not simply connected.

We are now in position to prove Borsuk's nonretraction of the ball in $\mathbb{R}^2$. The second proof shows that the result is just a consequence of the Borsuk–Ulam Theorem 6.4.16.

Corollary 7.1.38 (Borsuk's nonretraction of the ball). *In Euclidean plane $\mathbb{R}^2$, the unit circle*

$$S^1 = \{(x_1, x_2) \in \mathbb{R}^2 : x_1^2 + x_2^2 = 1\}$$

is not a retract of the closed unit ball B^2.

Proof. (1) *First proof.* Arguing by contradiction, assume that S^1 is a retract of B^2. Then Corollary 7.1.35 implies that $\pi_1(S^1)$ is isomorphic to a subgroup of $\pi_1(B^2)$. By convexity of the ball, $\pi_1(B^2)$ is trivial (see Example 4.4.2). Hence, $\pi_1(S^1)$ is trivial, which contradicts $\pi_1(S^1) \cong \mathbb{Z}$.

(2) *Second proof.* If $r : B^2 \longrightarrow S^1$ is a retraction, then it is antipode-preserving on S^1, which contradicts Theorem 6.4.16. □

Example 7.1.39. (1) The circle S^1 (and every subset $Y \subset \mathbb{R}^2$ homeomorphic to S^1) is not a retract of the n-dimensional Euclidean space $\mathbb{R}^n$, for $n \geq 2$. Otherwise $\pi_1(S^1) \cong \mathbb{Z}$ would be a subgroup of $\pi_1(\mathbb{R}^n) \cong 0$, a contradiction.
(2) Likewise, S^1 is not a retract of the bounded and unbounded solid cylinders $B^2 \times [0,1]$ and $B^2 \times \mathbb{R}$.

Example 7.1.40. If the torus $\mathbb{T}^2 \cong S^1 \times S^1$ were a retract of $B^2 \times B^2$, then by Theorem 7.1.32, $\pi_1(S^1 \times S^1)$ would be a subgroup of $\pi_1(B^2 \times B^2)$. This is impossible because the first group is homeomorphic to $\mathbb{Z} \times \mathbb{Z}$ (Example 6.1.8) whereas the second group is trivial. For the same reasons, the torus is not a retract of the solid torus $B^2 \times S^1$.

7.2 Deformation retract

7.2.1 Deformation retraction and strong deformation retraction

Definition 7.2.1. A nonempty subset $A \subset X$ is *a deformation retract* of X if A is a retract of X and $\mathrm{Id}_X \simeq i \circ r$, where $r : X \longrightarrow A$ is a retraction and $i : A \hookrightarrow X$ the inclusion map. Equivalently, there exists a continuous map $H : X \times I \longrightarrow X$ such that for every $x \in X$,

$$H(x,0) = x \quad \text{and} \quad H(x,1) = r(x).$$

The map r is called *a deformation retraction* and H a deformation homotopy.

Remark 7.2.2. A deformation retraction is a continuous map $r : X \longrightarrow A$ such that $r \circ i = \mathrm{Id}_{|A}$ and $i \circ r \simeq \mathrm{Id}_X$.

Remark 7.2.3. If $A, B \subset X$ and A, B are deformation retracts of X, the union $A \cup B$ is not necessarily a deformation retract of X. Think of the annulus $X = A(0;1,2)$ with the boundary circles $A = C(0;1)$ and $B = C(0;2)$. Then X deformation retracts on each of A and B with the radial retraction whereas AUB is not a retract of X since it is not connected but X is. For the same reason, in a connected space X, a two-point subset cannot be a retract of X (see Example 7.1.11).

Theorem 7.2.4. *Let $A \subset X$ be a deformation retract of X. Then A and X are homotopically equivalent.*

Proof. The homotopy equivalence maps are merely the inclusion map $i : A \hookrightarrow X$ and the retraction $r : X \longrightarrow A$. Let H be the deformation homotopy given in Definition 7.2.1. We have

$$\forall x \in X, \quad (i \circ r)(x) = r(x) = H(x,1)\ H(x,0) = x = \mathrm{Id}_X(x),$$
$$\forall x \in A, \quad (r \circ i)(x) = r(x) = x = \mathrm{Id}_{|A}(x). \qquad \square$$

Remark 7.2.5. In general, the converse of Theorem 7.2.4 does not hold true since the contractible spaces given by the square I^2 and the comb space X are homotopically equivalent but by Example 7.1.30, X is not even a retract of I^2.

Theorem 7.1.32 describes the fundamental group of a retract as a subgroup of the fundamental group of the space because the induced homomorphism is injective. Assuming that X and A are path-connected, Theorem 7.2.4 and Corollary 4.5.6 yield the following consequence.

Corollary 7.2.6. *Let $A \subset X$ be a deformation retract of X and $a \in A$. Then the fundamental groups $\pi_1(A,a)$ and $\pi_1(X,a)$ are isomorphic.*

Proposition 7.2.7. *The following statements are equivalent:*
(1) *X is contractible;*
(2) *X is homotopically equivalent to a point $x_0 \in X$;*
(3) *X deformation retracts onto a point $x_0 \in X$.*

Proof. The equivalence between (1) and (2) is Proposition 4.9.17. For some point $x_0 \in X$, let $r : X \longrightarrow \{x_0\}$ be the constant map retraction and $i : \{x_0\} \hookrightarrow X$ the inclusion map. Then $(r \circ i)(x_0) = r(x_0) = x_0$ and $(i \circ r)(x_0) = i(x_0) = x_0$. Hence, $i \circ .r = e_{x_0} = r$. If X is contractible, then $e_{x_0} \simeq \mathrm{Id}_X$ by Definition 4.9.1. Therefore, (1) implies (3). The implication (3) $\Longrightarrow$ (2) is Theorem 7.2.4. $\square$

By the way, we obtain an alternative proof of Corollary 4.9.12.

Corollary 7.2.8. *A contractible space is path-connected.*

Proof. Let X be contractible and $\{x_0\}$ a deformation retraction of X. Then $\gamma(t) = H(x,t)$ is a path joining $x_0 \in X$ to any point x of the space X. Hence, X is path-connected. $\square$

Remark 7.2.9. (1) It is clear that the notion of deformation retraction is stronger than the retraction given by Definition 7.1.1.
(2) Example 7.1.3 shows that a space X always retracts onto one of its point, says x_0.
(3) If X is not path-connected, then $\{x_0\}$ cannot be a deformation retract of X by Corollary 7.2.8.

Let A be a nonempty subspace of X.

Definition 7.2.10. (1) A is *a strong deformation retract* of X if there exists a continuous map $H : X \times I \longrightarrow X$ such that

$$\begin{cases} H(x,0) = x, & \forall x \in X, \\ H(x,1) \in A, & \forall x \in X, \\ H(x,t) = x, & \forall x \in A, \forall t \in I. \end{cases}$$

(2) The continuous map r given by $r(x) = H(x,1)$ for $x \in X$ is called *a strong deformation retraction* and H is a strong deformation homotopy.

Some books (e. g., [45, p. 2]), use merely Definition 7.2.10 to define a deformation retract.

Remark 7.2.11. (1) The map $r : X \longrightarrow A$ given by $r(x) = H(x,1)$ satisfies $r \circ i = \mathrm{Id}_{|A}$ (r is a retraction) and $i \circ r \simeq \mathrm{Id}_X$ with homotopy H.

(2) We have

$$\text{Strong deformation retraction} \implies \text{Deformation retraction} \implies \text{Retraction.}$$

Example 7.2.12. By Example 4.9.2 and Proposition 7.2.7, Euclidean space $\mathbb{R}^n$ deformation retracts onto the origin. In fact, for every point $x_0 \in \mathbb{R}^n$, $H(x,t) = tx_0 + (1-t)x$ is even a strong deformation homotopy.

7.2.2 Construction of deformation retractions

Example 7.2.13. By Example 7.1.9, the circle S^1 is a retract of the torus but not a deformation retract since they have not isomorphic fundamental groups (Theorem 6.1.5 and Corollary 6.1.11).

Example 7.2.14. The radial retraction shows that $A = S^{n-1}$ is a retract of space $X = \mathbb{R}^n \setminus \mathring{B}^n$ for all $n \geq 1$, where $\mathring{B}^n$ is the open unit ball. From Example 4.2.9 together with Example 1.5.13 and Example 1.5.18, we know that A and X are homotopically equivalent. Actually, A is a strong deformation retract of X. A suitable deformation homotopy is given by $H(x,t) = tr(x) + (1-t)x$, where r is the radial retraction onto the sphere. The same homotopy shows that S^{n-1} is a strong deformation retract of the punctured space $\mathbb{R}^n \setminus \{0\}$. It can be checked that $H(x,t) \neq 0$ for all $(x,t) \in (\mathbb{R}^n \setminus \{0\}) \times I$; see also Example 7.1.5.

Example 7.2.15. The comb space deformation retracts onto $\{0\} \times I$. A suitable homotopy $H : X \times I \longrightarrow X$ is given by (see [92, Example 14.1.10])

$$H((x,y),t) = \begin{cases} (x,(1-3t)y), & 0 \leq t \leq 1/3, \\ ((2-3t)x,0), & 1/3 \leq t \leq 2/3, \\ (0,(3t-2)y), & 2/3 \leq t \leq 1. \end{cases}$$

Proposition 7.2.16. *The sphere $S^n \cong S^n \times \{0\}$ is a strong deformation retract of the bounded cylinder $S^n \times I$ and the unbounded cylinder $S^n \times \mathbb{R}$.*

Proof. Let $r : S^n \times [0,1] \longrightarrow S^n \times \{0\}$ be the retraction $r(x,t) = (x,0)$ and $i : S^n \times \{0\} \hookrightarrow S^n \times [0,1]$ the inclusion map. Then, for $(x,t,s) \in (S^n \times I) \times I$, $H(x,t,s) = (x,(1-s)t)$ is a strong deformation homotopy. By Theorem 7.2.4, the sphere S^n and the cylinder $S^n \times I$ are homotopically equivalent, confirming the result of Example 4.2.6. □

The following result is a generalization of Proposition 7.2.16.

Proposition 7.2.17. *If $A \subset X$ is a strong deformation retract of the space X, then for any space Y, $Y \times A$ is a strong deformation retract of the space $Y \times X$.*

Proof. Let $H : X \times I \longrightarrow X$ be a strong deformation homotopy of X onto A. Then $s(x) = \widetilde{H}(x,1)$ is a strong retraction from $Y \times X$ onto $Y \times A$, where $\widetilde{H} : (Y \times X) \times I \longrightarrow Y \times X$ is given by

$$\widetilde{H}(y,x,t) = (y, H(x,t)).$$

□

Example 7.2.18. (1) Since $\mathbb{R}^n$ is contractible, for every subspace $A \subset \mathbb{R}^n$, $A \times \{0\}$ is a strong deformation retract of $A \times \mathbb{R}^n$.

(2) Also B^n is contractible. Then $S^{n-1} \times \{0\}$ is a strong deformation retract of both the cylinder $S^{n-1} \times \mathbb{R}^n$ (Proposition 7.2.16) and the solid torus $S^{n-1} \times B^n$.

Proposition 7.2.19. (1) *The punctured plane $\mathbb{R}^2 \setminus \{0\}$ is a strong deformation retract of the space $\mathbb{R}^3$ with the z-axis $(0z)$ removed, i. e., $\mathbb{R}^3 \setminus (0z)$.*

(2) *The cylinder $S^1 \times \mathbb{R}$ is a strong deformation retracts of $\mathbb{R}^3 \setminus (0z)$.*

(3) *The circle $S^1 \cong S^1 \times \{0\}$ is a strong deformation retract of $\mathbb{R}^3 \setminus (0z)$.*

Proof. (1) The retraction is the orthogonal projection $r(x,y,z) = (x,y,0)$ onto the (x,y)-plane.

(2) The retraction is the radial retraction in the plane parallel to the (x,y)-plane and passing through the point (x,y,z), i. e., $r(x,y,z) = (\frac{(x,y)}{\|(x,y)\|}, z)$.

(3) The retraction is the composition of the orthogonal projection and the radial retraction, i. e., $r(x,y,z) = (\frac{(x,y)}{\|(x,y)\|}, 0)$.

In all cases, the strong deformation homotopy is

$$H(x,y,z,t) = tr(x,y,z) + (1-t)(x,y,z), \quad \text{for } (x,y,z) \in \mathbb{R}^3 \setminus (0z) \text{ and } t \in I.$$

In part (1), $H(x,y,z,t) = (x,y,(1-t)z)$ is well-defined since $(x,y) \neq (0,0)$. In parts (2) and (3), we only need to verify that for all $(x,y,z) \in \mathbb{R}^3 \setminus (0z)$, and $t \in I$, $H(x,y,z,t) \in \mathbb{R}^3 \setminus (0z)$. On the contrary,

$$\frac{t}{\|(x,y)\|}x + (1-t)x = \frac{t}{\|(x,y)\|}y + (1-t)y = 0,$$

which implies $0 \leq \frac{t}{\|(x,y)\|} = t - 1 \leq 0$, a contradiction. □

Remark 7.2.20. In Proposition 7.2.19(2), a direct proof that $\mathbb{R}^2 \setminus \{0\}$ and $\mathbb{R}^3 \setminus (0z)$ are homotopically equivalent is immediate. For this, write

$$X = \mathbb{R}^3 \setminus (0z) \cong Y = \mathbb{R}^2 \setminus (0,0) \times (0z) \cong S^1 \times (0,\infty) \times (0z),$$

by appealing to Example 1.5.25. By Proposition 4.9.20, we conclude that X, Y and S^1 are homotopically equivalent.

The strong deformation homotopy in Proposition 7.2.19 is a general construction. We have the following.

Proposition 7.2.21. *Let $A \subset X$ be subspaces of a topological vector space E and $r : X \longrightarrow A$ a retraction. Suppose that X satisfies the following condition:*

$$\textit{for all } x \in X, \textit{ the segment } [x, r(x)] \textit{ lies in } X.$$

Then A is a strong deformation retract of X.

Proof. Just consider the convex homotopy $H(x,t) = tr(x) + (1-t)x$, where r retracts X onto A. □

Example 7.2.22. (1) If A is retract subset of a convex space X, then X strongly deformation retracts onto A.
(2) Let $X = \mathbb{R}^n \setminus \{0\}$, $A = S^{n-1}$ and $r(x) = x/\|x\|$, the radial retraction. So, we recover the result of Example 7.2.14.

Since $(\mathbb{R}^3 \setminus (0z)) \cap S^2 \cong S^2 \setminus \{N_p, S_p\} \cong \mathbb{R}^2 \setminus (0,0)\}$, Proposition 7.2.19(1) tells us that $\mathbb{R}^3 \setminus (0z)$ deformation retracts onto $S^2 \setminus \{N_p, S_p\}$. A more general version is given by the following.

Proposition 7.2.23. *Let $\ell_1, \ell_2, \dots, \ell_n$, $n \geq 1$ be n lines through the origin of Euclidean space $\mathbb{R}^3$ and $X = \mathbb{R}^3 \setminus \{\ell_1, \ell_2, \dots, \ell_n\}$. Then X deformation retracts onto $X \cap S^2$.*

Proof. Under the radial retraction, the space X strongly deformation retracts onto the subspace

$$X \cap S^2 = S^2 \setminus \{p_1, -p_1, p_2, -p_2, \dots, p_n, -p_n\},$$

where the diametral points $p_k, -p_k$ are the intersection points of the lines ℓ_k $(1 \leq k \leq n)$ with the sphere S^2. The strong deformation homotopy is assured by Proposition 7.2.21. □

Proposition 7.2.24. *S^1 embeds as a deformation retract of $S^3 \setminus S^1$.*

Proof. (1) The circle and the sphere identify with

$$S^1 = \{z \in \mathbb{C} : |z| = 1\} \cong \{0\} \times S^1 \cong S^1 \times \{0\},$$
$$S^3 = \{(z_1, z_2) \in \mathbb{C}^2 : |z_1|^2 + |z_2|^2 = 1\},$$

where $|z|$ refers to the modulus of the complex number z. For $0 \leq t \leq 1$, define the family of maps $h_t : S^1 \longrightarrow S^3$ by $h_t(z) = (tz, \sqrt{1-t^2}z)$. The map h_t is well-defined since $t^2|z|^2 + (1-t^2)|z|^2 = 1$ for all $z \in S^1$. h_t is clearly injective, hence an embedding by Theorem 1.5.36(1). Therefore, S^1 embeds in S^3 and $S^3 = \bigcup_{0\leq t\leq 1} h_t(S^1)$. Note that for each $t \in [0,1]$, the map h_t is just the rotation of angle $\theta = \tan^{-1}(\frac{\sqrt{1-t^2}}{t}) \in [0, \pi/2]$.

Let $X = S^3 \setminus h_0(S^1) = \bigcup_{0<t\leq 1} h_t(S^1)$ and $A = h_1(S^1)$. We claim that X deformation retracts onto A.

(2) For $z = (z_1, z_2) \in X$, let $t \in (0,1]$ and $u \in S^1$ be such that $z = h_t(u)$. The map $r : X \longrightarrow A$ given by

$$r(z) = r(tu, \sqrt{1-t^2}u) = h_1(u) = (u, 0)$$

is well-defined, continuous and for $z = h_1(u) = (u,0) \in A$, we have $r(z) = (u,0) = z$. Hence, r is a retraction.

(3) Define the continuous map $H : X \times I \longrightarrow X$ as

$$H(z,s) = H(tu, \sqrt{1-t^2}u, s) = ((s+t-ts)u, \sqrt{1-s^2}\sqrt{1-t^2}u).$$

Then, for every $z \in X$,

$$H(z,0) = (tu, \sqrt{1-t^2}u) = z \quad \text{and} \quad H(z,1) = (u,0) = r(z).$$

Therefore, H is a homotopy between r and the identity map on X. □

Definition 7.2.25. (1) The image of S^1 by an embedding f is called *a knot*, that is a Jordan curve in $\mathbb{R}^3$ (see Definition 1.5.5).
(2) The difference set $S^3 \setminus f(S^1)$ is called *a knot complement*.

Remark 7.2.26. In Theorem 10.4.24, we will prove that $S^n \setminus S^k$ is homotopically equivalent to the sphere S^{n-k-1}.

Remark 7.2.27. (1) Here is a informal proof of Proposition 7.2.24. Consider the one-point compactification $S^n \cong \mathbb{R}^n \cup \{\infty\}$ (Proposition 1.5.28). Then

$$S^3 \setminus S^1 \cong (\mathbb{R}^3 \cup \{\infty\}) \setminus (\mathbb{R} \cup \{\infty\}) \cong \mathbb{R}^3 \setminus \mathbb{R} \cong \mathbb{R}^3 \setminus (0z).$$

By Proposition 7.2.19, $\mathbb{R}^3 \setminus (0z)$ deformation retracts on S^1, whence the result.
(2) Along the same lines,

$$\mathbb{R}^3 \setminus (S^1 \cup (0z)) \cong \mathbb{R}^3 \setminus (S^1 \cup \mathbb{R}) \cong (\mathbb{R}^3 \setminus \mathbb{R}) \setminus (S^1) \cong (S^3 \setminus S^1) \setminus S^1.$$

Remark 7.2.28. Following Remark 7.2.27, if X deformation retracts onto $Y \subset X$ (with a retraction r and a homotopy H) and X is homeomorphic to X' (with some homomorphism h), then X' deformation retracts onto $h(Y)$. The corresponding retractions and homotopies are given by the composites $s = h \circ r \circ h^{-1}$ and $K = H \circ r \circ (h^{-1} \times \mathrm{Id}_{|I})$, respectively.

Remark 7.2.29. (1) One should be careful when manipulating logical operations involving homeomorphisms and homotopy equivalences. Here are two examples:
(a) By convexity (see Example 4.2.3), $\mathbb{R}^2$ and $\mathbb{R}$ are homotopically equivalent spaces. However, $\mathbb{R}^2 \setminus \{0\}$ is not homotopically equivalent to $\mathbb{R} \setminus \{0\}$ since the first punctured space is path-connected whereas the second one has two path-components (see Proposition 4.2.19).
(b) By Example 1.5.18, $\mathring{B}^2 \setminus \{0\}$ is homeomorphic to the hollowed plane $\mathbb{R}^2 \setminus B^2$. Taking the union with $\mathring{B}^2$, the space $\mathring{B}^2 \setminus \{0\} \cup \mathring{B}^2 = \mathring{B}^2$ is a connected space whereas $(\mathbb{R}^2 \setminus B^2) \cup \mathring{B}^2 = \mathbb{R}^2 \setminus S^1$ has two components. This result was proved by Camille Jordan in 1887 (see Proposition 10.4.17).

(2) However, if the spaces X, X' are homeomorphic (homotopically equivalent, respectively) with $\overline{X} \cap \overline{Y} = \emptyset$ and $\overline{X'} \cap \overline{Y} = \emptyset$, then $X \cup Y$ and $X' \cup Y$ are homeomorphic (homotopically equivalent, respectively), where Y is any space.

Remark 7.2.30. Does homotopy equivalence between two spaces imply that one space is a deformation retract of the other? A theorem due to Ralph H. Fox (1943) (see [34]) states that two spaces X and Y have the same homotopy type if and only if they are homeomorphic to deformation retracts of a single space Z involving the mapping cylinder (see [45, Corollary 0.21]). Yet one of the two spaces does not necessarily deformation retract on the other, as Hatcher's example shows in [45, p. 2]. However, for a Borsuk pair (X, A), the answer is positive (see also Proposition 7.3.5).

Proposition 7.2.31. *The space $(S^3 \setminus S^1) \setminus S^1$ is a deformation retract of $\mathring{B}^2 \setminus \{0\} \times \mathring{B}^2 \setminus \{0\}$.*

Proof. As in Proposition 7.2.24, consider the identifications $S^1 = \{z \in \mathbb{C} : |z| = 1\}$ and $S^3 = \{(z_1, z_2) \in \mathbb{C}^2 : |z_1|^2 + |z_2|^2 = 1\}$, where $|z|$ refers to the modulus of the complex number z. Let $X = (S^3 \setminus S^1) \setminus S^1$ and $Y = \mathring{B}^2 \setminus \{0\} \times \mathring{B}^2 \setminus \{0\}$. By Proposition 7.2.24, $S^3 = \bigcup_{0 \le t \le 1} h_t(S^1)$ and $(S^3 \setminus S^1) \setminus S^1 = \bigcup_{0<t<1} h_t(S^1)$, where $h_t(z) = (tz, \sqrt{1-t^2}z)$. Then $z = (z_1, z_2) \in X$ if and only if there exist $t \in (0,1)$ and $u \in S^1$ such that $z_1 = tu$ and $z_2 = \sqrt{1-t^2}u$. Hence, $|z_1| = t \in (0,1)$ and $|z_2| = \sqrt{1-t^2} \in (0,1)$, i. e., $z \in Y$. Therefore, $X \subset Y$. We prove the claim of the proposition.

(1) For $z = (z_1, z_2) \in Y$, let $r : Y \longrightarrow X$ be given by

$$r(z) = \frac{1}{\sqrt{|z_1|^2 + |z_2|^2}} z.$$

The map r is well-defined for $|r(z)_1|^2 + |r(z)_2|^2 = 1$ and r is continuous. In addition, for $z \in X$, we have $|z_1|^2 + |z_2|^2 = 1$. Then $r(z) = z$, proving that r is a retraction. Notice that $r(z) = z/\|z\|$, where $\|z\|$ is Euclidean norm in $\mathbb{R}^4$.

(2) Let $H : Y \times \backslash\{0\}) \times I \longrightarrow Y$ be given by

$$H(z,t) = \left(t + \frac{1-t}{\sqrt{|z_1|^2 + |z_2|^2}}\right)z.$$

Then H is well-defined since for all $z = (z_1, z_2) \in Y$, $t \in I$ and $i = 1, 2$, we have

$$\begin{aligned} 0 &< \left(t + \frac{1-t}{\sqrt{|z_1|^2 + |z_2|^2}}\right)|z_i| \\ &= t|z_i| + \frac{(1-t)|z_i|}{\sqrt{|z_1|^2 + |z_2|^2}} \\ &< t + (1-t) = 1. \end{aligned}$$

Finally, H is continuous, $H(z,0) = \frac{1}{\sqrt{|z_1|^2+|z_2|^2}} z = r(z)$ and $H(z,1) = z$. Hence, H is the required homotopy. □

Assume an equivalence relation $\sim$ is given in X. The equivalence relation $\sim_A$ on A is induced by $\sim$ via the inclusion map i, i. e.,

$$x \sim_A y \iff i(x) \sim i(y), \quad \text{for all } x, y \in A.$$

The subspace A ($A/_{\sim_A}$, respectively) is endowed with the subspace topology of X (of $X/_{\sim}$, respectively).

Proposition 7.2.32. (1) *Let A be a retract of a space X. Then $A/_{\sim_A}$ is a retract of $X/_{\sim}$.*
(2) *Assume that X is compact and $X/_{\sim}$ Hausdorff. If A is a deformation retract (a strong deformation retract, respectively) of X, then $A/_{\sim_A}$ is a deformation retract (a strong deformation retract, respectively) of $X/_{\sim}$.*

Proof. (1) Let $\pi : X \longrightarrow X/_{\sim}$, be the projection map, $r : X \longrightarrow A$ a retraction and $s : X/_{\sim} \longrightarrow A/_{\sim_A}$ the map such that the following diagram commutes ($s \circ \pi = \pi_{|A} \circ r$):

$$\begin{array}{ccc} X & \xrightarrow{\quad r \quad} & A \\ {\scriptstyle \pi}\downarrow & & \downarrow{\scriptstyle \pi_{|A}} \\ X/_{\sim} & \xrightarrow[\quad s \quad]{} & A/_{\sim_A} \end{array}$$

We claim that s is a retraction.
(a) Let $\pi_{|A}(a) \in A/_{\sim_A}$ for some $a \in A$. Since r is surjective, there is some $x \in X$ such that $a = r(x)$. Then $\pi_{|A}(a) = \pi_{|A}(r(x)) = s(\pi(x))$, proving that s is surjective.
(b) By the commutativity of the diagram,

$$s(\pi_{|A}(a)) = s(\pi(a)) = \pi_{|A}(r(a)) = \pi_{|A}(a),$$

for all $a \in A$. Therefore, s restricted to $A/_{\sim_A}$ is the identity.

(c) The continuity of s derives from Theorem 2.1.9 since $s \circ \pi = \pi_{|A} \circ r$, π is an identification map, and $\pi_{|A}$, r are continuous functions.

(2) Let $H : X \times I \longrightarrow X$ be a homotopy and $K : X/_\sim \times I \longrightarrow X/_\sim$ the map such that the diagram commutes ($K \circ (\pi \times \mathrm{Id}) = \pi \circ H$):

$$\begin{array}{ccc} X \times I & \xrightarrow{\quad H \quad} & X \\ {\scriptstyle \pi\times\mathrm{Id}}\big\downarrow & & \big\downarrow{\scriptstyle \pi} \\ X/_\sim \times I & \xrightarrow[\quad K \quad]{} & X/_\sim \end{array}$$

where $(\pi \times \mathrm{Id})(x,t) = (\pi(x), t)$ is the Cartesian product map of π and the identity Id on the interval I. The map K clearly satisfies the properties of a homotopy. To prove that K is continuous, let $C \subset X/_\sim$ be a closed set. Then $\pi^{-1}(C)$ is closed in X. Therefore,

$$H^{-1}(\pi^{-1}(C)) = (\pi \circ H)^{-1}(C) = \big(K \circ (\pi \times \mathrm{Id})\big)^{-1}(C)$$

is closed in X, hence compact in the compact space X. Since the product $\pi \times \mathrm{Id}$ is continuous and $X/_\sim$ is Hausdorff,

$$(\pi \times \mathrm{Id})\big(K \circ (\pi \times \mathrm{Id})\big)^{-1}(C) = (\pi \times \mathrm{Id})\big((\pi \times \mathrm{Id})^{-1} \circ K^{-1}\big)(C) = K^{-1}(C)$$

is compact, hence closed in the Hausdorff space $X/_\sim$. □

As application of Proposition 7.2.32, we present a result stronger than Proposition 4.2.17.

Proposition 7.2.33. (1) *The core circle $C(\mathbb{M}^2)$ embeds as a strong deformation retract of the Möbius band $\mathbb{M}^2$.*

(2) *We have $\pi_1(\mathbb{M}^2) \cong \mathbb{Z}$.*

Proof. (1) Let $X = [0,1]^2$ with $(x,0) \sim (1-x,1)$. We claim that the subset $A = \{1/2\} \times [0,1]$ is a strong deformation retract of X. Indeed, consider the homotopy $H : X \times I \longrightarrow X$ given by $H(x,y,t) = (t/2 + (1-t)x, y)$. Since $t/2 + (1-t)x \le 1 - t/2 \le 1$ for all $t, x \in [0,1]$, H is well- defined. Moreover, H satisfies

$$\begin{aligned} H(x,y,0) &= (x,y), \quad H(x,y,1) = (1/2,y) \in A, \quad \text{for all } (x,y) \in X, \\ H(1/2,y,t) &= (1/2,y), \quad \text{for all } (1/2,y) \in A. \end{aligned}$$

Hence, H is a strong deformation homotopy. By Proposition 7.2.32, the quotient space $A/_\sim = \pi(A)$ is a strong deformation retract of $X/_\sim = \mathbb{M}^2$. However, $\pi(A)$ is by definition the core circle (see Example 3.2.5), which is homeomorphic to a circle. Indeed, the equivalence relation becomes $(\frac{1}{2}, 0) \sim (\frac{1}{2}, 1)$ onto the subspace A, which means that we have identified the endpoints of the segment $[0,1]$, leading to a quotient space homeomorphic to S^1.

(2) By Theorem 6.1.4 and Corollary 7.2.6, we conclude that $\pi_1(\mathbb{M}^2) \cong \pi_1(\pi(A)) \cong \mathbb{Z}$. □

Remark 7.2.34. Proposition 7.2.33 can be derived from Example 3.5.6 for $n = 1$ combined with Exercise 22(1), this chapter.

7.2.3 Punctured spaces

Proposition 7.2.35. *The punctured torus (a torus with a point removed)* $\mathbb{T}'^2 = \mathbb{T}^2 \setminus \{*\}$ *deformation retracts onto a wedge sum of two circles.*

Proof. (1) Let (x_0, y_0) be an interior point in the square I^2, $X = I^2 \setminus \{(x_0, y_0)\}$ and $A = \partial I^2$. In Example 3.2.3, the torus is defined by the equivalence relation $\sim$ resulting from the identification of the opposite sides of the square I^2. Consider the radial retraction $r : X \longrightarrow A$. By Proposition 7.2.32(1), $A/_\sim$ is a retract of $X/_\sim$, where $X/_\sim = \mathbb{T}'^2$ is the punctured torus. Furthermore, the map $H : X \times I \longrightarrow X$ defined by $H((x,y),t) = (1-t)(x,y) + tr(x,y)$ is a strong deformation homotopy between the retraction r and the identity on X. We claim that $A/_\sim$ is a deformation retract of $X/_\sim$.

(2) Consider the diagram

$$\begin{array}{ccc} X \times I & \xrightarrow{\quad H \quad} & X \\ {\scriptstyle \pi\times\mathrm{Id}}\big\downarrow & & \big\downarrow{\scriptstyle \pi} \\ \mathbb{T}'^2 \times I & \xrightarrow[\quad K \quad]{} & \mathbb{T}'^2 \end{array}$$

The restriction of the projection map to space X is defined by

$$\pi(x,y) = \begin{cases} \{(x,y)\}, & \text{if } (x,y) \in \mathring{I}^2 \setminus \{(x_0,y_0)\}, \\ \{(x,0),(x,1)\}, & \text{if } y(1-y) = 0, \\ \{(0,y),(1,y)\}, & \text{if } x(1-x) = 0. \end{cases}$$

Then the map $(\pi \times \mathrm{Id})^{-1} : \mathbb{T}'^2 \times I \longrightarrow X \times I$ is explicitly expressible as

$$(\pi \times \mathrm{Id})^{-1}([(x,y)],t) = \begin{cases} ((x,y),t), & \text{if } (x,y) \in \mathring{I}^2 \setminus \{(x_0,y_0)\}, \\ \{((x,1-y),t)\}, & \text{if } y(1-y) = 0, \\ \{((1-x,y),t)\}, & \text{if } x(1-x) = 0. \end{cases}$$

Hence, $K = \pi \circ H \circ (\pi \times \mathrm{Id})^{-1}$ is well-defined, i. e., $K \circ (\pi \times \mathrm{Id}) = \pi \circ H$ (the diagram commutes).

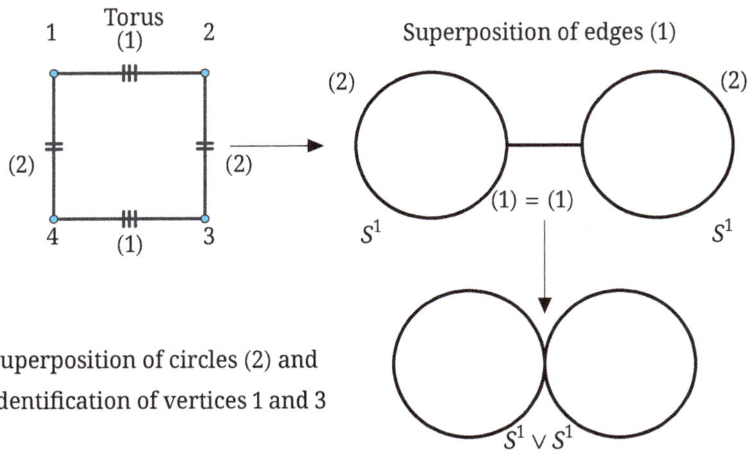

Figure 7.1: The boundary of the square retracts onto the wedge of two circles.

(3) (a) To prove that π is a closed map, let $C \subset X$ be closed and $C' = \pi(C)$. Since $C \subset X \subset I^2$, C is bounded, hence compact in I^2. Since X is open in I^2, C is compact in X. By continuity of the projection π, the range C' is compact in $\mathbb{T}'^2$. However, the punctured space $\mathbb{T}'^2 = \mathbb{T}^2 \setminus \{*\}$ is Hausdorff because $\mathbb{T}^2 \cong S^1 \times S^1$ is Hausdorff. Hence, C' is closed in $\mathbb{T}'^2$.

(b) We claim that K is continuous. Indeed, for a closed subset $C \subset \mathbb{T}'^2$, $\pi^{-1}(C)$ is closed in X and so $H^{-1}(\pi^{-1}(C)) = (\pi \circ H)^{-1}(C)$ is closed in $X \times I$. Therefore,

$$(\pi \times \mathrm{Id})(\pi \circ H)^{-1}(C) = K^{-1}(C)$$

is closed in $\mathbb{T}'^2 \times I$ because π is closed.

(4) Furthermore, it is easily checked that K is a deformation homotopy.

(5) Finally, notice that $\partial I^2/_\sim$ is homeomorphic to the boundary of the square with opposite sides identified, that is a pair of two circles joined at a corner, $S^1 \vee S^1$. See Figure 7.1, where 1, 2, 3, 4 denote the vertices and (1), (2) the edges. First, edges (1) are identified and then circles (2) are superposed.

(6) Therefore, the wedge $S^1 \vee S^1$ of circles is a deformation retract of the punctured torus $X/_\sim = \mathbb{T}'^2$. □

Remark 7.2.36. (1) The proof of Proposition 7.2.35 still works for the punctured Möbius band $\mathbb{M}'^2 = \mathbb{M}^2 \setminus \{*\}$ with the difference that because of the orientation of the opposite sides the space deformation retracts onto a figure eight instead of the wedge sum of two circles.

(2) Since the space $X = I^2 \setminus \{(0,0)\}$ is not compact, we did not apply Proposition 7.2.32 directly but the relative compactness of X allowed us to prove Proposition 7.2.35. The same procedure can be employed to prove the next result.

Proposition 7.2.37. *The complement of the core circle $C(\mathbb{M}^2)$ in the Möbius band $\mathbb{M}^2$ strongly deformation retracts on the boundary circle $\partial\,\mathbb{M}^2$.*

Proof. Consider the subsets $A_1 = \{(x,y) \in I^2 : 0 \le x < 1/2,\ 0 \le y \le 1\}$ and $A_2 = \{(x,y) \in I^2 : 1/2 < x \le 1,\ 0 \le y \le 1\}$ of the square. Then notice that the set $\mathbb{M}^2 \setminus C(\mathbb{M}^2)$ is just the image of the disjoint union $A_1 \cup A_2$ under the projection map π defining the Möbius band in Example 3.2.5. In addition, the set A_1 strongly deformation retracts onto the left-hand side of the square $\{0\} \times [0,1]$. The retraction $r : A_1 \longrightarrow \{0\} \times [0,1]$ is the horizontal projection $r(x,y) = (0,y)$ and the strong deformation homotopy is $H((x,y),t) = ((1-t)x,y)$ for all $(x,y) \in A_1$ and $t \in [0,1]$. Likewise A_2 strongly deformation retracts onto the right-hand side of the square, namely $\{1\} \times [0,1]$. Then the claim follows as in the proof of Proposition 7.2.35. □

The next proposition mentioned in [61, Exercise 2.4.1] tells more than Proposition 7.2.37.

Proposition 7.2.38. *The complement of the core circle $C(\mathbb{M}^2)$ in the Möbius band $\mathbb{M}^2$ is homeomorphic to the cylinder $S^1 \times (0,1]$.*

Proof. Let $X_1 = [0,1/2)$, $X_2 = (1/2,1]$, $X = X_1 \cup X_2$ and $Y = S^1 \times (0,1]$. Then $\mathbb{M}^2 \setminus C(\mathbb{M}^2) = \pi(X) = X/_\sim$. Consider the diagram

$$\begin{array}{ccc} X & \xrightarrow{\ \pi\ } & X/_\sim \\ & \searrow^{f} & \downarrow{\bar f} \\ & & Y \end{array}$$

where the map f is defined by

$$f(x,y) = (\operatorname{sgn}(1-2x)e^{i\pi y}, |1-2x|).$$

Then the map f is continuous, surjective and closed. It remains to check that f agrees with the equivalence relation, which defines the Möbius band. We distinguish two cases.

(1) If (x,y) and (x',y') both lie in X_1, then

$$\begin{aligned} f(x,y) = f(x',y') \quad &\Longleftrightarrow \quad (e^{i\pi y}, 1-2x) = (e^{i\pi y'}, 1-2x') \\ &\Longleftrightarrow \quad y - y' = 2n \in \mathbb{Z} \quad \text{and} \quad x = x'. \end{aligned}$$

Since $y,y' \in [0,1]$, $1 \ge |y-y'| = 2|n|$ implies $n = 0$. Thus,

$$f(x,y) = f(x',y') \quad \Longleftrightarrow \quad x = x' \quad \text{and} \quad y = y'.$$

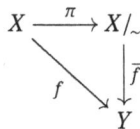

(2) If $(x,y) \in X_1$ and $(x',y') \in X_2$, then

$$\begin{aligned} f(x,y) = f(x',y') &\iff (e^{i\pi y}, 1-2x) = (-e^{i\pi y'}, 2x'-1) \\ &\iff (e^{i\pi y}, 1-2x) = (e^{i\pi(y'-1)}, 2x'-1) \\ &\iff y - y' = -1 + 2n \in \mathbb{Z} \quad \text{and} \quad x' = 1-x. \end{aligned}$$

Here again, $y, y' \in [0,1]$ and $1 \geq |y-y'| = |2n-1|$ yield $n = 0$ or $n = 1$. Hence,

$$\begin{aligned} & f(x,y) = f(x',y') \iff \\ & (x' = 1-x) \quad \text{and either} \quad (y = 0 \text{ and } y' = 1) \quad \text{or} \quad (y = 1 \text{ and } y' = 0). \end{aligned}$$

The same reasoning applies if (x,y) and (x',y') both lie in X_2 or $(x,y) \in X_2$ and $(x',y') \in X_1$. Therefore, we have proved that

$$f(x,y) = f(x',y') \iff x \sim x',$$

where $\sim$ defines the Möbius band. By Corollary 2.3.6, the induced map $\overline{f}$ is a homeomorphism. □

Corollary 7.2.39. *$\mathbb{M}^2 \setminus C(\mathbb{M}^2)$ is path-connected.*

Remark 7.2.40. Figures 7.2, 7.3 and 7.4 show the process of constructing a cylinder without the base circle by cutting the core circle of the Möbius band. We have numbered the vertices of the square $[0,1] \times [0,1]$, which are involved in the identification.

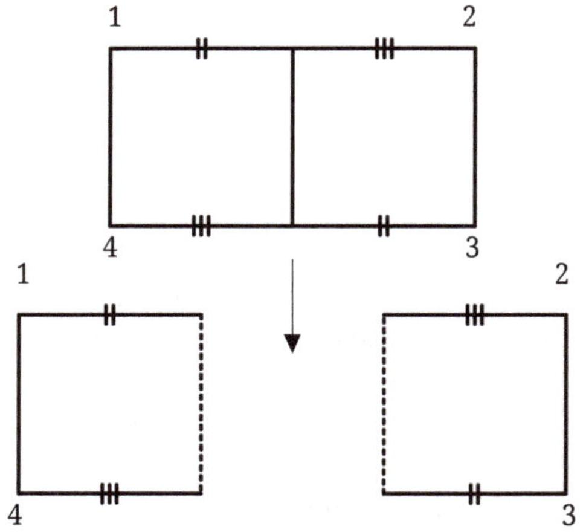

Figure 7.2: Cutting the Möbius strip.

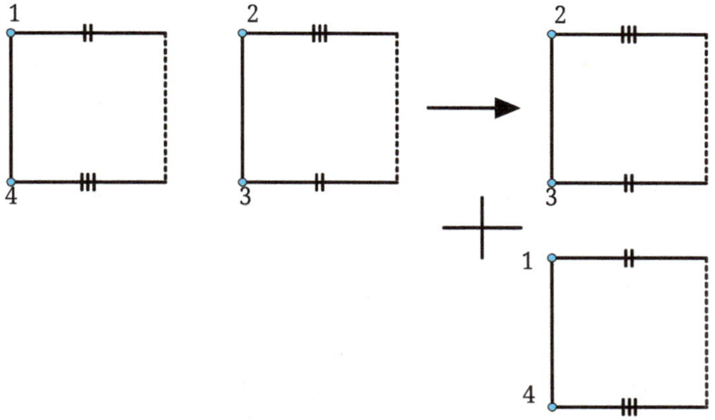

Figure 7.3: Reconstitution of the detached halves.

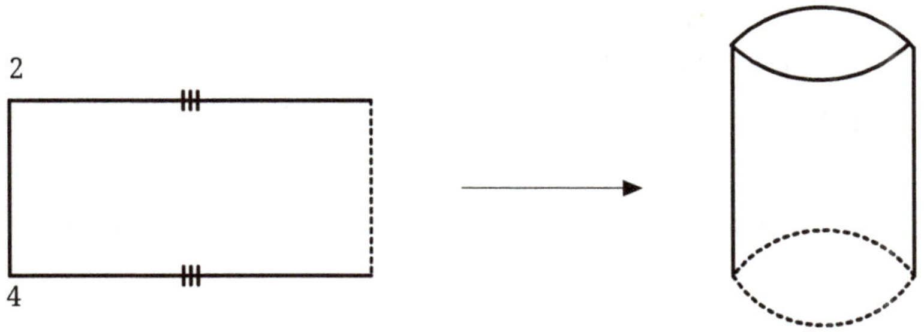

Figure 7.4: Building a cylinder with a base circle removed.

Proposition 7.2.41. *The* $(n-1)$*-projective space* $\mathbb{R}\mathrm{P}^{n-1}$ *is a deformation retract of the punctured n-projective space, i. e.,* $\mathbb{R}\mathrm{P}^n$ *with one point removed.*

Proof. Recall that $\mathbb{R}\mathrm{P}^n$ is obtained from the sphere S^n by identifying antipodal points ($x \sim_1 -x$) (Definition 3.1.1). Let π_1 and π_2 be the canonical projections given by Definition 3.1.1 and Proposition 3.1.10, respectively. Removing one point from $\mathbb{R}\mathrm{P}^n$ means that two antipodal points have been already removed from S^n. Let $X = S^n \setminus \{N_p, S_p\}$. For $x = (x_1, x_2, \dots, x_n, x_{n+1}) = (x', x_{n+1}) \in X$, let $h_N(x) = \frac{1}{1-x_{n+1}} x'$ be the stereographic projection of a point $x \in X$ via the northern pole (Example 1.5.26). Since $h_N(S_p) = 0$, X is homeomorphic to $X' = \mathbb{R}^n \setminus \{0\}$ with homeomorphism h_N. Consider the commutative diagram

$$\begin{array}{ccc} X & \xrightarrow{h_N} & X' \\ {\scriptstyle \pi_1}\downarrow & & \downarrow{\scriptstyle \pi_2} \\ Y & \xrightarrow{r} & Z \end{array}$$

where $Y = \mathbb{RP}^n \setminus \{*\}$ and $Z = \mathbb{RP}^{n-1}$. Suppose that $\pi_1(x) = (\pi_2 \circ h_N)(x)$ for some $x \in X$. Since $r \circ \pi_1 = \pi_2 \circ h_N$, we have

$$(r \circ \pi_1)(x) = (\pi_2 \circ h_N)(x) = \pi_1(x).$$

Then the map r is continuous by the factorization Theorem 2.1.9 and retracts the space Y onto Z. Finally, the map $H : Y \times I \longrightarrow Y$ defined by $H(\pi_1(x), t) = \pi_1(x_0, x_1, \dots, (1-t)x_n)$ is the required deformation homotopy. Indeed, for all $x \in X$, $H(\pi_1(x), 0) = \pi_1(x)$ and

$$H(\pi_1(x), 1) = \pi_1(x', 0) = \pi_2(x') = (\pi_2 \circ h_N)(x) = r(\pi_1(x)).$$ □

According to Theorem 3.2.8, a Möbius band $\mathbb{M}^2$ can be viewed as a subset of a projective plane $\mathbb{RP}^2$. We have the following.

Proposition 7.2.42. *The open Möbius band $\mathring{\mathbb{M}}^2$ is a deformation retract of the punctured projective plane $\mathbb{RP}^2 \setminus \{*\}$.*

Proof. Let $\mathring{\mathbb{M}}^2$ be given as in Remark 3.2.6 and consider the commutative diagram:

$$\begin{array}{ccc} S^2 \setminus \{N_p, S_p\} & \xrightarrow{\quad h \quad} & S^1 \times \mathbb{R} \\ {\scriptstyle \pi_1}\downarrow & & \downarrow{\scriptstyle \pi} \\ \mathbb{RP}^2 \setminus \{*\} & \xrightarrow{\quad r \quad} & \mathring{\mathbb{M}}^2 \end{array}$$

By the stereographic projection (Example 1.5.26) and Theorem 1.5.38, the map h is a homeomorphism. Arguing as in Proposition 7.2.41, we deduce that the map r defined by $r \circ \pi = \pi \circ h$ is a retraction. □

Next, we collect the essential properties of the Möbius band and the projective space. Results 3 and 10 will be proved in Chapter 9 and Chapter 10, respectively. ∂ means that the boundary of the Möbius band is glued to the circle $S^1 \subset B^2$. π is the projection map defining the projective plane.

Corollary 7.2.43. *We have:*
(1) $\mathbb{RP}^2 \cong B^2 \cup_\partial \mathbb{M}^2 \cong B^2 \cup_{z^2} S^1$ *(Theorem* 3.2.8*);*
(2) $\mathbb{RP}^n \cong B^n \cup_\pi \mathbb{RP}^{n-1}$ *(Example* 3.4.19*);*
(3) $\mathbb{RP}^{n-1}$ *is not a retract of* $\mathbb{RP}^n$ *(Corollary* 9.2.20*);*
(4) $\mathbb{RP}^{n-1}$ *is a deformation retract of* $\mathbb{RP}^n \setminus \{*\}$ *(Proposition* 7.2.41*);*
(5) $\mathring{\mathbb{M}}^2$ *is a deformation retract of* $\mathbb{RP}^2 \setminus \{*\}$ *(Proposition* 7.2.42*);*

(6) $C(\mathbb{M}^2)$ *is homeomorphic to* S^1 *(Example* 3.2.5*);*
(7) $\partial\,\mathbb{M}^2$ *is homeomorphic to* S^1 *(Proposition* 3.2.7*);*
(8) $\mathbb{M}^2$ *deformation retracts on its core circle* $C(\mathbb{M}^2)$ *(Proposition* 7.2.33*);*
(9) $\mathbb{M}^2 \setminus C(\mathbb{M}^2)$ *deformation retracts on the boundary circle* $\partial\,\mathbb{M}^2$ *(Proposition* 7.2.37*);*
(10) $\mathbb{M}^2$ *does not retract on its boundary circle* $\partial\,\mathbb{M}^2$ *(Corollary* 10.3.29*).*

Remark 7.2.44. By Theorem 7.1.17, the image of a retract of a space X under a homeomorphism is a retract of the homeomorphic space. However, parts (5), (6) and (8) of Corollary 7.2.43 show that if A, B are homeomorphic subspaces of X with A, a retract of X, then not necessarily X retracts onto B.

7.3 The homotopy extension property

We aim to present a practical result, which is the topological version of the algebraic fact stating that the quotient of a group G over the identity element is isomorphic to G (see Example B.2.11(2)). This works for some good pair (X, A) as defined in the next definition (see [25, Definition (7.1), Chapter XVI], [45, p. 14], and [49, Definition 9.1]). For those classes of pairs, a substantial result of stability type will be given in Theorem 7.4.8.

Definition 7.3.1. Let $A \subset X$. We say that the pair (X, A) has the homotopy extension property (HEP for brevity) if for every space Y, any continuous map from the slice $(X \times \{0\}) \cup (A \times I)$ to Y can be extended to a continuous map from $X \times I$ to Y.

The homotopy extension property is called the Borsuk property (and (X, A) is a Borsuk pair), as well. When (X, A) is a Borsuk pair, the inclusion $i : A \hookrightarrow X$ is called a cofibration (see, e. g., [94, Chapter 1, Section 4, p. 29]). Most properties of cofibrations are discussed in [12, Chapter 7] and [94].

From Theorem 7.1.24, we deduce the following.

Proposition 7.3.2. *A pair* (X, A) *satisfies the homotopy extension property if and only if* $(X \times \{0\}) \cup (A \times I)$ *is a retract of the cylinder* $X \times I$.

Actually, we have more in the following.

Proposition 7.3.3. *The following statements are equivalent:*
(1) (X, A) *is a Borsuk pair;*
(2) $(X \times \{0\}) \cup (A \times I)$ *is retract of the cylinder* $X \times I$*;*
(3) $(X \times \{0\}) \cup (A \times I)$ *is a deformation retract of the cylinder* $X \times I$.

Proof. Since by Proposition 7.3.2, statements (1) and (2) are equivalent, we only check that (2) implies (3). Let $r : X \times I \longrightarrow (X \times \{0\}) \cup (A \times I)$ be a retraction represented as $r(x, t) = (r_1(x, t), r_2(x, t))$, where r_1 and r_2 are the components of r, i. e., $r_1(x, t) \in X$

and $r_2(x,t) \in I$ Then a required deformation homotopy is given by (see [12, Exercise 6, Section 7.2])

$$H(x,t,s) = (r_1(x,st), t(1-s) + sr_2(x,t)).$$

We have $H(x,t,0) = (r_1(x,0),t) = (x,t)$ and $H(x,t,1) = (r_1(x,t), r_2(x,t)) = r(x,t)$, as claimed. In [25, Theorem 7.4, Chapter XV], the deformation homotopy is written equivalently as

$$H(x,t,s) = ((p_X \circ r)(x,st), (1-s)t + s(p_I \circ r)(x,t)),$$

where $p_X(x,t) = x$ and $p_I(x,t) = t$ are the projections onto the first and the second variable, respectively. □

Remark 7.3.4. In Remark 7.2.30, we have discussed the connection between the homotopy equivalence and the retraction. When the pair (X,A) is a Borsuk pair (Definition 7.3.1) and the inclusion map $i : A \hookrightarrow X$ is a homotopy equivalence, X deformation retracts on A. The following result can be found also in [45, Corollary 0.20].

Proposition 7.3.5. *Let (X,A) be a Borsuk pair. Then A is a deformation retract of X if and only if the cofibration $i : A \hookrightarrow X$ is a homotopy equivalence.*

Proof. Necessity is Exercise 17 of this chapter. So, we only check the sufficient condition. Let $g : X \longrightarrow A$ be a homotopy inverse of the inclusion map i, i. e., $g \circ i \simeq \mathrm{Id}_{|A}$ and let $H : A \times I \longrightarrow A$ be such a homotopy, i. e., $H(\cdot,0) = g_A$ and $H(\cdot,1) = \mathrm{Id}_{|A}$. Since (X,A) is a Borsuk pair, by Definition 7.3.1, H has a continuous extension $\widetilde{H} : X \times I \longrightarrow A$. Hence, $\widetilde{H}(x,0) = g(x)$ for all $x \in X$. For $x \in X$, let $r(x) = \widetilde{H}(x,1)$. Then $r(x) = H(x,1) = x$ for all $x \in A$. Hence, $r : X \longrightarrow A$ is a retraction and $\widetilde{H}$ is a homotopy from r to g. Finally, since by assumption $i \circ g$ and Id_X are homotopic, then so are $i \circ r$ and Id_X, proving that r is a deformation retraction. □

Now consider a Borsuk pair (X,A), where $A \subset X$ is a closed subset.

Proposition 7.3.6. *Let $B \subset A \subset X$ be such that (X,A) and (A,B) have the homotopy extension property. Then the pair (X,B) has the homotopy extension property.*

Proof. Let $r : X \times I \longrightarrow (X \times \{0\}) \cup (A \times I)$ and $s : A \times I \longrightarrow (A \times \{0\}) \cup (B \times I)$ be retractions. The map $\rho : X \times I \longrightarrow (A \times \{0\}) \cup (B \times I)$ given by

$$\rho(x,t) = \begin{cases} r(x,t), & \text{if } r(x,t) \in X \times \{0\}, \\ (s \circ r)(x,t), & \text{if } r(x,t) \in A \times I \end{cases}$$

is continuous because on the intersection $A \times \{0\} = X \times \{0\} \cap A \times I$, we have $r(a,0) = (a,0)$, which implies $(s \circ r)(a,0) = s(a,0) = (a,0)$. For $(x,t) \in X \times \{0\}$, $\rho(x,t) = r(x,t) = (x,t)$ and for $(x,t) \in B \times I \subset A \times I$, $r(x,t) = (x,t)$, which implies $(s \circ r)(x,t) = s(x,t) = (x,t)$, proving that ρ is the required retraction. □

Proposition 7.3.7. *Let (X, A) and (Y, B) have the homotopy extension property. Then the pair $(X \times Y, A \times B)$ has the homotopy extension property.*

Proof. Let $r = (r_1, r_2) : X \times I \longrightarrow (X \times \{0\}) \cup (A \times I)$ be a retraction, where r_1 and r_2 refer to the components of r, i. e., $r_1(x, t) \in X$ and $r_2(x, t) \in I$. Let $s = (s_1, s_2) : Y \times I \longrightarrow (Y \times \{0\}) \cup (B \times I)$ be a retraction, too. Then the map $\rho : (X \times Y) \times I \longrightarrow ((X \times Y) \times \{0\}) \cup ((A \times B) \times I)$ given by

$$\rho(x, y, t) = \left(r_1(x, t), s_1(y, t), \frac{r_2(x, t) + s_2(y, t)}{2} \right)$$

is the required retraction (see [11, 7.3.2]). □

Proposition 7.3.8. *Let (X, A) have the homotopy extension property. For some space Y, let $f, g : A \longrightarrow Y$ be homotopic maps such that f extends continuously to the whole space X. Then the same holds for g.*

Proof. Let $H : A \times I \longrightarrow Y$ be a homotopy between f and g and $\tilde{f} : X \longrightarrow Y$ a continuous extension of f. Then $H(a, 0) = f(a)$ and $H(a, 1) = g(a)$ for all $a \in A$. Define the map $F : (X \times \{0\}) \cup (A \times I) \longrightarrow Y$ by

$$F(x, 0) = \tilde{f}(x) \quad \text{for all } x \in X \quad \text{and} \quad F(a, t) = H(a, t) \quad \text{for all } (a, t) \in A \times I.$$

On the intersection $(X \times \{0\}) \cap (A \times I) = A \times \{0\}$, we have $F(a, 0) = \tilde{f}(a) = f(a) = H(a, 0)$ for all $a \in A$. Hence, F is continuous. Since (X, A) has the HEP, F has a continuous extension $\tilde{F} : X \times I \longrightarrow Y$. Let $\tilde{g}(x) = \tilde{F}(x, 1)$. Then, for all $a \in A$, $\tilde{g}(a) = \tilde{F}(a, 1) = F(a, 1) = H(a, 1) = g(a)$. Hence, $\tilde{g}$ is the required extension of the map g. □

As a consequence, we have the following.

Corollary 7.3.9. *Let (X, A) have the homotopy extension property. Then, for any space Y, every null-homotopic map $f : A \longrightarrow Y$ extends to a continuous map $\tilde{f} : X \longrightarrow Y$.*

Theorem 7.3.10. *Let (X, A) have the homotopy extension property and $A \subset X$ is contractible. Then the quotient space X/A is homotopically equivalent to X.*

Proof. Let $H : A \times I \longrightarrow A$ be the homotopy between the identity on A and a point $a \in A$. The homotopy equivalence maps are the projection map

$$f = \pi : X \longrightarrow X/A \quad \text{given by } f(x) = [x] = \begin{cases} \{A\}, & x \in A, \\ \{x\}, & x \notin A, \end{cases}$$

which is continuous and the map

$$g : X/A \longrightarrow X \quad \text{given by } g(y) = \begin{cases} a, & y = \{A\}, \\ x, & y = \{x\} \neq A. \end{cases}$$

(1) The map g is continuous. Let $U \subset X$ be an open set.

(a) If $a \in U$, then

$$V := g^{-1}(U) = \{\{x\}, x \in U \setminus A\} \cup \{A\}.$$

Since $\pi^{-1}(V) = U \setminus A \cup A = U$ is open, V is open in X/A.

(b) If $a \notin U$, then $\{A\} \notin g^{-1}(U)$ for otherwise $g(\{A\}) = a \in U$. Hence,

$$V := g^{-1}(U) = \{\{x\},\ x \in U \setminus A\} = U \setminus A,$$

which is open since A is closed.

(2) We have $f \circ g = \mathrm{Id}_{|X/A}$. Since $a \in A$, we have

$$\begin{aligned}(f \circ g)(y) &= \begin{cases} f(a), & y = \{A\}, \\ f(x), & y = \{x\} \neq A \end{cases} \\ &= \begin{cases} [a], & y = \{A\}, \\ \{x\}, & y = \{x\} \neq A \end{cases} \\ &= \begin{cases} \{A\}, & y = \{A\}, \\ y, & y \neq \{A\} \end{cases} \\ &= y = \mathrm{Id}_{|X/A}(y).\end{aligned}$$

Moreover,

$$(g \circ f)(x) = g([x]) = \begin{cases} a, & x \in A, \\ x, & x \notin A. \end{cases}$$

(3) The composite $g \circ f$ is homotopically equivalent to the identity on X. Define the function $K : X \times I \longrightarrow X$ by

$$K(x,t) = \begin{cases} H(x,t), & x \in A, \\ x, & x \notin A. \end{cases}$$

Then

$$K(x,0) = \begin{cases} H(x,0) = x, & x \in A, \\ x, & x \notin A \end{cases}$$

i. e., $K(x,0) = \mathrm{Id}_{|X}(x)$ for all $x \in X$ and

$$K(x,1) = \begin{cases} H(x,1) = a, & x \in A, \\ x, & x \notin A, \end{cases}$$

i. e., $K(x,1) = (g \circ f)(x)$ for all $x \in X$. K is further continuous. It can be checked that for an open subset $U \subset X$ we have

$$K^{-1}(U) = \begin{cases} U \times I, & U \cap A = \emptyset, \\ ((U \setminus A) \times I) \cup H^{-1}(U \cap A), & U \cap A \neq \emptyset. \end{cases}$$

Since A is closed and U is open in X, we deduce that $K^{-1}(U)$ is an open set. Therefore, K is a homotopy between $g \circ f$ and the identity on X, which completes the proof. □

Remark 7.3.11. Let the pair (X, A) have the HEP and $i : A \hookrightarrow X$ the inclusion map.

(1) Recall that by Proposition 4.9.5, the cone $C(A)$ is contractible. Using Exercise 27, in this chapter one can check that $(C(A) \cup X, C(A))$ has the HEP. Using the definitions of the quotient space X/A and the cone $C(A)$ together with Theorem 7.3.10, we obtain

$$X/A \cong (C(A) \cup_i X)/C(A) \simeq C(A) \cup_i X. \tag{7.2}$$

(2) By Example 3.5.8, the mapping cone K_i of the inclusion map i is homeomorphic to the union $C(A) \cup_i X$. This with (7.2) yield

$$K_i \cong C(A) \cup_i X \simeq X/A. \tag{7.3}$$

Example 7.3.12. By Example 3.4.24(2), the sphere is an n-skeleton. This with [45, Proposition 0.16] allow us to take $A = S^k$ and $X = S^n$ $(0 \le k \le n-1)$. By (7.3), Theorem 3.3.6(1), we have

$$K_i \cong C(S^k) \cup_i S^n \cong B^{k+1} \cup_i S^n \simeq S^n/S^k.$$

Example 7.3.13. By Exercise 21, Chapter 6, for all $k < n$ the inclusion map $i : S^k \hookrightarrow S^n$ is homotopic to a constant map $X \longrightarrow \{y_0\} \in Y$ (i. e., S^k is contractible in S^n). By Example 3.5.7(1), the mapping cone K_{cst} of a constant map cst is the wedge sum $C(X) \cup_{\mathrm{cst}} Y \cong \Sigma(X) \vee Y$. Using Example 7.3.12 together with Theorem 3.3.6(2) and Theorem 7.4.8, we get

$$K_i \cong K_{\mathrm{cst}} \cong \Sigma(A) \vee X = \Sigma(S^k) \vee S^n \quad \Longleftrightarrow \quad S^n/S^k \simeq S^{k+1} \vee S^n. \tag{7.4}$$

Example 7.3.14. Let us consider some interesting particular cases of (7.4):

(1) The case $n = 1$, $k = 0$:

$$S^1/S^0 \cong S^1 \vee S^1$$

(circle with two points identified, see Remark 6.2.8).

(2) The case $n = 2$, $k = 0$:

$$S^2/S^0 \cong S^2 \vee S^1$$

(sphere with two points identified, same as for the circle).

(3) The case $n = 2$, $k = 1$:

$$S^2/S^1 \cong S^2 \vee S^2$$

(sphere with the equator identified, see Example 2.5.4).

A standard example of an HEP pair is given by the following.

Proposition 7.3.15. *The pair (B^n, S^{n-1}) has the homotopy extension property.*

Proof. To apply Proposition 7.3.2, let $X = B^n \times I$ and $A = (B^n \times \{0\}) \cup (S^{n-1} \times I)$.

(1) *First proof.* A retraction $r : X \longrightarrow A$ can be defined by the radial retraction from a point $P_0 = (0, a)$ for some $a > 1$ as follows:

$$r(P) = \begin{cases} (P_0P) \cap (B^n \times \{0\}), & \text{if } P \text{ lies within the cone } (B^n, P), \\ (P_0P) \cap (S^{n-1} \times I), & \text{otherwise,} \end{cases}$$

where $P = (x, t) \in B^n \times I$ (for $n = 2$, see Figure 7.5). Since the cylinder is convex, one may take $H(t, x) = tr(x, t) + (1 - t)(x, t)$ as the homotopy between the identity on $B^n \times I$ and the retraction r.

(2) *Second proof.* For $(x, t) \in B^n \times I$, one may construct such a retraction explicitly as follows (see [38, Chapitre IV, Exemples 2.4(iii)]):

$$r(x, t) = \begin{cases} (\frac{x}{\|x\|}, \frac{2\|x\|-2+t}{\|x\|}), & 1 - t/2 \leq \|x\| \leq 1, \\ (\frac{2x}{2-t}, 0), & 0 \leq \|x\| \leq 1 - t/2. \end{cases}$$

(3) *Third proof.* (See [37, Lemma 1.3.9]). Let $f : B^n \longrightarrow S^{n-1}$ be given by

$$f(x) = \begin{cases} 0, & 0 \leq \|x\| \leq 1/2, \\ 2\|x\| - 1, & 1/2 \leq \|x\| \leq 1. \end{cases}$$

Define the epigraph and the graph of f:

$$Y = \{(x, t) \in B^n \times I : f(x) \leq t\} \quad \text{and} \quad Y_0 = \{(x, f(x)) : x \in B^n\}.$$

Then $r : Y \longrightarrow Y_0$ defined by $r(x, t) = (x, f(x))$ is a retraction and for $(x, t, s) \in Y \times I$, $H(x, t, s) = (x, (1 - s)t + sf(x))$ a deformation homotopy. □

Remark 7.3.16. More generally, it is proved in [45, Proposition 0.16] that every *CW* pair (X, A), where X a *CW* complex and $A \subset X$ a closed subcomplex has the homotopy extension property. (See also [37, Theorem 1.3.15]).

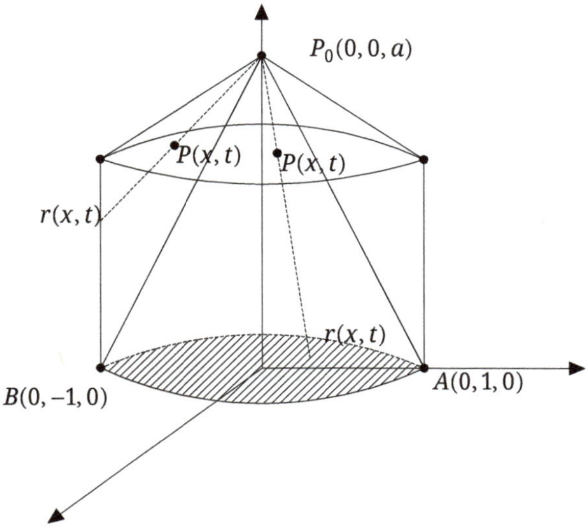

Figure 7.5: The pair (B^2, S^1) has the HEP.

7.4 Retract of an adjunction space

Let $f : A \subset X \longrightarrow Y$ be a continuous map, $A \subset X$ a closed subset and $Z_f = X \cup_f Y = (X \cup Y)/_\sim$ the adjunction space of X and Y (see Definition 3.4.1). According to Corollary 3.4.35, Z_f is homeomorphic to the partition, which consists of $\pi(Y)$ and $\pi(X \setminus A)$, where $\pi : X \cup Y \longrightarrow Z_f$ is the canonical projection map (see Section 3.4). We have the following.

Theorem 7.4.1. (1) *If A is a retract of X, then $\pi(Y)$ is a retract of Z_f.*
(2) *If A is a deformation retract of X, then $\pi(Y)$ is a deformation retract of Z_f.*

Proof. (1) For some retraction $r : X \longrightarrow A$, define the map $s : Z_f \longrightarrow \pi(Y)$ by

$$s(z) = \begin{cases} z, & z \in \pi(Y), \\ (\pi \circ f \circ r)(x), & z \in \pi(X \setminus A) \text{ with } z = \pi(x) \text{ and } x \in X \setminus A. \end{cases}$$

Note that $\pi_{|X \setminus A}$ is injective so x is unique in the definition of the map s. Then s is the required retraction. Furthermore,

$$(s \circ g)(x) = ((\pi \circ f) \circ r)(x) = (\pi \circ f) \circ r(x), \ x \in X,$$

i. e., the following first diagram commutes, where $g = \pi_{|X}$ is introduced in Section 3.4:

$$\begin{array}{ccc} A & \xleftarrow{\quad r \quad} & X \\ \Big\downarrow{\scriptstyle \pi\circ f} & & \Big\downarrow{\scriptstyle g} \\ \pi(Y) & \xleftarrow[\quad s \quad]{} & Z_f \end{array} \qquad\qquad \begin{array}{ccc} A & \xhookrightarrow{\quad i \quad} & X \\ \Big\downarrow{\scriptstyle f} & & \Big\downarrow{\scriptstyle g} \\ Y & \xhookrightarrow{\quad j \quad} & Z_f \end{array}$$

(2) Let $H : X\times I \longrightarrow X$ be a homotopy such that for all $x \in X, H(x,0) = x$ and $H(x,1) = r(x)$, where $r : X \longrightarrow A$ is a retraction of X onto A. Define the functions $K_1 : \pi(X\setminus A)\times I \longrightarrow Z_f$ and $K_2 : \pi(Y)\times I \longrightarrow Z_f$ by $K_2(z,t) = z$ for all $z \in \pi(Y)$ and

$$K_1(z,t) = \begin{cases} (\pi\circ f)(H(x,t)), & H(x,t) \in A, \\ g(H(x,t)), & H(x,t) \notin A, \end{cases}$$

for $z \in \pi(X\setminus A), z = \pi(x), x \in X\setminus A$. Then

$$\begin{aligned} K_1(z,0) &= \begin{cases} g(x) = \pi(x) = z, & H(x,0) \in A, \\ g(f(x)) = g(x) = z, & H(x,0) \notin A, \end{cases} \\ &= z, \end{aligned}$$

for all $z \in \pi(X\setminus A)$. Since $r(x) \in A$, we have $g(r(x)) = (\pi\circ f\circ r)(x)$. Hence,

$$\begin{aligned} K_1(z,1) &= \begin{cases} g(r(x)) = s(g(x)) = s(z), & H(x,1) \in A, \\ (\pi\circ f\circ r)(x)) = (s\circ g)(x) = s(z), & H(x,1) \notin A, \end{cases} \\ &= s(z), \quad \text{for all } z \in \pi(X\setminus A). \end{aligned}$$

Finally, define $K : Z_f\times I \longrightarrow Z_f$ by

$$K(z,t) = \begin{cases} K_1(z,t), & z \in \pi(X\setminus A), \\ K_2(z,t), & z \in \pi(Y). \end{cases}$$

Then K is continuous and $K(z,0) = z$ and $K(z,1) = s(z)$ for all $z \in Z_f$. Hence, K is a homotopy between the identity on Z_f and the retraction s. We conclude that $K\circ l = K_1$ and $K\circ m = K_2$, where $l : \pi(X\setminus A)\times I \hookrightarrow Z_f\times I$ and $m : \pi(Y)\times I \hookrightarrow Z_f\times I$ are the inclusion maps. □

Since by Theorem 7.1.24, the retraction is equivalent to an extension property, the following result is stronger than part (1) of Theorem 7.4.1.

Theorem 7.4.2. *Let $A \subset X$ be a closed subset. A continuous map $f : A \subset X \longrightarrow Y$ has a continuous extension $\widetilde{f} : X \longrightarrow \pi(Y)$ if and only if $\pi(Y)$ is a retract of Z_f.*

Here, the extension is understood in the sense $\widetilde{f}(a) = [f(a)]$ for all $a \in A$. By Proposition 3.4.28, $\widetilde{f}$ is an extension of f up to homeomorphism.

Proof. (1) Let $r : Z_f \longrightarrow \pi(Y)$ be a retraction and $\widetilde{f} = r \circ \pi_{|X}$. Then $\widetilde{f}$ is continuous and for all $a \in A$, $\pi_{|X}(a) = [a] = [f(a)]$. Hence, $\widetilde{f}(a) = r(\pi_{|X}(a)) = r([f(a)]) = [f(a)]$ since $[f(a)] \in \pi(Y)$. Therefore, $\widetilde{f}$ is an extension of f.

(2) Let $f : A \longrightarrow Y$ have a continuous extension $\widetilde{f} : X \longrightarrow \pi(Y)$. Since by Corollary 3.4.35, $\pi(X \setminus A)$ and $\pi(Y)$ form a partition of Z_f, define the function $r : Z_f \longrightarrow \pi(Y)$ by

$$r(z) = \begin{cases} \widetilde{f}(z), & z \in \pi(X \setminus A), \\ z, & z \in \pi(Y). \end{cases}$$

By definition, $r(z) = z$ for all $z \in \pi(Y)$. Moreover, for all $x \in X$,

$$\begin{aligned}(r \circ \pi_{|X})(x) = r([x]) &= \begin{cases} \widetilde{f}([x]), & x \notin A, \\ [f(x)], & x \in A \end{cases} \\ &= \widetilde{f}(x)\end{aligned}$$

because $[x] = \{x\}$ for all $x \in X \setminus A$. Also, for all $y \in Y$, $(r \circ \pi_{|Y})(y) = r([y]) = [y] = \pi_{|Y}(y)$ by definition of r. This shows that $r \circ \pi_{|X} = \widetilde{f}$ and $r \circ \pi_{|Y} = \pi_{|Y}$ are continuous, which implies by the factorization theorem 2.1.9 the continuity of the map r. Therefore, r is the required retraction. □

Let $f : A \subset X \subset X' \longrightarrow Y$ be a continuous map, where A is closed in X and X is closed in X'.

Theorem 7.4.3. (1) *If X is a retract of X', then so is $X \cup_f Y$ a retract of $X' \cup_f Y$.*
(2) *Assume further $X' \cup Y$ is compact and $X' \cup_f Y$ Hausdorff. If X is a deformation retract of X', then $X \cup_f Y$ is a deformation retract of $X' \cup_f Y$.*

Proof. Let $r : X' \longrightarrow X$ be a retraction and $H : X' \times I \longrightarrow X'$ the associated homotopy joining r to the identity on X'. Define $s : X' \cup Y \longrightarrow X \cup Y$ and $K : (X' \cup Y) \times I \longrightarrow (X' \cup Y)$ by

$$s(z) = \begin{cases} r(z), & z \in X', \\ z, & z \in Y \end{cases} \quad \text{and} \quad K(z,t) = \begin{cases} H(z,t), & z \in X', \\ z, & z \in Y. \end{cases}$$

Then s and K are continuous and $s(x) = r(x) = x$ for all $x \in X$ so s retracts $X' \cup Y$ onto $X \cup Y$. In addition,

$$K(z,0) = \begin{cases} H(z,0) = z, & z \in X', \\ z, & z \in Y \end{cases} = z. \qquad K(z,1) = \begin{cases} H(z,1) = r(z), & z \in X', \\ z, & z \in Y \end{cases} = s(z).$$

Hence, $X \cup Y$ is a deformation retract of $X' \cup Y$. Passing to the quotient spaces (Proposition 7.2.32), we deduce that $X \cup_f Y$ is a deformation retract of $X' \cup_f Y$. □

As a consequence, we recapture part (2) of Theorem 7.4.1.

Corollary 7.4.4. *Under the assumptions of Theorem 7.4.3, if A is a deformation retract of X, then $\pi(Y)$ is a deformation retract of Z_f.*

Remark 7.4.5. (1) Let $X = B^2, A = S^1, Y = \mathbb{T}^2$ the torus and $f = \pi \circ h$, where $h : B^2 \longrightarrow I^2$ is a homeomorphism and π the projection map of I^2 onto the torus. Since $f(S^1) \cong S^1 \vee S^1$ and $f(\mathring{B}^2) \cong \mathring{B}^2$, the adjunction space $Z_g = B^2 \cup_g (S^1 \vee S^1)$ is homeomorphic to the torus $\mathbb{T}^2$, where g is the restriction of f on S^1 (see Example [12, 4.5.6]).

(2) By Proposition 7.2.35, $S^1 \vee S^1$ is a deformation retract of the punctured torus $\mathbb{T}'^2 = \mathbb{T}^2 \setminus \{*\}$. Theorem 7.4.3 implies that $Z_g = B^2 \cup_g (S^1 \vee S^1)$ is a deformation retract of $B^2 \cup_g \mathbb{T}'^2$. By part (1), the torus $\mathbb{T}^2$ is a deformation retract of $B^2 \cup_g \mathbb{T}'^2$.

Let $f : A \subset X \longrightarrow Y$ be a continuous map, where A is a closed subset, $\pi : X \cup Y \longrightarrow Z_f$ the projection map and $g = \pi_{|Y} \circ f : A \longrightarrow Z_f$. Here, we ignore the identification between Y and $\pi_{|Y}(Y)$ given by Proposition 3.4.28 as in [49, Proposition 7.1]. The following result provides quite interesting applications in retraction theory.

Theorem 7.4.6. *The image $\pi(Y)$ is a retract of Z_f if and only if the continuous map $g = \pi_{|Y} \circ f$ has a continuous extension $\widetilde{g} : X \longrightarrow Z_f$.*

Proof. (1) If $\widetilde{g}$ exists, then define $r : Z_f \longrightarrow \pi_{|Y}(Y)$ by

$$r(z) = \begin{cases} z, & z \in \pi_{|Y}, \\ \widetilde{g}(x), & z \notin \pi_{|Y}, \end{cases}$$

where x is uniquely given by $z = \pi(x)$, $x \in X \setminus A$ (see the description of the projection map π in Section 3.4). Clearly, r is continuous and is the required retraction.

(2) Conversely, let $r : Z_f \longrightarrow \pi_{|Y}(Y)$ be a retraction and set $\widetilde{g} = r \circ \pi_{|X}$:

$$X \xrightarrow{\quad \pi_{|X} \quad} Z_f \xrightarrow{\quad r \quad} \pi_{|Y}(Y).$$

For all $x \in A$, we have $\pi_{|Y}(f(x)) \in \pi_{|Y}(Y)$, and thus

$$\widetilde{g}(x) = (r \circ \pi_{|X})(x) = r(\pi_{|X}(x)) = r(\pi_{|Y}(f(x)) = \pi_{|Y}(f(x) = g(x).$$

Then $\widetilde{g}$ is a continuous extension of g. □

Theorem 7.4.2 and Theorem 7.4.6 result in the following one.

Theorem 7.4.7. *Let (X, A) be a pair. Then the following statements are equivalent:*

(1) *A continuous map $f : A \subset X \longrightarrow Y$ has a continuous extension $\widetilde{f} : X \longrightarrow \pi(Y)$;*

(2) *$\pi(Y)$ is a retract of Z_f;*
(3) *The continuous map $g = \pi_{|Y} \circ f : A \longrightarrow Z_f$ has a continuous extension $\widetilde{g} : X \longrightarrow Z_f$.*

The following result shows the stability of adjunction spaces under homotopy when the homotopy extension property holds for a pair.

Theorem 7.4.8. *Let (X, A) be a Borsuk pair and $f, g : A \longrightarrow Y$ two homotopic maps. Then the adjunction spaces $X \cup_f Y$ and $X \cup_g Y$ are homotopically equivalent.*

Proof. Let $H : A \times I \longrightarrow Y$ be a homotopy between f and g and $C_H = (X \times I) \cup_H Y$ the mapping cylinder. Since $X \times \{0\} \cup A \times I$ is a retract of $X \times I$, let r be such a retraction.

(1) By Proposition 7.3.3, $(X\times\{0\})\cup(A\times I)$ is a deformation retract of $X\times I$. Theorem 7.4.3 implies that $(X \times \{0\} \cup A \times I) \cup_H Y$ is a deformation retract of $(X \times I) \cup_H Y$.

(2) We claim that $(X \times \{0\}) \cup_f Y$ is a deformation retract of $(X \times I) \cup (A \times \{0\}) \cup_H Y$. For this, we first write $X \times \{0\} = (X \times \{0\}) \cup (A \times \{0\})$ and $(X \times I) \cup (A \times \{0\}) = X \times I$. Using part (1) and Theorem 7.4.3, the claim follows by observing that on the product $A \times \{0\}$, the homotopy H restricts to f.

(3) We repeat the above process with the space $(X \times \{1\}) \cup_g Y$. By Theorem 7.2.4, the spaces $(X\times\{0\})\cup_f Y$, $(X\times\{0\})\cup_g Y$ and $(X\times I)\cup_H Y$ are homotopically equivalent, whence the proof of the theorem with the identification $X \times \{0\} \cong X$. □

Example 7.4.9. Let $f : S^{n-1} \longrightarrow Y$ be null-homotopic. By Example 3.4.14 and Theorem 7.4.8, the adjunction space $B^n \cup_f Y$ is homotopically equivalent to the wedge sum $S^n \vee Y$. In particular, $B^n \sim_f S^m \cong S^n \vee S^m$ whenever $f : S^{n-1} \longrightarrow S^m$ is null-homotopic.

We examine a special and important situation of maps on spheres. Recall (Definition 3.4.21) that a n-cell is a space homeomorphic to the ball B^n. So, let $f : A = S^{n-1} \subset B^n \longrightarrow X$ be a continuous map, where X is a path-connected space. Let $\pi : B^n \cup X \longrightarrow Z_f = B^n \cup_f X$ be the projection map, where the adjunction space is obtained by attaching an n-cell to space X. Then Theorem 6.4.1, Theorem 7.4.6 and Theorem 7.4.2 may be collected in the following one.

Theorem 7.4.10. *Let $n \geq 2$. The following statements are equivalent:*
(1) *$f : S^{n-1} \longrightarrow X$ is null-homotopic;*
(2) *f extends continuously to $\widetilde{f} : B^n \longrightarrow X$;*
(3) *$\pi(X)$ is a retract of Z_f;*
(4) *The map $g = \pi \circ f$ has a continuous extension $\widetilde{g} : B^n \longrightarrow Z_f$.*

Recall that the Moore space is given by $M_n = B^2 \cup_f S^1$, where $f(z) = z^n : S^1 \longrightarrow S^1$ is the power map (see Definition 3.4.17). By Proposition 6.2.3, $\deg f = n \neq 0$. According to Example 5.2.4 and Proposition 6.1.1, f is not null-homotopic. Hence, f does not admit a continuous extension $\widetilde{f} : B^2 \longrightarrow S^1$. By Theorem 7.4.10, we have the following.

Corollary 7.4.11. *The image $\pi(S^1)$ is not a retract of M_n.*

7.5 Absolute retract

In this section, we assume that all spaces are normal (see Definition 1.1.18).

Definition 7.5.1. A continuous map $r : X \longrightarrow Y$ is *an r-map* provided there exists a continuous map $s : Y \longrightarrow X$ such that $r \circ s = \mathrm{Id}_Y$ (r is right invertible).

If $r : X \longrightarrow Y$ is an r-map, then r is surjective and $s : Y \longrightarrow s(Y)$ is a bijection. Thus, we have the following extension property.

Proposition 7.5.2. *Let X be a space and $Y \subset X$ a subspace. A map $r : X \longrightarrow Y$ is an r-map if and only if for any space Z, every continuous map $f : Y \longrightarrow Z$ has a continuous extension $\widetilde{f} : X \longrightarrow Z$.*

Proof. (1) Let $r : X \longrightarrow Y$ be an r-map and $f : Y \longrightarrow Z$ a continuous map, where Z is an arbitrary space. The continuous map $\widetilde{f} = f \circ r$ satisfies $\widetilde{f}(x) = f(x)$ for all $x = s(y)$ and $y \in Y$. Since $s : Y \longrightarrow s(Y)$ is bijective and $s(Y) \subset X, \widetilde{f} : X \longrightarrow Z$ may be regarded as an extension of the map f.

(2) Conversely, if the condition holds, take $Z = Y$ and $f = \mathrm{Id}_Y$ to obtain the r-map $r = \widetilde{\mathrm{Id}_Y} : X \longrightarrow Z = Y$ and the inclusion map $s : Y \hookrightarrow X$. □

Example 7.5.3. A retraction is an r-map.

Definition 7.5.4. A space X is *an absolute retract* ($X \in \mathrm{AR}$) if for every space Y and for any embedding $h : X \longrightarrow Y$, the set $h(X)$ is a retract of Y. (Thus, X can be embedded as a retract of any space.)

Proposition 7.5.5. *A retract of an absolute retract is an absolute retract.*

Proof. Let $A \subset X$ be a retract of an absolute retract X. Let Y be a space and $h : A \longrightarrow Y$ an embedding. By Theorem 7.1.17, $h(A)$ is a retract of $h(X)$ and $h(X)$ is a retract of Y since $X \in \mathrm{AR}$. Proposition 7.1.26 concludes the proof. □

Theorem 7.5.6. *If a metric space E is an r-image of some normed space (image by some r-map), then E is an* AR *space.*

Proof. Suppose one is given an r-map $r : N \longrightarrow E$ from a normed space N onto the metric E and let $h : E \longrightarrow Y$ be an embedding, where Y is an arbitrary space. Define $f = s \circ h^{-1} : h(E) \longrightarrow N$, where $r \circ s = \mathrm{Id}_E$. Since a normed space has the extension property, there exists an extension $\widetilde{f} : Y \longrightarrow N$. Then the map $\varrho : Y \longrightarrow h(E)$ given by $\varrho = r \circ \widetilde{f}$ is the required retraction. □

The converse is true and relies on the following embedding theorem due to Arens–Fells. A proof can be found in [21, Theorem 5.16].

Lemma 7.5.7. *Given a metric space E, there exists a normed space N and an isometry $h : E \longrightarrow h(E) \subset N$.*

Theorem 7.5.8. *Let E be an* AR *metric space. Then E is the r-image of some normed space.*

Proof. By Lemma 7.5.7, there exists a normed space N and an embedding $h : E \longrightarrow h(E) \subset N$. Since $E \in$ AR, $h(E)$ is a retract of N, proving the claim. □

Definition 7.5.9. A space X possesses *the extension property* provided that for every space Y and every closed subset $B \subset Y$, any continuous map $f : B \longrightarrow X$ has a continuous extension $\tilde{f} : Y \longrightarrow X$. We write $X \in$ ES.

Example 7.5.10. By Dugundgi's extension theorem (Lemma A.5.2), every normed space has the extension property.

Definition 7.5.9 is different from the characterization given in Theorem 7.1.24 though it is connected to Definition 7.5.4 via the following result (see [21, Theorem 5.20]).

Theorem 7.5.11. *We have $X \in$ ES $\Leftrightarrow X \in$ AR.*

Proof. (1) We have ES $\Longrightarrow$ AR. Given $X \in$ ES, let Y be an arbitrary space, $h : X \longrightarrow Y$ an embedding, and $B = h(X)$. Then, for $f = h^{-1}$, there exists an extension $\tilde{f} : Y \longrightarrow X$. Hence, the map $r = h \circ \tilde{f}$ is a retraction from Y onto $h(X)$, proving that $X \in$ AR.

(2) We have AR $\Longrightarrow$ ES. Assume that $X \in$ AR. Let B be a closed subset of Y and $f : B \longrightarrow X$ a continuous map. Since $X \in$ AR, in view of Theorem 7.5.8, there exists a normed space N and an r-map $r : N \longrightarrow X$, i. e., there exists $s : X \longrightarrow N$ such that $r \circ s = \mathrm{Id}_X$. Define $f_1 : B \longrightarrow N$ by $f_1 = s \circ f$. Since $E \in$ ES, the map $\tilde{f}_1 : Y \longrightarrow N$ is an extension of f_1 onto Y. Finally, the map $\tilde{f} : Y \longrightarrow X$ given by $\tilde{f} = r \circ f_1$ is an extension of f onto Y. □

Proposition 7.5.12. *Every absolute retract space X is contractible.*

Proof. Let Y be an arbitrary space, $f, g : Y \longrightarrow X$ two continuous maps, and $h_0 : (Y \times \{0\}) \cup (Y \times \{1\}) \longrightarrow X$ the map given by

$$h_0(y, 0) = f(y), \quad \text{and} \quad h_0(y, 1) = g(y), \quad y \in Y.$$

Since $X \in$ AR, by Theorem 7.5.11 $X \in$ ES. Hence, h_0 has a continuous extension $h : Y \times [0, 1] \longrightarrow X$. Clearly, this is equivalent to saying that f, g are homotopic. Then, if X is an AR space, any two continuous maps $f, g : Y \longrightarrow X$ are homotopic. By Theorem 4.9.7, the space X is contractible. □

Therefore, the inclusions in Remark 4.9.27 extend to:

$$\begin{aligned} \text{convex} &\subset \text{absolute retract} \cong \text{has the extension property} \\ &\subset \text{contractible} \\ &\subset \text{simply connected} \\ &\subset \text{path-connected} \subset \text{connected}. \end{aligned}$$

Before closing this section, all equivalent definitions of a contractible space are summarized in the following theorem. They come from this section (Theorem 7.1.25 and Proposition 7.2.7) and from Chapter 4 (Theorem 4.9.7, Corollary 4.9.9, Corollary 4.9.11, Proposition 4.9.17).

Theorem 7.5.13. *For a space X, the following statements are equivalent:*
(1) *X is contractible;*
(2) *X is contractible to a point $x_0 \in X$;*
(3) *The identity map $\mathrm{Id}_{|X}$ is null-homotopic;*
(4) *For every space Y, any two continuous maps $f, g : X \longrightarrow Y$ are homotopic;*
(5) *For every space Y, any two continuous maps $f, g : Y \longrightarrow X$ are homotopic;*
(6) *For every space Y, any continuous map $f : X \longrightarrow Y$ is null-homotopic;*
(7) *For every space Y, any continuous map $f : Y \longrightarrow X$ is null-homotopic;*
(8) *X is homotopically equivalent to a one-point set;*
(9) *X is a retract of the cone $C(X)$ over X;*
(10) *X deformation retracts onto a point $x_0 \in X$.*

7.6 Brouwer's fixed-point theorem for the disc

Definition 7.6.1. We say that a space X has the *fixed-point property* if every continuous function $f : X \longrightarrow X$ has a fixed point, that is an element $x \in X$ such that $f(x) = x$.

The central result of this section is to prove that the closed unit ball has the fixed-point property. In the beginning of the twentieth century, Brouwer has studied this widely-used theorem in [13–15].

Theorem 7.6.2 (Brouwer's fixed-point theorem for the disc). *Every continuous function $f : B^2 \longrightarrow B^2$ has a fixed point.*

Proof. Assume that $f(x) \neq x$ for all $x \in B^2$. Then we can define a map $r : B^2 \longrightarrow S^1$ such that for given $x \in B^2$, $r(x)$ is the point of intersection of the half-line $[f(x), x)$ originated from $f(x)$ through x with the boundary $S^1 = \partial B^2$ of the closed unit ball B^2 (see Figure 7.6). Notice that the convexity of the ball implies that the ray $[f(x), x)$ lies entirely in B^2. Since $r(x)$ lies in the half-line $[f(x), x)$, we have

$$r(x) = \lambda x + (1 - \lambda)f(x) = f(x) + \lambda(x - f(x)),$$

where $\lambda = \lambda(x) > 0$ is such that $\|r(x)\| = 1$ (to express that $r(x) \in \partial B^2$), i. e.,

$$\|r(x)\|^2 = 1 \quad \Longleftrightarrow \quad \langle r(x), r(x)\rangle = 1,$$

where $\langle\cdot\rangle$ refers to the inner product in $\mathbb{R}^2$. By substitution, we obtain a second-order algebraic equation for the unknown $\lambda = \lambda(x)$:

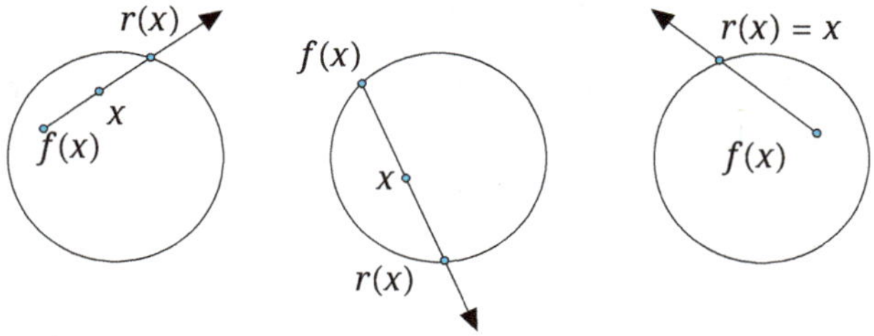

Figure 7.6: Construction of a retraction on the disk.

$$\lambda^2\|x-f(x)\|^2+2\lambda\langle f(x),x-f(x)\rangle+\|f(x)\|^2-1=0. \tag{7.5}$$

The reduced discriminant is

$$\triangle'=|\langle f(x),x-f(x)\rangle|^2+\|x-f(x)\|^2(1-\|f(x)\|^2).$$

Since $\|f(x)\|\le 1$, $\triangle'\ge 0$. We have

$$\begin{aligned}\triangle'=0 &\iff \|f(x)\|=1 \quad\text{and}\quad \langle f(x),x-f(x)\rangle=0\\ &\iff \lambda^2\|x-f(x)\|^2=0\\ &\iff \lambda=0 \quad\text{or}\quad f(x)=x,\end{aligned}$$

which is impossible. Hence, $\triangle'>0$ and the equation has a unique positive root:

$$\lambda_1=\lambda_1(x)=\frac{-\langle f(x),x-f(x)\rangle+\sqrt{\triangle'}}{\|x-f(x)\|^2}.$$

Since f is continuous, then so is the function $\lambda=\lambda(x)$. Hence, the map r is continuous and satisfies $r(x)=x$ for all $x\in S^1$, by definition of r. Indeed, since $r(x)-x=(1-\lambda)(f(x)-x)$ and $f(x)\neq x$ for all $x\in B^2$,

$$r(x)=x \iff \lambda=1.$$

Note that the expression of λ_1 yields the following equivalences:

$$\begin{aligned}\lambda_1=1 &\iff \triangle'=(\|x-f(x)\|^2+\langle f(x),x-f(x)\rangle)^2\\ &\iff \|x-f(x)\|^2+2\langle f(x),x-f(x)\rangle+\|f(x)\|^2-1=0\\ &\iff \|x-f(x)\|^2+2\langle f(x),x\rangle-\|f(x)\|^2-1=0\\ &\iff \|x\|^2+\|f(x)\|^2-2\langle f(x),x\rangle+2\langle f(x),x\rangle-\|f(x)\|^2-1=0\\ &\iff \|x\|=1\\ &\iff x\in S^1.\end{aligned}$$

This shows that r is a retraction of B^2 onto S^1, which contradicts Corollary 7.1.38. □

Remark 7.6.3. To prove Brouwer's fixed-point theorem, we have used both the compactness and the convexity of the ball B^2. The convexity of the ball is employed to say that the ray $[f(x), x)$ lies entirely in B^2, and thus B^2 can be replaced by a convex subset C. The compactness of B^2 is related to the nonretraction of the ball. The case of a general dimension is discussed in Section 11.1.

Remark 7.6.4. We have used Corollary 7.1.38 to prove Theorem 7.6.2. The following result shows that the converse holds true, that is to say Theorem 7.6.2 implies Corollary 7.1.38.

Theorem 7.6.5 (Borsuk's nonretraction of the ball). *The sphere S^1 is not a retract of the ball B^2.*

Proof. Suppose to the contrary that S^1 is a retract of B^2 and let $r : B^2 \longrightarrow S^1$ be a retraction. Then the map $f = -r$ is continuous from B^2 to $S^1 \subset B^2$. Note that

$$r(x) \in S^1 \iff -r(x) \in S^1,$$

thus f is well-defined. By Theorem 7.6.2, f has at least one-fixed point $x \in B^2$, i. e.,

$$f(x) = x \iff -r(x) = x.$$

Since $r(B^2) = S^1$, $-x \in S^1$ and so $x \in S^1$. Hence, $r(x) = x$, which gives $x = -x$, i. e., $x = 0$, which contradicts $x \in S^1$. □

In the next two results, we show that the fixed-point property is preserved under homeomorphism and retraction.

Theorem 7.6.6. *Let X and Y be homeomorphic spaces. Then X has the fixed-point property if and only if Y has the fixed point property.*

Proof. Let $h : X \longrightarrow Y$ be a homeomorphism and $f : Y \longrightarrow Y$ a continuous map. Then the map

$$g = h^{-1} \circ f \circ h : X \longrightarrow X$$

is continuous. If X has the fixed-point property, then there exists $x \in X$ such that

$$\begin{aligned} g(x) = x \quad &\iff \quad h^{-1} \circ f \circ h(x) = x \\ &\iff \quad f \circ h(x) = h(x) \\ &\iff \quad f(y) = y, \quad \text{with } y = h(x), \end{aligned}$$

that is f has a fixed point. □

Theorem 7.6.7. *If a space X has the fixed-point property and $r : X \longrightarrow Y$ is an r-map, then Y has the fixed-point property.*

Proof. Let $r : X \longrightarrow Y$ be an r-map, $s : Y \longrightarrow X$ its right-inverse, and $f : Y \longrightarrow Y$ a continuous map. Then the composite map $g = s \circ f \circ r : X \longrightarrow X$ is continuous. By supposition, there is some $x_0 \in X$ such that $g(x_0) = x_0$. Let $y_0 = r(x_0)$. Then $s(f(y_0)) = x_0$ and by left-composition with r, we get $f(y_0) = r(x_0) = y_0$. Hence, y_0 is a fixed point for the map f. □

Corollary 7.6.8. *If a space X has the fixed-point property and $A \subset X$ is a retract, then A has the fixed-point property.*

Example 7.6.9. Every retract of the closed unit ball has the fixed-point property.

Example 7.6.10. Every set homeomorphic to the closed unit ball has the fixed-point property. This holds for every closed ball $B(x_0, R)$, for

$$B(x_0, R) = \{x_0\} + RB(0, 1).$$

Any closed ball can be obtained from the closed unit ball using two simple geometric transformations: the shifting translation and the homothety (homogeneous dilation).

Remark 7.6.11. S^1 has not the fixed point property since rotations on the circle are fixed point- free. Actually, since B^2 has the fixed-point property, Corollary 7.6.8 shows that S^1 is not a retract of B^2, providing a second proof of Corollary 7.1.38.

Remark 7.6.12. The fixed-point property is not preserved under homotopy. For example, the continuous function $f : (0, 1) \longrightarrow (0, 1)$ given by $f(x) = x(1 - x)$ is null-homotopic for $(0, 1)$ is contractible (see Corollary 4.9.11) but has no fixed point.

Exercises (Chapter 7)

1. Prove that the radial retraction $r : \mathbb{R}^{n+1} \setminus \{0\} \longrightarrow S^n$ given by $r(x) = x/\|x\|$ is an open map.

2. Consider the ball retraction $\tilde{r}$ given in a normed space X by Example 7.1.4.
(1) Prove that $\tilde{r}$ is a 2-Lipschitz map (see Definition 1.2.6).
(2) Assume that X is a Hilbert space.
  (a) Show that $\langle x - \tilde{r}(x), \tilde{r}(y) - \tilde{r}(x)\rangle \le 0$ for all $x, y \in X$, where $\langle\cdot\rangle$ is the inner product in H.
  (b) Deduce that $\tilde{r}$ is a 1-Lipschitz map. (*Hint:* Write $x - y = \tilde{r}(x) - \tilde{r}(y) + x - \tilde{r}(x) + \tilde{r}(y) - y$).)

3. Prove that any closed arc of the unit disk B^2 is a retract of B^2 (*Hint:* Use the Tietze extension theorem, Theorem A.5.1 in the Appendix).

4. (1) Prove that the closed northern hemisphere $S^n_+ = \{x = (x_1, x_2, \dots, x_n, x_{n+1}) : x_{n+1} \geq 0\}$ is a retract of the half-ball B^{n+1}_+. (*Hint:* See [27, Example 3.1.12].)
(2) Deduce that, up to homeomorphism, the closed n-ball B^n is a retract of the half $(n+1)$-ball B^{n+1}_+.

5. (1) Prove that the closed northern hemisphere $S^n_+ = \{x = (x_1, x_2, \dots, x_n, x_{n+1}) : x_{n+1} \geq 0\}$ is a retract of the sphere S^n. (*Hint:* See [12, Exercise 2, Section 4.4].)
(2) Deduce that S^n_+ is simply connected. Is the result predictable?

6. Is the continuous image of a contractible space a contractible space? Prove or disprove.

7. Is a contractible subspace of X a retract of X? Prove or disprove.

8. Let $A \subset X$ be a deformation retract (strong deformation retract, respectively) of X and $B \subset Y$ be a deformation retract of Y. Prove that $A \times B$ is a deformation retract (strong deformation retract, respectively) of $X \times Y$ with the product topology. (*Hint:* See Proposition 7.1.8.)

9. If $A \subset X$ deformation retract of X, then prove that A/A is a deformation retract of X/A.

10. Let A be a retract of X and $U \subset A$.
(1) (a) Prove that U is open in A if and only if $r^{-1}(U)$ is open in X, where $r : X \longrightarrow A$ is a retraction. (*Hint:* Check that $r^{-1}(U) \cap A = U$.)
(b) Deduce the proof of Proposition 7.1.15.
(2) Given a second proof of (1) using Corollary 2.1.6.

11. Prove that the homotopy equivalence is an equivalence relation.

12. Prove that the composition of two homotopy equivalence maps is a homotopy equivalence map.

13. Prove that a triangle of the plane is homotopically equivalent to S^1. What about a triangle with one vertex removed?

14. Prove that the equator S^{n-1} is a deformation retract of the doubly punctured sphere $S^n \setminus \{N_p, S_p\}$. (*Hint:* One may use Remark 6.3.6.)

15. Prove that $\mathbb{RP}^1$ is a deformation retract of $X = \mathbb{RP}^2 \setminus \{Oz\}$. (*Hint:* Use the retraction $r([x]) = [(x', 0)/\|x'\|]$ and the homotopy $H([x], t) = [(x', tx_3)/\|(x', tx_3)\|]$ for $x = (x_1, x_2, \dots, x_n, x_{n+1}) = (x', x_{n+1}) \in \mathbb{R}^{n+1}$ and $t \in [0, 1]$) (of course, the result is weaker than Proposition 7.2.41 but explicitly determines the maps r and H).

16. We say that a space X is *locally contractible* at a point $x \in X$ if for every neighborhood $U \in \mathcal{N}_x$ in X, there exists $V \subset U$ such that the inclusion $V \hookrightarrow U$ is null-homotopic. A space is locally contractible if it is locally contractible at every point.

(1) Prove that a locally contractible space is locally simply connected and locally path-connected.
(2) Prove that the comb space is not locally contractible (even though it is contractible by Example 4.9.3). (*Hint:* Use Exercise 22 of Section 1.4.)
(3) Prove that the circle is locally contractible but not contractible.

17. Prove that if a space X deformation retracts onto a point $x \in X$, then X is locally contractible at x.

18. Let A be a deformation retract of a space X. Prove that the inclusion $i : A \hookrightarrow X$ is a homotopy equivalence map.

19. Let $A \subset X$ be a closed subset. Prove that a subset A is a deformation retract of a space X if and only if for any space Y, every continuous function $f : A \longrightarrow Y$ has a continuous extension and any two continuous maps $f, g : X \longrightarrow Y$ such that $f_{|A} = g_{|A}$ are homotopic.

20. Prove that a space X is contractible if and only if there exists $x_0 \in X$ such that $\{x_0\}$ is a deformation retract of X.

21. (1) By Example 7.2.15, the comb space deformation retracts onto $\{0\} \times I$. Show that it does not deformation retract strongly onto $\{0\} \times I$. (*Hint:* See [92, Example 14.1.10].)
(2) By Example 4.9.3, the comb space is contractible (though not locally contractible according to Exercise 16 above). Show that it does not deformation retract strongly onto the point $(0, 1)$. (*Hint:* See [61, Exercise 3.8.1].) (In [45, Exercise 7, p. 18], the author provides a contractible space, which does not deformation retract strongly onto any point of it.)

22. Let $f : X \longrightarrow Y$ be a continuous map and C_f be the mapping cylinder (Definition 3.5.1). Prove that:
(1) Y is a strong deformation retract of C_f, which implies Theorem 4.2.18;
(2) X is a retract of C_f if and only if f has a left homotopy inverse;
(3) X is deformable onto C_f if and only if f has a right homotopy inverse (we say that X is deformable onto A if the identity Id_X is homotopic to a continuous map from X to A);
(4) X is a deformation retract of C_f if and only if f is a homotopy equivalence.

(*Hint:* See [25, Section 4, Chapter XIII]), [45, Corollary 0.21] or [49, Proposition 12.1].)

23. We say that a subspace $A \subset X$ is a weak retract of X if there exists a continuous map $r : X \longrightarrow A$ such that $r \circ i$ is homotopic to the identity on A. Let A be the comb space and $X = I^2$. Prove that A is a weak retract of X although A is not a retract of X (see Example 7.1.30).

24. Let $A \subset X$ be such that the pair (X, A) has the homotopy extension property (Definition 7.3.1). Prove that:

(1) A is a weak retract of X if and only if A is a retract of X;
(2) if X is contractible, then the quotient space X/A is homotopically equivalent to the suspension $\Sigma(X)$ of X (compare with Theorem 7.3.10).

25. Let $A \subset X$ and $f : A \longrightarrow Y$ a continuous map. Show that A is contractible if and only if X is a retract of the adjunction space $C(A) \cup_f X$, where $C(A)$ is the cone over X. (*Hint:* See [45, Exercise 26, p. 157].)

26. Let $A \subset X$ be such that the pair (X, A) has the homotopy extension property (Definition 7.3.1). Show that $(C_f, A \cup B)$ has the homotopy extension property, where $f : B \longrightarrow A$ is a continuous map and C_f is the mapping cylinder.

27. Let $f : B \subset A \subset X \longrightarrow Y$ be a continuous map, where $X \cup Y$ is a disjoint union. Assume that (X, A) has the homotopy extension property. Prove that the pair $(X \cup_f Y, A \cup_f Y)$ has the homotopy extension property. (*Hint:* Consider the following commutative diagram, where $Z = (X \cup Y) \times I$, $W = [(X \cup Y) \times \{0\}] \cup [(A \cup Y) \times I]$ and Z_f, W_f are the corresponding quotient spaces for the equivalence relation $b \sim f(b)$, $b \in B$:

$$\begin{array}{ccc} Z & \xrightarrow{\tilde{r}} & W \\ {\scriptstyle \pi\times\mathrm{Id}}\downarrow & & \downarrow{\scriptstyle \pi\times\mathrm{Id}} \\ Z_f & \xrightarrow{s} & W_f \end{array}$$

Here, $\tilde{r}$ and s are retraction maps.)

28. A subspace $A \subset X$ is *a neighborhood retract* ($A \in \mathrm{NR}$ for brevity) of X if there exists an open neighborhood $U \subset X$ of A such that A is a retract of U. If A can be further embedded in $\mathbb{R}^n$, A is said to be *an Euclidean neighborhood retract* ($A \in \mathrm{ENR}$ for brevity).
(1) If $U \in \mathrm{NR}$ is an open subset of X, prove that U inherits all local topological properties of X.
(2) Prove that an Euclidean neighborhood retract is the intersection of a closed subset and an open subset of $\mathbb{R}^n$.
(3) Let $A \subset \mathbb{R}^n$ be a locally compact and locally contractible subset (see definition in Exercise 16 of Chapter 7). Prove that $A \in \mathrm{ENR}$. (*Hint:* See [24, Proposition 8.12].)
(4) Let $A \subset \mathbb{R}^n$ be a closed locally contractible subset (see definition in Exercise 16 of Chapter 7). Prove that $A \in \mathrm{ENR}$. (*Hint:* See [90, Lemma 5.1.17].)
(5) Let $A \in \mathrm{ENR}$ be compact. Prove that A is locally contractible in the weak sense, that is for every $x \in A$ and any neighborhood $U \in \mathcal{N}_x$, there exists a neighborhood $V \subset U \in \mathcal{N}_x$ such that the inclusion map $i : V \hookrightarrow U$ is null-homotopic. (*Hint:* See [45, Theorem A.7].)
(6) Show that a compact space $A \in \mathrm{ENR}$ if and only if it can be embedded as a retract of a finite simplicial complex. (*Hint:* [45, Corollary A.8].)

29. Let (X, A) be a pair such that:

(a) A is closed;

(b) $X \times [0, 1]$ is a normal space;

(c) $(X \times \{0\}) \cup (A \times I)$ is a neighborhood retract.

Prove that (X, A) is a Borsuk pair. (*Hint:* See [64, Lemma 7.1].)

30. A space X is *an absolute neighborhood retract* ($X \in$ ANR for brevity) if for any space Y and for any embedding $h : X \longrightarrow Y$, there exists an open subset $V \subset Y$ such that the image $h(X)$ is a retract of V.

(1) Let $X \in$ ANR and $U \subset X$ an open subset. Prove that $U \in$ ANR, a result due to O. Hanner (*Hint:* See [9, Chapter IV, Theorem (10.1)] or [40, Proposition (1.2)(i), Section 11].)

(2) Let $X \in$ ANR and $A \subset X$ a neighborhood retract. Prove that $A \in$ ANR. (*Hint:* See [40, Proposition (1.2)(ii), Section 11].)

(3) Clearly, every AR space is ANR. By Proposition 7.5.12, an AR space is contractible. Conversely, assume that a normal space X is contractible and ANR. Prove that X is an absolute retract.
(*Hint:* Let $f : B \longrightarrow X$ be a continuous map, where B is a closed subset of a metric space Y. Let U be an open set such that $B \subset U \subset Y$ and $\tilde{f} : U \longrightarrow X$ a continuous extension. Check that the map $F : Y \longrightarrow X$ defined by

$$F(x) = \begin{cases} H(\tilde{f}(x), \theta(x)), & x \in \overline{V}, \\ x_0, & x \in Y \setminus V, \end{cases}$$

is a continuous extension of f, where $x_0 \in X, B \subset V \subset \overline{V} \subset U$, H is a Homotopy and the function θ is given by Urysohn's theorem. Details can be found in [21, Proposition 5.42]; see also [40, Theorem (2.1), Section 11]).

(4) Let X be a metric space such that $X = X_1 \cup X_2$ and let $X_0 = X_1 \cap X_2$.
   (a) Suppose that X_1, X_2 and X_0 are AR spaces (ANR spaces, respectively). Prove that $X \in$ AR.
   (b) Suppose that X and X_0 are AR spaces (ANR spaces, respectively). Prove that $X_1, X_2 \in$ AR. (*Hint:* See [9, Chapter IV, Theorem (6.1)].)

31. Prove that the sphere $S^n \in$ ANR. (*Hint:* See [46, Theorems 2–36].)

32. Let X, Y be disjoint normal spaces such that $X \in$ AR. Let A be a closed subset of X, $f : A \longrightarrow Y$ be a continuous map and $Z_f = (X \cup_f Y)$, the adjunction space (see Definition 3.4.1) of X and Y. Prove that:

(1) Z_f is normal; (*Hint:* Use Urysohn's lemma, Theorem A.4.1 and Tietze's theorem, Theorem A.5.1.)

(2) f can be extended to X. This provides an alternative proof of the implication AR $\Longrightarrow$ ES in Theorem 7.5.11.

33. Let A be a retract of a space X.

(1) Prove that if X is compact (locally compact, respectively), then so is the subset A.

(2) Prove that if X is connected (locally connected, respectively), then so is the subset A. (*Hint:* Use Theorem 1.4.7, Theorem 1.3.8 and see [50, Proposition 9.2 and Proposition 10.1].)

34. Let X be a normed space. A nonempty subset $\mathscr{P}$ of X is called a cone if $\mathscr{P}$ is convex, closed and satisfies the conditions:

(1) $\alpha x \in \mathscr{P}$ for all $x \in \mathscr{P}$ and $\alpha \geq 0$,

(2) $x, -x \in \mathscr{P}$ imply $x = 0$.

For $R > 0$, define the complement of the conical shell to be the set $A = \{x \in \mathscr{P} : \|x\| \geq R\}$. Prove that A is a retract of X. (*Hint:* See [104, Theorem 2.1].)

8 Fundamental groups of some surfaces

This chapter is devoted to the computation of the fundamental groups of some spaces, surfaces and manifolds. The theory of the fundamental group (Chapter 4) and the theory of retraction (Chapter 7) are essential in conducting most cases of calculation. Further to these theories, the theorem of Seifert-van Kampen discussed in Section 8.1 turns out to be an efficient tool in regard to the computation of the fundamental groups of some spaces defined as union of spaces. However, the core of this chapter is Section 8.2, where we calculate with details the fundamental groups of the wedge sum of circles, the spheres and the manifolds, the n-petaled rose, the figure 8 and the figure Θ. We describe the fundamental groups of the mapping cone, the attaching cells, the connected sum of surfaces and the Klein bottle, as well. All computed fundamental groups are collected in a table, in Section 8.3.

8.1 Seifert–van Kampen theorem

Given the fundamental groups of two open sets U, V, which cover a space X, we present a general theorem which allows us to calculate the fundamental group of the union $X = U \cup V$. This is a result due to H. Seifert (1931) and E. R. van Kampen (1933). Actually, Herbert Seifert proved a first version in his thesis in 1930. We give two versions of the theorem as well as some corollaries useful for applications. For the proof of this standard theorem, we refer the reader, e. g., to [10, Theorem 9.4, Chapter III], [45, Theorem 1.20], [67, Theorem 2.1] or [76, Theorem 70.1, Theorem 70.2]. A first version was already given with proof in Theorem 6.3.1. We set some hypotheses.

Let U, V be open sets and $X = U \cup V$ such that U and V together with their intersection $X_0 = U \cap V$ are path-connected spaces. Let $x_0 \in X_0$, H a group, and

$$\phi_1 : \pi_1(U, x_0) \longrightarrow H \quad \text{and} \quad \phi_2 : \pi_1(V, x_0) \longrightarrow H$$

two homomorphisms. Consider the following pushout diagram, where the homomorphisms are induced by the inclusion maps

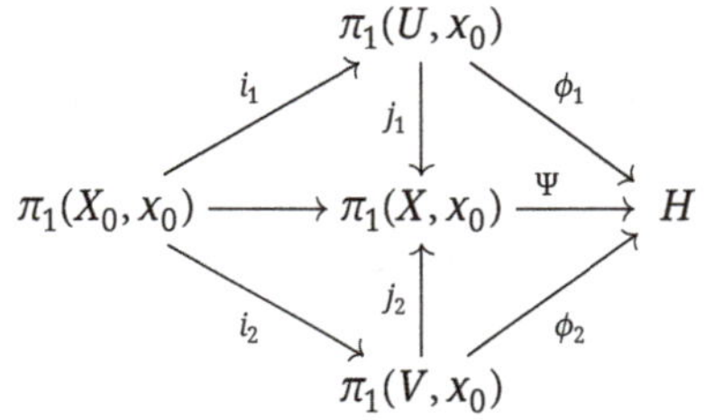

The classical version of the Seifert–van Kampen theorem is the following.

https://doi.org/10.1515/9783111517384-008

Theorem 8.1.1. *If $\phi_1 \circ i_1 = \phi_2 \circ i_2$, then there is a unique homomorphism*

$$\Psi : \pi_1(X, x_0) \longrightarrow H$$

such that $\Psi \circ j_1 = \phi_1$ and $\Psi \circ j_2 = \phi_2$.

Let $G_1 * G_2$ denote the free product of the groups G_1 and G_2 (Definition B.5.1). The group-theoretic version of the Seifert–van Kampen theorem is the following.

Theorem 8.1.2. *Under the hypotheses of the Seifert–van Kampen theorem, the fundamental groups $\pi_1(X, x_0)$ and*

$$\pi_1(U, x_0) *_{\pi_1(U \cap V)} \pi_1(V, x_0) \cong (\pi_1(U, x_0) * \pi_1(V, x_0))/N$$

are isomorphic, where N is the least normal subgroup that contains (generated by the set of) the elements $i_1(f)i_2(f)^{-1}$, $f \in \pi_1(U \cap V)$.

The notation $*_N$ refers to the free product of the two fundamental groups with amalgamation over the subgroup N (see Definitions B.5.1 and B.6.1).

Remark 8.1.3. Actually, the classical version sates that there is a surjection j from $\pi_1(U, x_0) * \pi_1(V, x_0)$ to $\pi_1(X, x_0)$ whose kernel is the subgroup N. By the First Homomorphism Theorem B.2.14, we deduce that

$$\pi_1(X, x_0) \cong (\pi_1(U, x_0) * \pi_1(V, x_0))/ \operatorname{Ker} j \cong (\pi_1(U, x_0) * \pi_1(V, x_0))/N.$$

We can as well state Theorem 8.1.2 by writing

$$\pi_1(X, x_0) \cong \pi_1(U, x_0) * \pi_1(V, x_0)$$

subject to the relations $i_1(f) = i_2(f)$ for all $f \in \pi_1(U \cap V)$.

Remark 8.1.4. This second version is often represented by the diagram

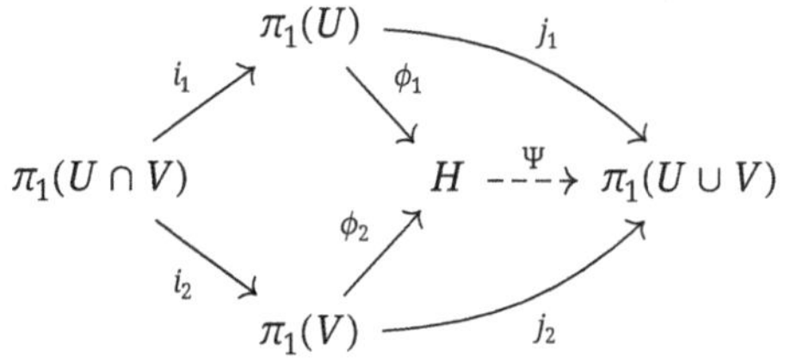

showing the pushout (Definition 3.4.4) $\pi_1(U \cup V)$, which is isomorphic to $H = \pi_1(U) *_{\pi_1(U \cap V)} \pi_1(V)$. Since all the mentioned spaces are path-connected, we do not mention explicitly the base point.

Corollary 8.1.5. *Under the hypotheses of the Seifert–van Kampen theorem, assume that* $U \cap V$ *is simply connected. Then* $\pi_1(X, x_0)$ *and* $\pi_1(U, x_0) * \pi_1(V, x_0)$ *are isomorphic.*

In particular, we recover Corollary 6.3.3.

Corollary 8.1.6. *Under the hypotheses of the Seifert–van Kampen theorem, assume that* U *and* V *are simply connected. Then* $X = U \cup V$ *is simply connected.*

Remark 8.1.7. Corollary 8.1.6 is a result of pure topology. A direct proof can be founded, e. g., in [6, Theorem (5.12)] or [36, Lemma 10.4)].

Remark 8.1.8. Corollary 8.1.6 no longer holds true if $U \cap V$ is not path-connected. Indeed, for $a = (1, 0)$, the open sets $U = S^1 \setminus \{a\}$ and $V = S^1 \setminus \{-a\}$ are homeomorphic to the real line by the stereographic projection, hence simply connected. Yet $X = U \cup V = S^1$ is not simply connected. Here, $U \cap V$ has two components.

Corollary 8.1.9. *Under the hypotheses of the Seifert–van Kampen theorem, assume that* V *is simply connected. Then* $\pi_1(X, x_0)$ *and* $\pi_1(U, x_0)/N$ *are isomorphic, where* N *is the least normal subgroup of* $\pi_1(U, x_0)$ *containing the image of the induced homomorphism*

$$i_* : \pi_1(U \cap V, x_0) \longrightarrow \pi_1(U, x_0).$$

8.2 Computations of some fundamental groups

8.2.1 Wedge sum of circles

As a first application of the Seifert–van Kampen theorem (Corollary 8.1.5), we begin with the wedge sum of a finite number of circles. We construct two open sets whose intersection is simply connected.

Proposition 8.2.1. *We have*

$$\pi_1(\overbrace{S^1 \vee S^1 \vee \cdots \vee S^1}^{n}) \cong \overbrace{\mathbb{Z} * \mathbb{Z} * \ldots \mathbb{Z}}^{n}.$$

Proof. Let $X = S^1 \vee S^1$ be the wedge sum of two circles $S^1_1 \vee S^1_2$ (union of two copies of circles), say $S^1_1 = C(-a; 1)$ and $S^1_2 = C(a; 1)$, where $a = (1, 0)$. Let $b = (2, 0) \in S^1_2$. Then the sets $U = S^1_1 \setminus \{-b\} \cup S^1_2$ and $V = S^1_1 \cup S^1_2 \setminus \{b\}$ are open in X (as complements of one-element sets), path-connected (since they intersect at the origin point), and cover the space X. Notice that U deformation retracts onto S^1_2 with the retraction r given by

$$r(x) = \begin{cases} 0, & x \in S^1_1 \setminus \{b\}, \\ x, & x \in S^1_2 \end{cases}$$

and homotopy $H(x, t) = tr(x) + (1 - t)x$, that is to say

$$H(x,t) = \begin{cases} (1-t)x, & x \in S_1^1 \setminus \{b\}, \\ x, & x \in S_2^1. \end{cases}$$

r and H are continuous by Lemma A.2.1. Hence, $\pi_1(U)$ and $\pi_1(V)$ are isomorphic to $\mathbb{Z}$. Since $U \cap V$ is simply connected (Proposition 4.9.6), by Theorem 6.1.4 and Corollary 8.1.5, the fundamental group $\pi_1(X)$ is isomorphic to $\pi_1(U) * \pi_1(V)$, the free product of two copies (with two generators) of the integers. By induction, we obtain the general result. □

By Proposition 7.2.35, a punctured torus and the wedge sum of two circles have isomorphic homotopy groups. Following [39, Proof of Proposition 3, Chapter 2], we prove that the same holds for the perforated torus.

Proposition 8.2.2. *The fundamental group of the perforated torus (a torus with a small ball removed)* $\mathbb{T}^2 \setminus \{B_\varepsilon\}$ *is isomorphic to a free group with two generators.*

Proof. By Example 3.2.3, the torus is defined by the equivalence relation $\sim$ resulting from the identification of the opposite sides of the square $I^2 = [0,1]^2$. The perforated torus is obtained by removing a small square $S_\varepsilon = [0,\varepsilon) \times [0,\varepsilon)$ from I^2 (see Figure 8.1). Then $\mathbb{T}^2 \setminus \{B_\varepsilon\} = \pi(I^2 \setminus S_\varepsilon)$. By definition of $\sim$, we can see that

$$\mathbb{T}^2 \setminus \{B_\varepsilon\} \cong ([0,1-\varepsilon) \times S^1) \cup (S^1 \times [0,1-\varepsilon)) =: U \cup V,$$

and the intersection $U \cap V$ is simply connected as a rectangle. The cylinders U and V are homotopy equivalent to S^1 by Example 4.2.6. Finally, Corollary 8.1.5 implies that $\pi_1(\mathbb{T}^2 \setminus \{B_\varepsilon\}) \cong \mathbb{Z} * \mathbb{Z}$. □

Example 8.2.3. Let $\mathbb{T}^2$, $\mathbb{T}'^2$ and $\mathbb{T}^2 \setminus \mathring{B}_\varepsilon^2$ be the torus, the punctured torus and the perforated torus, respectively. By Example 6.1.8, Proposition 7.2.35 and Proposition 8.2.2,

$$\pi_1(\mathbb{T}^2) \cong \mathbb{Z} \times \mathbb{Z} \quad \text{and} \quad \pi_1(\mathbb{T}'^2) \cong \pi_1(\mathbb{T}^2 \setminus \mathring{B}_\varepsilon^2) \cong \mathbb{Z} * \mathbb{Z}.$$

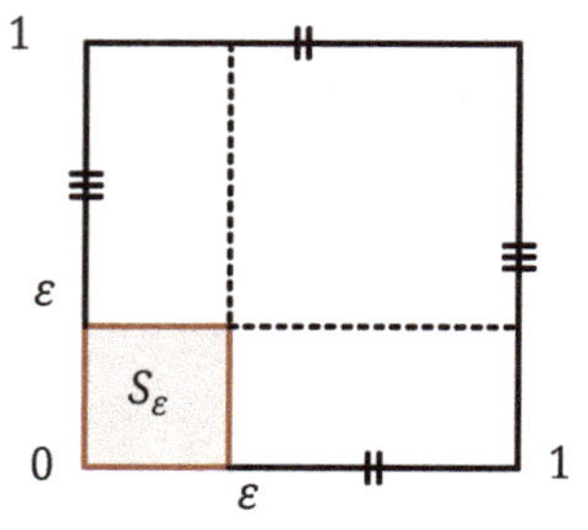

Figure 8.1: Perforated torus as union of two cylinders.

We can see that $S^1 \times S^1$ and $S^1 \vee S^1$ are not homotopically equivalent since their fundamental groups are not isomorphic. Indeed (see [33, Section 6.6, Theorem] or [76, Lemma 60.5]), we have the following.

Proposition 8.2.4. *The fundamental group of $S^1 \vee S^1$ is not Abelian.*

Corollary 8.2.5. *The fundamental group of the wedge sum $\mathbb{T}^2 \vee \mathbb{T}^2$ is not Abelian.*

Proof. By Example 7.1.9(3), there exists some retraction $r : \mathbb{T}^2 \vee \mathbb{T}^2 \longrightarrow S^1 \vee S^1$. Then the induced homomorphism $r_* : \pi_1(\mathbb{T}^2 \vee \mathbb{T}^2) \longrightarrow \pi_1(S^1 \vee S^1)$ is one-to-one. Then the claim follows from Proposition 8.2.4 and Theorem 7.1.32. □

Definition 8.2.6. The bouquet (wedge sum) $S^1 \vee S^1$ of two circles will be referred to as *a handle*.

Remark 8.2.7. In the literature (see, e. g., [89, p. 6] or [101, p. 152]), a handle is defined as the perforated torus $\mathbb{T}^2 \setminus \mathring{B}^2_\varepsilon$ involved in the construction of the connected sum of the torus and some other surfaces; see Proposition 8.2.2). By Proposition 7.2.35 and Proposition 8.2.2, the punctured torus, the perforated torus and the bouquet of two circles have isomorphic fundamental groups. This justifies the designation of the punctured (or perforated) torus by a handle in Definition 8.2.6.

Example 8.2.8. A mug (of coffee) Mg is the union of a handle with two supports attached to a cylinder (see Figure 8.2). Arguing as in Proposition 8.2.16, the handle deformation retracts onto another handle with one support. Hence,

$$\pi_1(\mathrm{Mg}) \cong \pi_1((S^1 \times [0,1]) \vee S^1) \cong \pi_1(S^1) * \pi_1(S^1) \cong \mathbb{Z} * \mathbb{Z}.$$

An alternative way to see this result is to apply Corollary 8.2.26 with $n = 1$. Indeed, the handle may thought of as a 1-cell attachment. Then

$$\pi_1(B^1 \cup_f (S^1 \times [0,1])) \cong \pi_1(S^1 \times [0,1]) * \mathbb{Z} \cong \mathbb{Z} * \mathbb{Z}.$$

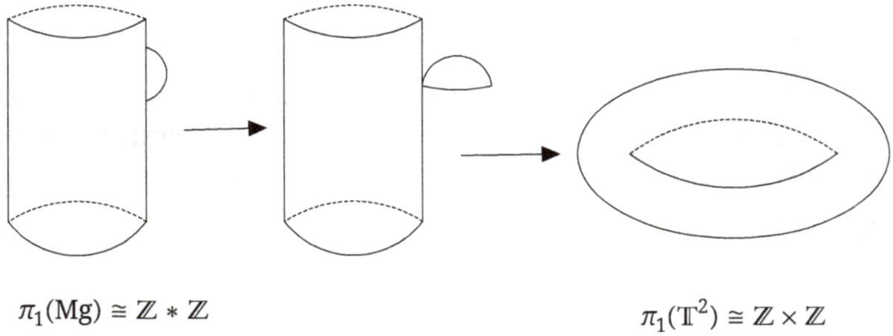

Figure 8.2: Coffee mug and doughnut.

8.2.2 Wedge sum of manifolds

The proof of Proposition 8.2.1 can be adapted to the case of the wedge sum of spheres $S^n \vee S^m$ for n and m or equal to 1. In the following theorem, the formula is extended to the more general setting of the wedge sum of manifolds, including the spheres. Once again, this is an application of the Seifert–van Kampen theorem (Corollary 8.1.5).

Theorem 8.2.9. *Let $M_1, M_2, \ldots, M_n$ be n disjoint manifolds. Then*

$$\pi_1(M_1 \vee M_2 \ldots \vee M_n) \cong \pi_1(M_1) * \pi_1(M_2) \ldots * \pi_1(M_n).$$

Proof. It is sufficient to prove the theorem for two manifolds M and N. Then the proof can be completed by induction.

Let $(x_0, y_0) \in M \times N$. By Definition 7.2.10, let $U_{x_0} \in \mathcal{N}(x_0)$ and $V_{y_0} \in \mathcal{N}(y_0)$ be open neighborhoods homeomorphic to open balls. Since balls of Euclidean space strongly deformation retract onto their centers (Example 4.9.2), so do the neighborhoods U_{x_0} and V_{y_0}. Thus, let $H_{x_0} : U_{x_0} \times I \longrightarrow U_{x_0}$ and $H_{y_0} : V_{y_0} \times I \longrightarrow V_{y_0}$ be continuous deformation retracts of U_{x_0} and V_{y_0} onto x_0 and y_0, respectively. Let $U = M \vee V_{y_0}$ and $V = U_{x_0} \vee N$. Then U and V are open sets, which cover the wedge sum $X = M \vee N$. The intersection $U \cap V$ deformation retracts on $\{x_0, y_0\}$, and thus it is simply connected. In addition, U and V deformation retract onto M and N, respectively, because U_{x_0} and V_{y_0} deformation retract onto x_0 and y_0, respectively. By Corollary 8.1.5,

$$\pi_1(M \vee N) \cong \pi_1(U) * \pi_1(V) \cong \pi_1(M) * \pi_1(N).$$ □

Example 8.2.10. We have

$$\pi_1(\mathbb{T}^2 \vee \mathbb{T}^2) \cong (\mathbb{Z} \times \mathbb{Z}) * (\mathbb{Z} \times \mathbb{Z}),$$
$$\pi_1(\mathbb{T}'^2 \vee \mathbb{T}'^2) \cong (\mathbb{Z} * \mathbb{Z}) * (\mathbb{Z} * \mathbb{Z}),$$

where $\mathbb{T}'^2 = \mathbb{T}^2 \setminus \{*\}$ is the punctured torus (see Proposition 7.2.35).

In Example 6.1.8, the fundamental group of the torus is computed using the formula for the fundamental group of a product given by Theorem 4.6.1. In the next proposition, $\pi_1(\mathbb{T}^2)$ is once again computed using the Seifert–van Kampen theorem.

Proposition 8.2.11. *We have*

$$\pi_1(\mathbb{T}^2) \cong \mathbb{Z} \times \mathbb{Z}.$$

Proof. In the interior of the square I^2, draw two balls $B \subset C$ such that B is open and C is closed. Then take $A = I^2 \setminus C$ (see Figure 8.3). B is homeomorphic to ball, the identifications of A yields a perforated torus and $A \cap B$ is homeomorphic to an annular region. The

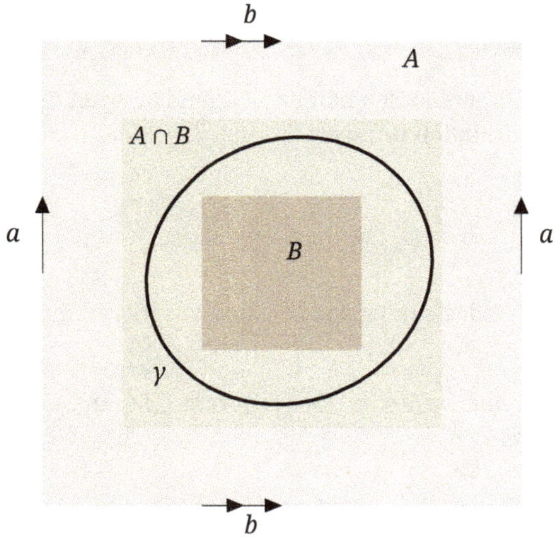

Figure 8.3: Fundamental group of the torus.

fundamental group of B is trivial. By Theorem 1.5.37, Example 4.2.5 and Example 4.2.8, $\pi_1(A \cap B) \cong \mathbb{Z}$. Finally, Proposition 8.2.2 yields $\pi_1(A) \cong \mathbb{Z} * \mathbb{Z}$. By applying Corollary 8.1.9, we get $\pi_1(\mathbb{T}^2) \cong \pi_1(A \cup B) \cong (\mathbb{Z} * \mathbb{Z})/N$, where N is the smallest normal subgroup which contains the generator of $i_*(A \cap B)$, where $i : A \cap B \hookrightarrow A$ is the inclusion map and $i_* : \mathbb{Z} \longrightarrow \mathbb{Z} * \mathbb{Z}$ is the induced homomorphism. In order to determine the normal subgroup N, let $[\gamma]$ be the generator of $\pi_1(A \cap B)$, where γ is a loop in the intersection $A \cap B$ (see Figure 8.3). By definition of the torus, the representation of γ is $aba^{-1}b^{-1}$. Hence, $[\gamma] = [a][b][a^{-1}][b^{-1}]$. However, $[a][b][a^{-1}][b^{-1}]$ is precisely the commutator of $[a]$ and $[b]$, where a, b are the generators of the first and the second $\mathbb{Z}$. Then N contains the commutator $[\mathbb{Z} * \mathbb{Z}, \mathbb{Z} * \mathbb{Z}]$. Since N is the smallest normal subgroup, which contains this commutator, $N = [\mathbb{Z} * \mathbb{Z}, \mathbb{Z} * \mathbb{Z}]$ and using Proposition B.4.5, we obtain that $\mathbb{Z} * \mathbb{Z}/N$ is an Abelian group with two generators, i. e.,

$$\pi_1(\mathbb{T}^2) \cong \mathbb{Z} * \mathbb{Z}/N \cong \mathbb{Z} \times \mathbb{Z}. \qquad \square$$

Remark 8.2.12. By Corollary 8.2.5, the group $\mathbb{Z} * \mathbb{Z}$ is not Abelian (non-Abelian group with two generators). The Abelianization of $\mathbb{Z} * \mathbb{Z}$ is the group $\mathbb{Z} \times \mathbb{Z}$ (a free Abelian group with two generators a and b) (see Definition B.4.4), which is the fundamental group of the torus (a doughnut for the coffee, a bagel) (see Example 10.2.29). By Proposition B.4.5, G/N is Abelian. Using merely group theory, it can be proved that the Abelianization of $G_1 * G_2$ is isomorphic to $G_1 \oplus G_2$. However, the coffee mug and the torus have isomorphic homology groups.

Theorem 8.2.9 is valid when the manifold is replaced by a pointed space (see, e. g., [62, Theorem 10.7]) according to the following.

Definition 8.2.13. A space X is pointed if there exist a point $x_0 \in X$ and an open neighborhood $U \in \mathscr{N}(x_0)$, which strongly deformation retracts onto x_0.

8.2.3 Wedge sum of spheres

Theorem 8.2.9 is now illustrated for the wedge sum of combination of spheres and circles.

Example 8.2.14. (1) Since $\pi_1(S^1) \cong \mathbb{Z}$ and $\pi_1(S^2) \cong \{0\}$ (Theorem 6.1.4 and Corollary 6.3.4), by Theorem 8.2.9, we have

$$\pi_1(S^1 \vee S^1) \cong \mathbb{Z} * \mathbb{Z}, \quad \pi_1(S^1 \vee S^2) \cong \mathbb{Z} \quad \text{and} \quad \pi_1(S^2 \vee S^2) \cong \{0\}.$$

(2) More generally, let $X = \bigvee_{k=1}^{k=n} S_k$ be the subspace of $\mathbb{R}^3$ consisting of the bouquet of n spheres $S_k = S^2(w_k, 1/k)$ centered at $w_k = (0,0,1/k)$ with radius $1/k$. Then X is simply connected.

Example 8.2.15. By Example 2.5.4, S^n/S^{n-1} is homeomorphic to $S^n \vee S^n$. More generally, Example 7.3.12 and Example 7.3.13 show that S^n/S^k is homotopically equivalent to $S^n \vee S^{k+1}$ for $k < n$. By Corollary 6.3.4 and Theorem 8.2.9,

$$\pi_1(S^n/S^k) \cong \begin{cases} \mathbb{Z} * \mathbb{Z}, & \text{if } n = 1,\ k = 0, \\ \mathbb{Z}, & \text{if } n > 1,\ k = 0, \\ \{0\}, & \text{if } n > k > 0. \end{cases}$$

The following proposition provides an example of two subspaces of Euclidean space, which are homotopically equivalent to the wedge sum of two spheres or the wedge sum of a circle and a sphere.

Proposition 8.2.16. *Consider the subspaces of* $\mathbb{R}^3$

$$A_1 = S^2, \quad A_2 = \{(0,0,z) \in \mathbb{R}^3 : -1 \le z \le 1\}, \quad A_3 = \{B^2 \cap (z = 0)\},$$

$X = A_1 \cup A_2$, *and* $Y = A_1 \cup A_3$. *Then:*

(1) X *and* Y *are homotopically equivalent to* $S^2 \vee S^1$ *and* $S^2 \vee S^2$, *respectively;*

(2) $\pi_1(X) \cong \mathbb{Z}$ *and* $\pi_1(Y) \cong 0$.

Proof. (1) Note that X/A_2 is homeomorphic to the sphere S^2 with two points identified, hence homeomorphic to S^2/S^0. By Example 7.3.14 and Example 8.2.15, S^2/S^0 is homotopically equivalent to $S^2 \vee S^1$. Clearly, the quotient space Y/A_2 obtained by squeezing the

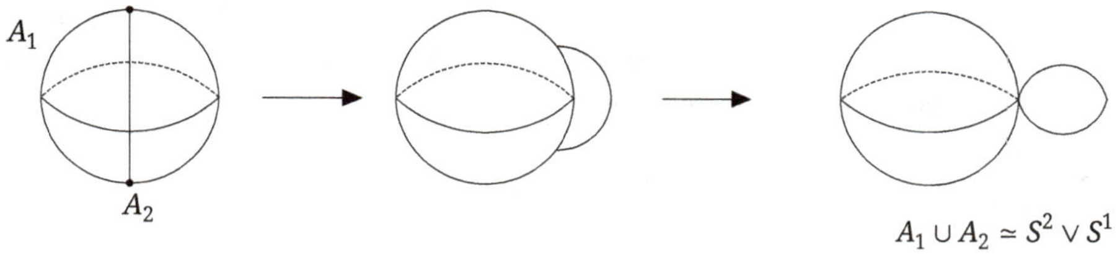

Figure 8.4: Homotopy equivalences from the sphere (1).

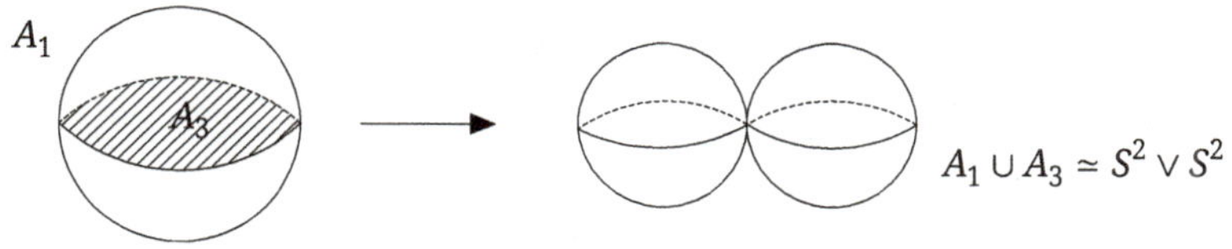

Figure 8.5: Homotopy equivalences from the sphere (2).

sphere along the equator is homeomorphic to $S^2 \vee S^2$ (see details for the circle in Remark 6.2.8). Since A_2 and A_3 are contractible spaces, by [45, Proposition 0.17], X and Y are homotopically equivalent to X/A_2 and Y/A_3, respectively, whence the first part (see Figure 8.4 and Figure 8.5).

(2) Part (2) is a consequence of part (1) and Theorem 8.2.9. □

Proposition 8.2.17. *The space $X = \mathbb{R}^3 \setminus S^1$ deformation retracts on the sphere S^2 with two points identified. Hence, $\pi_1(X) \cong \mathbb{Z}$.*

Proof. (1) *First proof.* Let us make the identification $S^1 \cong S^1 \times \{0\}$ and first ignore the plane (xy). Points outside this plane and points on (xy) lying outside S^1 deformation retract on the sphere S^2. The open disk within the circle S^1 deformation retracts on the open segment $(-1,1) \cong (-1,1)\times\{0\} \cong \{0\}\times(-1,1)$ with the retraction defined by $r(0,0) = 0$ and

$$r(x,y) = \frac{\sqrt{x^2+y^2}}{\max(|x|,|y|)}(x,0).$$

Then X deformation retracts on the space $X = A_1 \cup A_2$ given by Proposition 8.2.16, whence the claim of the proposition.

(2) *Second proof.* Since the unit circle is a one-point compactification of $\mathbb{R}$ (Proposition 1.5.28), we have the homeomorphisms

$$X = \mathbb{R}^3 \setminus S^1 \cong \mathbb{R}^3 \setminus (\mathbb{R} \cup \infty) \cong (\mathbb{R}^3 \setminus (0z)) \setminus \{*\},$$

where $\{*\}$ is a point lying outside the $(0z)$-axis. Let $r : \mathbb{R}^3 \setminus (0z) \longrightarrow S^2 \setminus \{N_p, S_p\}$ be the radial retraction as in the proof of Corollary 7.2.23. Notice that all points distinct from $\{*\}$, which lie on the line through the point $\{*\}$ and the origin retract on the punctured sphere $S^2 \setminus \{N_p, S_p\}$. Hence, $r(X) = r(\mathbb{R}^3 \setminus (0z))$, showing that X deformation retracts on $S^2 \setminus \{N_p, S_p\}$. Proposition 8.1.2 completes the proof. □

8.2.4 Fundamental groups of multipunctured plane, n-leafed rose, Figure 8, Figure Θ

Euclidean plane with n holes is first shown to be homotopically equivalent to n-leafed rose, i. e., the wedge sum of circles (see Figure 8.6 for $n = 5$).

Proposition 8.2.18. *If* $X = \mathbb{R}^2 \setminus \{x_1, x_2, \dots, x_n\}$, $n \geq 1$, *then*

$$\pi_1(X) \cong \pi_1(\overbrace{S^1 \vee S^1 \vee \cdots \vee S^1}^{n}) \cong \overbrace{\mathbb{Z} * \mathbb{Z} * \cdots * \mathbb{Z}}^{n}.$$

Proof. First, let $n = 2$. By Proposition 4.2.13 and Remark 4.2.14, we know that $\mathbb{R}^2 \setminus \{x_1, x_2\}$ and a disk with two holes $X = B^2 \setminus \{B_1, B_2\}$ are homeomorphic, hence homotopically equivalent to the figure 8. Again by Proposition 4.2.13, Figure 8 is homotopically equivalent to the wedge sum of two circles $S^1 \vee S^1$. By Proposition 8.2.1,

$$\pi_1(\mathbb{R}^2 \setminus \{x_1, x_2\}) \cong \pi_1(S^1 \vee S^1) \cong \mathbb{Z} * \mathbb{Z}.$$

For the general case, we claim that $\overbrace{S^1 \vee S^1 \vee \cdots \vee S^1}^{n}$ is a deformation retract of X. This can be checked by induction using the case of two circles. Therefore, these spaces are homotopically equivalent and once again the result follows from Proposition 8.2.1. □

The following result is a consequence of Proposition 4.2.13, Proposition 4.2.15, Proposition 8.2.1 and Proposition 8.2.18. It can be derived directly from Corollary 8.1.5 since the Figure 8 is the union of two circles whose intersection is a point.

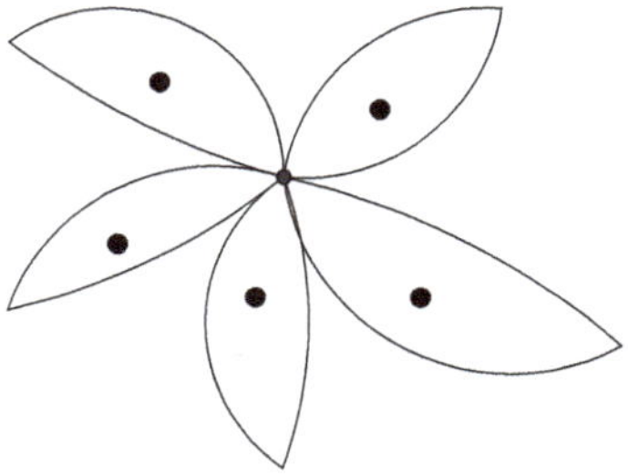

Figure 8.6: 5-leafed rose.

Corollary 8.2.19. *The fundamental groups of the bouquet of two circles, the Figure 8 and the figure theta are isomorphic to* $\mathbb{Z} * \mathbb{Z}$.

Corollary 8.2.20. $\mathbb{R}^3$ *and* $\mathbb{R}^2$ *are not homeomorphic.*

Proof. By contradiction, let $h : \mathbb{R}^3 \longrightarrow \mathbb{R}^2$ be a homeomorphism. Then $\mathbb{R}^3 \setminus \{0\}$ and $\mathbb{R}^2 \setminus \{h(0)\}$ are homeomorphic. However, $\mathbb{R}^3 \setminus \{0\}$ is simply connected (Corollary 6.3.9) while $\mathbb{R}^2 \setminus \{h(0)\}$ has fundamental group isomorphic to $\mathbb{Z}$ (Proposition 8.2.18). □

Using Proposition 7.2.19 and Proposition 8.2.17, we obtain

Corollary 8.2.21. *We have*

$$\pi_1(\mathbb{R}^3 \setminus (0z)) \cong \pi_1(S^1) \cong \pi_1(\mathbb{R}^3 \setminus S^1) \cong \mathbb{Z}.$$

By the stereographic projection, $S^2 \setminus \{x_1, x_2, \dots, x_n\}$ is homeomorphic to $\mathbb{R}^2 \setminus \{x_1, x_2, \dots, x_{n-1}\}$ $(n \geq 1)$.

Corollary 8.2.22. *We have*

$$\pi_1(S^2 \setminus \{x_1, x_2, \dots, x_n\}) \cong \overbrace{\mathbb{Z} * \mathbb{Z} * \dots * \mathbb{Z}}^{n-1}.$$

Corollary 8.2.22 combined with Proposition 7.2.23 lead to the following consequence.

Corollary 8.2.23. *Let* $\ell_1, \ell_2, \dots, \ell_n$ $(n \geq 1)$ *be* n *lines through the origin and* $X = \mathbb{R}^3 \setminus \{\ell_1, \ell_2, \dots, \ell_n\}$. *Then*

$$\pi_1(X) \cong \overbrace{\mathbb{Z} * \mathbb{Z} * \dots * \mathbb{Z}}^{2n-1}.$$

The case when the lines do not necessarily intersect is discussed when $n = 2$, $(\ell_1) : x = z = 0$, $(\ell_2) : x = z = 1$, $(\ell_3) : x = y = 0$ and $(\ell_4) : x + y = 1, z = 0$.

Corollary 8.2.24. *We have*

$$\pi_1(\mathbb{R}^3 \setminus \{\ell_1, \ell_2\}) \cong \pi_1(\mathbb{R}^3 \setminus \{\ell_3, \ell_4\}) \cong \mathbb{Z} * \mathbb{Z}.$$

Proof. Let $X = \mathbb{R}^3 \setminus \{\ell_1, \ell_2\}$ and $Y = \mathbb{R}^3 \setminus \{\ell_3, \ell_4\}$. We distinguish between two cases.

(1) The lines (ℓ_1) and (ℓ_2) are parallel. Let (P) be a plane perpendicular to (ℓ_1) and (ℓ_2), for instance the plane $(P) : y = 0$, and let $p_i = (P) \cap (\ell_i)$ $(i = 1, 2)$. The projection parallel to (ℓ_1), (ℓ_2) deformation retracts the space X onto the space $X' = (P) \setminus \{p_1, p_2\}$. By Proposition 8.2.18, $\pi_1(X') \cong \mathbb{Z} * \mathbb{Z}$.

(2) The lines (ℓ_3) and (ℓ_4) are skew lines (not parallel but do not intersect). Making rotations reduces to the first case. For instance, one can take the rotation $R_1(x, y, z) = (x', y', z)$ with $(x' + iy') - 1 = \frac{\sqrt{2}}{2}((x + iy) - 1)(1 - i)$ to rotate (ℓ_4) to the line $(\ell_4)' : x = 1$, $z = 0$. Also, rotate the line $(\ell_4)'$ to the line $(\ell_4)'' : x = 1, y = 0$ using the rotation

$R_2(1,y,0) = (1,0,yi)$. Here, $i^2 = -1$. The line $(\ell_4)''$ is parallel to (ℓ_3). Hence, $\pi_1(Y) \cong \pi_1(\mathbb{R}^3 \setminus \{\ell_3, \ell_4''\}) \cong \mathbb{Z} * \mathbb{Z}$, by the first part. □

8.2.5 Mapping cone and attaching cells

Let $f : X \longrightarrow Y$ be a continuous map and $K_f \cong C(X) \cup_f Y \cong [(X \times I) \cup Y]/\sim$ the mapping cone of f (see Definition 3.5.2). K_f is a partition consisting of the following equivalence classes:

(1) $\{\{(x,0), f(x)\} : x \in X\}\}$ (including $\{\{y, (f^{-1}(y), 0)\} : y \in f(X)\}$);
(2) the one-element $\{\{x_0, 1\}\}$ for some $x_0 \in X$ (apex of the cone);
(3) $\{(x,t), x \in X, 0 < t < 1\}$;
(4) the one-element sets $\{\{y\}, y \notin f(X)\}$.

The following result is brought from [5, Proposition B.4]. For the sake of completeness, the proof is given with some explanations.

Theorem 8.2.25. *We have*

$$\pi_1(K_f) \cong \pi_1(Y)/N,$$

where N is the least normal subgroup of $\pi_1(Y)$ containing $f_*(\pi_1(X))$ and f_* is the homomorphism induced by f.

Proof. Let π be the projection map and consider the open sets

$$U = \pi((X \times [0, 2/3)) \cup Y) \quad \text{and} \quad V = \pi(X \times (1/3, 1]).$$

The space V is a cone with the same apex as K_f and basis $X \times \{1/3\}$. By Proposition 4.9.5, it is contractible to its apex. The space U is homotopically equivalent to Y (see Figure 8.7). Then $U \cap V$ is a conical section (section of the cone between $t = 1/3$ and $t = 2/3$), which is homotopically equivalent to X. The proof is pretty much the same as for the cylinder and its basis (see Remark 4.2.7). Finally, $K_f = U \cup V$ and all spaces are path-connected. By Corollary 8.1.9, the spaces $\pi_1(U \cup V) \cong \pi_1(K_f)$ and $\pi_1(U) \cong \pi_1(Y)/N$ are isomorphic, where N is the least normal subgroup of $\pi_1(U)$ containing the image of the induced homomorphism

$$i_* : \pi_1(U \cap V) \longrightarrow \pi_1(U).$$

Since

$$f_* : \pi_1(X) \longrightarrow \pi_1(Y)$$

and

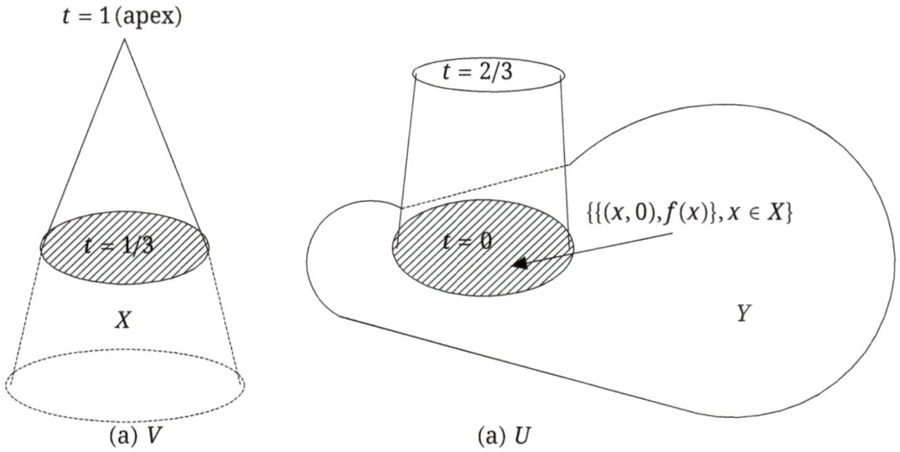

Figure 8.7: Decomposition of the mapping cone.

$$\pi_1(U \cap V) \cong_{h_1} \pi_1(X) \quad \text{and} \quad \pi_1(U) \cong_{h_2} \pi_1(Y),$$

where h_1, h_2 are the associated isomorphisms for all $s \in \pi_1(U \cap V)$,

$$i_*(s) = (h_2^{-1} \circ f_* \circ h_1)(s) \quad \Longrightarrow \quad \operatorname{Im} i_* = \operatorname{Im}(h_2^{-1} \circ f_* \circ h_1) \cong \operatorname{Im} f_*.$$

Therefore,

$$\pi_1(K_f) \cong \pi_1(Y)/N,$$

where N is the least normal subgroup of $\pi_1(X)$ containing the image of the induced homomorphism, i. e., $\operatorname{Im} f_*$. □

Let $f : S^{n-1} \subset B^n \longrightarrow X$ be a continuous map, where X is a pointed path-connected space (Definition 8.2.13). The following result provides a convenient way to compute the fundamental group of the adjunction space $Z_f = B^n \cup_f X$ in terms of that of the space X. This further shows that in the case when either $n = 1$ or $n \geq 3$, the map f is not involved in the calculation (see [31, Proposition 3.17]).

Corollary 8.2.26 (Attaching n-cells). *We have*

$$\pi_1(B^n \cup_f X) \cong \begin{cases} \pi_1(X) * \mathbb{Z}, & n = 1, \\ \pi_1(X), & n \geq 3, \\ \pi_1(X)/N, & n = 2, \end{cases}$$

where N is the least normal subgroup of $\pi_1(X)$ containing $f_*(\mathbb{Z})$ and f_* is the homomorphism induced by f.

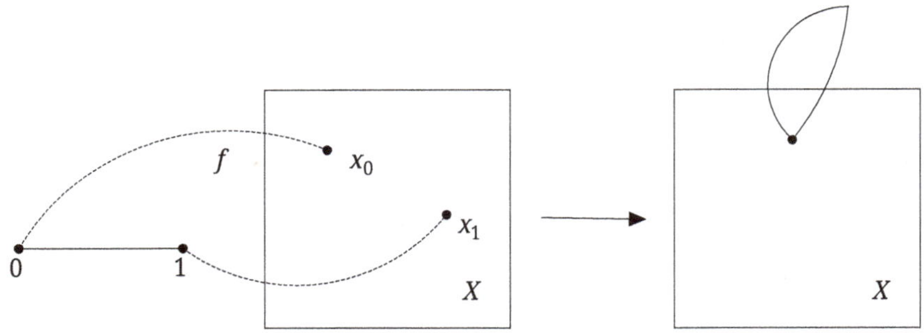

Figure 8.8: Space with an arc handle.

Proof. (1) For $n = 1$, let $f(-1) = x_0$ and $f(1) = x_1$, where $x_0, x_1 \in X$. Since X is path-connected, there is a path in X joining x_0 and x_1, hence an extension of f. By Theorem 6.4.1, f is null-homotopic. Theorem 7.4.8 shows that Z_f is homotopically equivalent to $B^1 \cup_{\text{cste}} X$, where cste is some constant function, say $f(x) = x_0 \in X$, for $x \in X$. However, the adjunction space $B^1 \cup_f X$ is just the wedge sum $X \vee S^1$ of X with a one-dimensional handle (Figure 8.8). Its fundamental group is given by Theorem 8.2.9.

(2) The case $n \geq 3$ is now discussed in two ways.

(a) *First proof.* Observe that

$$Z_f = B^n \cup_f X \cong C(S^{n-1}) \cup_f X.$$

Then $f_* : \pi_1(S^{n-1}) \longrightarrow \pi_1(X)$, where $\pi_1(S^{n-1}) \cong \{0\}$ for all $n \geq 3$. By applying Theorem 8.2.25 with $X \rightarrow S^{n-1}$ and $Y \rightarrow X$, we get $\pi_1(Z_f) \cong \pi_1(X)/\{0\} \cong \pi_1(X)$ by Example B.2.11, whence the result.

(b) *Second proof (direct proof).* Let $U = \mathring{B}^n$ be the open ball and $V = Z_f \setminus \{x_0\} = B^n \setminus \{x_0\} \cup_f X$, where $x_0 \in \mathring{B}^n$ is some given point. Then $\{U, V\}$ is an open covering of Z_f and U is contractible. Since S^{n-1} is a deformation retract of $B^n \setminus \{x_0\}$, by Theorem 7.4.3, $S^{n-1} \cup_f X = X$ is a deformation retract of V. Hence, V and X are homotopically equivalent. The intersection $U \cap V = \mathring{B}^n \setminus \{x_0\}$ is homotopically equivalent to the sphere S^{n-1}, which is contractible for $n \geq 3$. By Corollary 8.1.5, we obtain for $n \geq 3$,

$$\pi_1(Z_f) \cong \pi_1(U) * \pi_1(V) \cong \pi_1(X).$$

(3) For $n = 2$, $U \cap V$ is homotopically equivalent to S^1, where U and V are the same as in (2)(b). Using Seifert–van Kampen Theorem 8.1.2, we get

$$\pi_1(X) \cong \{0\} *_{\mathbb{Z}} \pi_1(X) \cong \pi_1(X)/N,$$

where N is the least normal subgroup of $\pi_1(X)$ that contains the elements $f_*(\mathbb{Z})$ and f_* is the homomorphism induced by f. □

Corollary 8.2.27. *Let M be an n-manifold and $M^\circ = M \setminus \mathring{B}^n$. Then, for $n \geq 3$, M and M° have isomorphic fundamental groups.*

Proof. Choose a continuous map $f : S^{n-1} \subset B^n \longrightarrow M^\circ$, let $M = B^n \cup_f M^\circ$ and apply Corollary 8.2.26. □

Example 8.2.28. Let $f : S^{n-1} \subset B^n \longrightarrow B^n = X$ be a continuous map. Then

$$\pi_1(B^n \cup_f B^n) \cong \{0\}, \quad \text{for all } n \geq 2.$$

Example 8.2.29. Let $f : S^{n-1} \subset B^n \longrightarrow S^{n-1} = X$ be a continuous map. Then

$$\pi_1(B^n \cup_f S^{n-1}) \cong \begin{cases} \{0\}, & n \geq 3, \\ \mathbb{Z}/N, & n = 2, \end{cases}$$

where N is the least normal subgroup of $\pi_1(S^1)$ containing $f_*(\mathbb{Z})$.

For example, if $n = 2$ and $f(z) = z^p$, by Proposition 6.2.3 $\deg f = p$. Hence, f_* is the multiplication by p in which case $f_*(\mathbb{Z}) = p\mathbb{Z}$. We deduce the fundamental group of the Moore space $M_p = B^2 \cup_{z^p} S^1$ introduced in Definition 3.4.17:

$$\pi_1(M_p) \cong \mathbb{Z}/p\mathbb{Z}. \tag{8.1}$$

In the particular case of the identity on S^1, we have $\pi_1(B^2 \cup_z S^1) \cong \{0\}$ because $\mathbb{Z}/\mathbb{Z}$ is trivial (see Example B.2.11), which can be obtained from Definition 3.4.17 since $B^2 \cup_z S^1 \cong B^2$.

Example 8.2.30. Since by Proposition 3.4.18, $\mathbb{RP}^2 \cong B^2 \cup_{z^2} S^1$ is a Moore space, we recover the fundamental of the projective plane already obtained in Corollary 6.1.11:

$$\pi_1(\mathbb{RP}^2) \cong \mathbb{Z}/2\mathbb{Z}.$$

Example 8.2.31. Let $f : S^{n-1} \subset B^n \longrightarrow \mathbb{RP}^n = X$ be the projection (and covering) map given by $f(x) = \{x, -x\}$ (see Proposition 5.1.7). By Example 3.4.19,

$$\mathbb{RP}^n \cong B^n \cup_f \mathbb{RP}^{n-1}, \quad \text{for all } n \geq 2.$$

By Corollary 8.2.26,

$$\pi_1(\mathbb{RP}^n) \cong \pi_1(\mathbb{RP}^{n-1}), \quad \text{for } n \geq 3 \quad \text{and} \quad \pi_1(\mathbb{RP}^2) \cong \pi_1(\mathbb{RP}^1)/N \cong \mathbb{Z}/N, \quad \text{for } n = 2,$$

where N is the least normal subgroup of $\mathbb{Z}$ containing $f_*(\mathbb{Z})$. For $n = 2$, f can be identified with the square map $f(z) = z^2$ and the latter map has degree 2 by Proposition 6.2.3. Then the induced map f_* is just the multiplication by 2. Hence, $\pi_1(\mathbb{RP}^2) \cong \mathbb{Z}/2\mathbb{Z}$. By induction, we find out a different method to compute the fundamental groups of the projective space (see Corollary 6.1.11 and Corollary 9.2.19):

$$\pi_1(\mathbb{R}\mathrm{P}^n) \cong \begin{cases} \mathbb{Z}/2\mathbb{Z}, & n \geq 2, \\ \mathbb{Z}, & n = 1. \end{cases}$$

8.2.6 Connected sum of manifolds and the Klein bottle

Let M_1, M_2 be two n-dimensional manifolds and $M_1\#M_2$ denote their connected sum (see Definition 3.7.1).

Theorem 8.2.32. *For $n \geq 3$,*

$$\pi_1(M_1\#M_2) \cong \pi_1(M_1) * \pi_1(M_2).$$

Proof. For $i = 1, 2$, let $h_i : S^{n-1} \longrightarrow M_i$, $M_i' = M_i \setminus \mathring{B}_i$, where $\mathring{B}_i$ are the n-dimensional open balls to be deleted from the manifolds M_i to form the connected sum, and let $B_i^\varepsilon \subset \mathring{B}_i$ be small balls therein. Consider the open sets $U_i = (M_i \setminus B_i^\varepsilon) \cup_h \mathring{B}_{3-i}$. For $n \geq 3$, Corollary 8.2.27 implies that M_i, M_i' and $M_i \setminus B_i^\varepsilon$ have isomorphic fundamental groups and by Corollary 8.2.26, $\pi_1(U_i) \cong \pi_1(M_i \setminus B_i^\varepsilon)$. The intersection $U_1 \cap U_2$ is homotopically equivalent to an annular region, thus to the sphere S^{n-1}. Finally, $(U \cup V) \cap (M\#N)$ covers $M_1\#M_2 = M_1' \cup_h M_2'$. Since S^{n-1} is simply connected for $n \geq 3$, Seifert–van Kampen Theorem 8.1.2 implies that

$$\begin{aligned} \pi_1(M_1\#M_2) &\cong \pi_1(U_1 \cup U_2) \\ &\cong \pi_1(U_1) * \pi_1(U_2) \\ &\cong \pi_1(M_1) * \pi_1(M_2). \end{aligned}$$

□

Remark 8.2.33. For $n \geq 3$, Theorem 8.2.9 and Theorem 8.2.32 show that

$$\pi_1(M_1\#M_2) \cong \pi_1(M_1 \vee M_2).$$

For $n = 2$, the amalgamation occurs on the infinite cyclic group $\pi_1(S^1) \cong \mathbb{Z}$ making the situation quite different. We discuss the connected sum of two projective planes, i. e., the Klein bottle.

Theorem 8.2.34. *The fundamental group of the Klein bottle $\mathbb{K}^2$ is the free product of $\mathbb{Z} * \mathbb{Z}$ with amalgamation over $\mathbb{Z}$, that is to say*

$$\pi_1(\mathbb{K}^2) \cong (\mathbb{Z} * \mathbb{Z})/N,$$

where N is the least normal subgroup of $\mathbb{Z} * \mathbb{Z}$ that contains the image of the induced inclusion homomorphism

$$i_* : \mathbb{Z} \longrightarrow \mathbb{Z} * \mathbb{Z}.$$

Equivalently, $\pi_1(\mathbb{K}^2) \cong \mathbb{Z} *_{\mathbb{Z}} \mathbb{Z}$ *or*

$$\pi_1(\mathbb{K}^2) \cong \langle a, b |\ a^2 b^2 = 1 \rangle.$$

Proof. (1) *First proof.* In order to make use of Corollary 8.1.9, let $x \in \mathbb{K}^2$, $U = \mathbb{K}^2 \setminus \{x\}$ and choose a small open disk $V = \overset{\circ}{B}{}^2(x, \varepsilon)$ in the Klein bottle (for instance $\mathbb{K}^2$ with the edges of the square I^2 removed). V is simply connected and $U \cap V = \overset{\circ}{B}{}^2 \setminus \{x\}$, which is homotopically equivalent to a circle (Example 4.2.8). By Remark 7.2.36, U is homotopically equivalent to the figure eight, i. e., a disk with two points removed. By Corollary 8.2.19, its fundamental group is $\pi(U) \cong \pi(\Theta) \cong \mathbb{Z} * \mathbb{Z}$. Therefore,

$$\pi_1(\mathbb{K}^2) \cong (\mathbb{Z} * \mathbb{Z})/N \cong \mathbb{Z} *_{\mathbb{Z}} \mathbb{Z},$$

where N is the least normal subgroup of $\pi_1(U) \cong \mathbb{Z} * \mathbb{Z}$ that contains $\operatorname{Im} i_*$ and

$$i_* : \mathbb{Z} \longrightarrow \mathbb{Z} * \mathbb{Z},$$

is the induced homomorphism of the inclusion map $i : \overset{\circ}{B}{}^2 \setminus \{x\} \hookrightarrow \mathbb{K}^2 \setminus \{x\}$.

(2) *Second proof.* Corollary 8.1.9 is once again employed with a slight modification in the choice of the open sets U and V. Take $U = \mathbb{K}^2 \setminus \overset{\circ}{B}{}^2_\varepsilon$ in the square representing the Klein bottle and $V = \overset{\circ}{B}{}^2 \supset \overset{\circ}{B}{}^2_\varepsilon$ a larger ball in $\mathbb{K}^2$. Then V is simply connected and $U \cap V$ is an open annulus whose fundamental group is isomorphic to $\mathbb{Z}$ (see Example 4.2.5). As in part (1), $\pi_1(U) \cong \mathbb{Z} * \mathbb{Z}$. Therefore,

$$\pi_1(\mathbb{K}^2) \cong \pi_1(U)/N \cong \mathbb{Z} * \mathbb{Z}/N \cong \mathbb{Z} *_{\mathbb{Z}} \mathbb{Z},$$

where N is the least normal subgroup of $\mathbb{Z} * \mathbb{Z}$ containing the image of the induced homomorphism

$$i_* : \mathbb{Z} \longrightarrow \mathbb{Z} * \mathbb{Z}$$

and $i : U \cap V \hookrightarrow V$. Up to homotopy equivalences, i is the inclusion of the circle into the figure eight.

(3) *Third proof.* Since by Proposition 3.8.10, $\mathbb{K}^2 \cong \mathbb{RP}^2 \# \mathbb{RP}^2$, let $M_1 = M_2 = \mathbb{RP}^2$ and consider the power map $h : S^1 \longrightarrow S^1$ given by $h(z) = z^2$. By Theorem 3.2.8, $\mathbb{RP}^2 \setminus B^2$ is homeomorphic to the Möbius band $\mathbb{M}^2$. Then $X = \mathbb{RP}^2 \# \mathbb{RP}^2 \cong (\mathbb{M}^2 \cup \mathbb{M}^2)/\sim_h$ (see Corollary 3.8.12). By Proposition 4.2.17, $\pi_1(\mathbb{M}^2) \cong \mathbb{Z}$. By Theorem 8.1.2,

$$\pi_1(\mathbb{K}^2) \cong \mathbb{Z} *_{\mathbb{Z}} \mathbb{Z} \cong (\mathbb{Z} * \mathbb{Z})/N,$$

where N is the least normal subgroup generated by the elements $i_1(f) i_2(f)^{-1}$, $f \in \pi_1(U \cap V)$, i. e., that contains the subgroup $2\mathbb{Z}$. The induced map $h_* : \mathbb{Z} \longrightarrow \mathbb{Z}$ is given by

$h_*(k) = 2k$ (see Definition 12.5.1) and Im $h_* = 2\mathbb{Z} \subseteq \mathbb{Z}$. Therefore, N is generated by a^2b^{-2}, where a, b are generators of first $\mathbb{Z}$ and second $\mathbb{Z}$, respectively. Finally,

$$\pi_1(\mathbb{K}^2) \cong \langle a, b |\, a^2b^{-2} = 1\rangle \cong \langle a, b |\, a^2 = b^2\rangle.$$

It is easy to see that the following equivalent representation holds:

$$\pi_1(\mathbb{K}^2) \cong \langle a, b |\, aba = b\rangle \cong \langle a, b |\, abab^{-1} = 1\rangle. \qquad \square$$

Remark 8.2.35. Theorem 8.2.34 shows that the adjunction space is not preserved under homotopy equivalence. Indeed, the Klein bottle is the adjunction space of two Möbius bands (Corollary 3.8.12) and $\mathbb{M}^2$ is homotopically equivalent to a circle (Proposition 4.2.17). However, the adjunction of two circles $S^1 \cup_f S^1$ has a fundamental group isomorphic to $\mathbb{Z}$ if, e. g., f is the identity map on S^1.

Remark 8.2.36. Let S be a connected compact surface represented symbolically by $\langle S |\ a_1a_2 \dots a_m\rangle$ (see Section 3.8). From Corollary 8.2.26 and using the representations of groups, one can deduce that the fundamental group of S is $\pi_1(S) \cong \langle a_1, a_2, \dots, a_m |\ a_1a_2 \dots a_m\rangle$ (see [62, Proposition 10.16] and [31, Corollary 3.19]). Thus, the fundamental group of the Klein bottle, the torus and the projective plane are:
(1) $\pi_1(\mathbb{K}^2) \cong \langle a, b |\, abab^{-1} = 1\rangle$;
(2) $\pi_1(\mathbb{T}^2) \cong \langle a, b |\, ab = ba\rangle$ (see also Remark 8.2.12);
(3) $\pi_1(\mathbb{RP}^2) \cong \langle a |\, aa = 1\rangle \cong \mathbb{Z}/2\mathbb{Z}$.

Remark 8.2.37. By Example 3.8.2, the code $abab^{-1}$ for the polygonal representation of the Klein bottle is equivalent to $aba^{-1}b$. Performing one permutation and one relabeling (see Section 3.8), this is equivalent to $baba^{-1}$ and then $abab^{-1}$. So, we can see that the formula for $\pi_1(\mathbb{K}^2)$ in the preceding remark is consistent with the result of Theorem 8.2.34.

We end this section with a result, which follows from the classification theorem 3.7.4 (see, e. g., [59, Corollary 26.2]).

Theorem 8.2.38. (1) *If S is homeomorphic to a connected sum of n 2-tori, then the Abelianization of $\pi_1(S)$, i. e., $\pi_1(S)/[\pi_1(S), \pi_1(S)]$, is isomorphic to $\mathbb{Z}^{2n}$.*
(2) *If S is homeomorphic to a connected sum of n projective planes, then the Abelianization of the fundamental group of S is isomorphic to $\mathbb{Z}^{n-1} \times \mathbb{Z}/2\mathbb{Z}$.*

Example 8.2.39. (1) The Abelianization of $\pi_1(\mathbb{T}^2\#\mathbb{T}^2)$ is isomorphic to $\mathbb{Z}^4$.
(2) The Abelianization of $\pi_1(\mathbb{RP}^2\#\mathbb{RP}^2)$ is isomorphic to $\mathbb{Z} \times \mathbb{Z}/2$.

Remark 8.2.40. Example 8.2.39 confirms that Theorem 8.2.32 does not in general hold for $n = 2$ even for the Abelianization of the groups. Indeed,

$$\begin{aligned}\text{Abelianization of } \pi_1(\mathbb{T}^2\#\mathbb{T}^2) &\cong \mathbb{Z}^4\\ &\cong \text{Abelianization of } (\mathbb{Z} * \mathbb{Z} * \mathbb{Z} * \mathbb{Z})\\ &\cong \pi_1(\mathbb{T}^2) \times \pi_1(\mathbb{T}^2)\\ &\cong \text{Abelianization of } \pi_1(\mathbb{T}^2) * \pi_1(\mathbb{T}^2).\end{aligned}$$

However,

$$\begin{aligned}\text{Abelianization of } \pi_1(\mathbb{RP}^2\#\mathbb{RP}^2) &\cong \mathbb{Z} \times \mathbb{Z}/2\\ &\ncong \mathbb{Z}/2 \times \mathbb{Z}/2,\end{aligned}$$

where

$$\begin{aligned}\mathbb{Z}/2 \times \mathbb{Z}/2 &\cong \pi_1(\mathbb{RP}^2) \times \pi_1(\mathbb{RP}^2)\\ &\cong \text{Abelianization of } \pi_1(\mathbb{RP}^2) * \pi_1(\mathbb{RP}^2).\end{aligned}$$

8.3 Summary of the main fundamental groups

In this last section of this chapter, we summarize the fundamental groups we have computed so far using results from previous chapters.

(1) By Example 1.5.22 and Example 4.9.2, the closed ball B^n, which is a contractible space, the northern hemisphere S^n_+ and southern hemisphere S^n_- have fundamental groups isomorphic to the trivial group.

(2) From Theorem 1.5.37 and Theorem 1.5.38, we have the following.

Corollary 8.3.1. *The spaces* $S^1 \times (0,\infty)$, $S^1 \times (0,1)$, $\mathring{B}^2 \setminus \{(0,0)\}$, $\mathbb{R}^2 \setminus \{(0,0)\}$ *and the open annulus* $A(0;1,2)$ *are homeomorphic, hence have isomorphic fundamental groups.*

(3) To check that S^1 and the closed annulus $A(0;1,2)$ have isomorphic fundamental groups, we refer to Example 4.2.5, where it is shown that the two sets are homotopically equivalent. Then we appeal to Corollary 4.5.5 and Theorem 6.1.4. The fundamental group of the spaces in Corollary 8.3.1 is isomorphic to the integers.

(4) By Corollary 6.3.4, for all $n \geq 2$, the sphere S^n is simply connected. Hence, it has a trivial fundamental group. Owing to the proof of Corollary 6.3.4, $S^n \setminus \{N_p\}$ and $\mathbb{R}^n$ are homeomorphic (with the stereographic projection). By Remark 6.3.6, we further know that for $n \geq 2$, $S^n \setminus \{N_p, S_p\}$ is homeomorphic to the punctured space $\mathbb{R}^n \setminus \{0\}$. The spaces $\mathbb{R}^n \setminus \{0\}$ is homotopically equivalent to S^{n-1} (with the radial projection, Example 7.1.5). Consequently, $S^n \setminus \{N_p, S_p\}$ and S^{n-1} have isomorphic fundamental groups.

(5) (a) $\mathbb{R}^n \setminus \{0\}$, $\mathring{B}^n \setminus \{0\}$ and $\mathbb{R}^n \setminus B^n$ are homeomorphic for $n \geq 2$ (see Example 1.5.13).

(b) By Example 7.2.14 and Example 1.5.13, $\mathbb{R}^n \setminus \{0\}$, $\mathring{B}^n \setminus \{0\}$, $\mathbb{R}^n \setminus \mathring{B}^n$ and S^{n-1} are homotopically equivalent. Indeed, by Example 1.5.18, $B^n \setminus \{0\}$ and $\mathbb{R}^n \setminus \mathring{B}^n$ are homeomorphic.

(c) Theorem 1.5.38 provides a homeomorphism between $S^{n-1} \times (0,1)$ and $\mathring{B}^n \setminus \{0\}$ in one hand, and a homeomorphism between $S^{n-1} \times (0,\infty)$ and $\mathbb{R}^n \setminus \{0\}$ in the other one.

Theorem 1.5.38 shows further that $S^{n-1} \times (0,1]$ and $B^n \setminus \{0\}$ are homeomorphic. Thus, their fundamental groups derive from those of the sphere S^n for all $n \geq 1$ combined with Proposition 7.2.16. See also Example 7.2.14.

(6) The fundamental groups of the quotient space S^n/S^k $(k < n)$ are computed in Example 8.2.15.

(7) For $n \geq 2$, the fundamental group of the projective space $\mathbb{R}P^n$ is given by Corollary 6.1.11. For $n = 1$, $\mathbb{R}P^1$ is homeomorphic to the circle (see Definition 3.1.1). By Corollary 6.3.11, the fundamental groups of the complex projective spaces $\mathbb{C}P^n$ reduce to those of the sphere for all $n \geq 1$.

(8) As for the torus and the solid torus, Example 6.1.9, Theorem 4.6.1 can be applied.

(9) The fundamental group of the Möbius strip is given by Proposition 4.2.17 and Proposition 7.2.33. The fundamental group of the Klein bottle is given by Theorem 8.2.34. The fundamental group of Moore space is given by formula (8.1).

(10) Addition topological structures of the Möbius band are discussed in Proposition 4.2.17, Proposition 7.2.33, Proposition 7.2.37 and Proposition 7.2.38.

(11) The fundamental groups of the punctured torus, the punctured Möbius band and the punctured projective plane follow from Proposition 7.2.41, Proposition 7.2.42, Proposition 7.2.35 and Remark 7.2.36.

(12) The computation of the fundamental groups of the torus and the Möbius band with holes are left as exercises (Exercises 6, 7, this chapter). See also Proposition 8.2.2.

(13) Propositions 7.2.19–7.2.31 provide further calculations of fundamental groups using the homotopy equivalence maps instead of homeomorphisms.

(14) Other fundamental groups were obtained using the Seifert–van Kampen theorem. For instance, we know that the double-punctured plane, the double-punctured disk, the figure eight, the figure theta, the n-leafed rose and the wedge sum (bouquet) of circles are homotopically equivalent (see Proposition 4.2.13, Proposition 4.2.15, Proposition 8.2.1, Proposition 8.2.18 and Corollary 8.2.19). The fundamental group of the Moore space is computed in Example 8.2.29.

(15) By Proposition 8.2.17 and Corollary 8.2.21, $\pi_1(\mathbb{R}^3 \setminus S^1) \cong \mathbb{Z}$. The following results are consequence of Proposition 7.2.19, Proposition 7.2.24 and Proposition 7.2.31, and Remark 7.2.27.

Corollary 8.3.2. *We have*
(1) $\pi_1(\mathbb{R}^3 \setminus (0z)) \cong \pi_1(\mathbb{R}^2 \setminus \{0\}) \cong \mathbb{Z}$;
(2) $\pi_1(S^3 \setminus S^1) \cong \mathbb{Z}$;
(3) $\pi_1(\mathbb{R}^3 \setminus (S^1 \cup (0z))) \cong \mathbb{Z} \times \mathbb{Z}$;
(4) $\pi_1((S^3 \setminus S^1) \setminus S^1) \cong \mathbb{Z} \times \mathbb{Z}$.

The obtained results are summarized in the following table, where $\{*\}$ and ℓ stand for a point and a line, respectively.

Topological space	**Fundamental group (up to isomorphism)**
A star-convex space, in particular a convex space, (open and closed balls B^n, $\mathbb{R}^n$ $n \geq 1$, $C(S^n)$ $n \geq 1, \ldots$) S^n $(n \geq 2)$, S^n_+, S^n_- $(n \geq 1)$, $S^n \setminus \{pt\}$ $(n \geq 1)$, For $n \geq 3$, $B^n \setminus \{0\}$, $\mathring{B}^n \setminus \{0\}$, For $n \geq 3$, $\mathbb{R}^n \setminus B^n$, $\mathbb{R}^n \setminus \mathring{B}^n$, For $n \geq 3$, $k \geq 0$, $\mathbb{R}^n \setminus \{k \text{ pts}\}$, For $n \geq 3$, $k \geq 1$, $S^n \setminus \{k \text{ pts}\}$, For $n \geq 3$, $S^{n-1} \times [0,1]$, $S^{n-1} \times (0,1]$, For $n \geq 3$, $S^{n-1} \times (0,1)$, $S^{n-1} \times (0,\infty)$, $\mathbb{CP}^1$, S^n/S^k $(1 \leq k < n)$	Trivial
S^1, $S^1 \times [0,1]$, $S^1 \times (0,1]$, $S^1 \times (0,1)$, $S^1 \times (0,\infty)$, $S^2 \setminus \{N_p, S_p\}$, $\mathbb{R}^2 \setminus \{0\}$, $\mathbb{R}^2 \setminus B^2$, $\mathbb{R}^2 \setminus \mathring{B}^2$, SO(2) $\mathring{B}^2 \setminus \{0\}$, $B^2 \setminus \{0\}$, $B^2 \times S^1$, $A(0;1,2)$, $\mathbb{RP}^1$, $\mathbb{CP}^n$, $n \geq 2$, $\mathbb{RP}^3 \setminus \mathbb{RP}^1$, $\mathbb{RP}^2 \setminus \{*\}$, $S^3 \setminus S^1$, $\mathbb{R}^3 \setminus \{0z\}$, $\mathbb{R}^3 \setminus S^1$, S^n/S^0 $(n > 1)$ $\mathbb{M}^2$, $\mathring{\mathbb{M}}^2$, $C(\mathbb{M}^2)$, $\mathbb{M}^2 \setminus C(\mathbb{M}^2)$, $\partial\mathbb{M}^2$	$\mathbb{Z}$
$\mathbb{T}^2$, $\mathbb{R}^3 \setminus \{(0z) \cup S^1\}$, $(S^3 \setminus S^1) \setminus S^1$	$\mathbb{Z} \times \mathbb{Z}$
$\mathbb{T}^n, n \geq 1$	$\overbrace{\mathbb{Z} \times \mathbb{Z} \times \cdots \times \mathbb{Z}}^{n}$
$\mathbb{RP}^n$, $n \geq 2$, SO(3)	$\mathbb{Z}/2\mathbb{Z}$
$M_n = B^2 \cup_{z^n} S^1$	$\mathbb{Z}/n\mathbb{Z}$
$S^1 \vee S^1$, 8, Θ, $\mathbb{T}^2 \setminus \{*\}$, $\mathbb{T}^2 \setminus B^2_\varepsilon$, $\mathbb{M}^2 \setminus \{*\}$, S^1/S^0	$\mathbb{Z} * \mathbb{Z}$
$S^2 \setminus \{(n+1) \text{ pts}\}$, $\mathbb{R}^2 \setminus \{n \text{ pts}\}$, $n \geq 1$, $\overbrace{S^1 \vee S^1 \vee \cdots \vee S^1}^{n}$, $\mathbb{T}^2 \setminus \{(n-1) \text{ pts}\}$	$\overbrace{\mathbb{Z} * \mathbb{Z} * \cdots * \mathbb{Z}}^{n}$
$\mathbb{K}^2$	$\mathbb{Z} *_{\mathbb{Z}} \mathbb{Z}$, $\langle a, b \mid a^2 = b^2 \rangle$

Exercises (Chapter 8)

1. Determine the fundamental group $\pi_1(\mathbb{RP}^n \times S^m)$ for n, $m \geq 2$.

2. Let X be the sphere S^2 with n points identified.

(1) Prove that X is homotopically equivalent to $S^2 \vee (\overbrace{S^1 \vee \cdots \vee S^1}^{n-1})$, $n \geq 1$.

(2) Deduce $\pi_1(X)$.

3. Determine the quotient space of the torus with n points identified and then find its fundamental group.

4. Let $f : S^1 \longrightarrow S^1$ be the nth power map given by $f(z) = z^n$.

(1) Determine the mapping cylinder C_f (the case $n = 2$ is Exercise 5 of Sections 3.3–3.7).

(2) Determine the mapping cone K_f.

5. Let B_1, B_2 and S_1, S_2 be copies of the unit ball and the unit circle of the plane, respectively. Let $f_1 : S_i \longrightarrow S_i$ be the power maps given by $f_i(z) = z^{n_i}$, for $i = 1, 2$ and define $f : S_1 \cup S_2 \longrightarrow S_1 \vee S_2$ by $f(z) = f_1(z) \vee f_2(z)$. Find the fundamental group of the adjunction space $X = (B_1 \cup B_2) \cup_f (S_1 \vee S_2)$.

6. Find the fundamental group of the torus $\mathbb{T}^2$ with n holes.

7. Find the fundamental group of the Möbius band $\mathbb{M}^2$ with n holes.

8. Let $X = \mathbb{R}^n \setminus \mathbb{R}^k$ $(n > k)$.

(1) Check that

$$X \cong (\mathbb{R}^{n-k} \setminus \{0\}) \times \mathbb{R}^k \simeq S^{n-k-1}.$$

(2) Deduce the fundamental group of the space X.

9. For $n > k$, find the fundamental group of the space $X = S^n \setminus S^k$. (*Hint:* One may see the results of the proof of Theorem 10.4.24.) Do not make confusion with the quotient S^n/S^k already discussed in Example 8.2.15.

10. Prove that $\mathbb{RP}^3 \setminus \mathbb{RP}^1$ deformation retracts onto S^1 and deduce the fundamental group $\pi_1(\mathbb{RP}^3 \setminus \mathbb{RP}^1)$.

11. Find the fundamental group of the space $X = \mathbb{R}^3 \setminus (S^1 \vee S^1)$.

12. Find the fundamental group of the subspace $X = A_1 \cup A_2 \cup A_3$ of $\mathbb{R}^3$ with the notation of Proposition 8.2.16.

13. Find the fundamental group of the space

$$X = \{(z_1, z_2) \in \mathbb{C}^2 : |z_1| = r_1,\ |z_2| = r_2,\ z_1^{p_1} = z_2^{p_2}\},$$

where $r_1, r_2 > 0$ and $p_1, p_2 \geq 1$.

14. Let $P_1, P_2 \subset \mathbb{R}^4$ be the planes $x = y = 0$ and $z = w = 0$, respectively. Compute $\pi_1(\mathbb{R}^4 \setminus \{P_1 \cup P_2\})$.

15. Let X be a connected and a locally path-connected space such that $X = U \cup V$ and $U \cap V = A \cup B$, where U, V are simply connected open subspaces and A and B are two connected components with disjoint closures. Prove that $\pi_1(X) \cong \mathbb{Z}$ (*Hint:* See [36, Exercise 1, Section 10, Chapter 3]).

9 Higher homotopy groups

In this chapter, the fundamental group $\pi_1(X)$ developed in Chapter 4 is extended to the higher-dimensional homotopy group $\pi_n(X)$, $n = 1, 2, \ldots$. In Section 9.1, the discussion is essentially restricted to the definitions and the main properties of $\pi_n(X)$. Based on the Freudenthal suspension theorem, the computations of the fundamental groups of the sphere are performed in Section 9.2. We will see that even for the sphere, the computations are not quite complete.

9.1 Definitions and main properties

Let $I^n = [0,1]^n$ be the n-cube and

$$\partial I^n = \{t = (t_1, t_2, \ldots, t_n) \in I^n \text{ and there exists } 1 \le k \le n \text{ such that } t_k = 0 \text{ or } t_k = 1\}$$

the boundary of I^n. Let X be a space and $x_0 \in X$ a point.

Definition 9.1.1. A *hyperloop* based at x_0 is a continuous map $f : I^n \longrightarrow X$ such that $f(\partial I^n) = x_0$.

Definition 9.1.2. Let f and g be hyperloops based at some point x_0. Define *the product* $f * g$ to be the hyperloop h given by

$$h(s_1, s_2, \ldots, s_n) = \begin{cases} f(2s_1, s_2, \ldots, s_{n-1}, s_n), & s_1 \in [0, 1/2], \\ g(2s_1 - 1, s_2, \ldots, s_{n-1}, s_n), & s_1 \in [1/2, 1]. \end{cases} \tag{9.1}$$

Since $(1, s_2, \ldots, s_{n-1}, s_n)$ and $(0, s_2, \ldots, s_{n-1}, s_n)$ lie in ∂I^n, we have

$$f(1, s_2, \ldots, s_{n-1}, s_n) = g(0, s_2, \ldots, s_{n-1}, s_n) = x_0.$$

By Lemma A.2.1, h is continuous at $(1/2, s_2, \ldots, s_{n-1}, s_n)$.

Definition 9.1.3. Let f and g be two hyperloops based at some point x_0. f and g are said to be homotopic if there exists a continuous map $H : I^n \times I \longrightarrow X$ such that:

(1) $H(s, 0) = f(s)$ and $H(s, 1) = g(s)$ for all $s \in I^n$;
(2) $H(s, t) = x_0$ for all $s \in \partial I^n$, $t \in I$.

In short, the homotopy H satisfies $H(\cdot, 0) = f(\cdot)$ and $H(\cdot, 1) = g(\cdot)$, $H(\partial I^n \times I) = x_0$. We shall write $f \simeq g$. As in the case $n = 1$, it is easily checked that $\simeq$ is an equivalence relation. Denote by $[f]$ the equivalence class of f and define in a natural way the class product as

$$[f] * [g] = [f * g].$$

https://doi.org/10.1515/9783111517384-009

As in Lemma 4.3.4, we can show that this operation is well-defined, i. e., $f' \simeq_p f$ and $g' \simeq_p g$ imply $[f' * g'] = [f * g]$. For this aim, consider two corresponding homotopies F, G and define the homotopy H between $f * k$ and $f' * g'$ by

$$H(\cdot, t) = F(\cdot, t) * G(\cdot, t), \quad t \in I.$$

Definition 9.1.4. The quotient space under the homotopy of the hyperloops based at x_0 is called the *nth higher homotopy group* of X relative to the base point x_0 and is denoted $\pi_n(X, x_0)$.

Remark 9.1.5. We can show that $\pi_n(X, x_0)$ is a group. The proof is analogous to that of $\pi_1(X, x_0)$. The identity element is $[e_{x_0}]$, where e_{x_0} is the constant map $e_{x_0}(s) = x_0$ for all $s \in I^n$. The inverse $[f]^{-1}$ is defined as $[\bar{f}]$, where $\bar{f}(s) = f(1 - s_1, s_2, \ldots, s_n)$ is the inverse loop of $f(s)$, for $s = (s_1, s_2, \ldots, s_n)$. The homotopy between $f * \bar{f}$ and e_{x_0} is given by

$$H(s,t) = \begin{cases} f(2s_1, s_2, \ldots, s_{n-1}, s_n), & s_1 \in [0, t/2], \\ f(t, s_2, \ldots, s_{n-1}, s_n), & s_1 \in [t/2, 1 - t/2], \\ f(2 - 2s_1, s_2, \ldots, s_{n-1}, s_n), & s_1 \in [1 - t/2, 1], \end{cases}$$

for $(s, t) \in I^n \times I$. The associativity of the operation $*$ is proved as for the case $n = 1$.

Remark 9.1.6. For $a = (1, 0, \ldots, 0) \in S^n$ and $x_0 \in X$, denote

$$\Omega(X, x_0) = \{f : I^n \longrightarrow X \text{ continuous and } f(\partial I^n) = x_0\},$$
$$\Sigma(X, x_0) = \{f : S^n \longrightarrow X \text{ continuous and } f(a) = x_0\}.$$

Let $[S^n, X]$ denote the set of homotopy classes of base point preserving continuous maps $f \in \Sigma(X, x_0)$. By collapsing the boundary ∂I^n of the n-cube I^n, the resulting quotient space $I^n/\partial I^n$ is homeomorphic to S^n. Indeed, since I^n is homeomorphic to the closed unit ball B^n, by Theorem 2.5.11, $I^n/\partial I^n \simeq B^n/S^{n-1} \simeq S^n$. Consider the diagram:

$$I^n \xrightarrow{\quad \pi \quad} I^n/\partial I^n \cong S^n \xrightarrow{\quad f \quad} X$$
$$\phi = f \circ \pi$$

The map $\phi : I^n \longrightarrow X$ given by $\phi = f \circ \pi$ is continuous and $\phi(\partial I^n) = x_0$. Hence, $\phi \in \Omega(X, x_0)$ is a hyperloop.

To the class $[\phi]$ of a hyperloop, we may assign, by a bijective correspondence, the homotopy class $[f]$, where $f : S^n \longrightarrow X$. Indeed, given $\phi : I^n \longrightarrow X$, one may assign $f : S^n \longrightarrow X$ by $\phi = f \circ \pi \in \Omega(X, x_0)$ exactly as in Remark 6.2.5, proving that ϕ is surjective. The map ϕ is injective since if $[f \circ \pi] = [g \circ \pi]$, then $f \circ \pi$ and $g \circ \pi$ are homotopic. Let $H : I^n \times I \longrightarrow X$ be such a homotopy. As for the existence of f in the surjection part, let

$\overline{H} : (I^n \times I)/(\partial I^n \times I) \longrightarrow X$, i. e., $\overline{H} : I^n/(\partial I^n \times I) \longrightarrow X$. Then $\overline{H} : S^n \times I \longrightarrow X$ is a homotopy between f and g (see [31, Exercise A.3(11)] for the last equivalence). Hence, $[f] = [g]$.

Finally, an operation $\overline{*}$ may be defined in $[S^n, X]$ as in (6.1). This motivates the following definition. The details for the case $n = 1$ were given in Remark 6.2.5.

Definition 9.1.7. We set

$$\pi_n(X, x_0) = [S^n, X]. \tag{9.2}$$

Note that B^n extends the interval $[0, 1]$ instead of the n-cube $[0, 1]^n$, which is a closed ball of Euclidean space homeomorphic to B^n (see Example 1.5.14). The n-cube gives rise to the concept of hyperloop. To complete the discussion, note that a third equivalent definition was given by W. Hurewicz in 1935 and consists of the recurrent relation

$$\pi_n(X, x_0) = \pi_{n-1}(\Omega(X, x_0), e_{x_0}), \quad n \geq 2. \tag{9.3}$$

Then most of properties of π_1 are transferred to π_n. For example, for the Cartesian product, we have the following.

Theorem 9.1.8. *For any spaces X, Y, $\pi_n(X \times Y, (x_0, y_0))$ and $\pi_n(X, x_0) \times \pi_n(Y, y_0)$ are isomorphic.*

Proof. We can use definition (9.3) to prove Theorem 9.1.8 by induction. Since the result has been proved for $n = 1$ in Theorem 4.6.1, we assume that it holds for $n - 1$. Notice that $\Omega(X \times Y, (x_0, y_0))$ is homeomorphic to $\Omega(X, x_0) \times \Omega(Y, y_0)$. Then

$$\begin{aligned}
\pi_n(X \times Y, (x_0, y_0)) &= \pi_{n-1}(\Omega(X \times Y, (x_0, y_0)), e_{(x_0, y_0)}) \\
&\cong \pi_{n-1}(\Omega(X, x_0) \times \Omega(Y, y_0), e_{(x_0, y_0)}) \\
&\cong \pi_{n-1}(\Omega(X, x_0), e_{x_0}) \times \pi_{n-1}(\Omega(Y, y_0), e_{y_0}) \\
&= \pi_n(X, x_0) \times \pi_n(Y, y_0).
\end{aligned}$$

□

By choosing a path joining two points x_0 and x_1 of X, the following result is easily checked.

Theorem 9.1.9. *If the space X is path-connected and x_0 and x_1 are points of X, then $\pi_n(X, x_0)$ is isomorphic to $\pi_n(X, x_1)$ for all $n \geq 1$.*

Viewing the fundamental group $\pi_1(X, x_0)$ as set of homotopy classes of maps $(S^1, x_0) \longrightarrow (X, x_0)$, Theorem 6.2.9 is reformulated as follows.

Theorem 9.1.10. *For a path-connected space X, the following statements are equivalent:*
(1) *Every map $f : S^1 \longrightarrow X$ is null-homotopic;*
(2) *Every map $f : S^1 \longrightarrow X$ extends continuously to a map $\widetilde{f} : B^2 \longrightarrow X$;*
(3) *X is simply connected.*

Remark 9.1.11. Theorem 9.1.10 can be extended to maps on the sphere S^n, $n \geq 1$. The case when $X = S^n$ is discussed in Theorem 12.5.12. In Corollary 9.2.13, we will prove that $\pi_n(S^n)$ is isomorphic to the set of integers. With Theorem 6.4.1, the following statements are equivalent for a path-connected space X:
(1) Every continuous map $f : S^n \longrightarrow X$ is null-homotopic;
(2) Every continuous map $f : S^n \longrightarrow X$ has a continuous extension $\tilde{f} : B^{n+1} \longrightarrow X$;
(3) X is n-aspherical.

Recall the following.

Definition 9.1.12. (1) Let $n = 1, 2, \ldots$. A space X is called *n-aspherical* if $\pi_n(X) \cong 0$.
(2) A path-connected space X is *aspherical* if it is n-aspherical for all $n > 1$.

Example 9.1.13. By Proposition 9.2.16, the circle S^1 is aspherical.

Example 9.1.14. Let $p : \widehat{X} \longrightarrow X$ be a covering map. Using the result of Theorem 9.2.18, X is aspherical if and only if $\widehat{X}$ is aspherical.

Remark 9.1.15. Let X be a path-connected space. By Corollary 6.2.12, X is 1-aspherical if and only if it is simply connected.

Proposition 9.1.16. *Every contractible space X is n-aspherical for all $n \geq 1$.*

Proof. For $n = 1$, every contractible space is simply connected (Corollary 4.9.15), hence 1-aspherical. For $n \geq 2$, let $H : X \times I \longrightarrow X$ be a continuous homotopy such that for all $x \in X$, $H(x, 0) = x$ and $H(x, 1) = x_0 \in X$. Then any continuous map $f : S^n \longrightarrow X$ has the continuous extension $\tilde{f} : B^{n+1} \longrightarrow X$ given by $\tilde{f}(x, t) = H(f(x), 1 - t)$. Therefore, every contractible space X is n-aspherical for all $n \geq 1$, hence aspherical. □

Definition 9.1.17. Let $n \geq 1$. A path-connected space X with $\pi_k(X) \cong \{0\}$ for all $k \in [1, n]$ is called *n-connected.*

Example 9.1.18. In Corollary 9.2.1, we will prove that the sphere S^n is $(n - 1)$-connected for all $n > 1$.

Remark 9.1.19. Following Remark 9.1.15, the following statements are equivalent for a path-connected space X:
(1) X is 1-aspherical;
(2) X is 1-connected;
(3) X is simply connected.

Remark 9.1.20. Let $n \geq 1$. According to Definition 9.1.12 and Definition 9.1.17, we have

$$\text{simply connected and aspherical} \subset n\text{-connected} \subset n\text{-aspherical}.$$

Remark 9.1.21. By Theorem 6.4.1 and Remark 9.1.11, the following statements are equivalent for a path-connected space X:

(1) X is n-connected;
(2) X is k-aspherical for all $1 \le k \le n$;
(3) For all $1 \le k \le n$, every continuous map $f : S^k \longrightarrow X$ has a continuous extension $\tilde{f} : B^{k+1} \longrightarrow X$.

Remark 9.1.22. By Proposition 7.5.12, every AR space is contractible hence, n-aspherical for all $n \ge 1$ by Proposition 9.1.16.

Remark 9.1.23. Remark 9.1.22 along with Theorem 7.5.11, Remark 9.1.15 and Proposition 9.1.16 give rise to the following proper inclusions, which complete those presented in Proposition 7.5.12.

$$\begin{aligned}\text{convex} \subset \text{absolute retract} &\cong \text{has the extension property}\\ &\subset \text{contractible}\\ &\subset \text{simply connected and aspherical}\\ &\subset n\text{-connected, for all } n \ge 1\\ &\subset n\text{-aspherical, for all } n \ge 1\\ &\subset \text{simply connected}\\ &\subset \text{path-connected} \subset \text{connected.}\end{aligned}$$

The following result is specific to higher homotopy groups. The reason is that the boundary ∂I^n is disconnected only for $n = 1$.

Theorem 9.1.24. *For $n \ge 2$, $\pi_n(X, x_0)$ is an Abelian group.*

Proof. According to the recurrence formula (9.3), it is sufficient to prove the result for $n = 2$. Let $f, g \in \Omega(X, x_0)$ and $h = f * g$ be given by (9.1), i. e.,

$$h(s_1, s_2) = \begin{cases} f(2s_1, s_2), & s_1 \in [0, 1/2],\\ g(2s_1 - 1, s_2), & s_1 \in [1/2, 1]. \end{cases}$$

By Theorem 4.4.4, the constant loop e_{x_0} satisfies $e_{x_0} * f \simeq f * e_{x_0} \simeq f$. This is used to construct at a first stage a homotopy joining h to the following loop:

$$k(s_1, s_2) = \begin{cases} f(2s_1, 2s_2 - 1), & s_1 \in [0, 1/2],\ s_2 \in [1/2, 1],\\ g(2s_1 - 1, 2s_2), & s_1 \in [1/2, 1],\ s_2 \in [0, 1/2],\\ e_{x_0}, & \text{otherwise.} \end{cases}$$

For fixed $s_1 \in [0, 1/2]$, we have $f(2s_1, s_2) \simeq e_{x_0} * f(2s_1, s_2)$, where the product is taken with respect to the second variable $s_2 \in [0, 1]$, i. e.,

$$f(2s_1, s_2) \simeq \begin{cases} e_{x_0}, & s_1 \in [0,1/2],\ s_2 \in [0,1/2], \\ f(2s_1, 2s_2 - 1), & s_1 \in [0,1/2],\ s_2 \in [1/2,1]. \end{cases}$$

Along similar ways, from the homotopy $g(2s_1 - 1, s_2) \simeq g(2s_1 - 1, s_2) * e_{x_0}$, we get

$$g(2s_1 - 1, s_2) \simeq \begin{cases} g(2s_1 - 1, 2s_2), & s_1 \in [1/2,1],\ s_2 \in [0,1/2], \\ e_{x_0}, & s_1 \in [1/2,1],\ s_2 \in [1/2,1]. \end{cases}$$

Inserting into $h(s_1, s_2)$ yields

$$h(s_1, s_2) \simeq \begin{cases} e_{x_0}, & s_1 \in [0,1/2],\ s_2 \in [0,1/2], \\ f(2s_1, 2s_2 - 1), & s_1 \in [0,1/2],\ s_2 \in [1/2,1], \\ g(2s_1 - 1, 2s_2), & s_1 \in [1/2,1],\ s_2 \in [0,1/2], \\ e_{x_0}, & s_1 \in [1/2,1],\ s_2 \in [1/2,1]. \end{cases}$$

This is exactly $k(s_1, s_2)$, which can be expressed as

$$\begin{aligned} k(s_1, s_2) &\simeq \begin{cases} f(2s_1, 2s_2 - 1), & s_1 \in [0,1/2],\ s_2 \in [1/2,1], \\ e_{x_0}, & s_1 \in [1/2,1],\ s_2 \in [1/2,1], \\ e_{x_0}, & s_1 \in [0,1/2],\ s_2 \in [0,1/2], \\ g(2s_1 - 1, 2s_2), & s_1 \in [1/2,1],\ s_2 \in [0,1/2] \end{cases} \\ &\simeq \begin{cases} (f * e_{x_0})(s_1, 2s_2 - 1), & s_1 \in [0,1],\ s_2 \in [1/2,1], \\ (e_{x_0} * g)(s_1, 2s_2), & s_1 \in [0,1],\ s_2 \in [0,1/2] \end{cases} \\ &\simeq \begin{cases} f(s_1, 2s_2 - 1), & s_1 \in [0,1],\ s_2 \in [1/2,1], \\ g(s_1, 2s_2), & s_1 \in [0,1],\ s_2 \in [0,1/2]. \end{cases} \end{aligned}$$

This is the second stage of the transformation through homotopies. Then the process is repeated twice. Once as in stage 1, by expressing f and g as products in terms of the first variable s_1 using the constant map e_{x_0} for fixed s_2. Then the product with the constant map allows us to group f and g in a final form. We get

$$k(s_1, s_2) \simeq \begin{cases} e_{x_0}, & s_1 \in [0,1/2],\ s_2 \in [1/2,1], \\ f(2s_1 - 1, 2s_2 - 1), & s_1 \in [1/2,1],\ s_2 \in [1/2,1], \\ g(2s_1, 2s_2), & s_1 \in [0,1/2],\ s_2 \in [0,1/2], \\ e_{x_0}, & s_1 \in [1/2,1],\ s_2 \in [0,1/2] \end{cases}$$

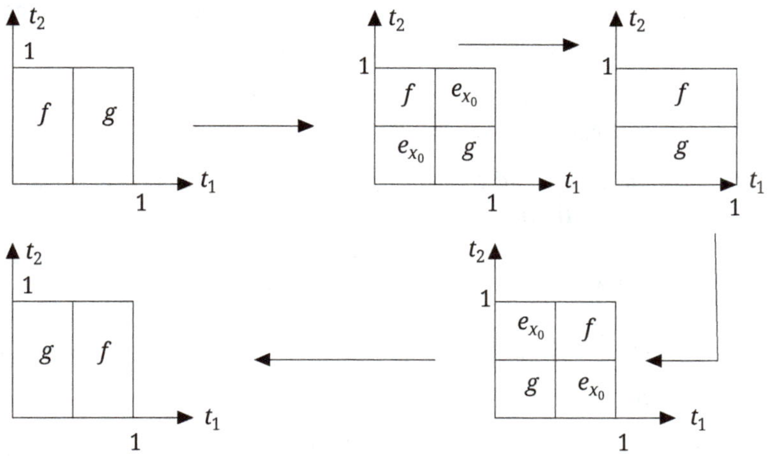

Figure 9.1: Step-by-step homotopies leading to commutativity.

$$\simeq \begin{cases} e_{x_0}, & s_1 \in [0,1/2],\ s_2 \in [1/2,1], \\ g(2s_1, 2s_2), & s_1 \in [0,1/2],\ s_2 \in [0,1/2], \\ f(2s_1 - 1, 2s_2 - 1), & s_1 \in [1/2,1],\ s_2 \in [1/2,1], \\ e_{x_0}, & s_1 \in [1/2,1],\ s_2 \in [0,1/2] \end{cases}$$

$$\simeq \begin{cases} (g * e_{x_0})(2s_1, s_2) \simeq g(2s_1, s_2), & s_1 \in [0,1/2],\ s_2 \in [0,1], \\ (e_{x_0} * f)(2s_1 - 1, s_2) \simeq f(2s_1 - 1, s_2), & s_1 \in [1/2,1],\ s_2 \in [0,1]. \end{cases}$$

The last formula is exactly that of $(g * f)(s_1, s_2)$. We conclude that $[f * g] = [g * f]$, proving the commutativity of the group operation $*$ in $\pi_n(X, x_0)$. A schematic representation of the five stages performed in the proof is summarized in Figure 9.1. □

Let $h : X \longrightarrow Y$ be a continuous map such that $y_0 = h(x_0)$. As in Definition 4.4.5, we have the following.

Proposition 9.1.25. *The map h defines an induced homomorphism* $h_* : \pi_n(X, x_0) \to \pi_n(Y, y_0)$ *by* $h_*([f]) = [h \circ f]$.

Proof. (1) If $[f] = [g]$, i. e., $f \simeq g$, there is a homotopy H between f and g. Then the composite map $h \circ H$ is a homotopy between $h\circ$ and $h \circ g$. Hence, h_* is well-defined.

(2) We have $h_*([f] * [g]) = h_*([f * g])$. By definition of the operation $*$, we get $h \circ (f * g) = (h \circ f) * (h \circ g)$. Hence, $h_*([f] * [g]) = h_*([f]) * h_*([g])$, proving that h_* is a homomorphism. □

The situation is exactly the same as for the case $n = 1$ (see Corollary 4.5.5, Remark 4.4.6 and Corollary 4.5.9). We have the following.

Corollary 9.1.26. *If* $\mathrm{Id}_X : (X, x_0) \longrightarrow (X, x_0)$ *is the identity, then the induced homomorphism* $(\mathrm{Id}_X)_* : \pi_n(X, x_0) \longrightarrow \pi_n(X, x_0)$ *is the identity.*

Corollary 9.1.27. *If* $h : (X, x_0) \longrightarrow (\{x_0\}, x_0)$ *is a constant map, then the induced homomorphism* $h_* : \pi_n(X, x_0) \longrightarrow \pi_n(\{x_0\}, x_0)$ *is trivial.*

Corollary 9.1.28. *For some* $x_0 \in X$ *and* $y_0 \in y$, *let* $f, g : (X, x_0) \longrightarrow (Y, y_0)$ *be two homotopic maps with homotopy H satisfying*

$$H(x_0, t) = f(x) = g(x), \quad \text{for all } t \in I.$$

Then for all n, the induced homomorphisms $f_*, g_* : \pi_n(X, x_0) \longrightarrow \pi_n(Y, y_0)$ *are identical.*

As a consequence, we have the following.

Corollary 9.1.29. *Let A be a deformation retract of X and* $i : (A, x_0) \hookrightarrow (X, x_0)$ *the inclusion map. Then the induced homomorphism* $i_* : \pi_n(A, x_0) \longrightarrow \pi_n(X, x_0)$ *is an isomorphism.*

The proof is essentially the same as in Theorem 7.2.4 and Corollary 4.5.5.

Corollary 9.1.30. *Let X, Y be two homotopically equivalent spaces. Then the fundamental groups* $\pi_n(X, x_0)$ *and* $\pi_n(Y, y_0)$ *are isomorphic.*

Proof. Let $f : X \longrightarrow Y$ and $g : Y \longrightarrow f$ be homotopy equivalence maps. Then Lemma 4.5.3, which does not depend on *n*, yields

$$(g \circ f)_* \cong \mathrm{Id}_X \quad \Longrightarrow \quad g_* \circ f_* \cong \mathrm{Id}_{\pi_n(X)}$$

and

$$(f \circ g)_* \cong \mathrm{Id}_Y \quad \Longrightarrow \quad f_* \circ g_* \cong \mathrm{Id}_{\pi_n(Y)}.$$ □

Since a contractible space is homotopically equivalent to a point, we have the following.

Corollary 9.1.31. *If X is contractible, then* $\pi_n(X)$ *is trivial for all* $n \geq 1$.

Example 9.1.32. We have $\pi_k(B^n) = 0$, for all $k, n \geq 1$.

The following result is analogous to Theorem 7.1.32.

Theorem 9.1.33. *Let A be a retract of X and* $a \in A$. *Then, for all positive integer n, the fundamental group* $\pi_n(A, a)$ *is isomorphic to a subgroup of the group* $\pi_n(X, a)$.

The next section is devoted to the computation of the *k*th fundamental groups of the sphere S^n. The computation of the groups $\pi_k(S^n)$ is not an easy task, especially for $k > n$.

Some cases can be computed (see [45]), others are still unknown (see [23, Appendix D.1]). We will discuss separately the case $k < n$, $k = n$ and $k > n$.

Early in 1930s, Hopf showed that $\pi_3(S^2)$ is isomorphic to the fundamental group of the circle by identifying $\mathbb{R}^4$ with $\mathbb{C}^2$ and then defining the Hopf map $H : S^3 \longrightarrow S^2$. The process yields that $\pi_2(S^2)$ is isomorphic to $\mathbb{Z}$, as shown in Section 9.2.2.

More generally, the cases $k \leq n$ can be determined starting from $\pi_2(S^2)$ and then using the following suspension theorem established by Freudenthal in 1937. For the proof, we refer, e. g., to [16, Theorem 6.15], [45, Section 4.1, p. 339] or [49, Theorem 2.1].

Theorem 9.1.34. *For $1 \leq k < 2n - 1$, we have*

$$\pi_{k+1}(\Sigma(S^n)) \cong \pi_k(S^n), \tag{9.4}$$

where $\Sigma(X)$ is the suspension space of X.

9.2 Higher homotopy groups of the sphere: $\pi_k(S^n)$

9.2.1 Case $k < n$

Corollary 9.2.1. *For all $1 \leq k < n$, the higher homotopy groups $\pi_k(S^n)$ are trivial.*

Proof. Since $\Sigma(S^n) \cong S^{n+1}$, relation (9.4) implies

$$\pi_{k+1}(S^{n+1}) \cong \pi_k(S^n), \quad \text{for all } 1 \leq k < 2n - 1. \tag{9.5}$$

Since $1 \leq k < n$, $1 \leq r < 2(n - k + r) - 1$ for all $r = 1, 2, \ldots, k$. By (9.5), we get

$$\pi_1(S^{n-k+1}) \cong \pi_2(S^{n-k+2}) \cong \cdots \cong \pi_k(S^n).$$

Since $n-k+1 \geq 2$, we know by Corollary 6.3.4 that the sphere S^{n-k+1} is simply connected. Therefore, $\pi_k(S^n) \cong \{0\}$ for all $1 \leq k < n$. □

In particular, we recover Corollary 6.3.4.

Corollary 9.2.2. *The fundamental groups $\pi_1(S^n)$ are trivial for all $n \geq 2$.*

Corollary 9.2.3. *The second higher homotopy groups $\pi_2(S^n)$ are trivial for all $n \geq 3$.*

We are now in position to present an alternative proof to Exercise 21 of Chapter 6.

Proposition 9.2.4. *For $k < n$, every continuous map $f : S^k \longrightarrow S^n$ is null-homotopic.*

Proof. Since $k < n$, by Definition 9.1.7 and Corollary 9.2.1, we have $0 \cong \pi_k(S^n) \cong [S^k, S^n]$. Hence, the induced homomorphism f_* is trivial and f is null-homotopic. □

Corollary 9.2.5. *For $k < n$, every continuous map $f : S^k \longrightarrow S^n$ is homotopic to a nonsurjective map.*

Remark 9.2.6. (1) An alternative proof of Corollary 9.2.1 can be performed using the identification $\pi_k(S^n) = [S^k, S^n]$. Then the proof relies on the idea that when $k < n$, one can homotope a map $f : S^k \longrightarrow S^n$ to a map $g : S^k \longrightarrow S^n$, which is not surjective, i. e., $g : S^k \longrightarrow S^n \setminus \{x_0\}$, for some $x_0 \in S^n$. Since $S^n \setminus \{x_0\}$ is homeomorphic to $\mathbb{R}^n$ and the latter space is contractible, g is null-homotopic showing that $[S^k, S^n]$ reduces to a single element. By the way, this proves Proposition 9.2.4 as well.
The reasoning is used to prove simple connectivity of the n-dimensional sphere in [45, Proposition 1.14] and [39, Theorem 1, Chapter 2]. See the second proof of Corollary 6.3.4, too.
The proof that f is homotopic to a nonsurjective map g comes from Theorem [25, Theorem 4.2, Chapter XVI], which states that for $k < n$, every continuous map $f : S^k \longrightarrow S^n$ is null-homotopic. This is a result based on the triangulation of the sphere S^k (see also [83, Theorem 7.5]).

(2) Conversely, in Proposition 9.2.4 the null-homotopy of such a map f is directly deduced from Corollary 9.2.1 (see also Exercise 25, Chapter 6). This means that there is equivalence between Corollary 9.2.1 and the null-homotopy of continuous maps $f : S^k \longrightarrow S^n$ for $k < n$, a fact resulting from the identification $\pi_k(S^n) = [S^k, S^n]$.

9.2.2 Case $k = n$

The key point is the computation of $\pi_2(S^2)$. For this purpose, we make use of the concept of the Hopf fibration. Let E, F and X be spaces.

Definition 9.2.7. A map $p : E \longrightarrow X$ is called *a fiber bundle* (or a Hopf fibration) with fiber F if p is continuous, surjective and for every $x \in X$, there exist $V \in \mathscr{N}(x)$ and a homeomorphism $h : V \times F \longrightarrow p^{-1}(V)$ such that $(p \circ h)(x, y) = x$ for all $(x, y) \in V \times F$.

Example 9.2.8. By Definition 5.1.1, a covering map $p : \widehat{X} \longrightarrow X$ is a fiber bundle with fiber $F = \{p^{-1}(x) : x \in X\}$. Then a homeomorphism $k : p^{-1}(V) \longrightarrow V \times F$ is given by $k(x) = (p_{|U_\alpha}(x), x)$.

For instance, $\pi : \mathbb{R} \longrightarrow S^1$ (see Theorem 5.1.13) and the projection map $\pi : S^n \longrightarrow \mathbb{RP}^n$ (see Example 3.5.6 and Proposition 5.1.7) are fiber bundles.

Example 9.2.9. With the identification $S^1 \cong [0,1]/_{\{0,1\}}$, the map $p : \mathbb{M}^2 \longrightarrow S^1$ defined by $p([x],[y]) = [y]$ (denoted g in the proof of Proposition 4.2.17) is a fiber bundle with fiber $F = S^1$. The homeomorphism h is $h([y],[x]) = [(x,y)]$ for $([y],[x]) \in V \times F$.

A further fiber bundle discovered by H. Hopf in 1935 (see [48]) is given by the following proposition.

Proposition 9.2.10. *With the identifications $S^3 \subset \mathbb{R}^4 \cong \mathbb{C} \times \mathbb{C}$ and $S^2 \subset \mathbb{R}^3 \cong \mathbb{C} \times \mathbb{R}$, the mapping $p : S^3 \longrightarrow S^2$ given by*

$$p(z_1, z_2) = (2z_1\overline{z_2}, |z_1|^2 - |z_2|^2)$$

is a fibration with fiber $F = S^1$.

Thus, the equations of S^3 and S^2 are $|z_1|^2 + |z_2|^2 = 1$ and $|z|^2 + |\lambda|^2 = 1$, respectively, where $(z_1, z_2) \in \mathbb{C}^2$ and $(z, \lambda) \in \mathbb{C} \times \mathbb{R}$. Here, $\overline{z}$ and $|z|$ denote the conjugate and the modulus of the complex number z, respectively.

Proof. (1) The map p is clearly continuous.

(2) The map p is well-defined. Squaring the components of $p(z_1, z_2)$ and adding term by term show that $(2z_1\overline{z_2})^2 + (|z_1|^2 - |z_2|^2)^2 = 1$ if and only if $|z_1|^2 + |z_2|^2 = 1$, i. e., $(z_1, z_2) \in S^3$.

(3) The map p is surjective. Let $Y = (Z, \lambda) \in S^2$ and consider two cases:

(a) When $Z = 0$, two subcases arise:
   (i) if $\lambda \geq 0$, then $p(\sqrt{\lambda}, 0) = Y$,
   (ii) if $\lambda \leq 0$, then $p(0, -\sqrt{\lambda}) = Y$.

(b) If $Z \neq 0$, let z_2 be such that $|z_2|$ solves the equation

$$|z_2|^4 + 4\lambda|z_2|^2 - |Z| = 0.$$

Then take $z_1 = \frac{Z}{2\overline{z_2}}$.

(4) The circle S^1 is a fiber. Notice that $p(z_1, z_2) = p(z_1', z_2')$ if and only if

$$\begin{cases} z_1\overline{z_2} = z_1'\overline{z_2'}, \\ |z_1|^2 - |z_2|^2 = |z_1'|^2 - |z_2'|^2. \end{cases} \tag{9.6}$$

The second relation of (9.6) both with $|z_1|^2 + |z_2|^2 = |z_1'|^2 + |z_2'|^2$ give $z_1' = \lambda_1 z_1$ and $z_2' = \lambda_2 z_2$, for some complex numbers λ_1, λ_2 whose moduli equal 1. Inserting into the first relation of (9.6) shows that $|\lambda_1| = |\lambda_2| = 1$. Hence, $(z_1', z_2') = z_0(z_1, z_2)$ with some $z_0 \in S^1$. Consequently, $p^{-1}(Z, \lambda) = \widehat{(Z, \lambda)}S^1$, which implies that $p^{-1}(Z, \lambda)$ is homeomorphic to S^1 for all $(Z, \lambda) \in S^2$. □

The proof of the following theorem can be found in [10, Chapter VII, Sections 5, 6] or [45, Section 4.2].

Theorem 9.2.11. *Given a fiber bundle $p : \widehat{X} \longrightarrow X$ with fiber F, the sequence*

$$\cdots \longrightarrow \pi_n(F) \xrightarrow{i_*} \pi_n(\widehat{X}) \xrightarrow{p_*} \pi_n(X) \longrightarrow \pi_{n-1}(F) \longrightarrow \cdots \pi_0(\widehat{X}) \longrightarrow \pi_0(X)$$

is exact, that is the image of a morphism is the kernel of the next one.

Corollary 9.2.12. *We have*

$$\pi_2(S^2) \cong \mathbb{Z}.$$

Proof. Applying Theorem 9.2.11 with $X = S^2$, $\widehat{X} = S^3$ and $F = S^1$, we get the exact sequence

$$\cdots \longrightarrow \pi_2(S^3) \xrightarrow{p_*} \pi_2(S^2) \xrightarrow{f} \pi_1(S^1) \xrightarrow{g} \pi_1(S^3) \longrightarrow \cdots$$

We already know that $\pi_1(S^1) \cong \mathbb{Z}$. By Corollary 9.2.1, $\pi_2(S^3) \cong \pi_1(S^3) \cong 0$. Hence, $\operatorname{Im} f = \mathbb{Z}$ (f is onto) and $\operatorname{Ker} f = p_*(0)$ (f is one-to-one), i. e., f is a bijection. Thus, $\pi_2(S^2) \cong \pi_1(S^1) \cong \mathbb{Z}$. □

We arrive at a generalization of Theorem 6.1.4.

Corollary 9.2.13. *For all $n \geq 1$, the fundamental group $\pi_n(S^n)$ is isomorphic to the set of integers.*

Proof. By (9.4), for $k = n$, the equality $\pi_{n+1}(S^{n+1}) \cong \pi_n(S^n)$ holds for $n < 2n - 1$, i. e., $n > 1$. Hence,

$$\pi_{n+1}(S^{n+1}) \cong \pi_n(S^n), \quad \text{for all } n \geq 2.$$

Since, by Proposition 9.2.12, $\pi_2(S^2)$ is isomorphic to $\mathbb{Z}$, the claim of the corollary follows. □

The case $n = 1$ is Theorem 6.1.4.

Example 9.2.14. By Definition 9.1.17, the sphere S^n is not n-connected.

As a consequence, we anticipate and solve Exercise 14 of Chapter 10, which is suggested to be treated using homology group theory. The proof follows from Corollary 9.2.1, Corollary 9.2.13 and Theorem 9.1.33.

Corollary 9.2.15. *For $0 < k < n$, S^k is not a retract of S^n.*

9.2.3 Case $k > n$

For the case $k > n$, we present two results, one for the unit circle and one for the sphere (Hopf's result).

Proposition 9.2.16. *We have $\pi_n(S^1) \cong 0$ for every $n > 1$.*

Proof. (1) *A first proof* relies on the identification $\pi_n(S^1) = [S^n, S^1] \cong [S^n, \mathbb{R}]$. By Corollary 5.2.19, any function $f : S^n \longrightarrow S^1$ has a unique lifting map $\widehat{f} : S^k \longrightarrow \widehat{S^1} = \mathbb{R}$. Again by the contractibility of $\mathbb{R}$, $\widehat{f}$ is null-homotopic and so is f. Hence, $\pi_n(S^1) = 0$ for all $n > 1$.

(2) *A second proof* appeals to Theorem 9.2.18. Since the map $p : \mathbb{R} \longrightarrow S^1$ given by $p(t) = e^{2\pi t}$ is a covering map (Theorem 5.1.13), by Theorem 9.2.18, p induces an isomorphism $p_* : \pi_n(\mathbb{R}) \rightarrow \pi_n(S^1)$. Since $\mathbb{R}$ is contractible, $\pi_n(\mathbb{R}) = 0$ and the claim follows. □

Proposition 9.2.17. *We have*
(1) $\pi_n(S^3) \cong \pi_n(S^2)$ *for all* $n \geq 3$;
(2) $\pi_3(S^2) \cong \mathbb{Z}$.

Proof. (1) By Proposition 9.2.10, let $p : S^3 \longrightarrow S^2$ be a fiber bundle with fiber $F = S^1$ (the Hopf map) and let S^2 be identified with $\mathbb{C}\mathrm{P}^1$. Let

$$S^3 = \{(z_1, z_2) \in \mathbb{C}^2 : |z_1|^2 + |z_2|^2 = 1\}.$$

Applying Theorem 9.2.11 with $X = S^2$, $\widehat{X} = S^3$ and $F = S^1$, we get the exact sequence

$$\cdots \longrightarrow \pi_n(S^1) \xrightarrow{i_*} \pi_n(S^3) \xrightarrow{p_*} \pi_n(S^2) \xrightarrow{g} \pi_{n-1}(S^1) \longrightarrow \cdots$$

where g is a homomorphism. Since $n \geq 3$, Proposition 9.2.16 gives $\pi_n(S^1) = \pi_{n-1}(S^1) = 0$. By exactness of the sequence, $\operatorname{Ker} p_* = \operatorname{Im} i_* = 0$ and so p_* is injective. Furthermore, $\operatorname{Im} p_* = \operatorname{Ker} g \cong \pi_k(S^2)$ and so p_* is surjective, hence an isomorphism. Therefore,

$$\pi_n(S^3) \cong \pi_n(S^2), \quad \text{for all } n \geq 3.$$

(2) By Corollary 9.2.13, $\pi_3(S^3) \cong \mathbb{Z}$. Hence, $\pi_3(S^2)$. □

A very interesting situation of isomorphic higher homotopy groups is given by the following theorem. This may help in computing some fundamental groups given some known covering spaces. The case $n = 1$ is Corollary 5.2.11, where the induced homomorphism is a monomorphism.

Theorem 9.2.18. *If* $p : \widehat{X} \longrightarrow X$ *is a covering map, then for all* $n \geq 2$, *the induced homomorphism*

$$p_* : \pi_n(\widehat{X}) \to \pi_n(X)$$

is an isomorphism.

Proof. Let $\pi_n(X) = [S^n, X]$ (see Remark 9.1.6) and $p_*([f]) = [p \circ f]$.

(1) The map p_* is surjective. Let $[f] \in \pi_n(X)$, where $f : S^n \longrightarrow X$ is a continuous map. Since for $n \geq 2$, S^n is simply connected by Corollary 6.3.4 and locally path-connected, Corollary 5.2.19 yields a unique lifting map $\widehat{f} : S^n \longrightarrow \widehat{X}$ such that $p \circ \widehat{f} = f$. Hence, $p_*([\widehat{f}]) = [f]$, as claimed.

(2) The map p_* is injective. Let $p_*([f]) = p_*([g])$, i. e., $[p \circ f] = [p \circ g]$ or equivalently $p \circ f \simeq p \circ g$. By Theorem 5.2.10, f and g are path-homotopic. Hence, $[f] = [g]$. □

As a consequence, in most cases the higher fundamental groups of the real projective space reduce to those of the sphere.

Corollary 9.2.19. *We have*

$$\pi_k(\mathbb{RP}^n) \cong \begin{cases} \mathbb{Z}, & \text{if } k = n = 1, \\ \mathbb{Z}/2\mathbb{Z}, & \text{if } k = 1,\ n > 1, \\ \pi_k(S^n), & \text{if } k, n > 1. \end{cases}$$

Proof. Since the sphere S^n is a covering space of the projective plane $\mathbb{RP}^n$ (see Proposition 5.1.7), $\pi_k(\mathbb{RP}^n) \cong \pi_k(S^n)$ for all $k, n > 1$. The claim is proved. □

Corollary 9.2.20. *For all $1 \leq k < n$, $\mathbb{RP}^k$ is not a retract of $\mathbb{RP}^n$.*

Proof. By contradiction, let r be such a retraction. Then $r \circ i = \mathrm{Id}_{\mathbb{RP}^k}$, where i is the inclusion map. Hence, $(r \circ i)_* = r_* \circ i_*$ is the identity on $\pi_k(\mathbb{RP}^k)$ and

$$\pi_k(\mathbb{RP}^k) \xrightarrow{i_*} \pi_k(\mathbb{RP}^n) \xrightarrow{r_*} \pi_k(\mathbb{RP}^{n-1}).$$

By Theorem 9.2.1 and Corollary 9.2.13 together with Corollary 9.2.19, for $1 < k < n$, $\pi_k(\mathbb{RP}^k) \cong \mathbb{Z}$ and $\pi_k(\mathbb{RP}^n) \cong 0$, i. e.,

$$\mathbb{Z} \xrightarrow{i_*} 0 \xrightarrow{r_*} \mathbb{Z}.$$

Hence, $r_* \circ i_*$ is the trivial homomorphism, a contradiction. If $k = 1$, we again reach a contradiction with the diagram

$$\mathbb{Z} \xrightarrow{i_*} \mathbb{Z}/2\mathbb{Z} \xrightarrow{r_*} \mathbb{Z}$$

because $\mathbb{Z}/2\mathbb{Z}$ is a finite group whereas $\mathbb{Z}$ is an infinite cyclic group. □

Since by Example 3.4.19, $B^n \cup_\pi \mathbb{RP}^{n-1} \cong \mathbb{RP}^{n-1}$, where $\pi(z) = \{z, -z\}$ is the canonical projection map, the following result is a consequence of Theorem 7.4.10 and Corollary 9.2.20.

Corollary 9.2.21. *For all $n \geq 1$, the covering map $\pi : S^n \longrightarrow \mathbb{RP}^n$ is not null-homotopic.*

Along with Theorem 9.2.18, we end this section with a result on the attaching n-cells to a space X. For some continuous map $f : S^{n-1} \subset B^n \longrightarrow X$, the adjunction space $B^n \cup_f X$ was introduced in Definition 3.4.1. The following result (see [10, Chapter VII, Theorem 11.7]) extends Corollary 8.2.26 to higher homotopy groups.

Proposition 9.2.22. *Let $n \geq 3$ and assume that X is path-connected and has a universal covering space ($\widehat{X}$ is simply connected). Then:*
(1) *$\pi_k(B^n \cup_f X) \cong \pi_k(X)$ for all $k < n$.*
(2) *For $k = n$, the homomorphism $i_* : \pi_n(X) \longrightarrow \pi_n(B^n \cup_f X)$ induced by the inclusion map is an epimorphism with kernel generated by $\{\phi([f]) : \phi \in \pi_n(X)\}$.*

Exercises (Chapter 9)

1. Prove that for $n \geq 1$, the operation $\overline{*}$ endows $[S^n, X]$ with a group structure.

2. Let $f : X \longrightarrow Y$ be a homotopy equivalence and $x_0 \in X$. Show that for all n, $\pi_n(X, x_0)$ and the fundamental group $\pi_1(Y, f(x_0))$ are isomorphic.

3. Show that for $n > 1$, every continuous map $S^n \longrightarrow S^1$ is null-homotopic.

4. Prove that $\pi_k(\mathbb{RP}^n) \cong \pi_k(S^n)$ for all $n \geq 1$ and $k > 1$ (argue as in Proposition 9.2.17). In particular, $\pi_2(\mathbb{RP}^2) \cong \mathbb{Z}$.

5. Prove that $\pi_{n+1}(S^n) \cong \mathbb{Z}_2 \cong \mathbb{Z}/2\mathbb{Z}$ for all $n \geq 3$.

6. Prove that $\pi_4(\mathbb{RP}^3)$ and $\pi_4(S^2)$. (*Hint:* Use Exercise 5.)

10 Elements of homology theory

This chapter is independent of the preceding chapters. We develop some aspects of the homology theory, which is a cornerstone in algebraic topology. There are several homology theories developed in the available literature. In this chapter, we have investigated three of them: the simplicial homology (Section 10.1), the singular homology (Section 10.2) and the relative homology (Section 10.3). In Section 10.2, the relationship between the first homotopy and the first homology groups due to Poincaré and Hurewicz is explained. Section 10.3 considers the axiomatic approach of the homology theory. Unlike the case of the higher fundamental groups, all homology groups of the spheres are computed in this section. The homology groups of the quotient spaces together with the local homology group are given. In homology theory, the Mayer–Vietoris exact theorem (Section 10.4) corresponds to the Seifert–van Kampen theorem in homotopy theory. With suitable exact sequences of morphisms, it helps in determining some homology groups of spaces. The retraction theory investigated in Chapter 7 is vital in determining the homology groups of some surfaces. The homology groups of the torus, the projective space and the Klein bottle are first discussed in the setting of the simplicial homology (Section 10.1) and then within the axiomatic approach of homology (Section 10.4). The homology groups of the wedge sum of spaces are computed, too. This chapter ends with the definitions of the Euler and Betti numbers commonly used in the classification of surfaces (Section 10.5).

10.1 Simplicial homology

The chain complexes allow us to define the homology groups. As example of application, we discuss the homology groups of the torus, the projective plane and the Klein bottle.

10.1.1 General properties

Definition 10.1.1. (1) *The (closed) standard p-simplex* (or p-Euclidean simplex) Δ_p is the convex hull (see Subsection A.1) in $\mathbb{R}^{p+1}$ of the canonical basis vectors $(e_0, e_1, \dots, e_p)$

$$\Delta_p = \left\{(t_0, \dots, t_p) \in \mathbb{R}^{p+1} : \ t_i \geq 0 \text{ and } \sum_{i=0}^{p} t_i = 1\right\}$$
$$= \left\{t_0 e_0 + \cdots + t_p e_p \in \mathbb{R}^{p+1} : \ t_i \geq 0 \text{ and } \sum_{i=0}^{p} t_i = 1\right\}.$$

(2) The number p is called *the dimension* of Δ_p.

https://doi.org/10.1515/9783111517384-010

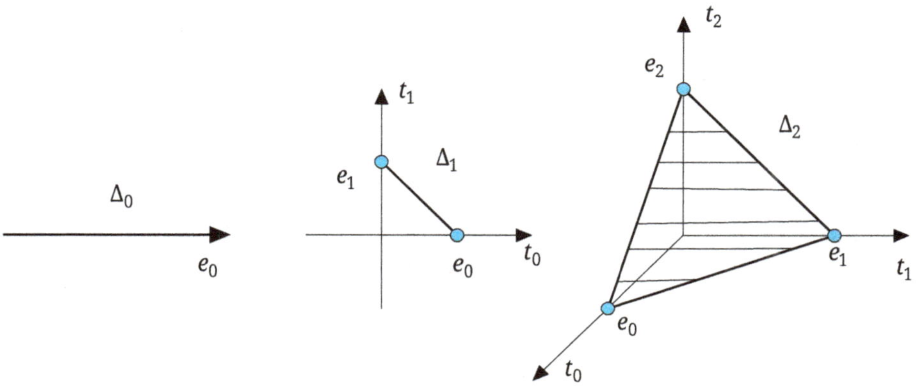

Figure 10.1: 0, 1 and 2 simplices.

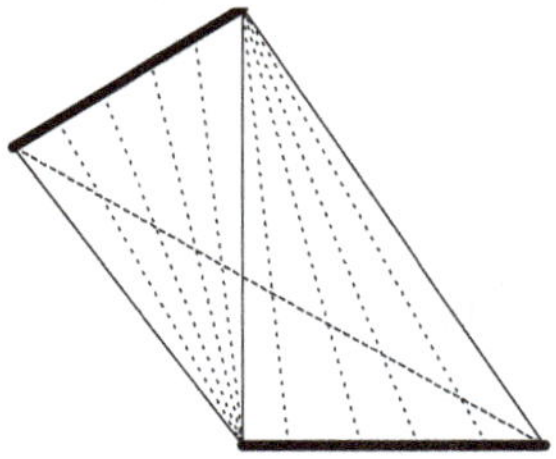

Figure 10.2: 3-Simplex as a join of two closed segments.

(3) The vectors $(e_0, e_1, \ldots, e_p)$ are called *the vertices* of the simplex.
(4) The numbers $(t_0, \ldots, t_p)$ are called *the barycentric coordinates* of the simplex.

The cases $p = 0$ (point), $p = 1$ (line segment) and $p = 2$ (triangle) are described in Figure 10.1. The tetrahedron has dimension $p = 3$ and the case $p = 4$ is called a 5-cell.

Remark 10.1.2. Instead of the canonical basis $(e_0, e_1, \ldots, e_p)$, we can consider an affinely independent family of vectors $(a_0, a_1, \ldots, a_p)$, that is the family of p vectors $(a_1 - a_0, a_2 - a_0, \ldots, a_p - a_0)$ is linearly independent.

Example 10.1.3. The join of two closed intervals is a 3-simplex (see Figure 10.2).

Remark 10.1.4. By Corollary A.1.10, a simplex is compact convex.

Definition 10.1.5. For $0 \leq i \leq p$, define the map $d_i : \mathbb{R}^{p+1} \longrightarrow \mathbb{R}^p$,

$$d_i(t_0, t_1, \ldots, \ldots, t_p) = (t_0, t_1, \ldots, t_{i-1}, t_{i+1}, \ldots, t_p),$$

obtained by omitting the ith coordinate t_i.

Definition 10.1.6. (1) The *ith face* of a standard p-simplex Δ_p is the image $d_i(\Delta_p)$ $(0 \leq i \leq p)$.
(2) The *boundary* of Δ_p is the union $\bigcup_{i=0}^{p} d_i(\Delta_p)$.

Example 10.1.7. (1) The faces of a triangle $\{t_0, t_1, t_2)$ are $[t_1, t_2]$, $[t_0, t_2]$ and $[t_0, t_1]$.
(2) The boundary of $\{t_0, t_1, t_2)$ is $[t_1, t_2] \cup [t_0, t_2] \cup [t_0, t_1]$.

Example 10.1.8. If $\{t_0, t_1, t_2, t_3)$ are non-coplanar vectors, then these vectors are vertices of a tetrahedron and the boundary consists of four triangles.

It is easy to check the following.

Proposition 10.1.9. *If we let*

$$\mathring{\Delta}_p = \left\{ (t_0, \dots, t_p) \in \mathbb{R}^{p+1} : \; t_i > 0 \text{ and } \sum_{i=0}^{p} t_i = 1 \right\},$$

then the boundary of Δ_p *is* $\Delta_p \setminus \mathring{\Delta}_p$.

Remark 10.1.10. (1) The boundary of a simplex is the set of points, which are not in the interior of the simplex.
(2) In general, the boundary of a simplex is not necessarily its topological boundary.

Definition 10.1.11. (1) The set $\mathring{\Delta}_p$ is called the *interior of the simplex*.
(2) The point $c = (1, \dots, 1)/(p+1)$ is called the *center* of the simplex (or center of mass).

Definition 10.1.12. (1) A permutation $\Delta_{a(p)} = \{a_0, a_1, \dots, a_p\}$ of a simplex Δ_p is odd (resp., even) if the number of inversions it contains is odd (resp., even). Recall that $\{a_i, a_j\}$ is an inversion if $i < j$ and $a_i > a_j$. Then the set of all $(p+1)!$ permutations is separated into two disjoints oriented classes (even and odd permutations).
(2) An *oriented simplex* is a simplex with one chosen orientation.

Definition 10.1.13. *A simplicial complex* (or a complex, for short) K is a finite union of simplices such that:
(1) If $\Delta_p \in K$ and if $\Delta_q \subset \Delta_p$, then $\Delta_q \in K$.
(2) If $\Delta_p, \Delta_q \in K$, then either $\Delta_p = \Delta_q$ or $\Delta_p \cap \Delta_q = \emptyset$.

Example 10.1.14. A triangle is a 1-simplicial complex built out of three 0-simplices (the vertices of the triangle) and three 1-simplices (the edges of the triangle) (Figure 10.3(a)).

Example 10.1.15. A square with one diagonal is a 2-simplicial complex built out of four 0-simplices (the vertices of the square), five 1-simplices (the edges of the square plus the diagonal), and two 2-simplices (the inscribed two triangles) (Figure 10.3(b)).

Example 10.1.16. The figure with two nested triangles (Figure 10.3(c)) is built out of six 0-simplices and twelve 1-simplices (see [17, Examples 7.1–7.3]).

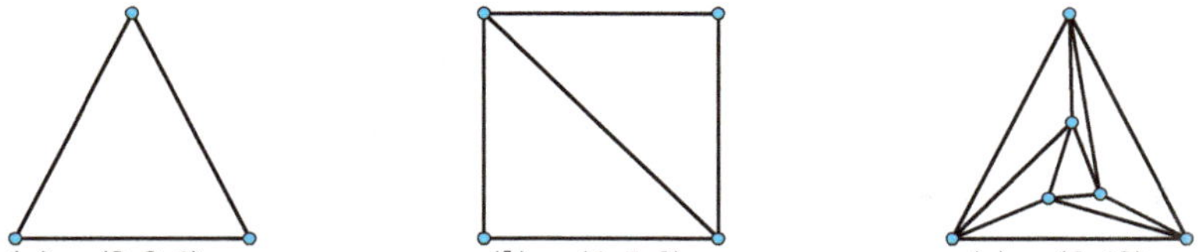

Figure 10.3: Simplicial complexes.

Example 10.1.17. (1) A finite graph (V, E) consists of a finite set V (of vertices) and a set E of unordered pairs of distinct elements of V (called edges).
(2) A finite graph (V, E) is an n-simplicial complex built out of n 0-simplices (the vertices of the graph) and $(n-1)$ 1-simplices (the edges of the graph).

Definition 10.1.18. (1) A space X is *a polyhedron* if there exists a simplicial complex K such that X is the union of all simplices of K.
(2) The dimension of a polyhedron X is the maximum of all dimensions of the simplices of K.
(3) A *finite polyhedron* is a polyhedron with finite dimension.
(4) A space X is *triangulable* if it is homeomorphic to a polyhedron. It is also called *polytope*.

By [19, Example 3.2.8], every sphere S^n is a polytope.

Definition 10.1.19. (1) *A simplicial map* is a continuous mapping

$$\sigma : K \longrightarrow L,$$

where K and L are simplicial complexes.
(2) In the case $K = \Delta_p$ is a p-simplex, σ is called a p-simplicial map.

Remark 10.1.20. (1) A p-simplex is a set. A p-simplicial map is a mapping.
(2) The plural of simplex is simplices (even though simplexes is used in some references, as well). The plural of complex is complexes.

10.1.2 Homology groups of a complex

Let K be a simplicial complex.

Definition 10.1.21. (1) Let $E_p = E_p(K)$ denote the free Abelian group generated by all p-simplices on the simplicial complex K. Define on E_p an equivalence relation by

$$[t_{\sigma(0)}, t_{\sigma(1)}, \ldots, t_{\sigma(p)}] \sim (-1)^{\varepsilon_\sigma}[t_0, t_1, \ldots, t_p],$$

where σ is a permutation of $\{0,1,\dots,p\}$ and ε_σ its signature. Recall that $\varepsilon_\sigma = \pm 1$ according as the permutation is even or odd.
(2) The quotient group $C_p(K) = E_p/_\sim$ is the set of oriented p-simplices.
(3) The elements of $C_p(K)$ are called *p-chains* and are denoted $\langle t_0, t_1, \dots, t_p\rangle$.

In other words, $C_p(K)$ is the set of all $\mathbb{Z}$-linear combinations of oriented simplices of the simplicial complex K.

Definition 10.1.22. For $p = 1, 2, \dots$, the *boundary homomorphism* $\partial_p : C_p(K) \longrightarrow C_{p-1}(K)$ is defined by

$$\partial_p\langle t_0, t_1, \dots, t_p\rangle = \sum_{i=0}^{p}(-1)^i\langle t_0, t_1, \dots, t_{i-1}, t_{i+1}, \dots, t_p\rangle.$$

Example 10.1.23. We have
(1) $\partial_1\langle t_0, t_1\rangle = \langle t_1 - t_0\rangle$;
(2) $\partial_2\langle t_0, t_1, t_2\rangle = \langle t_1, t_2\rangle - \langle t_0, t_2\rangle + \langle t_0, t_1\rangle$;
(3) $\partial_3\langle t_0, t_1, t_2, t_3\rangle = \langle t_1, t_2, t_3\rangle - \langle t_0, t_2, t_3\rangle + \langle t_0, t_1, t_3\rangle - \langle t_0, t_1, t_2\rangle$.

Lemma 10.1.24. *We have* $\partial_{p-1} \circ \partial_p = 0$.

Proof. We have

$$\begin{aligned}\partial_{p-1}(\partial_p\langle t_0, t_1, \dots, t_p\rangle) &= \partial_{p-1}\left(\sum_{i=0}^{p}(-1)^i\langle t_0, \dots, t_{i-1}, 0, t_{i+1}, \dots, t_p\rangle\right)\\ &= \sum_{i=0}^{p}(-1)^i\sum_{j=0}^{i-1}(-1)^j[t_0, \dots, t_{j-1}, 0, t_{j+1}, \dots, t_{i-1}, t_{i+1}, \dots, t_p]\\ &\quad + \sum_{i=0}^{p}(-1)^i\sum_{j=i+1}^{p}(-1)^{j-1}\langle t_0, \dots, t_{i-1}, t_{i+1}, \dots, t_{j-1}, 0, t_{j+1}, \dots, t_p\rangle,\end{aligned}$$

where the last terms cancel in pairs. □

Example 10.1.25. We have

$$\begin{aligned}\partial_1\partial_2\langle t_0, t_1, t_2\rangle &= \partial_1\langle t_1, t_2\rangle - \partial_1\langle t_0, t_2\rangle + \partial_1\langle t_0, t_1\rangle\\ &= (t_2 - t_1) - (t_2 - t_0) + (t_1 - t_0)\\ &= 0.\end{aligned}$$

Definition 10.1.26. (1) The sequence

$$\cdots \xrightarrow{\partial_{n+1}} C_n(K) \xrightarrow{\partial_n} C_{n-1}(K) \xrightarrow{\partial_{n-1}} C_{n-2}(K) \cdots \xrightarrow{\partial_2} C_1(K) \xrightarrow{\partial_1} C_0(K) \xrightarrow{\partial_0} 0,$$

is called *a chain complex*.
(2) For $n = 1, 2, \dots$, δ_n are the boundary maps.

Definition 10.1.27. (1) For $n = 1, 2, \ldots$, the *n cycles* of the chain complex are

$$Z_n(K) = \operatorname{Ker}\{\partial_n : C_n(K) \longrightarrow C_{n-1}(K)\}.$$

(2) For $n = 0, 1, \ldots$, the *n boundaries* of the chain complex are

$$B_n(K) = \operatorname{Im}\{\partial_{n+1} : C_{n+1}(K) \longrightarrow C_n(K)\}.$$

(3) The *nth homology group* of a complex K is defined by

$$H_n(K) = \begin{cases} Z_n(K)/B_n(K), & \text{if } n \geq 1, \\ C_0/\operatorname{Im}\partial_1, & \text{if } n = 0. \end{cases}$$

Remark 10.1.28. Since $\partial_n \circ \partial_{n+1} = 0$, $\operatorname{Im}\partial_{n+1} \subset \operatorname{Ker}\partial_n$, and thus the quotient group $Z_n(K)/B_n(K)$ makes sense.

Remark 10.1.29. It is clear that 0-chains are 0-cycles. Lemma 10.1.24 states that all n-boundaries are n-cycles.

Example 10.1.30. Assume that $\partial_n = 0$ for all n. Then $Z_n = C_n$ and $B_n = \{0\}$ so that $H_n \cong C_n$ for all n.

Example 10.1.31. Let $K = \{v\}$. Then $C_n(K) = \{0\}$ for all $n \geq 1$ and $C_0(K)$ is the free Abelian group generated by v. Hence, $H_n = 0$ for all $n \geq 1$,

$$C_0(K) = \operatorname{Ker}\partial_0 \cong \mathbb{Z} \quad \text{and} \quad \operatorname{Im}\partial_1 = \{0\}.$$

Therefore, $H_0 = \mathbb{Z}/\{0\} \cong \mathbb{Z}$.

Example 10.1.32. Consider the triangle K in Example 10.1.14. C_0 is the free Abelian group generated by the vertices $\{v_0, v_1, v_2\}$ and C_1 is generated by the edges $\{[v_0, v_1], [v_1, v_2], [v_2, v_0]\}$. Then both groups have rank 3. Since no more simplices exist, $C_n(K) = \{0\}$ for all $n \geq 2$. In particular, $\operatorname{Ker}\partial_2 = \operatorname{Im}\partial_2 = \operatorname{Im}\partial_3 = \{0\}$. Then $H_n \cong \{0\}$ for $n > 1$. Let us compute $H_1(K)$ and $H_0(K)$. By Definition 10.1.22, $\partial_1([v_i - v_j]) = v_i - v_j$ for $i, j \in \{0, 1, 2\}$. Then

$$\begin{aligned} \partial_1(\lambda_0[v_0, v_1] + \lambda_1[v_1, v_2] + \lambda_2[v_2, v_0]) &= \lambda_0(v_0 - v_1) + \lambda_1(v_1 - v_2) + \lambda_2(v_2 - v_0) \\ &= (\lambda_0 - \lambda_2)v_0 + (\lambda_1 - \lambda_0)v_1 + (\lambda_2 - \lambda_1)v_2. \end{aligned}$$

As a consequence, for an element be in the kernel of ∂_1, we must have $\lambda_0 = \lambda_1 = \lambda_2$. Hence, the kernel is generated by $[v_0, v_1] + [v_1, v_2] + [v_2, v_0]$, and thus

$$H_1(K) = \operatorname{Ker}\partial_1/\operatorname{Im}\partial_2 \cong \mathbb{Z}/\{0\} \cong \mathbb{Z}.$$

In addition, from the description above of ∂_1, we can see that the elements of $\operatorname{Im}\partial_1$ have the form

$$(\lambda_0 - \lambda_2)v_0 + (\lambda_1 - \lambda_0)v_1 - [(\lambda_0 - \lambda_2) + \lambda_1 - \lambda_0)]v_2.$$

Hence, $\operatorname{Im} \partial_1$ is isomorphic to $\mathbb{Z}^2$. Since C_{-1} is trivial, $C_0 = \operatorname{Ker} \partial_0$, which is isomorphic to $\mathbb{Z}^3$. We infer that

$$H_0(K) = \operatorname{Ker} \partial_0 / \operatorname{Im} \partial_1 \cong \mathbb{Z}^3/\mathbb{Z}^2 \cong \mathbb{Z}.$$

Example 10.1.33. Consider the one-dimensional complex K in Figure 10.4 (see [35, Figure 5.3]). Then C_0 is generated by the six vertices $\{v_0, v_1, v_2, v_3, v_4, v_5\}$ and C_1 is generated by the eight edges $\{[v_1, v_0], [v_0, v_3], [v_1, v_2], [v_2, v_3], [v_1, v_4], [v_4, v_3], [v_1, v_5], [v_5, v_3]\}$. Since no higher simplex exists, $C_n(K) = \{0\}$ for all $n \geq 2$ and ∂_2 is injective. Then $H_n \cong \{0\}$, for $n \geq 2$. Notice that by Definition 10.1.22, $\partial_1([v_i - v_j]) = v_i - v_j$, for $i, j \in \{0, 1, 2, 3, 4, 5\}$. Hence,

$$\begin{aligned}
&\partial_1(\lambda_0[v_1, v_0] + \lambda_1[v_0, v_3] + \lambda_2[v_1, v_2] + \lambda_3[v_2, v_3] \\
&\quad + \lambda_4[v_1, v_4] + \lambda_5[v_4, v_3] + \lambda_6[v_1, v_5] + \lambda_7[v_5, v_3]) \\
&= \lambda_0(v_1 - v_0) + \lambda_1(v_3 - v_0) + \lambda_2(v_2 - v_1) + \lambda_3(v_3 - v_2) \\
&\quad + \lambda_4(v_4 - v_1) + \lambda_5(v_3 - v_4) + \lambda_6(v_5 - v_1) + \lambda_7(v_3 - v_5) \\
&= -(\lambda_0 + \lambda_1)v_0 + (\lambda_0 - \lambda_2 - \lambda_4 - \lambda_6)v_1 + (\lambda_2 - \lambda_3)v_2 \\
&\quad + (\lambda_1 + \lambda_3 + \lambda_5 + \lambda_7)v_3 + (\lambda_4 - \lambda_5)v_4 + (\lambda_6 - \lambda_7)v_5.
\end{aligned}$$

Therefore $\operatorname{Im} \partial_1$ is generated by all six vertices of $C_0(K)$. Hence,

$$H_0(K) = C_0(K)/\operatorname{Im} \partial_1 \cong \mathbb{Z}^6/\mathbb{Z}^6 \cong \mathbb{Z}.$$

Also, a direct computation shows that the kernel of ∂_1 is generated by three elements. Therefore,

$$H_1(K) = \operatorname{Ker} \partial_1 / \operatorname{Im} \partial_2 \cong \mathbb{Z} \oplus \mathbb{Z} \oplus \mathbb{Z}/\{0\} \cong \mathbb{Z} \oplus \mathbb{Z} \oplus \mathbb{Z}.$$

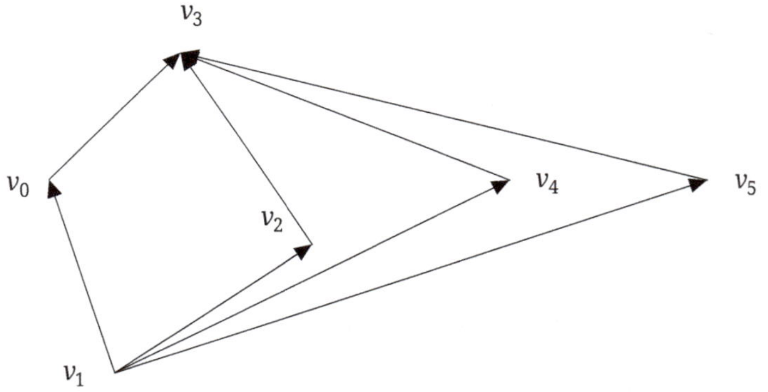

Figure 10.4: One-dimensional complex.

Proposition 10.1.34 (Homology groups of the torus). *The homology groups of the torus are*

$$H_n(\mathbb{T}^2) \cong \begin{cases} \mathbb{Z}^2, & n = 1, \\ \mathbb{Z}, & n = 0 \text{ or } n = 2, \\ 0, & \text{otherwise.} \end{cases}$$

Proof. The torus $K = \mathbb{T}^2 = S^1 \times S^1$ is represented by the square in Figure 10.5. It has one 0-simplex, the identified vertex v ($C_0(K) = \langle v \rangle$), three 1-simplices: two identified edges $\{a\}$, $\{b\}$ and the diagonal $\{c\}$ ($C_1(K) = \langle a, b, c \rangle$), two 2-simplices: the two triangles T_1, T_2 ($C_2(K) = \langle T_1, T_2 \rangle$), and zero 3-simplices (no tetrahedron exists), and of course no higher-dimensional simplices. Then $\operatorname{Im} \partial_3 = \{0\}$ and $C_n(K) = \{0\}$ for all $n \geq 3$. Hence, $H_n \cong \{0\}$ for $n \geq 3$. The nontrivial part of the chain sequence is

$$0 \xrightarrow{\partial_3} C_2(K) \xrightarrow{\partial_2} C_1(K) \xrightarrow{\partial_1} C_0(K) \xrightarrow{\partial_0} 0,$$

where $\partial_1 = 0$ because $\partial_1(a) = \partial_1(b) = \partial_1(c) = \langle v - v \rangle = 0$. Since $C_0(K) \cong \mathbb{Z}$, then

$$H_0(K) = C_0(K)/\operatorname{Im} \partial_1 \cong \mathbb{Z}/\{0\} \cong \mathbb{Z}.$$

Since $C_1(K) = \langle a, b, c \rangle$, then

$$H_1(K) = \operatorname{Ker} \partial_1/\operatorname{Im} \partial_2 = C_1(K)/\operatorname{Im} \partial_2 \cong \langle a, b, c \rangle/\operatorname{Im} \partial_2.$$

In addition, $\partial_2(T_1) = \langle c - a - b \rangle$ and $\partial_2(T_2) = \langle a + b - c \rangle$. Then $\operatorname{Im} \partial_2$ is generated by two elements. Consequently, the first homology group is

$$\begin{aligned} H_1(K) &\cong \langle a, b, c \rangle/\langle c - a - b, a + b - c \rangle \\ &\cong \langle a, b, c \rangle/\langle a + b - c \rangle \\ &\cong \mathbb{Z}^3/\mathbb{Z} \\ &\cong \mathbb{Z}^2. \end{aligned}$$

The second homology group is

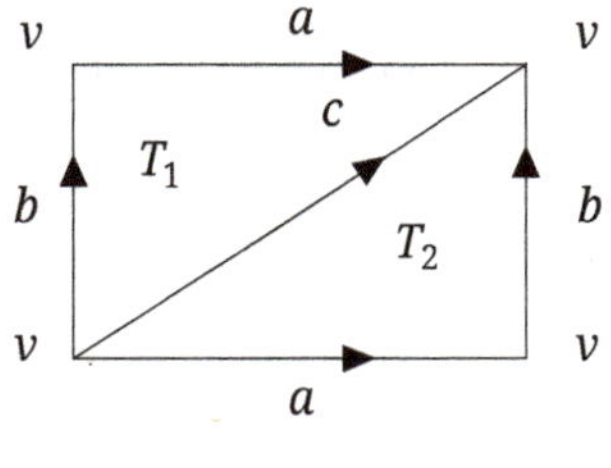

Figure 10.5: Torus as a simplicial complex.

$$H_2(K) = \operatorname{Ker} \partial_2 / \operatorname{Im} \partial_3 \cong \operatorname{Ker} \partial_2 / \{0\} \cong \operatorname{Ker} \partial_2.$$

Let $u = \lambda_1 T_1 + \lambda_2 T_2 \in \operatorname{Ker} \partial_2$. Then

$$\begin{aligned} \partial_2(u) = 0 \quad &\Longleftrightarrow \quad \lambda_1 \partial_2(T_1) + \lambda_2 \partial_2(T_2) = 0 \\ &\Longleftrightarrow \quad \lambda_1 \langle c - a - b \rangle + \lambda_2 \langle a + b - c \rangle = 0 \\ &\Longleftrightarrow \quad (\lambda_2 - \lambda_1)\langle a + b - c \rangle = 0 \\ &\Longleftrightarrow \quad \lambda_1 = \lambda_2 = \lambda \\ &\Longleftrightarrow \quad u = \lambda(T_1 + T_2). \end{aligned}$$

Hence, $\operatorname{Ker} \partial_2$ is generated by one element that is an infinite cyclic group. Finally,

$$H_2(K) \cong \operatorname{Ker} \partial_2 \cong \mathbb{Z},$$

and thus all homology groups are computed. □

Proposition 10.1.35. *The homology groups of the projective plane are*

$$H_n(\mathbb{RP}^2) \cong \begin{cases} \mathbb{Z}, & n = 0, \\ \mathbb{Z}/2\mathbb{Z}, & n = 1, \\ 0, & \textit{otherwise.} \end{cases}$$

Proof. The projective plane $K = \mathbb{RP}^2$ is represented by the square in Figure 10.6. It has two 0-simplices: the identified vertices v, w ($C_0(K) = \langle v, w \rangle$), three 1-simplices $\{a, b, c\}$: two identified edges $\{a\}$, $\{b\}$ and the diagonal $\{c\}$ ($C_1(K) = \langle a, b, c \rangle$), two 2-simplices: the two triangles T_1, T_2 ($C_2(K) = \langle T_1, T_2 \rangle$), and no higher-dimensional simplices. Then $\operatorname{Im} \partial_n = \{0\}$ and $C_n(K) = \{0\}$ for all $n \geq 3$. Hence, $H_n \cong \{0\}$ for $n \geq 3$. The nontrivial part of the chain sequence is

$$0 \xrightarrow{\partial_3} C_2(K) \xrightarrow{\partial_2} C_1(K) \xrightarrow{\partial_1} C_0(K) \xrightarrow{\partial_0} 0.$$

We have $\partial_1(a) = \partial_1(b) = \langle w - v \rangle$ and $\partial_1 c = 0$. Then $\operatorname{Im} \partial_1$ is generated by $\langle w - v \rangle$ and $\operatorname{Ker} \partial_1$ is generated by $\langle a - b, c \rangle$. Since $C_0(K) = \langle v, w \rangle$, we find

$$H_0(K) = C_0(K) / \operatorname{Im} \partial_1 \cong \langle v, w \rangle / \langle w - v \rangle \cong \mathbb{Z}.$$

Since $\partial_2 T_1 = \langle a - b + c \rangle$ and $\partial_2 T_2 = \langle b - a + c \rangle$, the image $\operatorname{Im} \partial_2$ is generated by $\langle a - b + c, b - a + c \rangle$. Hence,

$$\begin{aligned} H_1(K) &\cong \langle a - b, c \rangle / \langle a - b + c, b - a + c \rangle \\ &\cong \langle a - b + c, c \rangle / \langle a - b + c, 2c \rangle \\ &\cong \langle c \rangle / \langle 2c \rangle \\ &\cong \mathbb{Z}/2\mathbb{Z} \cong \mathbb{Z}_2. \end{aligned}$$

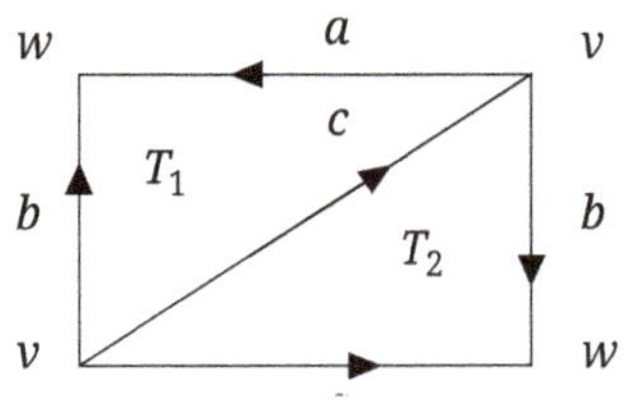

Figure 10.6: Projective plane as a simplicial complex.

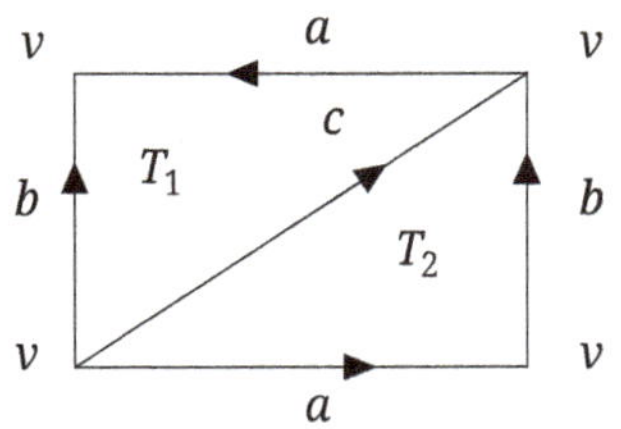

Figure 10.7: Klein bottle as a simplicial complex.

Furthermore, $\partial_2(\lambda_1 T_1 + \lambda_2 T_2) = 0$ implies $\lambda_1 = \lambda_2 = 0$. Hence, $\operatorname{Ker} \partial_2 = 0$ and the second homology group is

$$H_2(K) = \operatorname{Ker} \partial_2 / \operatorname{Im} \partial_3 \cong \{0\}/\{0\} \cong \{0\}. \qquad \square$$

Proposition 10.1.36 (Homology groups of the Klein bottle). *The homology groups of the Klein bottle are*

$$H_n(\mathbb{K}^2) \cong \begin{cases} \mathbb{Z}, & n = 0, \\ \mathbb{Z} \oplus \mathbb{Z}/2, & n = 1, \\ 0, & \textit{otherwise}. \end{cases}$$

Proof. The Klein bottle $K = \mathbb{K}^2$ is represented by the square in Figure 10.7. As for the torus (Proposition 10.1.34), $C_0(K) = \langle v \rangle)$, $C_1(K) = \langle a, b, c \rangle$, $C_2(K) = \langle T_1, T_2 \rangle$, but no higher-dimensional simplices exist for $n \geq 3$. Also, $\partial_1(a) = \partial_1(b) = \partial_1(c) = 0$, $\partial_2(T_1) = \langle a - b + c \rangle$ and $\partial_2(T_2) = \langle a + b - c \rangle$. Hence, $\partial_1 = 0$ and $C_n(K) = \{0\}$ for all $n \geq 3$, which implies $H_n \cong \{0\}$ for $n \geq 3$.

Similar to the torus, we have

$$H_0(K) = C_0(K)/\operatorname{Im} \partial_1 \cong \mathbb{Z}/\{0\} \cong \mathbb{Z}.$$

Since $C_1(K) = \langle a, b, c \rangle$ and $\operatorname{Im} \partial_2$ is generated by $\langle a - b + c, a + b - c \rangle$, the first homology group is

$$\begin{aligned}
H_1(K) &= \operatorname{Ker}\partial_1/\operatorname{Im}\partial_2\\
&\cong \langle a,b,c\rangle/\langle a-b+c, a+b-c\rangle\\
&\cong \langle a,b,c\rangle/\langle 2a, a+b-c\rangle\\
&\cong \langle a,b,a+b\rangle/\langle 2a\rangle\\
&\cong \langle a,b\rangle/\langle 2a\rangle\\
&\cong (\mathbb{Z}\langle a\rangle \oplus \mathbb{Z}\langle b\rangle)/(\mathbb{Z}\langle 2a\rangle)\\
&\cong \mathbb{Z}\oplus\mathbb{Z}/2.
\end{aligned}$$

As for the second homology group, we can check that as for the projective plane $\operatorname{Ker}\partial_2 = 0$ and the second homology group is

$$H_2(K) = \operatorname{Ker}\partial_2/\operatorname{Im}\partial_3 \cong \{0\}/\{0\} \cong \{0\}. \qquad \square$$

10.2 Singular homology

10.2.1 General properties

The singular homology is constructed from the generalization to chain complexes $C_n(X)$, where X is any space. Some properties are discussed as well.

Definition 10.2.1. *A singular n-simplex* on X is a continuous map

$$\sigma : \Delta_n \longrightarrow X,$$

where Δ_n is an n-simplex.

Remark 10.2.2. A singular 0-simplex can be identified with points of X whereas singular 1-simplex is identified with paths in X by using the affine map $t \mapsto (t, 1-t)$.

Remark 10.2.3. The following notation should not be mixed:
(1) Δ_n : n-simplex;
(2) $\sigma : \Delta_n \longrightarrow L$: n-simplicial map (L is a simplicial complex);
(3) $\sigma : \Delta_n \longrightarrow X$: singular n-simplex (X is a space).

In accordance with this definition, we define the faces and the boundary of a singular n-simplex.

Definition 10.2.4. Let σ be a singular n-simplex with $n \geq 1$. For $0 \leq i \leq n$, the *ith face* of σ is the singular $(n-1)$-simplex $\sigma_i : \Delta_{n-1} \longrightarrow X$, given by

$$\begin{cases} \sigma_0(t_1, \dots, t_n) = \sigma(0, t_1, \dots, t_n), \\ \sigma_i(t_0, \dots, t_{i-1}, t_{i+1}, \dots, t_n) = \sigma(t_0, \dots, t_{i-1}, 0, t_{i+1}, \dots, t_n) \quad (0 < i < n), \\ \sigma_n(t_0, \dots, t_{n-1}) = \sigma(t_0, \dots, t_{n-1}, 0). \end{cases}$$

Example 10.2.5. For $n = 2$, $\Delta_2 = \{t_0, t_1, t_2\}$ and the three faces are the singular 1-simplices $\sigma_0(t_1, t_2) = \sigma(0, t_1, t_2)$, $\sigma_1(t_0, t_2) = \sigma(t_0, 0, t_2)$ and $\sigma_2(t_1, t_2) = \sigma(t_0, t_1, 0)$.

Definition 10.2.6. The set $S_n(X)$ will denote the set of all singular n-simplices $\sigma : \Delta_n \longrightarrow X$.

Definition 10.2.7. With the notation of Example B.7.8 in the Appendix, let $E = S_n(X)$. Define $C_n(X) = L(S_n(X); \mathbb{Z})$ to be the free Abelian group of elements $\sum_{\sigma \in S_n(X;\mathbb{Z})} \lambda_\sigma e_\sigma$, where $\lambda_\sigma \in \mathbb{Z}$ and all but a finite number of elements in the sum are nonzero. Hence,

$$C_n(X) = \{c : S_n(X) \longrightarrow \mathbb{Z} \\ \text{such that } c(\sigma) = 0, \text{ for all but finitely many singular } n\text{-simplices } \sigma\}.$$

The elements of $C_n(X) = C_n(X, \mathbb{Z})$ are called *the singular n-chain complexes* with coefficients in the ring $\mathbb{Z}$.

Remark 10.2.8. Let $c \in C_n(X)$ and $\{\sigma^0, \sigma^1, \dots, \sigma^N\}$ be such that $c(\sigma^k) \neq 0$ for all $k \in [0, N]$. For such a k, define the so-called elementary singular n-chain by $c^k(\sigma) = 0$ for every $\sigma \neq \sigma^k$ and $c^k(\sigma^k) = 1$. Then a singular n-chain can be expressed as $c = \sum_{i=1}^N c(\sigma^k)c^k$. Hence, $C_n(X)$ is a free Abelian group with basis the set of all singular n-simplices of X. We may write as well $c = \sum_{\sigma \in S_n} n_\sigma \sigma$, where $n_\sigma \in \mathbb{Z}$ for each $\sigma \in S_n$.

Definition 10.2.9. *The boundary* $\partial_n \sigma$ of a singular n-simplex σ is defined by

$$\partial_n \sigma = \sum_{i=0}^{n} (-1)^i \sigma_i,$$

where σ_i, $0 \leq i \leq n$ are the faces of the singular n-simplex σ.

Example 10.2.10. For $n = 2$, $\partial_2 \sigma = \sigma(0, t_1, t_2) - \sigma(t_0, 0, t_2) + \sigma(t_0, t_1, 0)$ (compare with Example 10.1.23(2)).

Remark 10.2.11. Since the set of all singular n-simplices is a basis of $C_n(X)$, we can define a group homomorphism, called the boundary morphism as an extension of the boundary $\partial_n \sigma$ to $C_n(X)$,

$$\partial_n : C_n(X) \longrightarrow C_{n-1}(X).$$

Since no singular (-1)-simplex exists, $C_{-1}(X) = \{0\}$ and $\partial_0 : C_0(X) \longrightarrow C_{-1}(X)$ is the zero map. As for Lemma 10.1.24, we have the following.

Lemma 10.2.12. *For all* $n \geq 0$, $\partial_n \circ \partial_{n+1} = 0$.

Definition 10.2.13. (1) The sequence

$$\cdots \xrightarrow{\partial_{n+1}} C_n(X) \xrightarrow{\partial_n} C_{n-1}(X) \xrightarrow{\partial_{n-1}} C_{n-2}(X) \cdots \xrightarrow{\partial_2} C_1(X) \xrightarrow{\partial_1} C_0(X) \xrightarrow{\partial_0} 0,$$

is called *a singular chain complex* and is denoted $C_* = (C_n, \delta_n)$.

(2) The quotient group $H_n(X) = \operatorname{Ker} \partial_n / \operatorname{Im} \partial_{n+1}$ is called the singular nth homology group. We set $H_n(X) \cong \{0\}$, for $n < 0$.

Proposition 10.2.14. *Let $X = \{x\}$ be a one-element set. Then*

$$H_n(X) \cong \begin{cases} \{0\}, & \text{if } n \geq 1, \\ \mathbb{Z}, & \text{if } n = 0. \end{cases}$$

Proof. For each nonnegative integer n, there is only one singular n-simplex, namely the constant map. Each chain $C_n(\{x\})$ is an infinite cyclic group, hence isomorphic to $\mathbb{Z}$. The boundary of a singular n-simplex is given by $\partial_n \sigma^n = \sum_{i=0}^{n} (-1)^i \sigma_i^n$. Since $\sigma_i^n = \sigma^{n-1}$,

$$\partial_n \sigma^n = \sum_{i=0}^{n} (-1)^i \sigma^{n-1} = \sigma^{n-1} \sum_{i=0}^{n} (-1)^i.$$

As a consequence,

$$\partial_n \sigma^n = \begin{cases} 0, & \text{if } n \text{ is odd}, \\ \sigma^{n-1}, & \text{if } n \text{ is even}. \end{cases}$$

Consider the singular chain complex

$$\cdots \xrightarrow{\partial_{n+2}} C_{n+1}(\{x\}) \xrightarrow{\partial_{n+1}} C_n(\{x\}) \xrightarrow{\partial_n} C_{n-1}(\{x\}) \xrightarrow{\partial_{n-1}} \cdots$$

and discuss three cases:

(1) $n > 0$ is odd:

$$\begin{cases} \partial_{n+1} \sigma^{n+1} = \sigma^n & (\partial_{n+1} \text{ is isomorphism}), \\ \partial_n \sigma^n = 0 & (\partial_n = 0). \end{cases}$$

Hence, $\operatorname{Im} \partial_{n+1} = \operatorname{Ker} \partial_n = C_n$ and $H_n(\{x\}) = C_n / C_n \cong \{0\}$ by using Example B.2.11 in the Appendix.

(2) Case when $n > 0$ is even:

$$\begin{cases} \partial_{n+1} \sigma^{n+1} = 0, \\ \partial_n \sigma^n = \sigma^{n-1} & (\partial_n \text{ is isomorphism}). \end{cases}$$

By Example B.2.11,

$$\operatorname{Im}\partial_{n+1} = \operatorname{Ker}\partial_n = \{0\} \quad \text{and} \quad H_n(\{x\}) \cong \{0\}/\{0\} \cong \{0\}.$$

(3) Case when $n = 0$: Consider the portion of the singular chain complex

$$\cdots \xrightarrow{\partial_2} C_1(\{x\}) \xrightarrow{\partial_1} C_0(\{x\}) \xrightarrow{\partial_0} \{0\}.$$

The equality $\partial_1\sigma^1 = \{0\}$ implies $\operatorname{Im}\partial_1 = \{0\}$ and $\partial_0\sigma^0 = \{0\}$. Hence, $\operatorname{Ker}\partial_0 = C_0 \cong \mathbb{Z}$. Using again Example B.2.11, we find

$$H_0(\{x\}) \cong \mathbb{Z}/\{0\} \cong \mathbb{Z}. \qquad \square$$

Proposition 10.2.15. *Let $(X_i)_{i\in I}$ be the family of path-components of a space X. Then, for all n,*

$$H_n(X) \cong \bigoplus_{i\in I} H_n(X_i).$$

Proof. Let $\sigma : \Delta_n \longrightarrow X$ be a singular n-simplex. Since Δ_n is path-connected, $\sigma : \Delta_n \longrightarrow X_i$. Then every singular n-chain decomposes on X uniquely into a sum of n-chains on X_i. As a consequence,

$$C_n(X) \cong \bigoplus_{i\in I} C_n(X_i), \quad \operatorname{Ker}\partial_n \cong \bigoplus_{i\in I} \operatorname{Ker}\partial_n(X_i), \quad \text{and} \quad \operatorname{Im}\partial_n \cong \bigoplus_{i\in I} \operatorname{Im}\partial_n(X_i). \qquad \square$$

10.2.2 The zeroth homology group

We start with the following.

Proposition 10.2.16. *A nonempty space X is path-connected if and only if*

$$H_0(X) \cong \mathbb{Z}.$$

Proof. (1) Consider the portion of the singular chain complex

$$\cdots \xrightarrow{\partial_2} C_1(X) \xrightarrow{\partial_1} C_0(X) \xrightarrow{\partial_0} \{0\},$$

where $C_0 = \operatorname{Ker}\partial_0$ is the free Abelian group generated by the points of X. By Remark 10.2.8,

$$C_0(X) = \left\{c_0 : S_0(X) \longrightarrow \mathbb{Z}, c_0 = \sum_{k=1}^{N} n_{x_k} x_k, \text{where } x_k \in X \text{ for all } k \in [1,N], N \in \{1,2,\dots\}\right\}.$$

As for $C_1(X)$, it is generated by the path-components of X. Let $x_0 \in X$. Since X is path-connected, for each $k \in [1, N]$, there is a path joining x_k and x_0, that is by Remark 10.2.2, an element of C_1. In addition,

$$\partial_1 \sigma^i = \sigma^i(t) - \sigma^i(1-t) = x_i - x_0.$$

Define the map $\phi : C_0 = \operatorname{Ker} \partial_0 \longrightarrow \mathbb{Z}$ by

$$\phi\left(\sum_{k=1}^{N} n_k x\right) = \sum_{k=1}^{N} n_k.$$

Then ϕ is a surjective homomorphism. We claim that $\operatorname{Ker} \phi = \operatorname{Im} \partial_1$.

(a) Let $c_0 = \sum_{k=1}^{N} n_k x_k \in \operatorname{Ker} \phi$. We have $\sum_{k=1}^{N} n_k = 0$, $\sum_{k=1}^{N} n_k \sigma^k \in C_1(X)$ and

$$\begin{aligned}\partial_1\left(\sum_{k=1}^{N} n_k \sigma^k\right) &= \sum_{k=1}^{N} n_k \partial_1(\sigma^k) \\ &= \sum_{k=1}^{N} n_k (x_k - x_0) \\ &= \sum_{k=1}^{N} n_k x_k - x_0 \sum_{k=1}^{N} n_k \\ &= \sum_{k=1}^{N} n_k x_k \\ &= c_0.\end{aligned}$$

Hence, $c_0 \in \operatorname{Im} \partial_1$.

(b) Let $c_0 = \sum_{k=1}^{N} n_k x_k \in \operatorname{Im} \partial_1$. There is $c_1 = \sum_{k=1}^{N} n_k \sigma^k \in C_1(X)$ such that $c_0 = \partial_1(c_1)$. Then

$$\sum_{k=1}^{N} n_k x_k = \sum_{k=1}^{N} n_k x_k - x_0 \sum_{k=1}^{N} n_k,$$

which implies $\sum_{k=1}^{N} n_k = 0$, i. e., $c_0 \in \operatorname{Im} \partial_1$.

By First Homomorphism Theorem B.2.14 in the Appendix, we conclude that

$$H_0(X) = \operatorname{Ker} \partial_0 / \operatorname{Im} \partial_1 = \operatorname{Ker} \partial_0 / \operatorname{Ker} \phi \cong \mathbb{Z}.$$

(2) Conversely, if X is not path-connected, then by Proposition 10.2.15,

$$H_0(X) \cong \bigoplus_{i \in I} H_0(X_i),$$

where $(X_i)_{i \in I}$ are the path-components of X. Part (a) of the proof implies that $H_0(X)$ is the direct sum of more than one copy of $\mathbb{Z}$, whence the claim. □

Remark 10.2.17. The proof of Proposition 10.2.16 shows that any element of $C_0(X)$ is a generator of $H_0(X)$.

Remark 10.2.18. Proposition 10.2.16 confirms that the zero simplicial homology groups were isomorphic to $\mathbb{Z}$ in Examples 10.1.31, 10.1.32, 10.1.33 and Propositions 10.1.34, 10.1.35, 10.1.36.

From Proposition 10.2.16 and Proposition 10.2.15, we obtain a particular description of the zeroth homology group.

Corollary 10.2.19. *Let $(X_i)_{i\in I}$ be the family of path-components of a space X. Then*

$$H_0(X) \cong \bigoplus_{i\in I} \mathbb{Z} \cong \mathbb{Z}^I.$$

Remark 10.2.20. (1) Corollary 10.2.19 states that the rank of the zeroth group $H_0(X)$ equals the cardinality of the index set I, that is the number of path-components of X. Since, by Exercise 14, Section 4.9, $\pi_0(X)$ refers to the set of all path-components of X, Corollary 10.2.19 explains the connection between $\pi_0(X)$ and $H_0(X)$ for any space X.
(2) By Remark 10.2.17, the generators of the space $H_0(X)$ are elements of $C_0(X_i) : i \in I$.

10.2.3 Homology groups of a contractible space

The following result generalizes Proposition 10.2.14 and Proposition 10.2.16.

Proposition 10.2.21. *Let X a contractible space. Then*

$$H_n(X) \cong \begin{cases} \{0\}, & \text{if } n \geq 1, \\ \mathbb{Z}, & \text{if } n = 0. \end{cases}$$

Proof. Let $H : X \times I \longrightarrow X$ be a homotopy such that for all $x \in X$, $H(x,0) = x_0$ and $H(x,1) = x$ for some base point $x_0 \in X$. For $\sigma \in S_n(X)$, define $K\sigma \in S_{n+1}$ to be the cone over σ given by

$$K\sigma(t_0, t_1, \ldots, t_n, t_{n+1}) = \begin{cases} H(\sigma(\frac{1}{1-t_{n+1}}(t_0, t_1, \ldots, t_n)), 1 - t_{n+1}), & \text{if } t_{n+1} \neq 1, \\ \{x_0\}, & \text{if } t_{n+1} = 1. \end{cases}$$

By the definition of the homotopy H, K is continuous. We claim that

$$\partial_{n+1}(K\sigma) = \sigma - K\partial_n\sigma, \quad \text{for all } \sigma \in S_n(X). \tag{10.1}$$

Extending $K\sigma$ by linearity, we set $K(\sum n_\sigma \sigma) = \sum n_\sigma K(\sigma)$. Then (10.1) becomes

$$c = \partial_{n+1}Kc + K\partial_n c, \quad \text{for all } c \in S_n(X).$$

If $c \in \operatorname{Ker} \partial_n$, then $c = \partial_{n+1}Kc \in \operatorname{Im} \partial_{n+1}$, which implies $\operatorname{Ker} \partial_n = \operatorname{Im} \partial_{n+1}$, and thus $H_n(X) \cong \{0\}$ for all $n \geq 1$. Finally, since X is path-connected, by Proposition 10.2.16, $H_0(X) \cong \mathbb{Z}$, whence the claim of the proposition. □

Example 10.2.22. The path-components of the space

$$X = \mathbb{R} \setminus \{x_1, x_2, \ldots, x_n\} \cong S^1 \setminus \{x_1, x_2, \ldots, x_n, x_{n+1}\}, \quad n \geq 1$$

consist of the union of the open intervals $\bigcup_{k=1}^{n-1}(x_k, x_{k+1})$ and the half-lines $(-\infty, x_1)$, (x_n, ∞). All of them are contractible spaces. Hence, $H_n(X)$ are trivial groups for all $n \geq 1$.

Proposition 10.2.14 and Proposition 10.2.21 motivate the following definition of spaces, which look "homologically" like one-point sets.

Definition 10.2.23. A path-connected space X is called *acyclic* if $H_n(X) \cong \{0\}$ for all $n \geq 1$.

By Proposition 10.2.21, we can insert the acyclicity property into the diagram of Proposition 7.5.12 and Remark 9.1.23. This together with Proposition 9.1.16 lead to the following proper inclusions:

$$\begin{aligned}\text{convex} \subset \text{absolute retract} &\cong \text{has the extension property}\\ &\subset \text{contractible}\\ &\subset n\text{-aspherical for all } n \geq 1,\ \text{acyclic}\\ &\subset \text{simply connected}\\ &\subset \text{path-connected} \subset \text{connected}.\end{aligned}$$

10.2.4 One-Dimensional homology

Let $f : S^n \longrightarrow X$ be a continuous map such that $f(1, 0, \ldots, 0) = x_0$ for some $x_0 \in X$. Let $f_* : H_n(S^n) \longrightarrow H_n(X)$ be the induced map of f. By (9.2), $f \in \pi_n(X, x_0) \cong [S^n, X]$, the nth higher homotopy group of a space X at a base point x_0. Then assign to f the element $f_*(1)$ of the homology group $H_n(X)$. At this stage, we have anticipated and admitted that $H_n(S^n) \cong \mathbb{Z}$ (see Theorem 10.3.21). This defines a map ϕ by $\phi(f) = f_*(1)$, where 1 is the generator of $\mathbb{Z}$. For $n = 1$, one can find the proof of the following theorem, e. g., in [10, Chapter IV, Theorem 3.4], [17, Theorem 10.26], or [19, Theorem 6.3.4] using the commutator subgroup $[\pi_1(X), \pi_1(X)]$ of the group $\pi_1(X)$ (see Definition B.4.1), or [59, Theorem 29.16] for a direct proof.

Theorem 10.2.24 (Poincaré–Hurewicz theorem). (1) *The map ϕ is a homomorphism.*
(2) *Let X be path-connected. Then ϕ is an isomorphism between $H_1(X)$ and the Abelianization of $\pi_1(X)$.*

Example 10.2.25. If X is simply connected, then $H_1(X)$ is trivial.

Example 10.2.26. We have
(1) $H_1(S^1) \cong \mathbb{Z}$ and $H_1(S^n) \cong \{0\}$ for $n > 1$;
(2) $H_1(\mathbb{R}\mathrm{P}^n) \cong \mathbb{Z}/2\mathbb{Z} \cong \pi_1(\mathbb{R}\mathrm{P}^n)$ for all $n \geq 2$;
(3) $H_1(\mathbb{T}^2) \cong \mathbb{Z}^2 \cong \pi_1(\mathbb{T}^2)$;
(4) Homology groups $H_1(8)$, $H_1(\Theta)$, $H_1(\mathbb{T}^2 \setminus \{p\})$, $H_1(\mathbb{K}^2 \setminus \{p\})$, ... are isomorphic to the Abelianization of $\mathbb{Z} * \mathbb{Z}$, that is to group $\mathbb{Z} \times \mathbb{Z}$ (see Example 8.2.3 and Example 8.2.8).

Example 10.2.27. By Proposition 3.8.10 and Example 8.2.39, we get

$$H_1(\mathbb{K}^2) \cong \text{Abelianization of } \pi_1(\mathbb{K}^2) \cong \mathbb{Z}^2 \times \mathbb{Z}^2/2.$$

Remark 10.2.28. If the space X is $n-1$-connected (see Definition 9.1.17) and $n \geq 2$, Hurewicz' theorem states that $\pi_n(X)$ and $H_n(X)$ are isomorphic (see [10, Corollary 10.8, Chapter VII], [23, Theorem 4.5.3], [45, Theorem 4.37]). Notice that $\pi_k(X)$ and $H_k(X)$ are trivial for all $1 \leq k \leq n-1$. Since by Corollary 9.2.1, S^n is $(n-1)$-connected for all $n \geq 2$, we deduce that

$$\pi_{n-1}(S^n) \cong H_{n-1}(S^n) \cong 0,$$

the latter isomorphism is part of Theorem 10.3.21.

Example 10.2.29. Let Mg be the mug introduced in Example 8.2.8. By Theorem 10.2.24(2) and Proposition 10.1.34, we have

$$H_1(\mathrm{Mg}) \cong \text{Abelianization of } (\pi_1(\mathrm{Mg})) \cong \text{Abelianization of } (\mathbb{Z} * \mathbb{Z}) \cong \mathbb{Z} \times \mathbb{Z} \cong H_1(\mathbb{T}^2).$$

10.2.5 Induced homomorphism

Let

$$\Delta_n \xrightarrow{\sigma} X \xrightarrow{f} Y,$$

where σ is a singular n-simplex on X and f a continuous function between spaces. Then $f \circ \sigma$ is a singular n-simplex on Y. By linearity, we extend this composite function to a homomorphism

$$C_n(X) \xrightarrow{f_*} C_n(Y)$$

by

$$f_*\Big(\sum_{\sigma \in S_n} n_\sigma \sigma\Big) = \sum_{\sigma \in S_n} n_\sigma (f \circ \sigma).$$

With the notation of the simplex face in Definition 10.2.4, we easily check that

$$\begin{aligned}(f\circ\sigma)_i(t_0,\dots,t_{i-1},t_{i+1},\dots,t_n) &= (f\circ\sigma)(t_0,\dots,t_{i-1},0,t_{i+1},\dots,t_n)\\ &= f(\sigma_i(t_0,\dots,t_{i-1},t_i,t_{i+1},\dots,t_n)),\end{aligned}$$

i. e.,

$$(f\circ\sigma)_i = f\circ\sigma_i.$$

Using the extension of the map f_* to S_n yields

$$\partial_n f_* = f_*\partial_n$$

and the following diagram commutes:

$$\begin{array}{ccc} C_n(X) & \xrightarrow{\quad f_*\quad} & C_n(Y)\\ \partial_n\big\downarrow & & \big\downarrow\partial_n\\ C_{n-1}(X) & \xrightarrow{\quad f_*\quad} & C_{n-1}(Y)\end{array}$$

Hence, f_* maps $Z_n(X)$ into $Z_n(Y)$ and $B_n(X)$ into $B_n(Y)$. Furthermore f_* induces a quotient group homomorphism

$$H_n(X)\xrightarrow{f_*}H_n(Y).$$

The following proposition is a consequence of the definition of f_*.

Proposition 10.2.30. (1) *If* Id_X *is the identity map on a space* X, *then* $(\mathrm{Id})_{X*}$ *is the identity homomorphism on* $H_n(X)$.

(2) *If* $X\xrightarrow{f}Y\xrightarrow{g}Z$ *are continuous maps, then*

$$(g\circ f)_* = g_*\circ f_*.$$

The following result corresponds to Corollary 4.5.5 for higher homotopy groups. For the proof we refer to, e. g., [19, Theorem 6.4.1] or [59, Theorem 29.14].

Proposition 10.2.31. *If* $f,g: X\longrightarrow Y$ *are homotopic, then the induced homomorphisms* f_* *and* $g_*: H_n(X)\longrightarrow H_n(Y)$ *are identical.*

Useful for applications is the following corollary whose proof is immediate.

Corollary 10.2.32. (1) *If* $f: X\longrightarrow Y$ *is a homotopy equivalence map, then* $f_*: H_n(X)\longrightarrow H_n(Y)$ *is an isomorphism.*

(2) *If* X *and* Y *are homotopically equivalent, then for each* $n\ge 0$, *the corresponding homology groups* $H_n(X)$ *and* $H_n(Y)$ *are isomorphic for all* $n\ge 0$.

The latter corollary will be proved for the relative homology (see Proposition 10.3.10).

We end this section with a result on the attaching of n-cells to a space X. Let $B^n \cup_f X$ be the adjunction space for some continuous map $f : S^{n-1} \subset B^n \longrightarrow X$ (see Definition 3.4.1). The following result (see [51, Corollary 3.4]) corresponds to Proposition 9.2.22 concerning the higher homotopy groups. The proof is omitted.

Proposition 10.2.33. *Let $n \geq 1$. Then:*
(1) $H_p(B^n \cup_f X) \cong H_p(X)$, *for all $p \neq n, n-1$.*
(2) *The induced homomorphism $i_* : H_p(X) \longrightarrow H_p(B^n \cup_f X)$ of the inclusion map $X \hookrightarrow B^n \cup_f X$ is a monomorphism for $p = n$ and an epimorphism for $p = n-1$.*

This result can be used to compute all homology groups of the projective spaces with any dimension. Recall that $\mathbb{RP}^1$ is isomorphic to S^1. So, this together with Proposition 10.1.35 yield

$$H_p(\mathbb{RP}^1) \cong \begin{cases} \mathbb{Z}, & p = 0, p = 1, \\ 0, & \text{otherwise.} \end{cases} \quad H_p(\mathbb{RP}^2) \cong \begin{cases} \mathbb{Z}, & p = 0, \\ \mathbb{Z}_2, & p = 1, \\ 0, & \text{otherwise.} \end{cases}$$

These are the same homology groups computed in Proposition 10.1.35 for the simplicial complex $\mathbb{RP}^2$. In the general case, we have the following.

Corollary 10.2.34. *We have*

$$H_p(\mathbb{RP}^n) \cong \begin{cases} \mathbb{Z}, & p = 0 \text{ or } p = n \text{ odd}, \\ \mathbb{Z}_2, & p \text{ odd and } 0 < p < n, \\ 0, & \text{otherwise.} \end{cases}$$

Proof. By Example 3.4.19 and Proposition 3.1.13, $\mathbb{RP}^n \cong B^n \cup_f \mathbb{RP}^{n-1}$, where $f : S^{n-1} \longrightarrow \mathbb{RP}^{n-1}$ is the projection map $f(x) = [x] = \{x, -x\}$. We proceed by induction on n. By Proposition 10.2.33, for $p \neq n$ and $p \neq (n-1)$,

$$H_p(\mathbb{RP}^n) \cong H_p(\mathbb{RP}^{n-1}).$$

For $p = n$ or $p = n-1$, consider the homomorphism $p_* : H_n(S^n) \longrightarrow H_n(\mathbb{RP}^n)$ of the covering map. If n is odd, then by induction $H_n(\mathbb{RP}^n) \cong H_1(\mathbb{RP}^1) \cong \mathbb{Z}$. Hence, $H_n(\mathbb{RP}^{n+1}) \cong \operatorname{Im} p_* \cong \mathbb{Z}_2$ and $H_{n+1}(\mathbb{RP}^{n+1}) \cong \operatorname{Ker} p_* \cong 0$. If n is even, then by induction $H_n(\mathbb{RP}^n) \cong 0$, $H_n(\mathbb{RP}^{n+1}) \cong H_n(\mathbb{RP}^n) \cong \operatorname{Im} p_* \cong 0$ and $H_{n+1}(\mathbb{RP}^{n+1}) \cong / \operatorname{Ker} p_* \cong \mathbb{Z}$. □

A simpler result can be proved for the complex projective spaces (see, e. g., [52, Théorème 9.3.11] or [102, Theorem 5.3.3]).

Corollary 10.2.35. *For $n \geq 0$, we have*

$$H_p(\mathbb{C}\mathrm{P}^n) \cong \begin{cases} \mathbb{Z}, & p \text{ even and } 0 \leq p \leq 2n, \\ 0, & \text{otherwise.} \end{cases}$$

Proof. We make use of the homeomorphism $\mathbb{C}\mathrm{P}^n \cong B^{2n} \cup \mathbb{C}\mathrm{P}^{n-1}$ similar to the one for the real projective spaces (see Exercise 23, Sections 3.3–3.7). Then we proceed by induction. The result is clear for $n = 0$. By Proposition 10.2.33, $H_p(\mathbb{C}\mathrm{P}^n) \cong H_p(\mathbb{C}\mathrm{P}^{n-1})$ for any even integer p distinct from $2n$, that is $p < 2n$ or $p > 2n + 2$. The induced homomorphism $f_* : H_{2n-1}(S^{2n-1}) \longrightarrow H_{2n-1}(\mathbb{C}\mathrm{P}^{n-1})$ is trivial. Hence, $H_{2n-1}(\mathbb{C}\mathrm{P}^n) \cong 0$ and $H_{2n}(\mathbb{C}\mathrm{P}^n) \cong \mathrm{Ker} f_* \cong \mathbb{Z}$. □

10.3 Relative homology

10.3.1 Axiomatic approach

A third way to introduce homology is an axiomatic approach developed by S. Eilenberg, Samuel and N. E. Steenrod in 1945 (see [26]).

Definition 10.3.1. Let X be a space and $A \subset X$ a subset.
(1) The pair (X, A) is called a *topological pair*, or merely a pair.
(2) A map of pairs $f : (X, A) \longrightarrow (Y, B)$ is a continuous map $f : X \longrightarrow Y$ such that $f(A) \subset B$.
(3) The pairs (X, A) and (Y, B) are *homeomorphic* if $f : X \longrightarrow Y$ is a homeomorphism and $f(A) = B$.

Definition 10.3.2. (1) Two maps of pairs $f, g : (X, A) \longrightarrow (Y, B)$ are homotopic if there exists a homotopy $H(x, t)$ between f and g such that $H(A, t) \subset B$ for all $t \in I$.
(2) We say that the pairs (X, A) and (Y, B) are *homotopically equivalent* if there exist maps of pairs $f : (X, A) \longrightarrow (Y, B)$ and $g : (Y, B) \longrightarrow (X, A)$ such that $g \circ f \simeq \mathrm{Id}_{|(X,A)}$ and $f \circ g \simeq \mathrm{Id}_{|(Y,B)}$.

Definition 10.3.3. A *homology theory* associates:
(1) to each pair (X, A) a sequence of Abelian groups $H_n(X, A)$ ($n \in \mathbb{Z}$) and a sequence of homomorphisms $\partial_n : H_n(X, A) \longrightarrow H_{n-1}(A, \emptyset)$, called boundary homomorphisms;
(2) to each map of pairs $f : (X, A) \longrightarrow (Y, B)$ a sequence of homomorphisms $f_* : H_n(X, A) \longrightarrow H_n(Y, B)$ satisfying the Eilenberg–Steenrod axioms for homology.

The group $H_n(X, \emptyset)$ will be shortened to $H_n(X)$ and the composite $f \circ g$ to fg.

Definition 10.3.4. *The Eilenberg–Steenrod axioms* for homology are the following.

Axiom 1 (Identity axiom). If f is the identity, then so is f_*.

Axiom 2 (Composition axiom). If $f : (X,A) \longrightarrow (Y,B)$ and $g : (Y,B) \longrightarrow (Z,C)$ are maps of pairs, then $(gf)_* = g_* f_*$.

Axiom 3 (Commutativity axiom or boundary axiom). We have $\partial_n f_* = (f_{|A})_* \partial_n$, which means that the following diagram commutes:

$$\begin{array}{ccc} H_n(X,A) & \xrightarrow{f_*} & H_n(Y,B) \\ \downarrow{\scriptstyle \partial_n} & & \downarrow{\scriptstyle \partial_n} \\ H_{n-1}(A) & \xrightarrow{(f_{|A})_*} & H_{n-1}(B) \end{array}$$

Axiom 4 (Exactness axiom). Given a pair (X,A), the following long sequence is exact:

$$\cdots \xrightarrow{\partial_{n+1}} H_n(A) \xrightarrow{i_*} H_n(X) \xrightarrow{j_*} H_n(X,A) \xrightarrow{\partial_n} H_{n-1}(A) \xrightarrow{i_*} \cdots,$$

where i_* and j_* are the homomorphisms induced by the inclusions $i : A \to X$ and $j : (X,\emptyset) \to (X,A)$.

Axiom 5 (Homotopy axiom). Two homotopic maps of pairs $f,g : (X,A) \to (Y,B)$ have identical induced homomorphism.

Axiom 6 (Excision axiom). Let (X,A) be a pair, $U \subset X$ an open subset such that $\overline{U} \subset \mathring{A}$, and $i : (X \setminus U, A \setminus U) \to (X,A)$ the inclusion map. Then the induced homomorphism i_* is an isomorphism.

Axiom 7 (Dimension axiom). Let $x_0 \in X$. Then, for all $n \neq 0$, $H_n(\{x_0\}) = 0$.

Definition 10.3.5. (1) Let $G = H_0(\{x\})$, for $x \in X$. Then G is called the *group of coefficients of the homology*. This completes Axiom 7.
(2) For $n < 0$, we set $H_n(X,A) = 0$.

10.3.2 Main properties

Proposition 10.3.6. *Let A, B be an open partition of a space X (that is to say, A, B open sets, $X = A \cup B$, $A \cap B = \emptyset$). Then, for every integer n,*

$$H_n(X,A) \cong H_n(B).$$

Proof. Apply the excision axiom with $U = X \setminus B$. Then $\overline{U} = U = X \setminus B = A = \mathring{A}$, $A \setminus B = A \cap B = \emptyset$, and $X \setminus U = B$. Hence, $i : (B,\emptyset) \hookrightarrow (X,A)$ and

$$H_n(B,\emptyset) \cong H_n(B) \cong H_n(X,A).$$

□

Notice that the following result is more general than Proposition 10.3.6.

Theorem 10.3.7. *Let X_1, X_2 be subspaces of X and $\mathring{X}_1 \cup \mathring{X}_2 = X$. Then*

$$H_n(X, X_1) \cong H_n(X_2, X_1 \cap X_2).$$

Proof. Just apply excision axiom 6 with $A = X_1$ and $U = X \setminus X_2$. Indeed,

$$\overline{U} = \overline{X \setminus X_2} = X \setminus \mathring{X}_2 \subset \mathring{X}_1 = \mathring{A}.$$

Then

$$H_n(X \setminus (X \setminus X_2), X_1 \setminus (X \setminus X_2)) \cong H_n(X, X_1) \quad \Longleftrightarrow \quad H_n(X_2, X_1 \cap X_2) \cong H_n(X, X_1). \qquad \square$$

Theorem 10.3.7 turns out to be equivalent to excision axiom 6.

Theorem 10.3.8. *Theorem* 10.3.7 *implies excision axiom 6.*

Proof. Assume Theorem 10.3.7 and let (X, A) be a pair with $U \subset X$ an open subset such that $\overline{U} \subset \mathring{A}$. Set $X_1 = A$ and $X_2 = X \setminus U$. Then

$$\mathring{X}_1 = \mathring{A} \quad \text{and} \quad \mathring{X}_2 = \overline{X \mathring{\setminus} U} = X \setminus \overline{U}.$$

Hence,

$$\overline{U} \subset \mathring{A} \quad \Longrightarrow \quad X \setminus \mathring{X}_1 \subset \mathring{X}_2 \quad \Longrightarrow \quad X \subset \mathring{X}_1 \cup \mathring{X}_1 \quad \Longrightarrow \quad X = \mathring{X}_1 \cup \mathring{X}_1.$$

By Theorem 10.3.7, we get

$$H_n(X, A) \cong H_n(X \setminus U, A \cap (X \setminus U)) = H_n(X \setminus U, A \setminus U). \qquad \square$$

Proposition 10.3.9. *For every integer n, $H_n(X, X) \cong 0$.*

Proof. Since the inclusion map $i : X \to X$ is the identity Id_X, by Axiom 1, $i_* : H_n(X) \to H_n(X)$ is the identity map. By Axiom 4, the following long sequence is exact:

$$\cdots \xrightarrow{\partial_{n+1}} H_n(X) \xrightarrow{i_*} H_n(X) \xrightarrow{j_*} H_n(X, X) \xrightarrow{\partial_n} H_{n-1}(X) \xrightarrow{i*} \cdots.$$

As a consequence, $\operatorname{Im} i_* = H_n(X) = \operatorname{Ker} j_*$. Then $j_* = 0$, i. e., $\operatorname{Ker} \partial_n = 0$ and ∂_n is injective. Hence, $H_n(X, X) \cong \operatorname{Im} \partial_n = \operatorname{Ker} i_* = 0$. Finally, $\partial_n = 0$ and $H_n(X, X) = 0$, as claimed. $\square$

Proposition 10.3.10. *If the pairs (X, A) and (Y, B) are homotopically equivalent, then $H_n(X, A) \cong H_n(Y, B)$ for all n.*

Proof. By supposition, let f and g be given by Definition 10.3.2(2). Using Axioms 1 and 2, we have

$$g_* f_* \cong \mathrm{Id}_{H_n(X,A)} \quad \text{and} \quad f_* g_* \cong \mathrm{Id}_{H_n(Y,B)}. \qquad \square$$

Corollary 10.3.11. *If X and Y are homotopically equivalent, then $H_n(X) \cong H_n(Y)$ for all n.*

Also, we get Proposition 10.2.21 as a consequence.

Corollary 10.3.12. *Let X be a contractible space (Definition 4.9.1). Then $H_n(X) = 0$ for all $n \in \mathbb{Z} \setminus \{0\}$ and $H_0(X) = G$.*

Proof. Since the pairs $(X, \emptyset)$ and $(\{x\}, \emptyset)$ are homotopically equivalent for each $x \in X$, from Axiom 7 and Proposition 10.3.10, we infer that $H_n(X) \cong H_n(\{x\}) = 0$. □

Example 10.3.13. By Example 4.9.2, a ball B^n of Euclidean space $\mathbb{R}^n$ is contractible for all $n \geq 1$. Hence, $H_p(B^n) \cong 0$ for all $p, n \geq 1$.

Example 10.3.14. By Proposition 4.9.5, a cone $C(X)$ over a space X is contractible to its apex. Hence, $H_0(C(X)) \cong G$ and $H_n(C(X)) \cong 0$ for all $n \neq 0$.

Corollary 10.3.15. *Let A be a deformation retract of X (Definition 7.2.1). Then $H_n(X, A) \cong 0$ for all $n \neq 0$.*

Proof. By Theorem 7.2.4, X and A are homotopically equivalent and so are the pairs (X, A) and (X, X). Then the claim is a consequence of Proposition 10.3.9 and Proposition 10.3.10. □

In the case we merely know that A is a retract of X, a more general result holds.

Proposition 10.3.16. *Let A be a retract of X. Then, for all integer n,*

$$H_n(X) \cong H_n(A) \oplus H_n(X, A).$$

Proof. Let $r : X \longrightarrow A$ be a retraction and $i : A \hookrightarrow X$ the inclusion map. By Remark 7.1.2, $r \circ i = \mathrm{Id}_{|A}$ and by Axiom 1, $(r \circ i)_* = \mathrm{Id}_{H_n(A)}$. In addition, $i_* : H_n(A) \longrightarrow H_n(X)$ is injective while $r_* : H_n(X) \longrightarrow H_n(A) \subset H_n(X)$ is surjective (Exercise 6). We have

$$\begin{aligned} H_n(X) &= \operatorname{Ker} r_* \oplus \operatorname{Im} r_* \quad \text{(from algebra)} \\ &= \operatorname{Ker} r_* \oplus H_n(A) \quad (r_* \text{ is surjective}) \\ &\cong \operatorname{Ker} r_* \oplus \operatorname{Im} i_* \quad (i_* \text{ is injective}). \end{aligned} \tag{10.2}$$

By Axiom 4, the following short sequence is exact:

$$\xrightarrow{\partial_{n+1}} H_n(A) \xrightarrow{i_*} H_n(X) \xrightarrow{j_*} H_n(X, A) \xrightarrow{\partial_n}$$

In particular, $\operatorname{Ker} j_* = \operatorname{Im} i_*$. Since i_* is injective, $\operatorname{Im} \partial_{n+1} = 0$, i. e., $\partial_{n+1} = 0$. The last equality in (10.2) implies that $j_{*|\operatorname{Ker} r_*}$ is injective. Since $\operatorname{Im} j_* = \operatorname{Ker} \partial_n = H_n(X,A)$, $j_* : \operatorname{Ker} r_* \longrightarrow H_n(X,A)$ is a isomorphism. Therefore, $H_n(X) \cong H_n(X,A) \oplus H_n(A)$. □

Example 10.3.17. For every point $x_0 \in X$, we have

$$H_n(X) \cong \begin{cases} G \oplus H_0(X, \{x_0\}), & \text{if } n = 0, \\ H_n(X, \{x_0\}), & \text{if } n \neq 0. \end{cases}$$

Corollary 10.3.18. *Let A be a deformation retract of X. Then $H_n(X) \cong H_n(A)$ for every n.*

Proof. (1) *First proof.* We just apply Corollary 10.3.15 and Proposition 10.3.16.

(2) *Second proof* (direct proof). Let $r : X \longrightarrow A$ be a retraction and $i : A \hookrightarrow X$ the inclusion map. We have $(r \circ i)_* = \operatorname{Id}_{H_n(A)}$ and since A be a deformation retract of X, $i \circ r$ is homotopic to the identity on X. Hence, $(i \circ r)_* = \operatorname{Id}_{H_n(X)}$, and thus $i_* : H_n(A) \longrightarrow H_n(X)$ is an isomorphism. □

From Proposition 10.3.16 and Corollary 10.3.12, we get the following.

Corollary 10.3.19. *Let $A \subset X$ be a retract contractible subspace. Then*

$$H_n(X) = \begin{cases} G \oplus H_0(X,A), & n = 0, \\ H_n(X,A), & n \neq 0. \end{cases}$$

Arguing as in Proposition 10.2.15, we have the following.

Proposition 10.3.20. *Let $(X_i)_{i \in I}$ be the family of path-components of a space X and $A \subset X$. Then, for all positive n,*

$$H_n(X,A) \cong \sum_{i \in I} H_n(X_i, A \cap X_i).$$

10.3.3 Homology groups of the sphere

We put the stress on the computation of the homology groups of the sphere S^n, boundary of the closed ball B^{n+1} of Euclidean space $\mathbb{R}^{n+1}$. This is performed by using the excision axiom. Recall that

$$S^n_+ = \{x = (x_1, x_2, \dots, x_n, x_{n+1}) \in S^n : x_{n+1} \geq 0\} \quad \text{and}$$
$$S^n_- = \{x = (x_1, x_2, \dots, x_n, x_{n+1}) \in S^n : x_{n+1} \leq 0\}$$

denote the closed northern hemisphere and the closed southern hemisphere, respectively. Their intersection S^{n-1} is the equator of the sphere S^n. To avoid confusion with notation, homology groups of sphere S^n will be denoted $H_p(S^n)$ throughout this chapter.

Theorem 10.3.21 (Homology groups of the sphere). *We have*

$$H_p(S^n) \cong \begin{cases} \mathbb{Z} \oplus \mathbb{Z}, & p = n = 0, \\ \mathbb{Z}, & p = 0,\ n > 0, \\ \mathbb{Z}, & p = n > 0, \\ 0, & \textit{otherwise.} \end{cases}$$

Proof. Step 1. We check that for any integers $p \geq 0$, $n \geq 1$, $H_p(S^n_+, S^{n-1}) \cong H_p(S^n, S^n_-)$.

In order to apply Axiom 6, let $X = S^n$, $A = S^n_-$, and consider the open subset

$$U = \{x = (x_1, x_2, \ldots, x_n, x_{n+1}) \in S^n : x_{n+1} < -1/2\}.$$

Then $\overline{U} \subset \mathring{A}$. By the excision axiom, $H_p(S^n \setminus U, S^n_- \setminus U) \cong H_p(S^n, S^n_-)$. Since it is easy to check that S^n_+ and S^{n-1} are deformation retracts of $S^n \setminus U$ and $S^n_- \setminus U$, respectively, the claim results from Corollary 10.3.18.

Step 2. We claim that $H_p(S^n) \cong H_{p-1}(S^{n-1})$ for every $p \geq 2$ and $n \geq 1$.

By Axiom 4, the short sequence

$$0 \cong H_p(B^n) \xrightarrow{j_*} H_p(B^n, S^{n-1}) \xrightarrow{\partial_n} H_{p-1}(S^{n-1}) \xrightarrow{i_*} H_{p-1}(B^n) \cong 0$$

is exact. We have used the contractibility of the closed ball. Then, using Exercise 1(2) of this chapter, we obtain that $H_p(B^n, S^{n-1}) \cong H_{p-1}(S^{n-1})$ for all $p \geq 2$. By Example 1.5.22, B^n and S^n_+ are homeomorphic. Hence, $H_p(S^n_+, S^{n-1}) \cong H_{p-1}(S^{n-1})$. Since S^n_- is contractible (and homeomorphic to B^n), By Exercise 14, Corollary 10.3.12 and Corollary 10.3.11, we have

$$\begin{aligned} H_p(S^n, S^n_-) &\cong H_p(S^n) \oplus H_{p-1}(S^n_-) \\ &\cong H_p(S^n) \oplus H_{p-1}(B^n) \\ &\cong H_p(S^n) \end{aligned}$$

for every $p \geq 1$ and $n \geq 1$. Then the claim follows from Step 1.

Step 3. Conclusion.

(a) If $p = n \geq 1$, then by Step 2 and Proposition 10.3.20,

$$H_n(S^n) \cong H_{n-1}(S^{n-1}) \cong \cdots \cong H_1(S^1) \cong \mathbb{Z}.$$

(b) If $p \neq n$, distinguish four cases.

(i) If $p > n$, by Step 2 and Axiom 7,

$$H_p(S^n) \cong H_{p-1}(S^{n-1}) \cong \cdots \cong H_{p-n}(S^0) \cong 0.$$

(ii) If $2 \le p < n$, again by Step 2 and Proposition 10.2.16,

$$H_p(S^n) \cong H_{p-1}(S^{n-1}) \cong \cdots \cong H_1(S^{n-p+1}) \cong 0$$

because S^{n-p+1} is simply connected for $n - p + 1 \ge 2 \Longleftrightarrow n > p$.

(iii) If $p = 1 < n$, Corollary 6.3.4 and Hurewicz' Theorem 10.2.24 yield $H_1(S^n) \cong \pi_1(S^n) \cong 0$. Since the sphere is path-connected, $H_0(S^n) \cong \mathbb{Z}$ for $n \ge 1$.

(iv) Finally, for $n = 0$, the sets $\{-1\}$ and $\{1\}$ can be considered as path-components of $S^0 = \{-1, 1\}$. Then Proposition 10.2.15 gives $H_0(S^0) \cong H_0(\{-1, 1\}) \cong \mathbb{Z} \oplus \mathbb{Z}$. □

Corollary 10.3.22. *For $n \ge 1$,*

$$H_0(S^n) \cong \mathbb{Z} \quad \textit{and} \quad H_p(S^n) \cong \{0\} \quad \textit{for all } p \ne n,\ p > 0.$$

Example 10.3.23. By Example 4.2.9, the sphere S^{n-1} and $\mathbb{R}^n \setminus \{0\}$ are homotopically equivalent. Using Proposition 10.3.10 and Theorem 10.3.21, we obtain the homology groups of the punctured Euclidean space

$$H_p(\mathbb{R}^n \setminus \{0\}) \cong \begin{cases} \mathbb{Z}, & p = 0,\ p = n - 1, \\ 0, & \text{otherwise.} \end{cases}$$

Example 10.3.24. By Proposition 4.2.17, the Möbius band $\mathbb{M}^2$ and the unit circle are homotopically equivalent. Using Corollary 10.2.32 and Theorem 10.3.21, we find

$$H_n(\mathbb{M}^2) \cong \begin{cases} \mathbb{Z}, & n = 0, 1, \\ 0, & \text{otherwise.} \end{cases}$$

Example 10.3.25. By Proposition 7.2.24, S^1 embeds as a deformation retract of $S^3 \setminus S^1$. Then

$$H_n(S^3 \setminus S^1) \cong H_n(\mathbb{R}^3 \setminus \{(0z)\}) \cong \begin{cases} \mathbb{Z}, & n = 0, 1, \\ 0, & \text{otherwise.} \end{cases}$$

More generally, the homology groups of $S^n \setminus S^k$ will be computed in the more general framework of Jordan–Brouwer separation theorem (Theorem 10.4.24), where the set difference $S^n \setminus S^k$ is shown to have isomorphic homology groups as the sphere S^{n-k-1}.

10.3.4 Homology groups of a quotient space

We discuss the connection between the relative homology groups $H_n(X, A)$ and the homology groups of the quotient space X/A when (X, A) is "a good pair," more precisely

$$A = \overline{A} \subset \mathring{U} = U \subset X, \tag{10.3}$$

where A is a deformation retract of the neighborhood U. We have (see [45, Proposition 2.22]).

Proposition 10.3.26. *For all $n \neq 0$,*

$$H_n(X/A) \cong H_n(X, A).$$

Proof. Since A and U are homotopically equivalent, so are the pairs (X, A) and (X, U). Since A is deformation retract of U, A/A is a deformation retract of U/A by Exercise 9, Chapter 7. By Corollary 10.3.18, we have

$$H_n(X, A) \cong H_n(X, U) \quad \text{and} \quad H_n(X/A, A/A) \cong H_n(X/A, U/A). \tag{10.4}$$

Let $\pi : X \longrightarrow X/A$ be the surjective quotient map. By the excision axiom 6,

$$\begin{aligned} H_n(X, U) &\cong H_n(X \setminus A, U \setminus A), \\ H_n(\pi(X), \pi(U)) &\cong H_n(\pi(X) \setminus \pi(A), \pi(U) \setminus \pi(A)). \end{aligned} \tag{10.5}$$

By Proposition 2.5.1 and the choice of the closed set A and the open set U, the spaces $X \setminus A$ and $(X/A) \setminus (A/A)$ are homeomorphic. So, are $U \setminus A$ and $(U/A) \setminus (A/A)$. Since by surjectivity of π, we have $X/A = \pi(X)$, $U/A = \pi(U)$ and $A/A = \pi(A)$, we have

$$\begin{aligned} H_n(\pi(X) \setminus \pi(A), \pi(U) \setminus \pi(A)) &\cong H_n((X/A) \setminus (A/A), (U/A) \setminus (A/A)) \\ &\cong H_n(X \setminus A, U \setminus A). \end{aligned} \tag{10.6}$$

Combining (10.4), (10.5) and (10.6), we obtain the chain of homeomorphisms:

$$\begin{aligned} H_n(X, A) &\cong H_n(X, U) \\ &\cong H_n(X \setminus A, U \setminus A) \\ &\cong H_n(\pi(X) \setminus \pi(A), \pi(U) \setminus \pi(A)) \\ &\cong H_n(\pi(X), \pi(U)) \\ &\cong H_n(X/A, U/A) \\ &\cong H_n(X/A, A/A) \\ &\cong H_n(X/A, \{A\}) \\ &\cong H_n(X/A), \end{aligned}$$

where we have appealed to Example 10.3.17 for the last homeomorphism. □

Example 10.3.27. Let $X = B^n$ and $A = S^{n-1}$ be the closed unit ball and its boundary in $\mathbb{R}^n$, respectively. By Theorem 2.5.11, $B^n/S^{n-1} \cong S^n$. Using Proposition 10.3.21 and Proposition 10.3.26, we obtain the relative homology groups

$$H_p(B^n, S^{n-1}) \cong H_p(S^n) \cong \begin{cases} \mathbb{Z}, & p = n > 0, \\ 0, & \text{otherwise.} \end{cases}$$

Theorem 7.3.10 and Theorem 10.3.26 combined yield the following.

Corollary 10.3.28. *Let (X, A) have the homotopy extension property (Definition 7.3.1). If a closed contractible subspace $A \subset X$ satisfies (10.3), then for all n,*

$$H_n(X) \cong H_n(X, A).$$

Corollary 10.3.29. *The boundary circle of the Möbius band is not a retract of the Möbius band (although the Möbius band deformation retracts on its core circle, by Proposition 7.2.33).*

Proof. Given the pair $(\mathbb{M}^2, \partial\,\mathbb{M}^2)$, the following short sequence is exact:

$$H_2(\mathbb{M}^2, \partial\,\mathbb{M}^2) \xrightarrow{\partial_2} H_1(\partial\,\mathbb{M}^2) \xrightarrow{i_*} H_1(\mathbb{M}^2) \xrightarrow{j_*} H_1(\mathbb{M}^2, \partial\,\mathbb{M}^2) \xrightarrow{\partial_1} H_0(\partial\,\mathbb{M}^2) \xrightarrow{\partial_0} 0,$$

where i_* and j_* are the homomorphisms induced by the inclusions $i : \partial\,\mathbb{M}^2 \to \mathbb{M}^2$ and $j : (\mathbb{M}^2, \emptyset) \to (\mathbb{M}^2, \partial\,\mathbb{M}^2)$. By Proposition 10.3.26, $H_1(\mathbb{M}^2, \partial\,\mathbb{M}^2)$ and $H_1(\mathbb{M}^2/\partial\,\mathbb{M}^2)$ are isomorphic. However, by Exercise 18, Sections 3.1–3.2, the quotient space $\mathbb{M}^2/\partial\,\mathbb{M}^2$ is isomorphic to the projective plane $\mathbb{RP}^2$ whose first homology group is isomorphic to $\mathbb{Z}/2\mathbb{Z}$ and the second homology group is trivial (Proposition 10.1.35). By Example 10.3.24, $H_1(\mathbb{M}^2) \cong \mathbb{Z}$ whereas $H_1(\partial\,\mathbb{M}^2) \cong \mathbb{Z}$ is a consequence of Proposition 3.2.7 and Corollary 10.2.32. Hence, the short sequence reduces to

$$0 \xrightarrow{\partial_2} \mathbb{Z} \xrightarrow{i_*} \mathbb{Z} \xrightarrow{j_*} \mathbb{Z}/2\mathbb{Z} \xrightarrow{\partial_1} \mathbb{Z} \xrightarrow{\partial_0} 0.$$

Therefore, the following successive implications hold:

$$\begin{aligned} \operatorname{Ker} i_* \cong 0 \quad &\Longrightarrow \quad \operatorname{Im} i_* \cong \mathbb{Z} \quad \Longrightarrow \quad \operatorname{Ker} j_* \cong \mathbb{Z} \quad \Longrightarrow \quad j_* \cong 0 \\ &\Longrightarrow \quad \operatorname{Ker} \delta_1 \cong 0 \quad \Longrightarrow \quad \operatorname{Im} \delta_1 \cong \mathbb{Z}/2\mathbb{Z} \quad \Longrightarrow \quad \operatorname{Ker} \delta_0 \cong \mathbb{Z}/2\mathbb{Z}. \end{aligned}$$

Therefore, $\mathbb{Z} \cong \mathbb{Z}/2\mathbb{Z}$, a contradiction. □

Borsuk's nonretraction theorem (Corollary 7.1.38 and Theorem 7.6.5) is derived for any dimension.

Corollary 10.3.30. *For all $n \geq 1$, the sphere S^{n-1} is not a retract of the ball B^n.*

Proof. (1) *First proof.* If S^{n-1} were a retract of B^n, then by Proposition 10.3.16,

$$H_n(B^n) \cong H_n(S^{n-1}) \oplus H_n(B^n, S^{n-1}),$$

for all n. Combining Example 10.3.13, Theorem 10.3.21 and Example 10.3.27, we get

$$0 \cong \mathbb{Z} \oplus \mathbb{Z},$$

for all $n \geq 2$. For $n = 1$, by Proposition 10.2.15 and Proposition 10.2.14, we have $0 \cong 0 \oplus \mathbb{Z}$. In both cases, a contradiction is reached.

(2) *Second proof.* By contradiction, consider a possible retraction $r : B^n \longrightarrow S^{n-1}$. Then $r \circ i = \mathrm{Id}_{S^{n-1}}$, where $i : S^{n-1} \longrightarrow B^n$ is the inclusion map. Hence, the composition map $r_* \circ i_*$ is the identity map. We have the composition

$$H_{n-1}(S^{n-1}) \xrightarrow{i_*} H_{n-1}(B^n) \xrightarrow{r_*} H_{n-1}(S^{n-1}),$$

where $H_{n-1}(S^{n-1}) \cong \mathbb{Z}$ and $H_{n-1}(B^n) \cong \{0\}$, a contradiction. □

More generally, a compact subset of $\mathbb{R}^n$ (not necessarily convex) cannot be retracted onto its boundary if and only if it has nonempty interior (which is the case of a ball). The following result follows from Corollary 10.3.30 and extends it (see [25, 2.3, Chapter XVI], [32, Theorem, p. 95], and [86, Theorem 2.3]).

Proposition 10.3.31. *Let $U \subset \mathbb{R}^n$ be a nonempty bounded open subset. Then ∂U is not a retract of $\overline{U}$.*

Proof. Let $r : \overline{U} \longrightarrow \partial U$ be a possible retraction, $0 \in U$ and $B(0, R)$ a closed ball containing $\overline{U}$. Define the function $f : B(0, R) \longrightarrow B(0, R)$ by

$$f(x) = \begin{cases} \frac{R}{\|r(x)\|} r(x), & x \in \overline{U}, \\ \frac{R}{\|x\|} x, & x \in B \setminus U. \end{cases}$$

Since r is a retraction, $r(x) = x$ for all $x \in \partial U$. Hence, both definitions of f agree on the intersection $\partial U = \overline{U} \cap (B \setminus U)$ and so f is continuous. Finally, if $\|x\| = R$, then $f(x) = x$ by the second definition of f. Therefore, f is a retraction of the closed ball $B(0, R)$ onto its boundary, which contradicts Corollary 10.3.30. □

If in Proposition 10.3.31, U is further convex, then we will prove that $\overline{U}$ is homeomorphic to a closed unit ball (see Theorem 11.2.3).

10.3.5 Local homology groups

Definition 10.3.32. The local homology groups of a space X at a point x are $H_p(X, X \setminus \{x\})$.

Remark 10.3.33. Let $V \in \mathcal{N}_x$ be a closed neighborhood homeomorphic of x. Then there exists an open neighborhood $U \in \mathcal{N}_x$ such that $U \subset V$ and $X \setminus V \subset X \setminus U \subset X \setminus \{x\}$. Hence,

$$\overline{X \setminus V} \subset X \setminus U \subset X \setminus \{x\}.$$

Therefore, excision Axiom 6 applies with $U = X \setminus V$ and $A = X \setminus \{x\}$. We get

$$H_p(X, X \setminus \{x\}) \cong H_p(V, V \setminus \{x\}).$$

This shows that the local homology groups only depend on a local neighborhood of the point x.

Let M_n be an n-manifold (see Definition 3.6.1) and $x \in M_n$.

Proposition 10.3.34.

$$H_p(M_n, M_n \setminus \{x\}) \cong \begin{cases} \mathbb{Z}, & p = n > 0, \\ 0, & \textit{otherwise}. \end{cases}$$

Proof. Let $V \in \mathcal{N}_x$ be such that V is closed and homeomorphic to B^n. Using Proposition 10.3.10 and Remark 10.3.33 together with the homotopy equivalence between $B^n \setminus \{x\}$ and S^{n-1} (see Example 4.2.9), we get

$$\begin{aligned} H_p(M_n, M_n \setminus \{x\}) &\cong H_p(V, V \setminus \{x\}) \\ &\cong H_p(B^n, B^n \setminus \{x\}) \\ &\cong H_p(B^n, S^{n-1}) \\ &\cong \begin{cases} \mathbb{Z}, & p = n > 0, \\ 0, & \text{otherwise}. \end{cases} \end{aligned}$$

□

10.4 The Mayer–Vietoris exact sequence

10.4.1 Mayer–Vietoris theorem

The Mayer–Vietoris theorem helps in determining the homology groups of the union of certain spaces. This theorem plays the same role as the Seifert–van Kampen theorem does in the fundamental group theory (see Section 8.1). The setting is as follows.

Let $A, B \subset X$ be two subsets of a space $X = \mathring{A} \cup \mathring{B}$. Denote by i, j the inclusion maps of $A \cap B$ into A and B, respectively, and let k, l be the inclusion maps of A and B in X, respectively. Define the homomorphisms $\phi : H_n(A \cap B) \longrightarrow H_n(A) \oplus H_n(B)$ and $\psi : H_n(A) \oplus H_n(B) \longrightarrow H_n(X)$ by

$$\phi(x) = (i_*(x), j_*(x)) \quad \text{and} \quad \psi(u, v) = k_*(u) - l_*(v),$$

where f_* is the homomorphism induced by f for $f = i, j, k, l$. We will be content by the statement of the following theorem. The proof can be found in any specialized book in algebraic topology, e. g., [68, Theorem 5.1]).

Theorem 10.4.1. *The following sequence, called the* Mayer–Vietoris sequence, *is exact*

$$\cdots \xrightarrow{\triangle} H_n(A \cap B) \xrightarrow{\phi} H_n(A) \oplus H_n(B) \xrightarrow{\psi} H_n(X) \xrightarrow{\triangle} H_{n-1}(A \cap B) \xrightarrow{\phi} \cdots$$
$$\cdots \longrightarrow H_0(A) \oplus H_0(B) \xrightarrow{\psi} H_0(X) \longrightarrow 0,$$

where $\triangle$ *are homomorphisms.*

Example 10.4.2. Consider a bouquet of two circles $X = S^1 \vee S^1$ with $S^1 \cap S^1 = (0,0)$. Let U, V be the open sets obtained by removing from X the point $(-1,0)$ and the point $(1,0)$, respectively. Since S^1 with one point removed deformation retracts onto one point, say the origin, each of U and V deformation retracts on S^1. The intersection $U \cap V$ is homeomorphic to $(-1,1) \times (-1,1)$. Hence, $U \cap V$ is contractible. By Theorem 10.4.1, the following Mayer–Vietoris sequence is exact

$$\cdots \xrightarrow{\triangle} H_n(U \cap V) \xrightarrow{\phi} H_n(U) \oplus H_n(V) \xrightarrow{\psi} H_n(X)$$
$$\xrightarrow{\triangle} H_{n-1}(U \cap V) \xrightarrow{\phi} \cdots \longrightarrow H_0(U) \oplus H_0(V) \xrightarrow{\psi} H_0(X) \longrightarrow 0.$$

Given the homology groups of the sphere (Theorem 10.3.21), the sequence reads

$$\cdots \xrightarrow{\triangle} 0 \xrightarrow{\triangle} H_2(X) \xrightarrow{\phi} 0 \xrightarrow{\psi} \mathbb{Z} \oplus \mathbb{Z} \xrightarrow{\triangle}$$
$$\xrightarrow{\triangle} H_1(X) \xrightarrow{\phi} \mathbb{Z} \longrightarrow \mathbb{Z} \oplus \mathbb{Z} \xrightarrow{\psi} \mathbb{Z} \longrightarrow 0.$$

Therefore,

$$H_n(S^1 \vee S^1) \cong \begin{cases} \mathbb{Z}, & n = 0, \\ \mathbb{Z} \oplus \mathbb{Z}, & n = 1, \\ 0, & \text{otherwise}. \end{cases}$$

Remark 10.4.3. In particular, when X either is a figure eight or a figure theta, we have as in Corollary 8.2.19,

$$H_1(X) \cong H_1(S^1 \vee S^1) \cong \mathbb{Z} \oplus \mathbb{Z}.$$

Notice that by Theorem 8.2.9, $\pi_1(X) \cong \pi_1(S^1 \vee S^1) \cong \mathbb{Z} * \mathbb{Z}$, where $\mathbb{Z} \oplus \mathbb{Z}$ (also denoted $\mathbb{Z} \times \mathbb{Z}$) is the Abelianization group of $\mathbb{Z} * \mathbb{Z}$ as expected from Theorem 10.2.24.

10.4.2 Computation of some homology groups

In Proposition 10.1.34, we have already computed the homology groups of the torus $\mathbb{T}^2$ considered as a simplicial complex. Let us employ the Mayer–Vietoris theorem to compute once again $H_n(\mathbb{T}^2)$.

Proposition 10.4.4 (Homology groups of the torus (2)). *The homology groups of the torus are*

$$H_n(\mathbb{T}^2) \cong \begin{cases} \mathbb{Z}^2, & n = 1, \\ \mathbb{Z}, & n = 0 \text{ or } n = 2, \\ 0, & \text{otherwise.} \end{cases}$$

Proof. Consider the open sets U, V obtained by removing a different circle in $\mathbb{T}^2$. The obtained surfaces are homeomorphic to two disjoint open cylinders. The latter cylinders are homotopically equivalent to the circle S^1. The intersection $U \cap V$ of the cylinders is homotopically equivalent to the disjoint union of two circles. By Theorem 10.4.1, we have the Mayer–Vietoris exact sequence

$$\cdots \xrightarrow{\triangle} H_n(U \cap V) \xrightarrow{\phi} H_n(S^1) \oplus H_n(S^1) \xrightarrow{\psi} H_n(\mathbb{T}^2) \xrightarrow{\triangle} H_{n-1}(U \cap V) \xrightarrow{\phi} \cdots$$
$$\cdots \longrightarrow H_0(S^1) \oplus H_0(S^1) \xrightarrow{\psi} H_0(\mathbb{T}^2) \longrightarrow 0.$$

We know that $H_0(S^1) \cong H_1(S^1) \cong \mathbb{Z}$ and $H_n(S^1)$ is trivial for $n \geq 2$. Moreover, $H_0(U \cap V) \cong \mathbb{Z} \oplus \mathbb{Z}$ because $U \cap V$ has two path-components (Proposition 10.2.15). Combining Theorem 4.6.1 and Theorem 10.2.24, we have $H_1(U \cap V) \cong \mathbb{Z} \oplus \mathbb{Z}$. Also, $H_0(\mathbb{T}^2) \cong \mathbb{Z}$ because the torus is path-connected. Hence, the homology groups $H_n(\mathbb{T}^2)$ are all trivial for $n \geq 3$ and the Mayer–Vietoris sequence reduces to the short exact sequence

$$0 \longrightarrow H_2(\mathbb{T}^2) \xrightarrow{f} \mathbb{Z} \oplus \mathbb{Z} \xrightarrow{g_1} \mathbb{Z} \oplus \mathbb{Z}$$
$$\xrightarrow{g_2} \mathbb{Z} \oplus \mathbb{Z} \xrightarrow{g_3} \mathbb{Z} \oplus \mathbb{Z} \xrightarrow{g_4} \mathbb{Z} \oplus \mathbb{Z} \xrightarrow{h} \mathbb{Z} \longrightarrow 0,$$

where f, g_i, h are homomorphisms ($1 \leq i \leq 4$). We have $\operatorname{Ker} f = 0$ and $\operatorname{Im} h = \mathbb{Z}$. By the First Homomorphism Theorem B.2.14, $(\mathbb{Z} \oplus \mathbb{Z}) / \operatorname{Ker} h \cong \mathbb{Z}$, which implies that $\operatorname{Ker} h \cong \mathbb{Z}$. Hence, $\operatorname{Im} g_4 \cong \mathbb{Z}$ and $\operatorname{Ker} g_4 \cong \mathbb{Z}$. Continuing the process, we find that $\operatorname{Im} f \cong \operatorname{Ker} g_1 \cong \mathbb{Z}$. As a consequence, $H_2(\mathbb{T}^2)/\operatorname{Ker} f \cong \mathbb{Z}$. Therefore, we recover the computations already performed in Proposition 10.1.34. □

Remark 10.4.5. Combining Theorem 8.2.38 and the Hurewicz Theorem 10.2.24, we obtain the homology groups

$$H_1(\mathbb{T}^2 \# \mathbb{T}^2) \cong \mathbb{Z}^4 \quad \text{and} \quad H_1(\mathbb{RP}^2 \# \mathbb{RP}^2) \cong \mathbb{Z} \times \mathbb{Z}/2.$$

In [102, Theorem 6.4.17], it is proved that for any two oriented compact n-manifolds M_1 and M_2, we have

$$H_k(M_1 \# M_2) \cong \begin{cases} \mathbb{Z}, & k = n, \\ H_k(M_1) \oplus H_k(M_2), & 1 \leq k \leq n-1. \end{cases}$$

In particular, for two oriented compact surfaces S_1 and S_2, we have

$$H_1(S_1\#S_2) \cong H_1(S_1) \oplus H_1(S_2).$$

Such a formula is not valid with the fundamental groups (Theorem 8.2.32 holds only for n-dimensional manifolds with dimensions $n \geq 3$). Using the fundamental group of the torus (Theorem 10.1.34), we can write

$$H_1(\mathbb{T}^2\#\mathbb{T}^2) \cong H_1(\mathbb{T}^2) \oplus H_1(\mathbb{T}^2) \cong \mathbb{Z}^2 \oplus \mathbb{Z}^2 \cong \mathbb{Z}^4,$$

which corroborates Theorem 8.2.38.

Further to the simplicial homology groups computed in Proposition 10.1.36, higher homology groups of the Klein bottle are evaluated using an exact Mayer–Vietoris sequence.

Proposition 10.4.6 (Homology groups of the Klein bottle).

$$H_n(\mathbb{K}^2) \cong \begin{cases} \mathbb{Z}, & n = 0, \\ \mathbb{Z} \oplus \mathbb{Z}/2, & n = 1, \\ 0, & \textit{otherwise}. \end{cases}$$

Proof. By Example 3.8.13 and Corollary 3.8.12, the Klein bottle $\mathbb{K}^2$ may be decomposed as the union of two Möbius bands $\mathbb{M}^2$ whose intersection is homeomorphic to a circle. Then the Mayer–Vietoris exact sequence reads

$$\begin{aligned} H_2(S^1) &\xrightarrow{\phi} H_2(\mathbb{M}^2) \oplus H_2(\mathbb{M}^2) \xrightarrow{\psi} H_2(\mathbb{K}^2) \\ &\xrightarrow{\triangle} H_1(S^1) \xrightarrow{\phi} H_1(\mathbb{M}^2) \oplus H_1(\mathbb{M}^2) \xrightarrow{\psi} H_1(\mathbb{K}^2) \\ &\xrightarrow{\triangle} H_0(S^1) \xrightarrow{\phi} H_0(\mathbb{M}^2) \oplus H_0(\mathbb{M}^2) \xrightarrow{\psi} H_0(\mathbb{K}^2) \longrightarrow 0. \end{aligned}$$

Since the Klein bottle is path-connected, $H_0(\mathbb{K}^2) = Z$. Using Example 10.3.24 and Theorem 10.3.21, we get

$$\begin{aligned} 0 &\xrightarrow{\psi} H_2(\mathbb{K}^2) \xrightarrow{\triangle} \mathbb{Z} \xrightarrow{\phi} \mathbb{Z} \oplus \mathbb{Z} \\ &\xrightarrow{\psi} H_1(\mathbb{K}^2) \xrightarrow{\triangle} \mathbb{Z} \xrightarrow{\phi} \mathbb{Z} \oplus \mathbb{Z} \xrightarrow{\psi} \mathbb{Z} \longrightarrow 0. \end{aligned}$$

The exactness of the sequence $0 = \operatorname{Im} \psi = \operatorname{Ker} \triangle$ at the first line implies that $\triangle$ is injective. By [45, Example 2.47], $\phi(n) = (2n, -2n)$. Hence, ϕ is injective, too. As a consequence, $\operatorname{Im} \triangle = \operatorname{Ker} \phi = 0$. Thus, $\triangle = 0$, $H_2(\mathbb{K}^2) \cong 0$ and so are the higher homology groups. Then the long sequence reduces to the five-term exact sequence of homology groups

$$0 \xrightarrow{\triangle} \mathbb{Z} \xrightarrow{\phi} \mathbb{Z} \oplus \mathbb{Z} \xrightarrow{\psi} H_1(\mathbb{K}^2) \xrightarrow{\triangle} 0.$$

Since $(2n, -2n) \mapsto (0, 2n)$ is bijective, $\phi(\mathbb{Z}) \cong (0, 2\mathbb{Z})$. Using Exercise 2(1) of this chapter, we get

$$H_1(\mathbb{K}^2) \cong (\mathbb{Z} \oplus \mathbb{Z})/\phi(\mathbb{Z}) \cong (\mathbb{Z} \oplus \mathbb{Z})/(0 \oplus 2\mathbb{Z}) \cong (\mathbb{Z}/0) \oplus (\mathbb{Z}/2\mathbb{Z}) \cong \mathbb{Z} \oplus (\mathbb{Z}/2\mathbb{Z}).$$

Then the claim of the proposition follows. □

Proposition 10.4.7. *Let A, B be an open partition of a space X (see Proposition 10.3.6). Then, for every positive integer n,*

$$H_n(X) \cong H_n(A) \oplus H_n(B).$$

Proof. Consider the exact sequence

$$\xrightarrow{\triangle} 0 \xrightarrow{\phi} H_n(A) \oplus H_n(B) \xrightarrow{\psi} H_n(X) \xrightarrow{\triangle} 0$$

and apply Exercise 1(2), this chapter. □

Let A and B be two closed disjoint subsets of S^n and $X = S^n \setminus (A \cup B)$. Since $(S^n \setminus A) \cup (S^n \setminus B) = S^n$ and $(S^n \setminus A) \cap (S^n \setminus B) = X$, we obtain an open covering of the sphere, which is not necessarily a partition. Next, we compute the homology groups of the intersection X.

Proposition 10.4.8. *For every integer $p < n - 1$,*

$$H_p(X) \cong H_p(S^n \setminus A) \oplus H_p(S^n \setminus B).$$

Proof. Consider the Mayer–Vietoris exact sequence

$$\cdots \longrightarrow H_{p+1}(S^n) \xrightarrow{\triangle} H_p(X) \xrightarrow{\phi} H_p(S^n \setminus A) \oplus H_p(S^n \setminus B) \xrightarrow{\psi} H_p(S^n) \longrightarrow \cdots.$$

Since for $p < n - 1$, $H_{p+1}(S^n) \cong H_p(S^n) \cong 0$, Exercise 1(2) of this chapter provides the required isomorphisms. □

As in Theorem 8.2.9, a formula for the homology groups of the wedge sum of two spaces X and Y can be obtained (see, e. g., [10, Chapter IV, Proposition 7.3]).

Proposition 10.4.9. *For all positive n,*

$$H_n(X \vee Y) \cong H_n(X) \oplus H_n(Y).$$

Example 10.4.10. We have

$$H_1(\mathbb{T}^2 \vee \mathbb{T}^2) \cong H_1(\mathbb{T}^2) \oplus H_1(\mathbb{T}^2) \cong \mathbb{Z}^2 \oplus \mathbb{Z}^2 \cong \mathbb{Z}^4.$$

This together with Remark 10.4.5 show that $\mathbb{T}^2 \vee \mathbb{T}^2$ and $\mathbb{T}^2 \# \mathbb{T}^2$ have isomorphic homology groups.

Example 10.4.11. For every n,

$$H_n(S^2 \vee S^1) \cong \begin{cases} \mathbb{Z}, & n = 0, 1, 2, \\ 0, & n \geq 3. \end{cases}$$

Example 10.4.12. For every n,

$$H_n(S^1 \vee S^1 \vee S^2) \cong \begin{cases} \mathbb{Z}, & n = 0, 2, \\ \mathbb{Z} \oplus \mathbb{Z}, & n = 1, \\ 0, & n \geq 3. \end{cases}$$

Notice that these homology groups coincide with those of the torus already calculated in Proposition 10.1.34 and Proposition 10.4.4.

The following proposition is the homology version of the Freudenthal suspension theorem (see Equation (9.4)). Let $\Sigma(X)$ denote the suspension of the space X (see Definition 3.3.1(2)).

Proposition 10.4.13. *We have*

$$H_{n+1}(\Sigma(X)) \cong H_n(X), \quad \text{for } n \geq 1.$$

Proof. The suspension $\Sigma(X)$ is the quotient space obtained by collapsing the slice $X \times \{0\}$ to one point x_0 and the slice $X \times \{1\}$ to another point x_1. Let the open sets $U = \Sigma(X) \setminus \{x_0\}$ and $V = \Sigma(X) \setminus \{x_1\}$. Then $\Sigma(X) = U \cup V$. By definition $U \cong C(X)$ and $V \cong C(X)$, which are contractible because the cone $C(X)$ is contractible (Proposition 4.9.5). Moreover, X is a deformation retract of $U \cap V$. Applying the Mayer–Vietoris theorem, we get the following exact sequence:

$$\cdots \xrightarrow{\triangle} H_{n+1}(X) \xrightarrow{\phi} 0 \xrightarrow{\psi} H_{n+1}(\Sigma(X)) \xrightarrow{\triangle} H_n(X) \xrightarrow{\phi} 0 \cdots,$$

proving the claim of the proposition. □

Remark 10.4.14. Notice that the formula in Proposition 10.4.13 does not allow the computation of the first homology group which however can be obtained by the Hurewicz Theorem 10.2.24. This is illustrated by the following example concerning the homology groups of the sphere in the case $1 \leq p \leq n$ and supporting Theorem 10.3.21.

Theorem 10.4.15 (Homology groups of the sphere (2)). *We have*

$$H_p(S^n) \cong \begin{cases} \mathbb{Z} \oplus \mathbb{Z}, & \text{if } 0 = p = n, \\ \mathbb{Z}, & \text{if } 0 = p < n \text{ or } 1 \leq p = n, \\ 0, & \text{if } 1 \leq p < n. \end{cases}$$

Proof. Since by Theorem 3.3.6, $\Sigma(S^n) = S^{n+1}$, we obtain that $H_0(S^{n+1}) \cong \mathbb{Z}$ and for $1 \leq p \leq n$,

$$\begin{aligned} H_{p+1}(S^{n+1}) &\cong H_p(S^n) \\ &\cong H_{p-1}(S^{n-1}) \\ &\cong \dots \\ &\cong H_1(S^{n+1-p}) \\ &\cong \begin{cases} \mathbb{Z}, & 1 \leq p = n, \\ 0, & 1 \leq p < n. \end{cases} \end{aligned}$$

For the last equivalence, we have appealed to the Hurewicz Theorem 10.2.24 together with Theorem 6.1.4 and Corollary 9.2.1. Finally, by Proposition 10.2.15, $H_0(S^0) \cong H_0(\{-1, 1\}) \cong \mathbb{Z} \oplus \mathbb{Z}$. □

Example 10.4.16. By Example 4.2.6, the cylinder $C = S^n \times [0, 1]$ and the circle the sphere S^n are homotopically equivalent. Proposition 10.3.10 implies that $H_p(C) \cong H_p(S^n)$.

10.4.3 Jordan–Brouwer separation theorems

The homology groups of the complement of the sphere are easy to compute.

Proposition 10.4.17. *We have*

$$H_p(\mathbb{R}^n \setminus S^{n-1}) \cong \begin{cases} \mathbb{Z} \oplus \mathbb{Z} \oplus \mathbb{Z}, & p = 0,\ n = 1, \\ \mathbb{Z} \oplus \mathbb{Z}, & p = 0,\ n \geq 2, \\ \mathbb{Z}, & p = n - 1 > 0 \\ 0, & \textit{otherwise}. \end{cases}$$

Proof. The space $X = \mathbb{R} \setminus S^0$ has three path-components. For $n \geq 2$, the space $X = \mathbb{R}^n \setminus S^{n-1}$ has two path-components, namely the open unit ball and the complement of the closed ball. Thus, the zeroth homology groups are isomorphic to $\mathbb{Z}$ for $n \geq 2$. By Example 1.5.13, Example 1.5.18 and Example 7.2.14, the complement of the closed ball is homotopically equivalent to S^{n-1} whereas the open unit ball is contractible. Then for $p > 0$, we can make use of Proposition 10.4.7 to deduce that X has same homology groups as the sphere S^{n-1}:

$$H_p(\mathbb{R}^n \setminus S^{n-1}) \cong H_p(S^{n-1}), \quad \text{for all } p > 0$$

and the claim follows. □

Remark 10.4.18. Proposition 10.4.17 is the first form of the Jordan–Brouwer separation theorem (see Theorem 10.4.22, too). The general case of $\mathbb{R}^n \setminus S^k$ is Exercise 23. Notice that by Exercise 10 of Section 1.4, $\mathbb{R}^n \setminus S^k$ has a countable number of path-components.

Proposition 10.4.19. *The spaces S^n and S^m are homotopically equivalent if and only if $m = n$.*

Proof. Assume by contradiction that S^n and S^m are homotopically equivalent and $m \neq n$. By Proposition 10.3.10, the nth homology groups $H_n(S^n)$ and $H_n(S^m)$ are isomorphic. By Theorem 10.3.21, the first group is isomorphic to $\mathbb{Z}$ while the second is isomorphic to 0, a contradiction. The converse is obvious. □

We now prove a generalization of Corollary 8.2.20.

Corollary 10.4.20. *The spaces $\mathbb{R}^m$ and $\mathbb{R}^n$ are homeomorphic if and only if $m = n$.*

Proof. We have the equivalences

$$\begin{aligned} \mathbb{R}^m \cong \mathbb{R}^n \quad &\Longleftrightarrow \quad \mathbb{R}^m \setminus \{0\} \cong \mathbb{R}^n \setminus \{0\} \\ &\Longleftrightarrow \quad S^{m-1} \simeq S^{n-1} \quad \text{(by Example 4.2.9)} \\ &\Longleftrightarrow \quad m = n \quad \text{(by Proposition 10.4.19).} \end{aligned}$$

□

Remark 10.4.21. The result still holds for $\mathbb{R}^m$ and $\mathbb{R}^n$ replaced by open sets $U \subset \mathbb{R}^m$ and $V \subset \mathbb{R}^n$, respectively (see Exercise 1, Chapter 12).

Theorem 10.4.22 (Jordan–Brouwer separation theorem). *The sphere S^{n-1} separates S^n into two components having same homology groups as a point. More precisely,*

$$H_p(S^n \setminus S^{n-1}) \cong \begin{cases} \mathbb{Z} \oplus \mathbb{Z}, & p = 0, \\ 0, & p \neq 0. \end{cases}$$

Proof. Since $S^n \setminus S^{n-1}$ is the union of the northern and the southern open hemispheres, by Example 1.5.22 (see also Example 1.4.5), we have $S^n \setminus S^{n-1} \cong \mathring{B}_1^n \cup \mathring{B}_2^n$, the union of two disjoint open n-balls. Then the claim follows from Proposition 10.4.7 and Corollary 10.2.19. □

Since by Corollary 10.2.19, $H_0(X)$ designates the number of path-components, we have the following.

Corollary 10.4.23. *$S^n \setminus S^{n-1}$ have two path-components whose common boundary is S^{n-1}. By Definition 10.2.23, both of components are acyclic.*

More generally, we have the following.

Theorem 10.4.24. *For $0 \leq k < n$,*

$$H_p(S^n \setminus S^k) \cong \begin{cases} \mathbb{Z} \oplus \mathbb{Z}, & 0 = p = n-k-1, \\ \mathbb{Z}, & 0 = p \leq k < n-1, \\ \mathbb{Z}, & 0 < p = n-k-1, \\ 0, & \textit{otherwise}. \end{cases}$$

Proof. Let $X = S^n \setminus S^k$. Then $x = (x_1, x_2, \ldots, x_n, x_{n+1}) \in X$ if and only if $\sum_{i=1}^{i=n+1} x_i^2 = 1$ and, up to a shifting of coordinates, $\sum_{i=1}^{i=k+1} x_i^2 \neq 1$. Hence, $\sum_{i=k+2}^{i=n+1} x_i^2 \neq 0$, i. e., $0 < \sum_{i=k+2}^{i=n+1} x_i^2 \leq 1$, which means that $(x_{k+2}, \ldots, x_{n+1}) \in B^{n-k} \setminus \{0\}$ and $(x_1, \ldots, x_{k+1}) \in \mathbb{R}^{k+1}$. This with Theorem 1.5.38 yields

$$S^n \setminus S^k \cong (B^{n-k} \setminus \{0\}) \cap S^n \cong (S^{n-k-1} \times (0,1]) \cap S^n$$

and the latter space deformation retracts on S^{n-k-1}. Finally, by Remark 4.2.7 and Theorem 10.3.21, we obtain

$$H_p(S^n \setminus S^k) \cong H_p(S^{n-k-1}) \cong \begin{cases} \mathbb{Z} \oplus \mathbb{Z}, & p = n-k-1 = 0, \\ \mathbb{Z}, & p = 0,\ 0 \leq k < n-1, \\ \mathbb{Z}, & 0 < p = n-k-1, \\ 0, & \text{otherwise}. \end{cases}$$ □

Example 10.4.25. Since by Exercise 8 of Chapter 8, the set difference $\mathbb{R}^n \setminus \mathbb{R}^k$ is homotopically equivalent to S^{n-k-1} for $n > k$, it follows from Corollary 10.3.11 that $H_p(\mathbb{R}^n \setminus \mathbb{R}^k) \cong H_p(S^{n-k-1})$. This combined with Theorem 10.4.24 shows that $\mathbb{R}^n \setminus \mathbb{R}^k$ and $S^n \setminus S^k$ have isomorphic homology groups.

In order to present a more general version of Theorem 10.4.24, let us mention the following important lemma. The proof can be found in [10, Corollary 19.3], [72, Chapter 5, Section 3, Theorem 20], [75, Theorem 36.1], [94, Chapter 4, Section 7, Lemma 13] (see also Exercise 25(2), this chapter).

Lemma 10.4.26. *Let B_k be a k-cell for some $0 \leq k \leq n-1$ and $X = S^n \setminus B_k$. Then $H_0(X) \cong \mathbb{Z}$ and $H_p(X) \cong 0$ for all positive p (i. e., X is acyclic). In particular, B_k does not separate S^n.*

The generalized form of the Jordan separation theorem, which extends Theorem 10.4.24, is given by

Theorem 10.4.27. *Let $f : S^k \longrightarrow S^n$ be an embedding and $0 \le k \le n-1$. Then*

$$H_p(S^n \setminus f(S^k)) \cong \begin{cases} \mathbb{Z} \oplus \mathbb{Z}, & 0 = p = n-k-1, \\ \mathbb{Z}, & 0 = p < n-k-1, \\ \mathbb{Z}, & 0 < p = n-k-1, \\ 0, & \textit{otherwise.} \end{cases}$$

This means that as in Theorem 10.4.24, $H_p(S^n \setminus f(S^k)) \cong H_p(S^{n-k-1})$.

Proof. We argue by induction on k. By Remark 6.3.6 and Example 4.2.9, for $k = 0$, we have

$$S^n \setminus \{f(-1), f(1)\} \cong \mathbb{R}^n \setminus \{0\} \simeq S^{n-1}.$$

Assume the formula valid for $k-1$ and suppose first $p > 0$. Let

$$U_+ = S^n \setminus f(S^k_+) \quad \text{and} \quad U_- = S^n \setminus f(S^k_-),$$

where S^k_+ and S^k_- are the northern hemisphere and the southern closed hemisphere of S^k, respectively. By Example 1.5.22, these hemispheres are homeomorphic to the closed ball B^k. Then by Lemma 10.4.26, $H_p(U_+) \cong H_p(U_-) \cong 0$ for all positive p. Since $f(S^k_+) \cup f(S^k_-) = f(S^k)$ and $f(S^k_+) \cap f(S^k_-) = f(S^{k-1})$, we have

$$U_+ \cap U_- = S^n \setminus f(S^k) \quad \text{and} \quad U_+ \cup U_- = S^n \setminus f(S^{k-1}).$$

By Theorem 10.4.1, the following Mayer–Vietoris sequence is exact:

$$\begin{aligned} \cdots &\xrightarrow{\triangle} H_{p+1}(S^n \setminus f(S^k)) \xrightarrow{\phi} H_{p+1}(U_+) \oplus H_{p+1}(U_-) \xrightarrow{\psi} H_{p+1}(S^n \setminus f(S^{k-1})) \\ &\xrightarrow{\triangle} H_p(S^n \setminus f(S^k)) \xrightarrow{\phi} H_p(U_+) \oplus H_p(U_-) \xrightarrow{\phi} \cdots \longrightarrow H_0(U_+) \oplus H_0(U_-) \\ &\xrightarrow{\psi} H_0(S^n \setminus f(S^{k-1})) \longrightarrow 0. \end{aligned}$$

The exactness of the short sequence

$$0 \xrightarrow{\psi} H_{p+1}(S^n \setminus f(S^{k-1})) \xrightarrow{\triangle} H_p(S^n \setminus f(S^k)) \xrightarrow{\phi} 0$$

both with Exercise 1 implies that $H_{p+1}(S^n \setminus f(S^{k-1}))$ and $H_p(S^n \setminus f(S^k))$ are isomorphic. Since, by assumption

$$H_{p+1}(S^n \setminus f(S^{k-1})) \cong \begin{cases} \mathbb{Z}, & p+1 = n-k, \\ 0, & \text{otherwise,} \end{cases}$$

the formula holds for k. We now examine the case $p = 0$ and argue again by induction. The exact Mayer–Vietoris sequence takes the form

$$0 \xrightarrow{\psi} H_1(S^n \setminus f(S^{k-1})) \xrightarrow{\triangle} H_0(S^n \setminus f(S^k)) \xrightarrow{\phi} \mathbb{Z} \oplus \mathbb{Z} \xrightarrow{\psi} H_0(S^n \setminus f(S^{k-1})) \longrightarrow 0.$$

By the inductive hypothesis, $H_0(S^n \setminus f(S^{k-1})) \cong \mathbb{Z}$. The discussion of the case $p > 0$ yields

$$H_1(S^n \setminus f(S^{k-1})) = \begin{cases} \mathbb{Z}, & k = n-1, \\ 0, & k < n-1. \end{cases}$$

So, we distinguish two cases:

(1) If $k < n - 1$, we have a short exact sequence

$$0 \xrightarrow{\triangle} H_0(S^n \setminus f(S^k)) \xrightarrow{\phi} \mathbb{Z} \oplus \mathbb{Z} \xrightarrow{\psi} \mathbb{Z} \longrightarrow 0.$$

Hence, $H_0(S^n \setminus f(S^k)) \cong \mathbb{Z}$ follows from First Theorem of Homomorphism B.2.14.

(2) If $k = n - 1$, we have

$$\mathbb{Z} \xrightarrow{\triangle} H_0(S^n \setminus f(S^k)) \xrightarrow{\phi} \mathbb{Z} \oplus \mathbb{Z} \xrightarrow{\psi} \mathbb{Z} \longrightarrow 0.$$

Again, by Theorem B.2.14, $H_0(S^n \setminus f(S^k)) \cong \mathbb{Z} \oplus \mathbb{Z}$. □

Notice that the Jordan curve separation theorem takes the following form, extending Corollary 10.4.23.

Corollary 10.4.28. *Let $f : S^{n-1} \longrightarrow S^n$ be an embedding. Then the space $S^n \setminus f(S^{n-1})$ has two path-components and $f(S^{n-1})$ is their common topological boundary. Both of components are acyclic.*

Proof. Let $X = S^n \setminus f(S^{n-1})$. By Theorem 10.4.27,

$$H_p(X) \cong \begin{cases} \mathbb{Z} \oplus \mathbb{Z}, & p = 0, \\ 0, & p > 0. \end{cases}$$

The case $p = 0$ shows that X has two components C_1 and C_2 (see Remark 10.2.20). Since S^n is locally connected and X is open in S^n, by Corollary 1.4.55, C_1 and C_2 are path-connected and clopen in X. Hence, C_1 and C_2 are open in S^n. Let $\partial C_i = \overline{C_i} \setminus C_i$ be the boundary of each C_i ($i = 1, 2$). Since $C_1 \cap C_2 = \emptyset$ and C_2 is open, by Exercise 8, Section 1.1, $\overline{C_1} \cap C_2 = \emptyset$. Let $x \in \overline{C_1} \setminus C_1$. Then $x \notin (C_1 \cup C_2)$. Since $S^n = C_1 \cup C_2 \cup f(s^{n-1})$, we obtain that $x \in f(S^{n-1})$, proving that $\partial C_1 \subset f(S^{n-1})$. Also, $\partial C_2 \subset f(S^{n-1})$.

It remains to show that $f(S^{n-1}) \subset \partial C_i$, for $i = 1, 2$. We follow the proof of [10, Corollary 19.7]. Let $x \in f(S^{n-1})$, $x = f(p)$, $p \in S^{n-1}$. By contradiction, let U be a neighborhood of x in S^n such that for instance $U \cap C_1 = \emptyset$. Let $B_\varepsilon = B(p, \varepsilon)$ be a small open ball in S^{n-1}

such that $f(B_\varepsilon) \subset U$. Since $S^{n-1} \setminus B_\varepsilon \cong B^{n-1}$, by Lemma 10.4.26, $X = S^n \setminus f(S^{n-1} \setminus B_\varepsilon)$ is open and connected. But $X = C_1 \cup C_2 \cup f(B_\varepsilon) \subset C_1 \cup C_2 \cup U$ and $C_1 \cap (C_2 \cup U) = \emptyset$. Then $X = (C_1 \cap X) \cup ((C_2 \cup U) \cap X)$ is a disjoint union of two nonempty open sets, a contradiction. □

Finally, consider a knot complement $X = S^3 \setminus f(S^1)$ (see Definition 7.2.25).

Corollary 10.4.29. *Let $f : S^1 \longrightarrow S^3$ be an embedding. We have*

$$H_p(S^3 \setminus f(S^1)) \cong \begin{cases} \mathbb{Z}, & p = 0, 1, \\ 0, & p \neq 0, 1. \end{cases}$$

10.5 Euler and Betti numbers

Given a simplicial complex K (Section 10.1), denote by $N_j(K)$ the number of j-simplices in K ($j \geq 0$).

Definition 10.5.1. The *Euler (or characteristic) number* of K is defined as

$$\chi(K) = \sum_{j\geq 0}(-1)^j N_j(K) = \sum_{j\geq 0}(-1)^j \dim C_j(K).$$

Example 10.5.2. In Figure 10.3, the Euler number for the triangle is $\chi = 3 - 3 = 0$ and it is $\chi = 4 - 5 + 2 = 1$ for the square.

Proposition 10.5.3. *We have*

$$\chi(K) = \sum_{j\geq 0}(-1)^j \dim H_j(K).$$

Proof. Since $H_j = \operatorname{Ker}\partial_j / \dim \operatorname{Im} \partial_{j+1}$, we have

$$\dim H_j(K) = \dim \operatorname{Ker}\partial_j - \dim \operatorname{Im}\partial_{j+1} = \dim C_j - \dim \operatorname{Im}\partial_j - \dim \operatorname{Im}\partial_{j+1},$$

$\operatorname{Im}\partial_0 = \{0\}$, and $\sum_{j\geq 0}(-1)^j(\operatorname{Im}\partial_j + \operatorname{Im}\partial_{j+1}) = 0$. Then the summation over j yields

$$\sum_{j\geq 0}(-1)^j \dim H_j(K) = \sum_{j\geq 0}(-1)^j \dim C_j(K).$$ □

Example 10.5.4. By Example 10.3.13 and Theorem 10.3.21, we get

$$\chi(B^n) = 1 \quad \text{and} \quad \chi(S^n) = 1 + (-1)^n.$$

Corollary 10.5.5. *Homotopy equivalent simplicial complexes have the same Euler characteristic.*

Example 10.5.6. We have already computed the homology groups of the most basic surfaces: the sphere S^2, the cylinder $S^1 \times [0,1]$, the torus $\mathbb{T}^2$, the projective plane $\mathbb{RP}^2$, the Möbius band $\mathbb{M}^2$ and the Klein bottle $\mathbb{K}^2$. These surfaces can be represented by 0, 1 or 2 simplices. As a consequence, their characteristic numbers are given by

$$\begin{aligned} \chi(S^2) &= 1-0+1 = 2, \quad & \chi(S^1 \times (0,1)) &= 1-0+1 = 2, \\ \chi(\mathbb{T}^2) &= 1-2+1 = 0, \quad & \chi(\mathbb{RP}^2) &= 1-0+0 = 1, \\ \chi(\mathbb{M}^2) &= 1-1+0 = 0, \quad & \chi(\mathbb{K}^2) &= 1-1+0 = 0. \end{aligned}$$

Definition 10.5.7. The rank (or dimension) of the group $H_n(X)$ $(n \geq 0)$ is called the *nth Betti number* of X and is denoted $\beta_n(X)$ or β_n for brevity.

Proposition 10.5.3 provides a relation between Euler and Betti numbers, namely

$$\chi(K) = \sum_{n \geq 0} (-1)^n \beta_n(K).$$

Remark 10.5.8. (1) By Remark 10.2.20, $\beta_0(X)$ equals the number of path-components of X.
(2) Since $\dim H_n(S^n) = 1$ and $\dim H_k(S^n) = 0$, for $k \neq n$, by Theorem 10.3.21 one may regard $\dim H_n(X)$ as the number of n-dimensional holes in a space X $(n \geq 1)$ (see [94, p. 155]).

Example 10.5.9. (1) A simply connected space has one component and no holes. In particular, this holds for a contractible space (see Corollary 4.9.15 and Corollary 10.3.12).
(2) Concerning the circle $X = S^1$, the first Betti numbers are $\beta_0 = 1$ (one path-component), $\beta_1 = 1$ (one hole) and $\beta_2 = 0$ (no cavity). The same holds for the cylinder $S^1 \times [0,1]$.
(3) The situation is similar for the Möbius band $X = \mathbb{M}^2$ (see Example 10.3.24) because $\beta_0 = \beta_1 = 1$ and $\beta_n \neq 0$, for $n \geq 2$.
(4) As for the torus $X = \mathbb{T}^2$, $\beta_0 = 1$ (one path-component), $\beta_1 = 2$ (two holes surrounded by the small circle and the big circle, respectively) and $\beta_2 = 1$ (one cavity).
(5) The Klein bottle $X = \mathbb{K}^2$ has one path-component (see Proposition 10.4.6), one hole $(\dim H_1 = \dim(\mathbb{Z} \oplus \mathbb{Z}/2\mathbb{Z}) = 1)$, and no cavity.
As for the rank of the groups $\mathbb{Z}$ and $\mathbb{Z}/2\mathbb{Z}$, see Example B.7.5 and Example B.7.6.

Similar to the table of fundamental groups set up at end of Chapter 8, we collect some of the homology groups computed in this chapter. The last three homology groups are drawn from Exercise 21.

Topological space	Nontrivial homology group	Trivial homology group
X	$H_p(X)$	$\{0\}$
acyclic	$\mathbb{Z}, p = 0$	$p \geq 1$
S^n	$\mathbb{Z} \oplus \mathbb{Z}\ (p = n = 0)$, $\mathbb{Z}\ (p = 0, n > 0 \text{ or } p = n > 0)$	otherwise
$\mathbb{R}^n \setminus \{0\}$	$\mathbb{Z}\ (p = 0, p = n - 1)$	otherwise
$S^3 \setminus S^1, \mathbb{R}^3 \setminus \{(0z)\}$	$\mathbb{Z}\ (p = 0, 1)$	$p \geq 2$
$\mathbb{RP}^n$	$\mathbb{Z}\ (p = 0 \text{ or } p = n \text{ odd})$, $\mathbb{Z}_2\ (p \text{ odd and } 0 < p < n)$	otherwise
$\mathbb{CP}^n$	$\mathbb{Z}\ (p \text{ even and } 0 \leq p \leq 2n)$	otherwise
$\mathbb{T}^2$	$\mathbb{Z}^2\ (p = 1), \mathbb{Z}\ (p = 0 \text{ or } p = 2)$	$p \geq 3$
$\mathbb{K}^2$	$\mathbb{Z}\ (p = 0), \mathbb{Z} \oplus \mathbb{Z}/2\ (p = 1)$	$p \geq 2$
$S^1, \mathbb{M}^2$	$\mathbb{Z}\ (p = 0, 1)$	$p \geq 2$
$S^1 \vee S^1$	$\mathbb{Z}\ (p = 0), \mathbb{Z} \oplus \mathbb{Z}\ (p = 1)$	$p \geq 2$
$S^2 \vee S^1$	$\mathbb{Z}\ (p = 0, 1, 2)$	$p \geq 3$
$S^1 \vee S^1 \vee S^2$	$\mathbb{Z}\ (p = 0, 2), \mathbb{Z} \oplus \mathbb{Z}\ (p = 1)$	$p \geq 3$
$S^n \vee S^m$	$H_p(S^n) \oplus H_p(S^m)$	
$M_1^n \# M_2^n$	$\mathbb{Z}\ (p = n)$, $H_p(M_1) \oplus H_p(M_2)$ (for $1 \leq p \leq n - 1$)	
$\mathbb{T}^2 \# \mathbb{T}^2$	$\mathbb{Z}^4\ (p = 1)$	
$\mathbb{RP}^n \# \mathbb{RP}^n$	$\mathbb{Z} \times \mathbb{Z}/2\ (p = 1)$	
$\mathbb{R}^n \setminus S^{n-1}$	$\mathbb{Z} \oplus \mathbb{Z} \oplus \mathbb{Z}\ (p = 0, n = 1)$, $\mathbb{Z} \oplus \mathbb{Z}\ (p = 0, n \geq 2)$, $\mathbb{Z}\ (p = n - 1 > 0)$	otherwise
$S^n \setminus S^{n-1}$	$\mathbb{Z} \oplus \mathbb{Z}\ (p = 0)$	$p \neq 0$
$S^n \setminus S^k$	$\mathbb{Z} \oplus \mathbb{Z}\ (0 = p = n - k - 1)$ $\mathbb{Z}\ (0 = p \leq k < n - 1$ or $0 < p = n - k - 1)$	
$X \times S^n$	$H_p(X) \oplus H_{p-n}(X)$	
$S^n \times S^n$	$\mathbb{Z} \oplus \mathbb{Z}\ (p = n)$	
$S^m \times S^n, m \neq n$	$\mathbb{Z}\ (p = n \text{ or } p = m)$	

Exercises (Chapter 10)

1. (1) Consider the exact sequence of groups

$$0 \xrightarrow{f} A \xrightarrow{g} 0.$$

Prove that $A \cong 0$.

(2) Consider the exact sequence of groups

$$0 \xrightarrow{f} A \xrightarrow{g} B \xrightarrow{h} 0.$$

Prove that A and B are isomorphic.

2. (1) Consider the five-term exact sequence of groups

$$0 \longrightarrow A \xrightarrow{f} B \xrightarrow{g} C \longrightarrow 0.$$

(a) Prove that $C \cong B/f(A)$.
(b) Suppose that A is Abelian and g is right invertible. Prove that

$$A \cong f(A) \quad \text{and} \quad B \cong A \oplus C.$$

(2) Consider the exact sequence of groups

$$A \xrightarrow{f} B \xrightarrow{g} C \xrightarrow{h} D.$$

Prove that f is surjective if and only if h is injective.

3. Consider the exact sequence of homology groups of a space X,

$$0 \xrightarrow{\partial_{n+1}} H_n(X) \xrightarrow{\partial_n} H_{n-1}(X) \xrightarrow{\partial_{n-1}} \cdots \xrightarrow{\partial_0} 0.$$

Prove that

$$\sum_{k=1}^{n} (-1)^n \beta_k(X) = 0,$$

where $\beta_k(X)$ is the kth Betti number of X.

4. Determine the homology groups of the 2-simplicial complex K in Example 10.1.15.

5. Prove Lemma 10.2.12.

6. Let $A \subset X$ be a subset of a space X, $i : A \longrightarrow X$ the inclusion map and $r : X \longrightarrow A$ a retraction. Prove that for every $n \geq 0$:
(1) $r_* : H_n(X) \longrightarrow H_n(A)$ is a epimorphism;
(2) $i_* : H_n(A) \longrightarrow H_n(X)$ is a monomorphism.

7. Show that for every x_0, y_0 in a space X, the pairs (X, x_0) and (X, y_0) are homotopically equivalent.

8. (1) Assume that a space X is deformable onto a subspace $A \subset X$, that is to say there exists a homotopy $H : X \times [0,1] \longrightarrow X$ such that $H(\cdot, 0)$ is the identity and

$H(\cdot, 1) \subset A$ (this is the definition of a strong deformation without Condition 3 in Definition 7.2.10). Prove that for all integer n,

$$H_n(A) \cong H_n(X) \oplus H_{n+1}(X, A).$$

(2) Let X be contractible and $A \subset X$ a subspace. Deduce the homology groups $H_n(A)$ in terms of $H_{n+1}(X, A)$.

9. Let $f : X \longrightarrow Y$ be a continuous map homotopic to a constant. Prove that the induced homomorphism is trivial.

10. Let (X, A) be a pair such that X and A are path-connected. Prove that $H_0(X, A) = 0$.

11. Let (X, A) be a pair such that the induced homomorphism $i_* : H_n(A) \longrightarrow H_n(X)$ is an isomorphism. Determine the homology groups $H_n(X, A)$, $n \geq 0$. (*Hint:* See [19, Section 6.4, Exercise 4].)

12. Let (X, A, B) a topological triple $(B \subset A \subset X)$.
(1) Prove the existence of a long exact sequence

$$\cdots \longrightarrow H_n(A, B) \longrightarrow H_n(X, B) \longrightarrow H_n(X, A) \longrightarrow H_{n-1}(A, B) \longrightarrow \cdots$$

(2) Suppose that A is a retract of X. Prove that for all integer n,

$$H_n(X, B) \cong H_n(X, A) \oplus H_n(A, B).$$

13. Let M_n be an n-dimensional manifold and $\{x_1, x_2, \dots, x_k\}$ k mutually distinct points in M_n. Find the relative homology groups $H_p(M_n, M_n \setminus \{x_1, x_2, \dots, x_k\})$.

14. Let $A \subset X$ be contractible in X. Prove that for all $n \geq 1$,

$$H_n(X, A) \cong H_n(X) \oplus H_{n-1}(A).$$

(*Hint:* Use Exercise 21, Chapter 7 and see [45, Exercise 26, p. 157].) Compare with Corollary 10.3.28.

15. (1) For $p, n = 0, 1, \dots$ and $q = 1, 2, \dots$, compute the homology groups

$$H_p(\overbrace{S^n \vee S^n \vee \cdots \vee S^n}^{q}).$$

(2) Find the homology groups of S^n/S^k for $k < n$. (*Hint:* Use Example 8.2.15.)

16. (1) (a) Find the homology groups of the subspace

$$X = \{(x, y) \in \mathbb{R}^2 \setminus \{(0, 0)\} : xy = 0\}.$$

(b) Deduce that X is not homotopically equivalent to the Euclidean plane.

(2) Consider the subspace of Euclidean space $\mathbb{R}^3$:

$$Y = \{(x,y,z) \in \mathbb{R}^3 \setminus \{(0,0,0)\} : xy = 0\}.$$

(a) Using a Mayer–Vietoris sequence, prove that

$$H_1(Y) \cong \mathbb{Z} \oplus \mathbb{Z} \oplus \mathbb{Z}.$$

(b) Show that $Y' = Y \cup \{(0,0,0)\}$ deformation retracts on Euclidean plane $\mathbb{R}^2$ but Y' is not homeomorphic to $\mathbb{R}^2$.

(3) Find the first homology group of the space

$$Z = \{(x,y,z) \in \mathbb{R}^3 \setminus \{(0,0,0)\} : xyz = 0\}.$$

17. Find the homology groups of the space $X = S^2 \cup \{Oz\}$ in $\mathbb{R}^3$.

18. Find the relative homology groups $H_n(X,A)$, where $X = S^2$ and $A = \{x_1, \dots, x_N\} \subset X$.

19. Using a Mayer–Vietoris sequence, compute the homology groups of the punctured torus $\mathbb{T}'^2 = \mathbb{T}^2 \setminus \{*\}$. (*Hint:* The study of $\mathbb{K}^2$ in Proposition 10.4.6 may be helpful. Another possibility offered is to appeal to Proposition 7.2.35 and Example 10.4.2.)

20. Let A, B be two closed subsets S^n and $X = S^n \setminus (A \vee B)$. Write a formula for the homology groups $H_p(X)$. (*Hint:* Argue as in Proposition 10.4.8.)

21. Let p, n be positive integers, X a space and $x_0 \in S^n$.

(1) Prove that:

(a) $H_p(X \times S^n) \cong H_p(X) \oplus H_p(X \times S^n, X \times \{x_0\})$;

(b) $H_p(X \times S^n, X \times \{x_0\}) \cong H_{p-1}(X \times S^{n-1}, X \times \{x_0\})$;

(c) $H_p(X \times S^n) \cong H_p(X) \oplus H_{p-n}(X)$. (*Hint:* [45, Exercise 36, Section 2.2]).

(2) Compute the homology groups $H_p(S^m \times S^n)$.

22. (1) Let $C(X)$ be the cone over a space X and $x \in X$. Prove that

$$H_n(C(X), C(X) \setminus \{x\}) \cong H_{n-1}(X).$$

(2) Determine the local homology groups $H_n(B^n, B^n \setminus \{x\})$, for $x \in B^n$.

23. (1) Find a space homotopically equivalent to $X = \mathbb{R}^n \setminus S^k$, for $n > k + 1$.

(2) Deduce the homology groups of X.

24. Suppose $1 \le k \le n$.

(1) Find a space homeomorphic to $S^n \setminus \mathbb{R}^k$.
(*Hint:* Notice that $\mathbb{R}^k \times \mathbb{R}^{n-k-1} \setminus \{0\} \cong \mathbb{R}^{n+1} \setminus \mathbb{R}^k$).

(2) Deduce the homology groups of space $S^n \setminus \mathbb{R}^k$.

25. Suppose $1 \leq k \leq n$.

(1) (a) Prove that the nonzeroth homology groups of $S^n \setminus \mathring{B}^n$ are trivial. (*Hint:* Use Remark 3.7.8.)

(b) Determine the homology groups of $S^n \setminus \mathring{B}^k$.

(2) (a) Show that for all $p > 0$, $H_p(S^n \setminus B^k) \cong 0$. (*Hint:* See [72, Chapter 5, Section 3, Theorem 20] or [94, Chapter 4, Section 7, Lemma 13].)

(b) Deduce the homology groups of $S^n \setminus S^k$ and compare with Theorem 10.4.24. (*Hint:* See [72, Chapter 5, Section 3, Theorem 21] or [94, Chapter 4, Section 7, Corollary 14].)

26. Let $f : S^n \longrightarrow \mathbb{R}^{n+1}$ be a continuous function such that $f(x) = x_0$ for all x on the equator $\mathrm{Eq} = S^n \cap \mathbb{R}^n$, where $x_0 \in \mathbb{R}^{n+1}$ and f is injective on $S^n \setminus \mathrm{Eq}$.

(1) (a) Describe the set $f(S^n)$. (*Hint:* Use Example 2.5.4.)

(b) How many components does the space $X = \mathbb{R}^{n+1} \setminus f(S^n)$ have?

(2) Compute the homology groups of X.

27. Let M_1 and M_2 be two manifolds of dimension m and n, respectively. Prove that M_1 and M_2 are hoemomorphic if and only if $m = n$.

28. Suppose that a finite simplicial complex K is covered by two subcomplexes K_1 and K_2. Prove that the characteristic numbers satisfy

$$\chi(K) = \chi(K_1) + \chi(K_2) - \chi(K_1 \cap K_2).$$

29. (1) Let K_1 and K_2 be two compact surfaces. Prove that

$$\chi(K_1 \# K_2) = \chi(K_1) + \chi(K_2) - 2.$$

(2) Deduce the characteristic number of n tori and n projective planes. (*Hint:* See [67, Proposition 8.1].)

11 Fixed-point theorems

Brouwer's fixed-point theorem for the disk B^2 was proved in Section 7.6 as application of the fundamental group of the circle. At that stage, we did not know about the higher homotopy groups of spheres. In this chapter, we take up Brouwer's fixed-point theorem for any dimension as well as its extension to infinite-dimension spaces, namely Schauder's fixed-point theorem. Brouwer's fixed-point theorem is proved in Section 11.1 using the nonretraction of the ball B^n in $\mathbb{R}^n$. This is one of different ways to prove this theorem. Some results on the surjectivity of continuous functions are established in this section, too. Historically, Brouwer's fixed-point theorem was presented by Jacques Salomon Hadamard [43] in 1910 and by Brouwer [15] in 1912. In Section 11.2, Brouwer's fixed-point theorem is proved for any compact convex subset of $\mathbb{R}^n$ with nonempty interior instead of the closed unit ball. Various equivalent formulations are given in Section 11.3: the Hartmann–Stampacchia theorem, the Knaster–Kuratowski–Mazurkiewicz theorem and the mini-max Ky Fan theorem. Actually, these versions are different approaches to deal with this theorem.

Regarding the infinite dimension, Schauder's fixed-point theorem is proved in Section 11.4. The proof goes through up to some approximation techniques. The retraction and the contractibility of the sphere are established for the infinite-dimensional spaces, which is a major difference with the finite-dimensional case. Some versions of Schauder's fixed-point theorem are proposed with proofs, including a nonlinear alternative and Rothe–Schaefer fixed-point theorem.

This chapter closes with Mönch's fixed-point theorem (Section 11.5) and Lefschetz fixed-point theorem (Section 11.6).

11.1 Brouwer's fixed-point theorem (the finite dimension)

The proof of Corollary 7.1.38 using the fundamental group does not go through in higher dimensions. Indeed, by Corollary 6.3.4, we know that for $n > 2$, S^{n-1} is simply connected. Then both B^n and S^{n-1} have the trivial fundamental group for all $n > 2$. However, the result of Borsuk's non-retraction of the sphere was extended to higher dimensions $n > 2$ in Corollary 10.3.30 by making use of the relative homology theory. As a consequence, we obtain the following.

Theorem 11.1.1 (Brouwer's fixed-point theorem). *Let $f : B^n \longrightarrow B^n$ be a continuous map ($n \geq 1$). Then f has at least one fixed point.*

Proof. The method of proof used in the case of the closed unit disc B^2 still carries over when $n > 2$. A contradiction is reached with Corollary 10.3.30 by making use of a possible retraction $r : B^n \longrightarrow S^{n-1}$. The case $n = 1$ follows from the intermediate value theorem on a closed bounded interval of the real line. □

https://doi.org/10.1515/9783111517384-011

Remark 11.1.2. As mentioned in Remark 7.6.4, Corollary 10.3.30 and Theorem 11.1.1 are equivalent. The proof is the same as for Theorem 7.6.2 and is omitted.

Remark 11.1.3. Unlike Banach's fixed-point theorem for contraction mappings, Brouwer's fixed- point theorem does not provide uniqueness of solution as the identity map shows. Next is a further example of a nonlinear map with infinitely many fixed points.

Example 11.1.4. Define the function $f : B^n \longrightarrow B^n$ by

$$f(x) = \begin{cases} \sin(2\pi\|x\|)x/\|x\|, & \text{if } x \neq 0, \\ 0 & \text{if } x = 0. \end{cases}$$

Then f is continuous and if $t_0 = \sin(2\pi t_0)$, then $1/4 < t_0 < 1/2$ and

$$\|x\| = t_0 \quad \Longrightarrow \quad f(x) = x.$$

So, all points on the circle $\|x\| = t_0$ are self-mapped.

A third equivalent version of Brouwer's fixed-point theorem is given by the following.

Theorem 11.1.5 (Noncontractibility of the sphere). *For all $n > 1$, the sphere S^{n-1} is not contractible (although S^{n-1} is simply connected for $n > 2$).*

Proof. First proof. If S^{n-1} were contractible, there would exist some $x_0 \in S^{n-1}$ such that $\mathrm{Id}_{S^{n-1}} \simeq x_0$. Then there exists a continuous map (homotopy) $H : S^{n-1} \times I \longrightarrow S^{n-1}$ such that $H(x, 0) = x_0$ and $H(x, 1) = x$ for all $x \in S^{n-1}$. This allows us to define a map $r : B^n \longrightarrow S^{n-1}$ by

$$r(x) = \begin{cases} x_0, & 0 \leq \|x\| \leq 1/2, \\ H(x/\|x\|, 2\|x\| - 1), & 1/2 \leq \|x\| \leq 1. \end{cases}$$

Since for $\|x\| = \frac{1}{2}$, $H(x/\|x\|, 2\|x\|) = H(2x, 0) = x_0$, by the pasting lemma (Lemma A.2.1), r is continuous. For every $x \in S^{n-1}$, we have $\|x\| = 1$. Then $r(x) = H(x, 1) = x$. Therefore, r is a retraction of B^n onto S^{n-1}, violating Corollary 10.3.30.

Second proof. Since (B^n, S^{n-1}) is a Borsuk pair, if S^{n-1} were contractible, then by Theorem 7.3.10, $B^n/S^{n-1} \cong S^n$ would be homotopically equivalent to B^n. By Corollary 9.1.30, their higher fundamental groups π_k should be isomorphic for each k. However, by Corollary 9.2.13, $\pi_n(S^n) \cong \mathbb{Z}$ for all $n \geq 1$ whereas by Example 9.1.32, $\pi_k(B^n)$ is trivial for all $k, n \geq 1$. □

Example 11.1.6. Since the punctured space $\mathbb{R}^n \setminus \{0\}$ strongly deformation retracts onto S^{n-1}, by Theorem 11.1.5, for all $n \geq 2$, $\mathbb{R}^n \setminus \{0\}$ is not contractible.

In Proposition 7.2.16, we have shown that the sphere $S^n \cong S^n \times \{0\}$ is a strong deformation retract of the bounded cylinder $S^n \times I$. However, we have the following.

Corollary 11.1.7. *There is no retraction of the cylinder $S^n \times I$ on its basis $S^n \times \{0\}$, which maps the top of the cylinder $S^n \times \{1\}$ to a point x_0 of the basis.*

Proof. If not, let $r : S^n \times I \longrightarrow S^n \times \{0\} \subset S^n \times I$ be such a retraction and $H(x,t) = r(x,t)$. Then $H(x,0) = x$ and $H(x,1) = x_0$ for all $x \in S^n$. Hence, the identity Id_{S^n} is homotopic to the constant map x_0, which violates Corollary 11.1.5. □

Actually, Theorem 11.1.5 is equivalent to Corollary 10.3.30.

Theorem 11.1.8. *Theorem* 11.1.5 *implies Corollary* 10.3.30.

Proof. Otherwise, let $r : B^n \longrightarrow S^{n-1}$ be a retraction. For some $x_0 \in S^{n-1}$, consider the map $H(x,t) = (r \circ s)(x,t) : S^{n-1} \times I \xrightarrow{s} B^n \xrightarrow{r} S^{n-1}$, where $s(x,t) = tx + (1-t)x_0$. H is well-defined because

$$\|tx + (1-t)x_0)\| \leq t\|x\| + (1-t)\|x_0\| = 1, \quad \text{for all } (x,t) \in S^{n-1} \times I.$$

The map H is continuous as the composite of two continuous functions. Furthermore, it satisfies $H(x,1) = r(x) = x$ and $H(x,0) = r(x_0) = x_0$ for all $x \in S^{n-1}$. Hence, H is a homotopy between $\mathrm{Id}_{S^{n-1}}$ and x_0, i. e., S^{n-1} is contractible, contradicting Theorem 11.1.5. □

Remark 11.1.9. A quick way to prove the equivalence between Theorem 11.1.5 and Corollary 10.3.30 is to appeal to Theorem 3.3.6 and Theorem 7.1.25. Actually S^{n-1} is contractible if and only if S^{n-1} is a retract of the cone $C(S^{n-1}) = B^n$.

Remark 11.1.10. A third way to prove the equivalence between Theorem 11.1.5 and Corollary 10.3.30 is to apply both Theorem 6.4.1 and Theorem 7.4.6. For this purpose, let $f = \mathrm{Id}_{|S^{n-1}} : A = S^{n-1} \subset B^n = X \longrightarrow S^{n-1} = Y$ be the identity map. Then $\pi_{|Y}(Y) = \pi_{|S^{n-1}}(S^{n-1}) = S^{n-1}$ and the adjunction space is $Z_f = B^n \cup_{\mathrm{Id}} S^{n-1} = B^n$ (see Example 3.4.15). Then $\pi : X \cup Y = B^n \longrightarrow Z_f = B^n$ is the identity on B^n and $g = \pi_{|Y} \circ f$ is the identity on S^{n-1}. By Theorem 7.4.6,

$$\begin{aligned} S^{n-1} \text{ is a retract of } B^n &\iff g \text{ has a continuous extension } \tilde{g} : B^n \longrightarrow S^{n-1} \\ &\iff \mathrm{Id}_{|S^{n-1}} \text{ is null-homotopic (Theorem 6.4.1)} \\ &\iff S^{n-1} \text{ is contractible (Definition 4.9.1).} \end{aligned}$$

Theorem 11.1.11 (Böhl's theorem). *Let $f : B^n \to \mathbb{R}^n$ be a continuous function satisfying the boundary condition*

$$x \neq \lambda f(x), \quad \text{for all } x \in S^{n-1} \text{ and } \lambda \in (0,1). \tag{11.1}$$

Then f has a fixed point $x \in B^n$.

Proof. Let $\tilde{r} : \mathbb{R}^n \longrightarrow B^n$ be the ball retraction (see Example 7.1.4) and $g = \tilde{r} \circ f$ the composite of f and $\tilde{r}$:

$$\begin{array}{ccc} B^n & \xrightarrow{f} & \mathbb{R}^n \\ & \searrow^{g} & \downarrow \tilde{r} \\ & & B^n \end{array}$$

Then g is continuous. By Brouwer's fixed-point theorem, g has a fixed point $x \in B^n$,

$$g(x) = x \iff (\tilde{r} \circ f)(x) = x. \tag{11.2}$$

If $f(x) \notin B^n$, then $\|f(x)\| > 1$, and thus by definition of the ball retraction $\tilde{r}$, $\tilde{r}(f(x)) = \frac{f(x)}{\|f(x)\|}$. Hence,

$$\frac{f(x)}{\|f(x)\|} = x \implies \|x\| = 1,$$

i. e., $x \in S^{n-1}$. By substitution in (11.2), we get $x = \frac{1}{\|f(x)\|} f(x)$ with $\lambda = \frac{1}{\|f(x)\|} < 1$, infringing Condition (11.1). Therefore, $f(x) \in B^n$ and so $\tilde{r}(f(x)) = f(x) = x$, proving that f has a fixed point $x \in B^n$. □

Remark 11.1.12. Theorem 11.1.11 may be formulated as *a nonlinear alternative.*

Let $f : B^n \to \mathbb{R}^n$ be a continuous function. Then either:

(1) there exist $x \in S^{n-1}$ and $\lambda \in (0, 1)$ such that $x = \lambda f(x)$;

(2) or f has a fixed point $x \in B^n$.

Remark 11.1.13. (1) In Theorem 11.1.11, Condition (11.1) is added because f does not self-map the closed ball B^n.

(2) The ball B^n can be replaced by any closed ball with adequate modification in Condition (11.1).

Remark 11.1.14. Condition (11.1) is equivalent to

$$f(x) - \lambda x \neq 0, \quad \text{for all } x \in S^{n-1}, \text{ and } \lambda > 1, \tag{11.3}$$

and in particular for all nonnegative values of λ.

Corollary 11.1.15. *Let $f : B^n \to \mathbb{R}^n$ be a continuous function such that*

$$\langle x, f(x) \rangle \leq 0, \quad \textit{for all } x \in S^{n-1}.$$

Then there exists $x \in S^{n-1}$ such that $f(x) = 0$.

Proof. We claim that Condition (11.1) holds for $g(x) = f(x) + x$ in which case g has a fixed point and f a zero as required. Suppose to the contrary that there are $x_0 \in S^{n-1}$

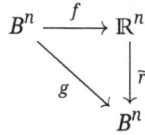

and $\lambda_0 \in (0,1)$ such that $x_0 = \lambda_0(f(x_0)+x_0)$, which implies $\|x_0\|^2 = \lambda_0\langle x_0, f(x_0)\rangle + \lambda_0\|x_0\|^2$. Using the hypothesis, we get $1 \le \lambda_0$, a contradiction. □

It turns out that Corollary 11.1.15 is also an equivalent form of Brouwer's fixed-point theorem. We have the following.

Proposition 11.1.16. *Corollary* 11.1.15 *implies Corollary* 10.3.30.

Proof. Suppose Corollary 11.1.15, assume by contradiction that some retraction $r : B^n \longrightarrow S^{n-1}$ exists, and define the dilation function $f(x) = x - ar(x)$, for some parameter $a > 1$. For all $x \in S^{n-1}$, we have $r(x) = x$ and

$$\langle x, f(x)\rangle = \|x\|^2 - a\langle x, r(x)\rangle = \|x\|^2 - a\|x\|^2 = 1 - a < 0.$$

Then, by Corollary 11.1.15, there exists $x_0 \in B^n$ such that $f(x_0) = 0$. However, for all $x \in B^n$,

$$\langle r(x), f(x)\rangle = \langle r(x), x\rangle - a\|r(x)\|^2 \le \|x\|\|r(x)\| - a \le 1 - a < 0.$$

Therefore, $f(x) \ne 0$ for all $x \in B^n$, and thus we reach a contradiction. □

The next result was formulated by Henri Poincaré (1886) and Piers Bohl (1904). In the one-dimensional case, this is the intermediate value theorem.

Corollary 11.1.17. *Let $f : B^n \to \mathbb{R}^n$ be a continuous function such that*

$$f(x) + \lambda x \ne 0, \quad \textit{for all } x \in S^{n-1} \textit{ and } \lambda > 0. \tag{11.4}$$

Then there exists $x \in B^n$ such that $f(x) = 0$.

Proof. Just apply Theorem 11.1.11 to the function $g(x) = -f(x) + x$, which satisfies Condition (11.3). □

Example 11.1.18. The function $f : B^n \longrightarrow B^n$ given in Example 11.1.4 has infinitely many zeros on the boundary S^{n-1}.

Corollary 11.1.19. *Let $f : B^n \to \mathbb{R}^n$ be a continuous function such that*

$$f(x) = x, \quad \textit{for all } x \in S^{n-1}.$$

Then $B^n \subseteq f(B^n)$.

Proof. Let $y_0 \in B^n$ and $g(x) = f(x) - y_0$. Then g satisfies Condition (11.4) of Corollary 11.1.17. If not, there are $x_0 \in S^{n-1}$ and $\lambda_0 > 0$ such that $g(x_0)+\lambda_0 x_0 = x_0 - y_0 + \lambda_0 x_0 = 0$, i. e., $y_0 = (1+\lambda_0)x_0$. Taking the norms yields the contradiction $1 \ge \|y_0\| = 1+\lambda_0 > 1$. Hence, there is some $\overline{x} \in B^n$ such that $0 = g(\overline{x})$, i. e., $f(\overline{x}) = y_0$, and thus $y_0 \in f(B^n)$. □

Corollary 11.1.20. *Let $f : B^n \longrightarrow B^n$ be continuous and $f(x) = x$ for all $x \in S^{n-1}$. Then f is surjective.*

Proof. Indeed, by Corollary 11.1.19, $B^n \subseteq f(B^n) \subseteq B^n$. □

Corollary 11.1.21. *Let $f : B^n \longrightarrow B^n$ be a continuous function such that*

$$f(x) = x, \quad \textit{for all } x \in S^{n-1}.$$

Then f is surjective.

The following result is known as the theorem of surjective maps.

Corollary 11.1.22. *Let $f : \mathbb{R}^n \to \mathbb{R}^n$ be a continuous function such that*

$$\lim_{\|x\|\to+\infty} \frac{\langle f(x), x\rangle}{\|x\|} = +\infty. \tag{11.5}$$

Then f is surjective.

Proof. To prove that f is onto, let $y_0 \in \mathbb{R}^n$ and $g(x) = f(x) - y_0$. Condition (11.5) implies the following one:

$$\lim_{\|x\|\to+\infty} \frac{\langle g(x), x\rangle}{\|x\|} = +\infty.$$

Hence, there exists some positive real number R such that

$$\langle g(x), x\rangle > 0, \quad \text{for all } x : \ \|x\| = R. \tag{11.6}$$

We claim that g fulfills Condition (11.4). Otherwise, there are $x_0 \in \partial B_R$ and $\lambda_0 > 0$ such that $g(x_0) + \lambda_0 x_0 = 0$, where B_R is the closed ball $B(0, R)$. Then $\langle g(x_0), x_0\rangle + \lambda_0\|x_0\|_0^2 = 0$, contradicting (11.6). By Corollary 11.1.17, there exists some solution $x_0 \in B_R$ of the equation $g(x_0) = 0$, i. e., $f(x_0) = y_0$, proving surjectivity of the function f. □

To end this section, we show that some special growth of the function f implies its surjectivity.

Corollary 11.1.23. *Let $f : \mathbb{R}^n \to \mathbb{R}^n$ be a continuous function such that there exists some constant $0 < k < 1$ such that*

$$\|f(x) - x\| \le k\|x\|, \quad \textit{for all } x \in \mathbb{R}^n. \tag{11.7}$$

Then f is surjective.

Proof. We check that Condition (11.7) implies Condition (11.5) of Corollary 11.1.22. Squaring both terms in Condition (11.7), expanding and evaluating the inner products in

$$\langle x - f(x), x - f(x)\rangle \le k^2\|x\|^2$$

yield the inequality

$$(1-k^2)\|x\|^2 + \|f(x)\|^2 \le 2\langle f(x), x\rangle.$$

Hence,

$$2\frac{\langle f(x), x\rangle}{\|x\|} \ge (1-k^2)\|x\| + \frac{\|f(x)\|^2}{\|x\|} \ge (1-k^2)\|x\|. \qquad \square$$

Remark 11.1.24. (1) Condition (11.7) is a linear growth condition for f. Indeed, by the triangle inequality,

$$(1-k)\|x\| \le \|f(x)\| \le (1+k)\|x\|, \quad \text{for all } x \in \mathbb{R}^n.$$

In particular, $f(0) = 0$ and $\lim_{\|x\|\to\infty} \|f(x)\| = +\infty$.

(2) Condition (11.7) can be easily extended to the following one:

$$\|f(x) - x\| \le k\|x\| + k', \quad \text{for all } x \in \mathbb{R}^n,$$

for some constants k, k' such that $0 < k < 1$.

11.2 Brouwer's fixed-point theorem for compact convex sets

We show that in Theorem 11.1.1 one can replace the closed unit ball by any compact convex subset of $\mathbb{R}^n$ with a nonempty interior.

Definition 11.2.1. A subset $C \subset \mathbb{R}^n$ is called *a convex body* if C is compact convex with a nonempty interior.

Theorem 11.2.2. *Brouwer's fixed-point theorem holds for any convex body C (and even for any set homeomorphic to such a set C).*

The proof of Theorem 11.2.2 is just a consequence of Brouwer's fixed-point theorem for the closed ball (Theorem 11.1.1), Theorem 7.6.6 and the following result.

Theorem 11.2.3. *Let $C \subset \mathbb{R}^n$ be a convex body. Then C is homeomorphic to the closed unit ball B^n and ∂C is homeomorphic to S^{n-1}.*

Proof. First proof. (1) A continuous correspondence. Without loss of generality, assume that $0 \in \mathring{C}$ for otherwise we can consider the shifting set $\widetilde{C} = C - \{x_0\}$ for some $x_0 \in \mathring{C}$ with $\widetilde{C}$ homeomorphic to C. Since C is bounded and has nonempty interior, there exist positive constants $R > r$ such that $B(0,r) \subset C \subset B(0,R)$. We construct a homeomorphism $C \longrightarrow B^n$ by

$$h(x) = \begin{cases} \frac{j(x)}{\|x\|}x, & \text{if } x \ne 0, \\ 0, & \text{if } x = 0, \end{cases}$$

where j is Minkowski's functional given by Lemma A.8.1 of the Appendix. By Property (3) of Lemma A.8.1, we have $\|h(x)\| = j(x) \leq 1$ for all $x \in C$. Thus, $h(C) \subset B^n$. Property (6) of Lemma A.8.1 yields the continuity of h because $x \neq 0$ and by property (1) of Lemma A.8.1, we have

$$\|h(x)\| = j(x) \leq \frac{\|x\|}{r}, \quad \text{for all } x \neq 0.$$

By the squeeze theorem $\lim_{x \to 0} h(x) = 0 = h(0)$, proving the continuity of h at 0, hence on the whole C.

(2) The map h is bijective. Let k be the map defined by

$$k(x) = \begin{cases} \frac{\|x\|}{j(x)} x, & \text{if } x \neq 0, \\ 0, & \text{if } x = 0. \end{cases}$$

We claim that $h \circ k = \mathrm{Id}_{|B^n}$. Indeed,

$$h(k(x)) = \begin{cases} \frac{j(\frac{\|x\|}{j(x)} x)}{\|k(x)\|} k(x), & \text{if } k(x) \neq 0, \\ 0, & \text{if } k(x) = 0 \end{cases}$$

with $\|k(x)\| = \frac{\|x\|^2}{j(x)}$ for all $x \neq 0$. Then $\frac{k(x)}{\|k(x)\|} = \frac{x}{\|x\|}$ for all $x \neq 0$. Since j is positively homogeneous (property (4) of Lemma A.8.1), we get

$$j(k(x)) = j\left(\frac{\|x\|}{j(x)} x\right) = \frac{\|x\|}{j(x)} j(x) = \|x\|, \quad \text{for all } x \neq 0.$$

Hence, for all $x \in B^n$,

$$\begin{aligned} h(k(x)) &= \begin{cases} x, & \text{if } x \neq 0, \\ 0, & \text{if } x = 0 \end{cases} \\ &= x. \end{aligned}$$

Similarly, we can check that $k \circ h = \mathrm{Id}_{|B^n}$. Hence, h is a bijection with continuous inverse k.

Second proof. Again assume that $0 \in \mathring{C}$.

(1) Let $h : \partial C \longrightarrow S^{n-1}$ be the radial retraction $h(x) = x/\|x\|$. We claim that h is a homeomorphism. Since ∂C is compact and S^n is Hausdorff, by Theorem 1.5.36, it suffices to prove that h is bijective. Clearly, h is continuous. Since C is compact with 0 as interior point, let $\mathring{B}(0, r) \subset C \subset \mathring{B}(0, R)$ for some $0 < r < R$.

(a) For $y \in \partial B(0, r)$, consider the ray $A = \{ty,\ t \geq 1\}$ originating from y. A is connected as a convex set. Moreover, $A \cap C \neq \emptyset$ since $y \in A \cap C$. Also, $A \cap (\mathbb{R}^n \setminus C) \neq \emptyset$ because it contains the element $z = \frac{R}{r} y$. Then $A \cap \partial C \neq \emptyset$ (see Exercise 5, Section 1.4). Hence, there exists some $t > 1$ such that $ty \in \partial C$.

(b) The map h is onto. For $y \in S^{n-1}$, $y' = ry \in \partial B(0,r)$. By part (a), there exists $t > 0$ such that $ty' \in \partial C$. Since

$$h(ty') = \frac{ty'}{t\|y'\|} = \frac{y'}{r} = \frac{ry}{r} = y,$$

h is surjective.

(c) h is one-to-one. By contradiction, let $x, x' \in \partial C$ be such that $x \neq x'$ and $\frac{x}{\|x\|} = \frac{x'}{\|x'\|}$. Assume $\|x\| < \|x'\|$. Then $x = \frac{\|x\|}{\|x'\|}x'$ lies in the open line segment $(0, x')$. Consider the set $E = B(0,r) \cup \{x'\}$. Since $E \subset C$ and C is convex, $\mathrm{Conv}(E) \subset C$. In addition, 0 and x' lie in $\mathrm{Conv}(E)$ for $0, x'$ lie in E. Hence, the closed line segment $[0, x']$ is a subset of $\mathrm{Conv}(E)$. Taking the interior points,

$$x \in (0, x') \subset \overbrace{\mathrm{Conv}(E)}^{\circ} \subset \mathring{C},$$

contradicting $x \in \partial C$. An alternative way to show the latter result is to appeal to Proposition A.1.2(2). Indeed, since $0 \in \mathring{C}$ and $x' \in \partial C$, $tx' = tx' + (1-t)0 \in \mathring{C}$ for all $t \in [0,1)$. Hence, $x = \frac{\|x\|}{\|x'\|}x' \in \mathring{C}$ for $\|x\| < \|x'\|$.

(2) Define the map $\tilde{h} : B^n \longrightarrow C$ by

$$\tilde{h}(x) = \begin{cases} \|x\| h^{-1}(x/\|x\|), & x \neq 0, \\ 0, & x = 0, \end{cases}$$

where h is the homeomorphism of part (1). We claim that $\tilde{h}$ is also a homeomorphism.

(a) The map $\tilde{h}$ is well-defined. Indeed, for $x \in B^n \setminus \{0\}$, we have $x/\|x\| \in S^{n-1}$, and thus $h^{-1}(x/\|x\|) \in \partial C \subset C$. By convexity of C, $0 \in \mathring{C}$ and $0 < \|x\| \leq 1$, we deduce that $\|x\| h^{-1}(x/\|x\|) \in C$.

(b) The map $\tilde{h}$ is bijective. Let $y \in C$. $x' = 0$ is the unique image of $y = 0$. If $y \in \partial C$, then $x' = h(y) \in S^{n-1}$ is its unique preimage. Indeed,

$$\tilde{h}(x) = \|x\| h^{-1}(x/\|x\|) = h^{-1}(x) = h^{-1}(h(y)) = y.$$

So, let $y \in \mathring{C} \setminus \{0\}$. By part (1), there exists a unique element $x \in \partial C$ such that $h(x) = y/\|y\|$, i. e., $x = \frac{\|x\|}{\|y\|}y$. Notice that $\|x\| \neq \|y\|$ because $x \neq y$. Furthermore, $\|x\| > \|y\|$ for otherwise $0 < \frac{\|x\|}{\|y\|} < 1$, which means that x lies in the open line segment $(0, y)$. But $(0, y) \subset \mathring{C}$ because $0, y$ lie in $\mathring{C}$ and C convex implies $\mathring{C}$ convex (see Proposition A.1.2(1)). Hence, $x \in \mathring{C}$, a contradiction. Finally, let $x' = \frac{\|y\|}{\|x\|}h(x) = \frac{y}{\|x\|}$. Then $\|x'\| = \frac{\|y\|}{\|x\|} < 1$ and

$$\begin{aligned} \tilde{h}(x') &= \frac{\|y\|}{\|x\|} h^{-1}\left(\frac{\|x\|}{\|y\|} x'\right) \\ &= \frac{\|y\|}{\|x\|} h^{-1}\left(\frac{\|x\|}{\|y\|} \frac{\|y\|}{\|x\|} h(x)\right) \end{aligned}$$

$$
\begin{aligned}
&= \frac{\|y\|}{\|x\|} h^{-1}(h(x)) \\
&= \frac{\|y\|}{\|x\|} x \\
&= \frac{\|y\|}{\|x\|} \frac{\|x\|}{\|y\|} y \\
&= y.
\end{aligned}
$$

Since in each case x' is unique, this proves bijectivity.

(c) The map $\widetilde{h}$ is continuous. Since for all $y \in \partial C$, $\|y\| \leq R$, we have $h^{-1}(x/\|x\|)$ for all $x \in \partial B^n$. Hence, $0 \leq \|\widetilde{h}(x)\| \leq R\|x\|$ for all $x \in B^n$. By the squeeze theorem, $\lim_{x \to 0} \widetilde{h}(x) = 0 = \widetilde{h}(0)$, proving continuity at $x = 0$. Therefore, $\widetilde{h}$ is a homeomorphism according to Theorem 1.5.36.

A second proof of part (2). Since ∂C is compact, the topological cone $C(\partial C)$ and the geometric cone $G(\partial C)$ are homeomorphic (see Proposition 3.3.12), where $G(\partial C) = \{ty : y \in \partial C,\ t \in [0,1]\}$ is the union of all line segments from the origin to the boundary points of C.

(a) We claim that $C = G(\partial C)$, which yields that C and $C(\partial C)$ are homeomorphic. First, notice that $G(\partial C) \subset C$. Indeed, for all $y \in \partial C$, we have $y \in \partial C \subset C$ and $0 \in \mathring{C} \subset C$. By convexity of C, we deduce that $ty \in C$ for all $t \in [0,1]$. Conversely, let $x \in C$ and consider three cases.

(i) If $x = 0$, then $x = ty \in G(\partial C)$ for $t = 0$ and arbitrary $y \in \partial C$.

(ii) If $x \in \partial C$, then $x = ty \in G(\partial C)$ for $t = 1$ and $y = x \in \partial C$.

(iii) If $x \in \mathring{C} \setminus \{0\}$, then $x/\|x\| \in S^{n-1}$. By part (1), there exists a unique $y \in \partial D$ such that $x = \frac{\|x\|}{\|y\|} y$. Arguing as in part (1)(b), we can check that $\|x\| < \|y\|$. Hence, $x \in G(\partial C)$.

(b) By part (1), ∂C and S^{n-1} are homeomorphic. Therefore, the cones over ∂C and S^{n-1} are homeomorphic (see Remark 2.4.1). Using part (a) above and Theorem 3.3.6, we conclude the proof. □

Remark 11.2.4. The proof of Theorem 11.2.3 shows the existence of homeomorphism of pairs between (B^n, S^{n-1}) and $(\overline{C}, \partial C)$. By Lemma A.8.1(7) of the Appendix, notice that for all $x \in \partial C$, the Minkowski functional is $j(x) = 1$. Then the homeomorphisms h in the first and the second proof coincide.

Since a simplex is a compact convex whose interior contains the center (see Remark 10.1.4 and Definition 10.1.11), by Theorem 11.2.3, it is homeomorphic to the closed unit ball B^n. Thus, we have the following.

Corollary 11.2.5. *Let Δ_n be an n-dimensional simplex. Then every continuous function $f : \Delta_n \longrightarrow \Delta_n$ has a fixed point.*

However, notice that Theorem 11.2.2 can be proved without appealing to Theorem 11.2.3 and Theorem 7.6.6 and without the condition $\mathring{C} \neq \emptyset$. We have the following.

Theorem 11.2.6. *Brouwer's fixed-point theorem holds for every nonempty compact convex subset $C \subset \mathbb{R}^n$.*

Proof. Since C is bounded, let $B(0,R)$ be a closed ball containing C in $\mathbb{R}^n$. Since C is closed and convex, by Theorem 7.1.21, it is a retract of $B(0,R)$. Actually by Lemma A.6.1, such a retraction $r : B(0,R) \longrightarrow C$ is just the projection on C. By Corollary 7.6.8, f has a fixed point in C. □

Remark 11.2.7. Suppose that C is as in Theorem 11.2.6, $0 \in \mathring{C}$ and C is symmetric with respect to the origin, i. e., $x \in S^n$ implies $-x \in S^n$. Then, arguing as in Theorem 7.6.5, we can prove that Brouwer's fixed-point theorem implies that ∂C is not a retract of C. However, we know from Proposition 10.3.31, that for any compact set $K \subset \mathbb{R}^n$ with nonempty interior, the boundary ∂K is not a retract of K.

Next, we show that Proposition 10.3.31 result can be derived from Brouwer's fixed point as well.

Proposition 11.2.8. *Let $U \subset \mathbb{R}^n$ be a nonempty bounded open subset. Then ∂U is not a retract of $\overline{U}$.*

Proof. Let $r : \overline{U} \longrightarrow \partial U$ be a possible retraction, $0 \in U$ and $B = B(0,R)$ a closed ball containing $\overline{U}$. Define the function f by

$$f(x) = \begin{cases} -\frac{R}{\|r(x)\|} r(x), & x \in \overline{U}, \\ -\frac{R}{\|x\|} x, & x \in B \setminus U. \end{cases}$$

Since r is a retraction, $r(x) = x$ for all $x \in \partial U$. Hence, both definitions of f agree on the intersection $\partial U = \overline{U} \cap (B \setminus U)$ and so $f : B \longrightarrow \partial B \subset B$ is continuous. By Brouwer's fixed-point Theorem 11.1.1 for the ball, f has a fixed point $x_0 = f(x_0)$. Hence, $x_0 \in \partial B$ and so $x_0 \in B \setminus U$ since $U \subset \mathring{B}$. By the second definition of f,

$$x_0 = -\frac{R}{\|x_0\|} x_0 \quad \Longrightarrow \quad x_0 = 0,$$

which contradicts $x_0 \in \partial B$. □

Remark 11.2.9. An analytic proof of Brouwer's fixed-point theorem using only vector fields in $\mathbb{R}^n$ (with neither reference to the homotopy theory nor the homology theory) was given by Milnor in 1978 and then simplified by Rogers in 1980. Actually, Rogers proves the nonretraction of the sphere (Corollary 10.3.30) first for continuously differentiable functions and then for continuous functions by Weierstrass–Stone approximation. We refer the interested reader to the papers [73, 80]. Weierstrass–Stone approximation is also used in [39, Lemma 2, Chapter 1] to approximate loops in S^1 by differentiable loops.

Remark 11.2.10. The proof of the nonexistence of a $\mathscr{C}^2$-retraction of the closed ball onto its boundary is not difficult (see, e. g., [56, Theorem 5.2.2]). Then the proof of Brouwer's fixed-point theorem goes through by approximating the function f by a $\mathscr{C}^2$ map on the compact unit ball and then by reasoning as in the proof of Theorem 11.1.1.

11.3 Equivalent forms of Brouwer's fixed-point theorem

Collecting the results obtained in Corollary 10.3.30, Theorem 11.1.1, Theorem 11.1.5 and Corollary 11.1.15, we obtain the following.

Corollary 11.3.1. *The following equivalent statements hold:*
(1) S^{n-1} *is not a retract of* B^n*;*
(2) S^{n-1} *is not contractible;*
(3) *Brouwer's fixed-point theorem holds for* B^n*;*
(4) *Corollary* 11.1.15 *holds.*

Three more equivalent versions are proposed in this section. The first one, in connection with variational inequalities, was proved by P. Hartman and G. Stampacchia in 1966 (see [44]).

Theorem 11.3.2 (Hartmann–Stampacchia theorem). *Let* $C \subset \mathbb{R}^n$ *be a nonempty compact convex subset and* $T : C \longrightarrow \mathbb{R}^n$ *a continuous map. Then there exists* $x_0 \in C$ *such that*

$$\langle Tx_0, x - x_0 \rangle \geq 0, \quad \textit{for all } x \in C. \tag{11.8}$$

Proof. (1) Let f given by $f(x) = x - \mathrm{Pr}_C(Tx)$. Then $f : C \longrightarrow C$ is continuous. By Brouwer's fixed-point Theorem 11.2.6, there exists some $x_0 \in C$ such that

$$x_0 = x_0 - \mathrm{Pr}_C(Tx_0) = \mathrm{Pr}_C(x_0 - Tx_0).$$

Using Lemma A.6.1, part (2), we have

$$\langle x_0 - Tx_0 - x_0, x - x_0 \rangle \leq 0, \quad \text{for all } x \in C,$$

which is (11.8).

(2) Conversely, assume that Theorem 11.3.2 holds and let $f : C \longrightarrow C$ be a continuous map. Define $T : C \longrightarrow C$ by $Tx = x - f(x)$ and set $x = f(x_0)$ in (11.8). We get

$$\begin{aligned} \langle Tx_0, f(x_0) - x_0 \rangle \geq 0 \quad &\Longleftrightarrow \quad \langle x_0 - f(x_0), f(x_0) - x_0 \rangle \geq 0 \\ &\Longleftrightarrow \quad -\|Tx_0\|^2 \geq 0 \\ &\Longleftrightarrow \quad f(x_0) = x_0. \end{aligned}$$

Then x_0 is a fixed point of f. □

The next equivalent version bears the name of the mathematicians Knaster, Kuratowski and Mazurkiewicz who first established the result in 1929 (see [58]). It can be also proved solely using topological reasoning (see, e. g., [72, Lemma 18, Chapter 4]).

Theorem 11.3.3 (Knaster–Kuratowski–Mazurkiewicz theorem). *Let* $x_1, \ldots, x_p \in \mathbb{R}^n$ *and* $F_1, \ldots, F_p$ *be closed subsets of* $\mathbb{R}^n$ *such that* $\operatorname{Conv}\{x_1, \ldots, x_p\} \subset \bigcup_{j=1}^{i=p} F_j$ *for all multiindex* $\{i_1, \ldots, i_p\} \in \{1, 2, \ldots\}^p$. *Then* $\bigcap_{i=1}^{i=p} F_i \neq \emptyset$.

Proof. (1) Let $C = \operatorname{Conv}\{x_{i_1}, \ldots, x_{i_N}\}$ be the convex hull of points $\{x_{i_1}, \ldots, x_{i_N}\}$ (see Subsection A.1). We use the reduction to absurdity and assume that $\bigcap_{i=1}^{i=N} F_i = \emptyset$. Since $C \subset F_1 \cup \cdots \cup F_N$ for all $x \in C$, there exists some $i_0 \in \{1, \ldots, N\}$ such that $x \notin F_{i_0}$. Consider the continuous functions on C given by $\phi_j(x) = d(x, F_i)$ for $i \in [1, N]$. Then $\sum_{i=1}^{i=N} \phi_i(x) > 0$. For $x \in C$, define the function

$$\phi(x) = \frac{\sum_{i=1}^{i=N} \phi_i(x) x_i}{\sum_{i=1}^{i=N} \phi_i(x)}.$$

Since $\phi : C \longrightarrow C$ is continuous, by Brouwer's fixed-point theorem, there is $x_0 \in C$ such that $\phi(x_0) = x_0$. By assumption, $C \subset \bigcup_{i=1}^{i=N} F_i$. Then there exists an index $j \in [1, N]$ such that $x_0 \in F_j$. By definition, $\phi_j(x_0) = 0$. Moreover,

$$\phi(x_0) = x_0 = \frac{\sum_{i=1}^{i=N} \phi_i(x_0) x_i}{\sum_{i=1}^{i=N} \phi_i(x_0)}.$$

Thus, $x_0 \in \operatorname{Conv}\{x_i,\ i \neq j\} \subset \bigcap\{F_i,\ i \neq j\}$ and there exists an index k such that $x_0 \in F_k$. Hence, $\phi_k(x_0) = 0$ and $x_0 \in \operatorname{Conv}\{x_i,\ i \neq j\}$. Repeating the same reasoning N times, we arrive at $x_0 \in \bigcap_{i=1}^{i=N} F_i$, violating our hypothesis.

(2) Conversely, assume that Theorem 11.3.3 holds. Let C be a nonempty compact convex of $\mathbb{R}^n$ and $f : C \longrightarrow C$ a continuous map. Let $g(x) = f(x) - x$. If $C = \{x_1\}$, then $f(x_1) = x_1$ and Brouwer's fixed-point theorem is proved. If $\{x_1, x_2\} \subset C$, consider for $N = 2$, the closed sets $F_1 = C$ and $F_2 = g^{-1}(0)$. Then $\operatorname{Conv}\{x_1, x_2\} \subset C \subset F_1 \cup F_2$. By the Knaster–Kuratowski–Mazurkiewicz theorem, $F_1 \cap F_2 \neq \emptyset$. As a consequence, there exists $x_0 \in C$ such that $f(x_0) = x_0$, proving Brouwer's fixed-point theorem. □

The following result has application in convex analysis. It was proved by Ky Fan in 1952 who extended the result to compact convex subsets of topological vector spaces in 1972 (see [29, 30]) It is equivalent to Brouwer's fixed-point theorem, too.

Theorem 11.3.4 (Mini-max Ky Fan theorem). *Let* $C \subset \mathbb{R}^n$ *be a nonempty compact convex subset and* $f : C \times C \to \mathbb{R}$ *a map such that:*

(1) *the map* $y \mapsto f(\cdot, y)$ *is lower semicontinuous (i. e., for all* $y_0 \in C$ *and all* $\varepsilon > 0$*, there exists a neighborhood* $U \in \mathcal{N}_{y_0}$ *such that* $f(\cdot, y) > f(\cdot, y_0) - \varepsilon$ *for all* $y \in U$*;*
(2) *the map* $x \mapsto f(x, \cdot)$ *is continuous and quasiconcave down (i. e., for all* $\lambda \in \mathbb{R}$*, the set* $\varphi^{-1}(\lambda, \infty)$ *is convex).*

Then

$$\min_{y\in C}\max_{x\in C} f(x,y) \le \max_{x\in C} f(x,x). \tag{11.9}$$

Proof. (1) The Knaster–Kuratowski–Mazurkiewicz theorem implies Ky Fan theorem. Consider a function f satisfying the hypotheses of Ky Fan theorem and set $M = \max_{x\in C} f(x,x)$. Define a family of sets by

$$F_x = \{y \in C : f(x,y) \le M\}, \quad x \in C.$$

Recall that $y \mapsto f(\cdot,y)$ lower semicontinuous means that for all $y_0 \in C$, $\lim_{y\to y_0} f(\cdot,y) \ge f(\cdot,y_0)$. This implies that the sets F_x are closed. Let $\{x_1,\dots,x_N\}$ be a finite set of elements of C. We claim that its convex hull is a subset of $\bigcup_{i=1}^{i=N} F_{x_i}$. Otherwise, there would exist $(\lambda_1,\dots,\lambda_N) \in [0,1]^N$ such that

$$\sum_{i=1}^{i=N}\lambda_i = 1 \quad \text{and} \quad \sum_{i=1}^{i=N}\lambda_i x_i \notin \bigcup_{i=1}^{i=N} F_{x_i}.$$

Then

$$f\left(x_i, \sum_{i=1}^{i=N}\lambda_i x_i\right) > M, \quad \text{for all } i \in [1,N].$$

Since f is quasiconcave down in the first argument,

$$f\left(\sum_{i=1}^{i=N}\lambda_i x_i, \sum_{i=1}^{i=N}\lambda_i x_i\right) > M,$$

contradicting the definition of M. Then the Knaster–Kuratowski–Mazurkiewicz theorem applies and we have $\bigcap_{x\in C} F_x \neq \emptyset$, i. e., there exists some $y_0 \in C$ such that $\max_{x\in C} f(x,y_0) \le M$, proving the Ky Fan theorem.

(2) Ky Fan theorem implies Brouwer's fixed-point theorem. Let $f : C \longrightarrow C$ be a continuous map from the compact convex set C to C. The function ϕ given by $\phi(x,y) = (f(y) - y, x - y)$ satisfies the assumptions of the Ky Fan theorem because it is continuous in x and affine in y. By assumption (11.9),

$$\min_{y\in C}\max_{x\in C} \phi(x,y) \le \max_{x\in C} \phi(x,x) = 0.$$

Then there exists some $y_0 \in C$ such that

$$\langle f(y_0) - y_0, x - y_0\rangle \le 0, \quad \text{for all } x \in C.$$

Since f self-maps C in C, one can take $x = f(y_0)$ to get $\|f(y_0) - y_0\|^2 \le 0$, i. e., $f(y_0) = y_0$, proving Brouwer's fixed-point theorem. □

The last three theorems combined together with Corollary 11.3.1 are collected in the following one.

Corollary 11.3.5. *The following equivalent statements hold:*
(1) *Brouwer's fixed-point theorem.*
(2) *Hartmann–Stampacchia theorem.*
(3) *Knaster–Kuratowski–Mazurkiewicz theorem.*
(4) *Mini-max Ky Fan theorem.*
(5) *The boundary of a closed ball is not a retract of the ball.*
(6) *The boundary of a closed ball is not contractible.*
(7) *Corollary* 11.1.15 *holds.*

Two more equivalent versions will be provided in Theorem 12.5.15.

11.4 Schauder's fixed-point theorem (the infinite dimension)

11.4.1 Defect of Brouwer's fixed-point theorem

Consider the space of square-summable sequences

$$\ell_2 = \left\{ x = (x_1, x_2, \dots, x_n, \dots) : \sum_{n=1}^{\infty} |x_n|^2 < \infty \right\}.$$

This is a Banach space with the norm

$$\|x\| = \left(\sum_{n=1}^{\infty} |x_n|^2 \right)^{\frac{1}{2}}.$$

Let $B = \{x \in l_2 : \|x\| \leq 1\}$ be the closed unit ball in ℓ_2, called the Hilbert ball. Define a map f by

$$f(x) = (\sqrt{1 - \|x\|^2}, x_1, x_2, \dots, x_n, \dots).$$

Then, for all $x \in \ell_2$,

$$\|f(x)\|^2 = (1 - \|x\|^2) + |x_1|^2 + |x_2|^2 + \dots + |x_n|^2 + \dots = (1 - \|x\|^2) + \|x\|^2 = 1,$$

i. e., $f(\ell_2) \subseteq \partial B$. In particular, $f(B) \subseteq B$. The function f is further continuous. Indeed for $x, y \in B$, we have $\|x\| \leq 1$, $\|y\| \leq 1$ and

$$\begin{aligned}\|f(x)-f(y)\|^2 &= \left|\sqrt{1-\|x\|^2}-\sqrt{1-\|y\|^2}\right|^2+|x_1-y_1|^2+|x_2-y_2|^2\\ &\quad+\cdots+|x_n-y_n|^2+\cdots\\ &= \left|\sqrt{1-\|x\|^2}-\sqrt{1-\|y\|^2}\right|^2+\|x-y\|^2.\end{aligned}$$

Since the following inequality holds,

$$|\sqrt{s}-\sqrt{t}| \le \sqrt{|s-t|},$$

for all $s, t \ge 0$,

$$\begin{aligned}\left|\sqrt{1-\|x\|^2}-\sqrt{1-\|y\|^2}\right|^2 &\le \left|\|x\|^2-\|y\|^2\right|\\ &= (\|x\|+\|y\|)\big|\|x\|-\|y\|\big|\\ &\le 2\big|\|x\|-\|y\|\big|.\end{aligned}$$

By the second triangle inequality,

$$\big|\|x\|-\|y\|\big| \le \|x-y\|, \quad \text{for all } x, y \in l_2.$$

Hence,

$$\|f(x)-f(y)\|^2 \le 2\|x-y\|+\|x-y\|^2.$$

This proves that f is uniformly continuous (Definition 1.2.6). Finally, the map f has no fixed point for otherwise there would exist some $x \in B$ such that $f(x) = x$. Since $f(B) \subseteq \partial B$, $x \in \partial B$ and so $\|x\| = 1$. Notice that $f(x) = x$ is equivalent to

$$\begin{cases} x_1 = \sqrt{1-\|x\|^2}\\ x_2 = x_1\\ x_3 = x_2\\ \cdots \quad \cdots\\ \cdots \quad \cdots\\ \cdots \quad \cdots\\ x_n = x_{n-1}\\ \cdots \quad \cdots \end{cases} \iff \begin{cases} x_1^2 = 1-\|x\|^2 = 0\\ x_2 = x_1\\ x_3 = x_2\\ \cdots \quad \cdots\\ \cdots \quad \cdots\\ \cdots \quad \cdots\\ x_n = x_{n-1}\\ \cdots \quad \cdots \end{cases}$$

$$\begin{aligned}&\iff x_1 = x_2 = \cdots = x_n = \cdots = 0 \quad \text{and} \quad \|x\|^2 = 1\\ &\iff x = 0 \quad \text{and} \quad \|x\| = 1,\end{aligned}$$

a contradiction. Therefore, f has no fixed point.

Conclusion. Brouwer's fixed-point theorem fails for the space ℓ_2 because the latter space has infinite dimension. Contrary to the finite-dimension case, a closed bounded set is not necessary compact in the infinite dimension case. To be more precise, by the Riesz theorem a space has finite dimension if and only if every closed bounded subset is compact (equivalently said, the closed unit ball is compact). As we will see now, a further difference between the finite dimension and the infinite dimension is the property of retraction and contractibility of the sphere. In the infinite-dimensional case, the situation is quite different (see Remark 11.4.2(1), too). In [9, Chapter I, Proposition (3.10)], it is proved that in a Hilbert space, the boundary of a closed ball B is a retract of B. Before proving this result in any normed space X, we discuss two concrete examples in infinite dimension.

11.4.2 Contractibility of the sphere in $\mathbb{R}^\infty$

We discuss the contractibility of the infinite sphere. Let

$$\mathbb{R}^{\mathbb{N}} = \mathbb{R} \times \mathbb{R} \times \cdots \times \mathbb{R} \times \cdots$$

be the countably infinite product of the real line. Define $\mathbb{R}^\infty$ to be the set of real sequences $(x_n)_n$ such that at most finite many terms are nonzero, i. e.,

$$\mathbb{R}^\infty = \{x = (x_n)_n : \text{ there exists } n_0 \in \mathbb{N} : \ x_n = 0, \text{ for all } n \geq n_0\}.$$

For $p \geq 1$, consider the space of p-summable sequences:

$$\ell_p = \left\{x = (x_n)_n : \sum_{n=1}^{\infty} |x_n|^p < \infty\right\}.$$

Endowed with the norm

$$\|x\| = \left(\sum_{n=1}^{\infty} |x_n|^p\right)^{\frac{1}{p}},$$

this is a Banach space. Then

$$\mathbb{R}^\infty = \bigcup_{n\geq 1} \mathbb{R}^n \quad \text{and} \quad \mathbb{R}^\infty \subset \ell_p \subset \mathbb{R}^{\mathbb{N}}.$$

It can be checked that $\mathbb{R}^\infty$ is closed in $\mathbb{R}^{\mathbb{N}}$ for the box topology and dense in $\mathbb{R}^{\mathbb{N}}$ for the product topology (Definition 1.1.40). Regarding the difference between these two topologies, we recommend [76, Section 19]. Define the infinite sphere

$$S^\infty = \{x \in \mathbb{R}^\infty : \ \|x\| = 1\}.$$

Then $S^\infty = \bigcup_{n\geq 1} S^n$ and

Theorem 11.4.1. *The space S^∞ is contractible.*

Proof. Let $f : S^\infty \longrightarrow S^\infty$ be given by $f(x) = (0, x)$. Then f is a continuous well-defined map. Let

$$H(t,x) = \frac{tx + (1-t)f(x)}{\|tx + (1-t)f(x)\|}.$$

For $t = 0$ or $t = 1$, $tx + (1-t)f(x) \neq 0$ for all $x \in S^\infty$. For $t \in (0,1)$,

$$\begin{aligned} tx + (1-t)f(x) = 0 \quad &\Longleftrightarrow \quad (tx_1, tx_2 + (1-t)x_1, \dots, tx_n + (1-t)x_{n-1}, \dots) = 0 \\ &\Longleftrightarrow \quad x = 0, \end{aligned}$$

which is impossible. Then H is a homotopy between f and the identity map on S^∞. Let the constant vector $c(x) = (1, 0, 0, \dots)$ and define the map

$$K(t,x) = \frac{tf(x) + (1-t)c(x)}{\|tf(x) + (1-t)c(x)\|}.$$

Then $tf(x) + (1-t)c(x) = 0$ if and only if $t = 1$ and $x = 0$, which is impossible. Thus, K is a homotopy between f and the constant map c. Therefore, Id_{S^∞} and c are homotopic, which completes the proof the theorem. □

Remark 11.4.2. (1) From Theorem 11.1.5 and Theorem 11.4.1, we can see that $\mathbb{R}^\infty$ has infinite dimension and so ℓ_p has infinite dimension.
(2) As in the finite-dimensional case, S^∞ is a deformation retract of the space $\mathbb{R}^\infty \setminus \{0\}$ via the radial retraction $r(x) = x/\|x\|$.

11.4.3 Retraction of the sphere in ℓ_2

Let B be the closed unit ball of the space $X = \ell_2$ of square-summable real sequences. According to Remark 11.4.2, ℓ_2 has infinite dimension. We construct explicitly a retraction of B onto ∂B. In Subsection 11.4.1, we have checked the existence of a fixed-point free continuous map f from B to B. Let $r(x) = f(x) + t(x)(x - f(x))$, where $t(x)$ is the positive root of the quadratic equation (see the proof of Brouwer's fixed-point Theorem 7.6.2):

$$t^2\|x - f(x)\|^2 + 2t\langle f(x), x - f(x)\rangle + \|f(x)\|^2 = 1.$$

Then we have the following.

Proposition 11.4.3. *The map $r : B \longrightarrow \partial B$ is a retraction.*

Proof. The map r is well-defined since f has no fixed point. Clearly, r is continuous and by definition $r(x) = x$, for all $x \in \partial B$, □

Corollary 11.4.4. *Any pair of continuous maps $f, g : \partial B \longrightarrow \partial B$ are homotopic.*

Proof. Such a homotopy $H : \partial B \times I \longrightarrow \partial B$ is given by

$$H(x,t) = r\big(tf(x) + (1-t)g(x)\big),$$

where r is the retraction given by Proposition 11.4.3. □

Next, we consider the general case of infinite-dimensional normed spaces. B refers to the closed unit ball and $S = \partial B$ the unit sphere.

11.4.4 Retraction and contractibility of the sphere in a normed space

Theorem 11.4.5 (Retraction of the sphere). *Let X be a normed space with $\dim X = \infty$. Then the sphere $S = \partial B$ is a retract of the closed ball B.*

We suggest a proof based on the following lemma due to V. L. Klee (1956) combined with Dugundgi's extension theorem (1951) (Lemma A.5.2).

Lemma 11.4.6 ([57]). *Let X be an infinite-dimensional normed space and $K \subset X$ a compact subset. Then X and $X \setminus K$ are homeomorphic.*

Proof of Theorem 11.4.5. By Lemma 11.4.6, let $h : X \setminus \{0\} \longrightarrow X$ be a homeomorphism, $\mathrm{Id}_{|S} : S \longrightarrow S$ the identity map and $i : S \hookrightarrow X \setminus \{0\}$ the inclusion map. The composite map $f = h \circ i \circ \mathrm{Id}_{|S}$ is continuous:

$$S \xrightarrow{\ \mathrm{Id}_{|S}\ } S \xrightarrow{\ i\ } X \setminus \{0\} \xrightarrow{\ h\ } X, \qquad f : S \longrightarrow X$$

Since S is closed, by Dugundgi's extension theorem (Lemma A.5.2), the map f has a continuous extension $\tilde{f} : B \longrightarrow X$. Consider the radial retraction $\varrho : X \setminus \{0\} \longrightarrow S$ given by $\varrho(x) = \frac{1}{\|x\|}x$ and let $r = \varrho \circ h^{-1} \circ \tilde{f}$:

$$B \xrightarrow{\ \tilde{f}\ } X \xrightarrow{\ h^{-1}\ } X \setminus \{0\} \xrightarrow{\ \varrho\ } S, \qquad r : B \longrightarrow S$$

Then r is continuous as the composite of continuous maps and for any $x \in S$,

$$
\begin{aligned}
r(x) &= (\varrho \circ h^{-1})(\tilde{f}(x)) = (\varrho \circ h^{-1})(f(x)) \\
&= (\varrho \circ h^{-1} \circ h \circ i \circ \mathrm{Id}_{|S})(x) = (\varrho \circ i \circ \mathrm{Id}_{|S})(x) \\
&= (\varrho \circ) i(x) = \varrho(x) \\
&= \frac{1}{\|x\|} x = x
\end{aligned}
$$

because $\|x\| = 1$. Therefore, $r : B \longrightarrow S = \partial B$ is a retraction, as claimed. □

Remark 11.4.7. Lemma 11.4.6 is specific to the infinite dimension. For $X = \mathbb{R}^n$ and $K = S^{n-1}$, X and $X \setminus K$ are not homeomorphic since X is path-connected whereas $X \setminus K$ has two path-components (see Proposition 10.4.17).

A second result specific to the infinite dimension is the following.

Corollary 11.4.8 (Contractibility of the sphere). *The sphere $S = \partial B$ is contractible.*

Proof. Since S is a retract of B, let $r : B \longrightarrow S$ be a retraction and define the function $H : S \times I \longrightarrow S$ by

$$H(x, t) = r((1 - t)x).$$

Since for $(x, t) \in S \times I$, $(1 - t)x \in B$, H is well-defined. Moreover, H is continuous and satisfies $H(x, 0) = r(x)$ and $H(x, 1) = r(0)$ for all $x \in S$. Hence, the identity $\mathrm{Id}_{|S}$ is homotopic to the constant $r(0)$, proving that S is contractible. □

To prove a fixed-point theorem in the infinite-dimension case, we will require more than the continuity on the map f, namely the compactness of f (see Subsection 1.3.4). Otherwise, the compactness of the domain of f is a sufficient condition. We present three formulations of Schauder's fixed-point theorem [88] and two corollaries.

11.4.5 Some versions of Schauder's fixed-point theorem

Theorem 11.4.9 (Version 1: Compact set). *Let C be a nonempty compact convex subset of a normed space X and let $f : C \longrightarrow C$ be a continuous map. Then f has at least one fixed point $x \in C$.*

Proof. We apply Schauder's Approximation Corollary A.7.3 in the Appendix by taking $K = C$ and $\varepsilon = 1/n$ $(n = 1, 2, \ldots)$. Then there exist a subspace $X_n \subset X$ with $\dim X_n < \infty$, $n = 1, 2, \ldots$ and a sequence of continuous maps $P_n : C \longrightarrow X_n$ such that for all $y \in C$,

$$\|P_n(y) - y\| < 1/n, \quad n = 1, 2, \ldots. \tag{11.10}$$

In addition,

$$P_n(C) \subset \mathrm{Co}(C) = C,$$

for C is convex. Let $K_n = \overline{\mathrm{co}}(P_n(C))$. Since C is closed and convex, $K_n \subset \overline{\mathrm{co}}(C) = C$. The set C is further compact. Hence, K_n is compact as a closed subset of C. Finally, $K_n \subset X_n$ because $P_n(C) \subset X_n$ for all $n = 1, 2, \ldots$. Consider the following diagram, where the composite function $g_n = P_n \circ f$ is continuous because P_n and f are continuous:

$$\begin{array}{ccc} K_n \subset C & \xrightarrow{\;f\;} & f(C) \subset C \\ & \searrow^{g_n} & \downarrow P_n \\ & & P_n(C) \subset K_n \end{array}$$

Since $\dim X_n < \infty$, Brouwer's fixed-point theorem implies that, for each $n = 1, 2, \ldots$, there exists $x_n \in K_n$ such that $g_n(x_n) = x_n$, i. e., $P_n(f(x_n)) = x_n$. Since $x_n \in K_n \subset C$, $f(x_n) \in f(C) \subset C$. Letting $y = f(x_n)$ in (11.10), we find

$$\|P_n(f(x_n)) - f(x_n)\| < 1/n, \quad n = 1, 2, \ldots.$$

Therefore,

$$\|x_n - f(x_n)\| < 1/n, \quad n = 1, 2, \ldots. \tag{11.11}$$

Since $(x_n)_n \subset C$ and the set C is compact, there exists a subsequence $(x_{n_k})_k \subset C$ and an element $x \in C$ such that $\lim_{k\to\infty} x_{n_k} = x$. Letting $n = n_k$ in (11.11) and taking the limit, as $k \to \infty$, we find $x = f(x)$, with $x \in C$, which completes the proof. □

Example 11.4.10. Let

$$\ell_2 = \left\{ x = (x_1, x_2, \ldots, x_n, \ldots) : \sum_{n=1}^{\infty} |x_n|^2 < \infty \right\}$$

be the Banach space of square-summable sequences (see Subsection 11.4.1).

Consider the Hilbert cube

$$Q^{\omega} = \{x = (x_1, x_2, \ldots, x_n, \ldots) \in \ell_2 : \; 0 \le x_n \le 1/n \text{ for all } n = 1, 2, \ldots\}.$$

Clearly, Q^{ω} is convex. Since $Q^{\omega} = \prod_{n=1}^{\infty} [0, 1/n]$, it is compact for the product topology in ℓ_2 by Tychonoff's theorem (see, e. g., [76, Corollary 37.3]). Therefore, by Theorem 11.4.9 every continuous function $f : Q^{\omega} \longrightarrow Q^{\omega}$ has a fixed point.

Example 11.4.11. Let E be an AR metric space. By Theorem, 7.5.8, E is the r-image of some normed space. Assume further E compact. According to [9, Chapter V, Theorem (1.1)], E is the r-image of a Hilbert cube. Therefore, Theorem 7.6.7 and Example 11.4.10 imply that every continuous map $f : E \longrightarrow E$ has a fixed point in E.

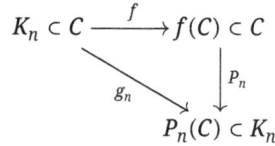

Theorem 11.4.12 (Version 2, compact map). *Let C be a nonempty closed convex subset of a Banach space and let $f : C \longrightarrow C$ be a continuous and compact map. Then f has at least one fixed point $x \in C$.*

Proof. The set $D = \overline{\mathrm{Co}}(f(C))$ is convex. Since f is compact, $f(C)$ is relatively compact. By Mazur's theorem (Lemma A.1.11), which requires the space be Banach, the set $\mathrm{Co}(f(C))$ is relatively compact, i. e., D is compact. Since C is closed and convex and by definition of D,

$$D \subseteq C \quad \Longrightarrow \quad f(D) \subseteq f(C) \subseteq D,$$

we conclude that

$$f(C) \subseteq C \quad \Longrightarrow \quad D = \overline{\mathrm{Co}}(f(C)) \subseteq \overline{\mathrm{Co}}(C) = C.$$

By Theorem 11.4.9, the map $f : D \longrightarrow D$ has a fixed point $x \in D \subseteq C$. □

The completeness of the space may be dropped. We have the following.

Theorem 11.4.13 (Version 3, compact map). *Let C be a nonempty closed convex subset of a normed space and let $f : C \longrightarrow C$ be a continuous and compact map. Then f has at least one fixed point $x \in C$.*

Proof. Let $(f_n)_n : C \longrightarrow X$ be a sequence of approximations of f such that

$$\|f_n(x) - f(x)\| < 1/n,$$

for each $n = 1, 2, \ldots$ with range in a finite-dimensional space X_n and $f_n(C) \subset \mathrm{Co}(f(C))$ (see Theorem A.7.2). Since $f(C) \subseteq C$ and C is convex, we have $\mathrm{Co}(f(C)) \subseteq C$. This implies that $f_n(C) \subseteq \mathrm{Co}(f(C)) \subseteq C$ and so $f_n(C) \subseteq C \cap X_n$. Since $f_n : C \cap X_n \longrightarrow C \cap X_n$ is continuous, by Brouwer's fixed point there exists $x_n \in C \cap X_n$ such that $f_n(x_n) = x_n$. Since f is compact, up to a subsequence $(x_{n_k})_k$, there exists $x = \lim_{k\to\infty} f(x_{n_k})$. Then the subsequence $(x_{n_k})_k$ converges to x as well because

$$\begin{aligned}
\|x_{n_k} - x\| &= \|x_{n_k} - f(x_{n_k}) + f(x_{n_k}) - x\| \\
&\leq \|f_n(x_{n_k}) - f(x_{n_k})\| + \|f(x_{n_k}) - x\| \\
&\leq 1/n_k + \|f(x_{n_k}) - x\|,
\end{aligned}$$

which tends to 0, as $k \to \infty$.

Note that if C were bounded, then $(x_n)_n$ would be bounded in X_n and $\dim(X_n) < \infty$. By the Bolzano–Weierstrass theorem, we obtain a subsequence $(x_{n_k})_k$, which converges to some limit $x \in C$ as C is closed. So, we may skip the compactness of f in this case.

To complete the proof, write

$$\begin{aligned}\|f(x)-x\| &= \|f(x)-f(x_{n_k})+f(x_{n_k})-f_{n_k}(x_{n_k})+f_{n_k}(x_{n_k})-x\| \\ &\le \|f(x)-f(x_{n_k})\| + \|f(x_{n_k})-f_{n_k}(x_{n_k})\| + \|f_{n_k}(x_{n_k})-x\|.\end{aligned}$$

Since f is continuous, as $k \longrightarrow \infty$, we have

$$\begin{aligned}\|f(x)-f(x_{n_k})\| &\longrightarrow 0, \\ \|f(x_{n_k})-f_{n_k}(x_{n_k})\| &< 1/n_k \longrightarrow 0, \\ \|f_{n_k}(x_{n_k})-x\| &= \|x_{n_k}-x\| \longrightarrow 0.\end{aligned}$$

Therefore, $\|f(x)-x\| = 0$. □

As a consequence, we derive the following fixed-point theorems. The first one is Böhl's theorem 11.1.11 in the infinite dimension.

Theorem 11.4.14. *Let B be the closed unit ball of a normed space X and $f : B \longrightarrow X$ a compact continuous map. Suppose that f satisfies the boundary condition:*

$$x \neq tf(x), \quad \text{for all } x \in \partial B \text{ and } t \in (0,1). \tag{11.12}$$

Then f has at least a fixed point $x \in B$.

Proof. Let $\widetilde{r} : X \longrightarrow B$ be the ball retraction (Example 1.4.23) and the composite map $g = \widetilde{r} \circ f$:

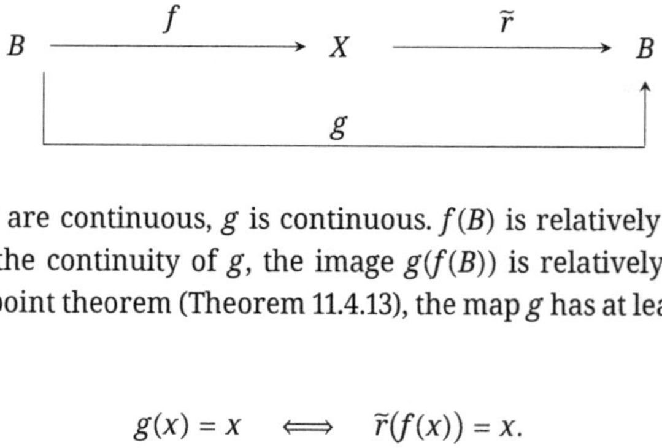

Since $\widetilde{r}$ and f are continuous, g is continuous. $f(B)$ is relatively compact because f is compact. By the continuity of g, the image $g(f(B))$ is relatively compact, too. By Schauder's fixed-point theorem (Theorem 11.4.13), the map g has at least one fixed point $x \in B$:

$$g(x) = x \quad \Longleftrightarrow \quad \widetilde{r}(f(x)) = x.$$

We claim that $f(x) \in B$ for otherwise $\|f(x)\| > 1$. By the definition of the retraction $\widetilde{r}$, we have $\widetilde{r}(f(x)) = \frac{1}{\|f(x)\|}f(x)$. Then $x = \frac{1}{\|f(x)\|}f(x)$ and $\|x\| = 1$, i. e., $x \in \partial B$, $\frac{1}{\|f(x)\|} < 1$ and $x = \frac{1}{\|f(x)\|}f(x)$, infringing (11.12). Hence, $f(x) \in B$, and thus $\widetilde{r}(f(x)) = f(x)$, which implies $f(x) = x$, $x \in B$. □

Remark 11.4.15. Theorem 11.4.14 may be formulated as *a nonlinear alternative.*

Let $f : B \longrightarrow X$ be a compact continuous map, where B is a closed ball in a normed space X. Then:

(1) either there exist $x \in \partial B$ and $t \in (0,1)$ such that $x = tf(x)$;

(2) or f has a fixed point $x \in B$.

The following result shows that f needs to map only the boundary of B into B. This is a theorem due to H. Schaefer (1955) [87].

Corollary 11.4.16 (Schaefer's fixed-point theorem). *Let $f : B \longrightarrow X$ be a compact continuous map such that*

$$f(\partial B) \subseteq B. \tag{11.13}$$

Then f has at least one fixed point in B.

Proof. We check that Condition (11.13) implies Condition (11.12) in Theorem 11.4.14. This means that Corollary 11.4.16 is a particular case of Theorem 11.4.14. Let Condition (11.13) be satisfied and assume that (11.12) does not hold. Then there exist $x \in \partial B$ and $t \in (0,1)$ such that $x = tf(x)$. Then

$$\lambda\|f(x)\| = 1 \quad \Rightarrow \quad \|f(x)\| = 1/t > 1 \quad \Rightarrow \quad f(x) \notin B.$$

Hence, $x \in \partial B$ and $f(x) \notin B$, which is a contradiction to Condition (11.13). □

Remark 11.4.17. Clearly, $f(\partial B) \subseteq B$ is weaker than the self-mapping condition $f(B) \subseteq B$ considered in Schauder's fixed-point theorem (Theorem 11.4.12). Corollary 11.4.16 was derived from Theorem 11.4.14 but provides a stronger result.

The next result shows that in Theorem 11.4.14, the closed ball B may be replaced by any closed subset with nonempty interior.

Theorem 11.4.18. *Let U be a nonempty open set of a Banach space X and $f : \overline{U} \longrightarrow X$ a compact continuous map such that $x \neq tf(x)$ for all $x \in \partial U$ and $t \in (0,1)$. Then f has a fixed point $x \in \overline{U}$.*

Proof. Define the set

$$S = \{x \in \overline{U} : x = tf(x), \text{ for some } t \in [0,1]\}.$$

Without loss of generality, assume $0 \in U$. S is a nonempty set since $0 \in S$ for $t = 0$. If there exists $x \in \partial U$ such that $x = f(x)$, then we are done. Then, using the hypothesis, $S \cap \partial U = \emptyset$. The set S is further closed because f is continuous. By Urysohn's lemma (Theorem A.4.1), there exists a continuous function $\theta : \overline{U} \longrightarrow \overline{U}$ such that $\theta(x) = 1$ for all $x \in S$ and $\theta(x) = 0$ for all $x \in \partial U$. Define the function $F(X) = \theta(x)f(x)\chi_{\overline{U}}$, where χ_A is the characteristic function of the set A defined by

$$\chi_A(x) = \begin{cases} 1, & x \in A, \\ 0, & x \notin A. \end{cases}$$

For $x \in \partial U = \partial \overline{U}$, $\theta(x) = 0$. So, by the pasting Lemma A.2.1, the mapping $F : X \longrightarrow X$ is continuous. In addition, by definition of F, $F(X) \subset \overline{\mathrm{Co}}(f(\overline{U}) \cup \{0\})$. Therefore, by Mazur's theorem (Lemma A.1.11), F is compact. By Schauder's fixed-point theorem (Theorem 11.4.13), there exists $x \in X$ such that $F(x) = x$. Since $0 \in U$, we have $x \in \overline{U}$. Hence, $x \in S$, which implies $\theta(x) = 1$, and then $f(x) = x$, which completes the proof. □

Corollary 11.4.19. *Let U be a nonempty open set of a Banach space and $f : \overline{U} \longrightarrow X$ a compact continuous map such that $f(\overline{U}) \cap \partial U = \emptyset$ for all $t \in (0,1)$. Then f has a fixed point $x \in \overline{U}$.*

As a consequence of Theorem 11.4.18, we obtain Altman's fixed-point theorem (see [3], 1958).

Corollary 11.4.20 (Altman's fixed-point theorem). *Let U be a nonempty open set of a Banach space and $f : \overline{U} \longrightarrow X$ a compact continuous map such that $\|f(x)\|^2 \leq \|x\|^2 + \|f(x) - x\|^2$ for all $x \in \partial U$. Then f has a fixed point $x \in \overline{U}$.*

Proof. Assume that $x = tf(x)$ for some $x \in \partial U$ and $t \in (0,1)$. Then squaring both sides and using the hypothesis yield $0 \leq 2t(t-1) < 0$, and a contradiction is reached. The claim of the corollary follows from Theorem 11.4.18. □

A particular case is given by Rothe's fixed-point theorem (see [82], 1937).

Corollary 11.4.21 (Rothe's fixed-point theorem). *Let U be a nonempty open set of a Banach space and $f : \overline{U} \longrightarrow X$ a compact continuous map such that $\|f(x)\| \leq \|x\|$ for all $x \in \partial U$. Then f has a fixed point $x \in \overline{U}$.*

The last result of this subsection is known as the Leray–Schauder nonlinear alternative (see [63], 1934). Notice that f is no longer assumed to be compact but only completely continuous (see Definition 1.3.36). X is any normed space.

Theorem 11.4.22 (Leray–Schauder fixed-point theorem). *Let X be a normed space and $f : X \longrightarrow X$ a completely continuous map. Then either the set*

$$S = \{x \in X : x = tf(x), \text{ for some } t \in [0,1]\}$$

is unbounded, or f has a fixed point $x_0 \in X$.

Proof. Notice that $0 \in S$ and assume that S is bounded. Let $M > 0$ be such that $S \subset \mathring{B}(0,M)$ and consider the function $\tilde{f} = \tilde{r}_M \circ f$, where $\tilde{r}_M$ is the retraction of the ball $B(0,M)$. That is,

$$\tilde{f}(x) = \begin{cases} f(x), & \|f(x)\| \leq M, \\ \frac{M}{\|f(x)\|} f(x), & \|f(x)\| \geq M. \end{cases}$$

Since f is completely continuous and $\widetilde{r}_M$ is continuous, $\widetilde{f}$ is a completely continuous map. By definition, $\|\widetilde{f}(x)\| \le M$. Hence, $\widetilde{f} : B(0,M) \longrightarrow B(0,M)$. Schauder's fixed-point theorem (Theorem 11.4.13) yields some fixed point $x_0 = \widetilde{f}(x_0)$. If $\|f(x_0)\| \ge M$, then $M\frac{f(x_0)}{\|f(x_0)\|} = x_0$, which implies $\|x_0\| = M$. But $x_0 = \frac{M}{\|f(x_0)\|}f(x_0)$ and $\frac{M}{\|f(x_0)\|} \le 1$ means that $x_0 \in S$. Hence, $x_0 \in \mathring{B}(0,M)$, a contradiction. Therefore, $\|f(x_0)\| < M$ and so $\widetilde{f}(x_0) = f(x_0) = x_0$, as claimed. □

Theorem 11.4.22 is generally used to prove the existence of a fixed point by showing that the set S is bounded. Here is an application.

Corollary 11.4.23. *Let X be a normed space and $f : X \longrightarrow X$ a completely continuous map satisfying the sublinear condition:*

$$\limsup_{\|x\|\to+\infty} \frac{\|f(x)\|}{\|x\|} < 1.$$

Then f has at least one fixed point in X.

Proof. We prove that the set

$$S = \{x \in X : x = tf(x), \text{ for some } t \in [0,1]\}$$

is bounded. By contradiction, assume that there exists a sequence $(x_n)_n \subset S$ such that $\|x_n\| \ge n$ for all positive integer $n = 1, 2 \dots$. Since $x_n = tf(x_n)$ for all n, $t \in (0,1]$ and we have

$$1 \le \frac{1}{t} = \frac{\|f(x_n)\|}{\|x_n\|}, \quad \text{for all } n.$$

Taking the limit as $n \to +\infty$ infringes the condition of the corollary. Therefore, S is bounded. By Theorem 11.4.22, we conclude that f has a fixed point in X. □

11.5 Mönch's fixed-point theorem

Some fixed-point theorems can be deduced from Schauder's fixed-point theorem such that the following result proved by H. Mönch in 1980 (see [74]). The compactness condition in Theorem 11.4.9 is relaxed.

Theorem 11.5.1. *Let Ω be an open convex subset of a Banach space X and $x_0 \in \Omega$. Let $f : \overline{\Omega} \longrightarrow \overline{\Omega}$ be a continuous map, which satisfies the hypothesis*

($\mathscr{H}$) *If $C \subseteq \overline{\Omega}$ is countable and $C \subseteq \overline{\mathrm{Co}}\,(f(C) \cup \{x_0\})$, then C is relatively compact.*

Then f has a fixed point.

Proof. Define a sequence of sets by $D_0 = \{x_0\}$ and $D_n = \overline{\text{Co}}\,(f(D_{n-1}) \cup \{x_0\})$, for $n = 1, 2, \ldots$. D_0 is compact. Inductively, if D_{n-1} is compact, then D_{n-1} is compact and by Mazur's theorem (Lemma A.1.11) D_n is compact. Hence, for each n, D_n is separable, and thus there exists a countable set C_n such that $\overline{C_n} = D_n$, for $n = 0, 1, 2, \ldots$. Let $D = \bigcup_{n=0}^{\infty} D_n$ and $C = \bigcup_{n=0}^{\infty} C_n$. Arguing once again by induction, we can show that $(D_n)_n$ is an ascending sequence of sets. Then

$$D = \bigcup_{n=0}^{\infty} \text{Co}(f(D_{n-1}) \cup \{x_0\}) = \text{Co}\,(f(D) \cup \{x_0\}).$$

Since $\bigcup_{n=0}^{\infty} \overline{C_n} \subseteq \overline{\bigcup_{n=0}^{\infty} C_n}$, we have

$$\overline{D} = \overline{\bigcup_{n=0}^{\infty} D_n} = \overline{\bigcup_{n=0}^{\infty} \overline{C_n}} = \overline{\bigcup_{n=0}^{\infty} C_n} = \overline{C}.$$

Since f is continuous, by Theorem 1.2.3, $f(\overline{D}) \subseteq \overline{f(D)}$. Hence,

$$f(D) \cup \{x_0\} \subseteq f(\overline{D}) \cup \{x_0\} \subseteq \overline{f(D) \cup \{x_0\}} \subseteq \overline{\text{Co}}\,(f(D) \cup \{x_0\}).$$

Therefore, $\overline{\text{Co}}\,(f(D) \cup \{x_0\}) = \overline{\text{Co}}\,(f(\overline{D}) \cup \{x_0\})$, which implies

$$\begin{aligned} C \subseteq \overline{C} = \overline{D} &= \overline{\text{Co}}\,(f(D) \cup \{x_0\}) \\ &= \overline{\text{Co}}\,(f(\overline{D}) \cup \{x_0\}) \\ &= \overline{\text{Co}}\,(f(\overline{C}) \cup \{x_0\}) \\ &= \overline{\text{Co}}\,(f(C) \cup \{x_0\}). \end{aligned}$$

Furthermore, C is countable. By hypothesis $(\mathscr{H})$, C is relatively compact. Then $\overline{D}$ is compact. Finally,

$$f(\overline{D}) \subseteq f(\overline{D}) \cup \{x_0\} \subseteq \overline{\text{Co}}\,(f(\overline{D}) \cup \{x_0\}) = \overline{D}.$$

By Schauder's fixed-point theorem (Theorem 11.4.9), we conclude that f has at least one fixed point in $\overline{D}$. □

11.6 Lefschetz fixed-point theorem

Definition 11.6.1. Let K be a finite polyhedron of dimension $K \leq n$ (see Definition 10.1.18) and $f : K \longrightarrow K$ a map. The Lefschetz number $\Lambda(f)$ of f is defined as

$$\Lambda(f) = \sum_{i=0}^{i=n} (-1)^i \,\text{tr}[f_*^i, H_i(K)],$$

where $f_*^i : H_i(K) \longrightarrow H_i(K)$ is the induced homomorphism of f over the homology group $H_i(K)$ and $\mathrm{tr}[f_*^i, H_i(K)]$ refers to its trace, $i = 1, \ldots n$ (see Definition B.7.1).

The following result is due to Lefschetz and Hopf. For the proof, we refer the reader to, e. g., [16, Theorem 7.8], [19, Theorem 4.9.8], [40, Theorem (2.4), Section 9], [83, Theorem 9.19].

Theorem 11.6.2 (Lefschetz–Hopf fixed-point theorem (Version 1: Finite polyhedron)). *Let K be a finite polyhedron and $f : K \longrightarrow K$ a continuous map. If $\Lambda(f) \neq 0$, then f has a fixed point in K.*

Remark 11.6.3. Since the fixed-point property is a topological invariant, the theorem holds for a finite polytope X (see Definition 10.1.18(4)).

Also, we recapture a version of Corollary 11.2.5 and also the following.

Corollary 11.6.4 (Brouwer's fixed-point theorem). *Let K be a finite polyhedron such that the homology groups $H_i(K)$ are trivial for all $i > 0$. Then every continuous map $f : K \longrightarrow K$ has a fixed point.*

Indeed, in such a situation the Lefschetz number is just $\Lambda(f) = 1$. A particular and important case is that of a closed ball in Euclidean space. Thus, Lefschetz fixed-point theorem is a generalization of Brouwer's fixed-point theorem.

Finally, notice that the condition that $H_i(K)$ are trivial for all $i > 0$ means that K has no holes of positive dimension (see Remark 10.5.8(2)). However, K is neither supposed to be convex as in Theorem 11.2.6 nor connected.

Since every compact locally contractible space of $\mathbb{R}^n$ is a retract of a finite simplicial complex (see [45, Theorem A.7]), the following corollary follows from Corollary 7.6.8 and Lefschetz–Hopf fixed-point theorem.

Corollary 11.6.5. *Let K be a compact locally contractible subspace of $\mathbb{R}^n$ such that $H_i(K)$ are trivial for all $i > 0$. Then every continuous map $f : K \longrightarrow K$ has a fixed point.*

Since AR space is ANR (see Exercise 27, Chapter 7) and an AR space is contractible (Proposition 7.5.12), we recapture Example 11.4.11.

Corollary 11.6.6. *Let K be a compact* AR *metric space. Then every continuous map $f : K \longrightarrow K$ has a fixed point.*

We close this chapter with a second version of Lefschetz fixed-point theorem (see [21, Theorem 1.26]).

Theorem 11.6.7 (Lefschetz–Hopf fixed-point theorem (Version 2: Compact map)). *Let X be an* ANR *space and $f : X \longrightarrow X$ a compact map. If $\Lambda(f) \neq 0$, then f has a fixed point in X.*

As a consequence, we have (see [40, Proposition (1.6), Section 0]).

Corollary 11.6.8 (Granas fixed-point theorem). *Let X be an* AR *space and $f : X \longrightarrow X$ a compact map. Then f has a fixed point in X.*

Indeed, in this case the Lefschetz number $\Lambda(f)$ is just equal to unity.

12 Applications

In this last chapter entirely devoted to applications, a fundamental theorem of algebra is first proved using homotopy and retraction theory (Section 12.1). Then the focus is essentially on the applications of some of the fixed-point theorems presented in Chapter 11. As application of Brouwer's fixed-point theorem, the Perron–Frobenius theorem is given in Section 12.2. Some retracts of Euclidean space are investigated in Section 12.3. Section 12.4 consists of some applications of Schauder's fixed-point theorem to the solvability of some initial and boundary value problems associated with nonlinear ordinary differential equations. Section 12.5 is concerned with the application of the homology theory developed in Chapter 10. It consists of three subsections. The degree of a map on the circle, which was introduced in Subsection 6.2, is extended here to the degree of a map on the sphere of arbitrary dimension. In particular, the extension of continuous maps defined on the sphere to the whole ball is discussed in detail. Nine equivalent versions of Brouwer's fixed- point theorem are summarized in this subsection. We study the degree of the reflection map and the antipodal map in Subsection 12.5.2. Some properties of the tangent vector fields are explored and the hairy ball theorem, also called the hedgehog theorem, is proved in Subsection 12.5.3. The final section (Section 12.6) thoroughly investigates the Borsuk–Ulam theorem and some of its applications in the fixed-point theory (the Lyusternik–Schnirelmann theorem and the ham sandwich theorem). Recall that the case of the dimension 2 was already discussed in Section 6.4. Several versions of the Borsuk–Ulam theorem which involve the antipode-preserving map are presented. In particular, we will prove that the Borsuk–Ulam theorem implies Brouwer's fixed-point theorem.

12.1 A fundamental theorem of algebra

Theorem 12.1.1. *A polynomial equation of degree $n > 0$,*

$$z^n + a_{n-1}z^{n-1} + \cdots + a_1 z + a_0 = 0$$

with real or complex coefficients has at least one (real or complex) root.

Proof. Suppose $a_0 \neq 0$ for otherwise we are done with $z = 0$. Let $f(z) = z^n + a_{n-1}z^{n-1} + \cdots + a_1 z + a_0$, $G(z,t) = z^n + t(a_{n-1}z^{n-1} + \cdots + a_1 z + a_0)$ and $H(z,t) = f(tz)$. If f does not vanish, then neither does H. Furthermore, for $|z| = R$, we have the estimates:

$$\begin{aligned}|G(z,t)| &\geq |z|^n - t|a_{n-1}z^{n-1} + \cdots + a_1 z + a_0| \\ &\geq R^n - (|a_{n-1}|R^{n-1} + \cdots + |a_1|R + |a_0|) \\ &> 0,\end{aligned}$$

https://doi.org/10.1515/9783111517384-012

for R sufficiently large. Hence, $G(z,t) \neq 0$ for all $(z,t) \in C_R \times I$, where C_R denotes the circle centered at the origin with a sufficiently large value of the radius R. Let $g = i \circ p_n$, where $p_n(z) = z^n$ is the nth power map and i is the inclusion map of the circle C_{nR} of the nonzero complex numbers. Viewed as maps from $C_R \times I$ to $\mathbb{C} \setminus \{0\}$, G and H are homotopies between (f, g) and $(f, f(0))$, respectively. By transitivity, g and $f(0)$ are homotopic from C_R to $\mathbb{C} \setminus \{0\}$, then g is null-homotopic. We show that this is not possible. Indeed, we have the diagram

$$\begin{array}{ccc} C_R & \xrightarrow{p_n} & C_{nR} \\ & \searrow^{g} & \downarrow{i} \\ & & \mathbb{C} \setminus \{0\} \end{array}$$

Then $g_* = i_* \circ (p_n)_*$. Since p_n is a covering map (Theorem 5.1.15), by Corollary 5.2.11, the induced homomorphism $(p_n)_*$ is injective. Moreover, C_R is a retract of $\mathbb{C} \setminus \{0\} \cong \mathbb{R}^2 \setminus \{0\}$. Hence, i_* is injective, i. e., there exists a retraction $r : \mathbb{C} \setminus \{0\} \longrightarrow C_R$ such that $I|_{C_R} = r \circ i$. As a consequence, g_* is injective. Therefore, the map $g : C_R \longrightarrow \mathbb{C} \setminus \{0\}$ is not null-homotopic, a contradiction. □

12.2 Perron–Frobenius theorem

A well-known result from algebra is proved using Brouwer's fixed-point theorem.

Theorem 12.2.1. *Let $A = (a_{ij})_{1 \le i,j \le n}$ be an $(n \times n)$ matrix with nonnegative entries $a_{ij} \geq 0$, for all $i, j = 1, \ldots, n$. Then A has a real nonnegative eigenvalue corresponding to a nonnegative eigenvector (i. e., with nonnegative components).*

Proof. Let

$$C = \left\{ x = (x_1, x_2, \ldots, x_n) \in \mathbb{R}^n : \; x_i \geq 0, \text{ for all } i = 1, 2, \ldots, n \text{ and } \sum_{i=1}^{n} x_i = 1 \right\}.$$

We have

(1) C is convex: if $x, y \in C$ and $t \in [0,1]$, then $tx_i + (1-t)y_i \geq 0$ for all $i = 1, 2, \ldots, n$ and

$$\sum_{i=1}^{n} tx_i + (1-t)y_i = t \sum_{i=1}^{n} x_i + (1-t) \sum_{i=1}^{n} y_i = t + (1-t) = 1.$$

(2) The set C is closed: let $(x^k)_k \subset C$ be a sequence which converges to some limit $x \in \mathbb{R}^n$. Then, for all $i = 1, 2, \ldots, n$, $\lim_{k\to\infty} (x_i)^k = x_i$. Since $(x_i)^k \geq 0$ for all $i = 1, 2, \ldots, n$ and for all $k = 1, 2, \ldots$, $x_i \geq 0$ for all $i = 1, 2, \ldots, n$. Furthermore,

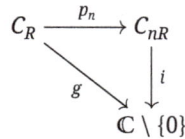

$$\sum_{i=1}^{n} x_i = \sum_{i=1}^{n} \lim_{k\to\infty} x_i^k = \lim_{k\to\infty} \sum_{i=1}^{n} x_i^k = \lim_{k\to\infty} 1 = 1.$$

Hence, $x \in C$.

(3) The set C is bounded: If we consider the norm $\|x\|_1 = \sum_{i=1}^{n} |x_i|$ in Euclidean space $\mathbb{R}^n$, then $C \subset \partial B(0,1) \subseteq B(0,1)$. Hence, C is bounded.

(4) If there exists $x \in C$ such that $Ax = 0$, then the theorem is proved with the eigenvalue $\lambda = 0$.

(5) If $Ax \neq 0$ for all $x \in C$, then we can define a map

$$f : C \longrightarrow C \quad \text{by}\, f(x) = \frac{1}{\|Ax\|_1} Ax.$$

The function f is well-defined.

(a) Since the matrix $A = (a_{ij})_{1\le i,j\le n}$ has positive entries, $(Ax)_i \geq 0$, for all $x \in C$. Hence, $\|Ax\|_1 = \sum_{i=1}^{n} (Ax)_i > 0$.

(b) We have

$$\begin{aligned}\sum_{i=1}^{n} (f(x))_i &= \sum_{i=1}^{n} \frac{1}{\|Ax\|_1}(Ax)_i \\ &= \frac{1}{\|Ax\|_1} \sum_{i=1}^{n} (Ax)_i \\ &= \frac{1}{\|Ax\|_1} \|Ax\|_1 = 1.\end{aligned}$$

Therefore, f maps C in C. Finally, f is continuous for Ax and $\|Ax\|_1$ are continuous. By Brouwer's fixed-point theorem, there exists $x_0 \in C$ such that

$$\begin{aligned} f(x_0) = x_0 \quad &\Longleftrightarrow \quad \frac{1}{\|Ax_0\|_1} Ax_0 = x_0 \\ &\Longleftrightarrow \quad Ax_0 = \lambda x_0, \quad \text{with } \lambda = \|Ax_0\|_1. \end{aligned}$$

□

12.3 Retracts of $\mathbb{R}^n$

We mainly prove two results mentioned without proof in [9]. The first one is [9, Chapter I (3.5)].

Proposition 12.3.1. *Let A be a retract of $\mathbb{R}^n$. Then $\mathbb{R}^n \setminus A$ is unbounded.*

We need an auxiliary result (see [9, Chapter I (3.6)]). This is an application of Brouwer's fixed- point theorem.

Lemma 12.3.2. *Let $X \subset \mathbb{R}^n$ be closed and C a bounded component of $\mathbb{R}^n \setminus X$. Then the identity on the boundary ∂C does not admit a continuous extension $f : \overline{C} \longrightarrow X$.*

Proof. Without loss of generality, assume that C is the component contained in the unit open ball $\mathring{B}^n(0,1)$. By contradiction, let f be as in the statement of the lemma. Define the map $\varphi : B^n(0,1) \longrightarrow \partial B^n(0,1)$ by

$$\varphi(x) = \begin{cases} -\frac{x}{\|x\|}, & x \in B^n \setminus C, \\ -\frac{f(x)}{\|f(x)\|}, & x \in \overline{C}. \end{cases}$$

Since $f(x) = x$ for all $x \in \overline{C} \setminus C$, both definitions of φ coincide, and thus by the pasting Lemma A.2.1 φ is continuous. By Brouwer's fixed-point theorem, there exists $x_0 \in B^n$ such that $\varphi(x_0) = x_0$. Then $\varphi(x_0) \notin C$ since $C \subset \mathring{B}^n \subset B^n$ and $\varphi(x_0) \in \partial B^n(0,1)$. Hence, $x_0 \notin C$, which entails by definition of the function φ,

$$\varphi(x_0) = -\frac{x_0}{\|x_0\|} = x_0.$$

Since $\|x_0\| = 1$, we get $x_0 = 0$ and a contradiction is reached. □

Proof of Proposition 12.3.1. Let $r : \mathbb{R}^n \longrightarrow A$ be a retraction and $U = \mathbb{R}^n \setminus A$, where $U = \bigcup_k C_k$ is a countable union of components (Exercise 10, Section 1.4). Suppose that $\partial C_k \subset A$ for some $k = 1, 2 \dots$. If U is bounded, each component C_k is bounded. In addition, the restriction r_A is the identity on A, hence on ∂C_k. This contradicts Lemma 12.3.2. If $\partial C_k \not\subset A$ for all $k = 1, 2 \dots$, then for all k there exists $x_k \in \partial C_k = \overline{C_k} \setminus C_k$ and $x_k \notin A$. Hence, for all k, $x_k \notin C_k$ and $x_k \in U = \bigcup_k C_k$. That is for all k, there exists $k' \neq k$ such that $x_k \in \overline{C_k} \cap C_{k'}$. Since U is open and $\mathbb{R}^n$ locally connected, all components are open. Therefore, $C_{k'}$ is a neighborhood of x_k, which intersects C_k. This is impossible since two components do not intersect. Finally, recall that the components of the open set U are path-connected (Exercise 10, Section 1.4). □

The second result is [9, Chapter I (3.7)].

Proposition 12.3.3. *Let A be a compact retract of $\mathbb{R}^n$, $n > 1$. Then the complement $\mathbb{R}^n \setminus A$ is path-connected.*

Proof. Let $r : \mathbb{R}^n \longrightarrow A$ be a retraction and set $U = \mathbb{R}^n \setminus A$. By contradiction, assume that U has more than one component, i. e., $U = \cup_k C_k$ is a countable union of components. Since $n > 1$ and A is compact, only one component C is unbounded. Let $U = \bigcup_k C_k \cup C$, where C_k is bounded for all $k = 1, 2, \dots$. Arguing as in the proof of Proposition 12.3.1, one can find some k such that $\partial C_k \subset A$. Then the restriction $f = r_{|\overline{C_k}}$ contradicts Lemma 12.3.2.
□

Example 12.3.4. (1) For all $n \geq 1$, $\mathbb{R}^n \setminus \mathring{B}^n$ is not a retract of $\mathbb{R}^n$.
(2) For all $n \geq 2$, S^{n-1} is not a retract of $\mathbb{R}^n$.

For $n = 1$, the situation is different ([9, Chapter I (3.8)]).

Proposition 12.3.5. *Let A be a compact retract of $\mathbb{R}$. Then the complement $\mathbb{R}^n \setminus A$ has two path-components.*

Proof. Arguing as in the proof of Proposition 12.3.3, the complement $U = \mathbb{R} \setminus A = (\bigcup_k C_k) \cup C$ is a countable union of intervals, where C is unbounded and all C_k are bounded. For instance,

$$C = (-\infty, c_0) \quad \text{and} \quad C_k = (a_k, b_k), \quad k \geq 1 \quad (c_0 < a_k \text{ and } a_k < b_k \text{ for all } k \geq 1).$$

Then $A = [c_0, a_1] \cup_k [b_k, a_{k+2}]$ is not connected, which contradicts the existence of a continuous function $r : \mathbb{R}^n \longrightarrow A$. □

12.4 Initial and boundary value problems

In this section, Schauder's fixed-point theorem is employed to solve some initial and boundary value problems for some ordinary differential equations (IVPs and BVPs for short, respectively).

12.4.1 Existence theory for initial value problems

Consider the Cauchy problem

$$\begin{cases} y'(x) = f(x, y(x)), & x \in J \subseteq \mathbb{R}, \\ y(x_0) = y_0, \end{cases} \tag{12.1}$$

where J is an interval of $\mathbb{R}$ with x_0 as left endpoint, i. e., $J = [x_0, x_0 + \delta]$, $J = [x_0, x_0 + \delta)$ or $J = [x_0, \infty)$. The nonlinear nonzero function $f : J \times \Omega \longrightarrow \mathbb{R}^n$ is continuous and y_0 is some point in $\Omega \subseteq \mathbb{R}^n$, an open set. Note that problem (12.1) is equivalent to the integral equation

$$y(x) = y_0 + \int_{x_0}^{x} f(t, y(t))dt, \quad x \in J.$$

Theorem 12.4.1 (Peano's theorem, local existence). *Problem* (12.1) *has at least one solution y defined on some interval* $[x_0, x_0 + a]$ *($a > 0$). y is called a local solution.*

Proof. Let $a > 0$ be such that $J_1 = [x_0, x_0 + a] \subseteq J$ and let $B = \mathring{B}(y_0, b)$ be an open ball contained in the open set Ω. Let the compact subset $K = J_1 \times \overline{B}$ in $\mathbb{R} \times \mathbb{R}^n$ and

$$M = \sup_{(t,y) \in K} \|f(x, y)\|.$$

Since K is compact and f is continuous, $0 < M < \infty$. Let

$$0 < \alpha \leq \min\left(a, \frac{b}{M}\right) \quad \text{and} \quad J_2 = [x_0, x_0 + \alpha] \subseteq J_1 \subseteq J.$$

Consider the space $X = \mathscr{C}(J_2, \overline{B})$ equipped with the sup-norm

$$\|y\|_X = \sup_{x \in J_2} \|y(x)\|.$$

Then $(X, \|\cdot\|_X)$ is a Banach space. Given the nonlinear integral equation, define an operator T on X by

$$Ty(x) = y_0 + \int_{x_0}^{x} f(t, y(t))dt.$$

Then T enjoys the following properties:

(1) The mapping T maps X in X. Let $y \in X$. Then, for all $x \in J_2$,

$$\begin{aligned}\|Ty(x) - y_0\| &\leq \int_{x_0}^{x} \|f(t, y(t))\|dt \\ &\leq M(x - x_0) \\ &\leq M\alpha \leq b.\end{aligned}$$

Hence, $Ty(x) \in \overline{B}(y_0, b)$ for all $x \in J_2$.

(2) The map T is continuous. Let $(y_n)_n \in X$ be a sequence, which converges to some limit $y \in X$. Since f is continuous, $\lim_{n\to\infty} f(t, y_n(t)) = f(t, y(t))$ for all $t \in J_2$. Hence, for all x in the compact interval J_2, we have

$$\lim_{n\to\infty} \int_{x_0}^{x} f(t, y_n(t))dt = \int_{x_0}^{x} f(t, y(t))dt.$$

Therefore, $\lim_{n\to\infty} Ty_n = Ty$ for the sup-norm in X. From (1) and (2), we deduce that $Ty \in X$.

(3) The map T is compact. Let $(y_n)_n \subset X$ and $\mathscr{H} = (Ty_n)_n \subset X$ By Corollary A.9.3, we need to check only the equicontinuity of $\mathscr{H}$, which is the case here with $F = \overline{B}$. Let $Y_n = Ty_n$.

(a) For all $x \in J$ and $n = 1, 2, \ldots$, $\|Y_n(x) - y_0\| \leq b$. Hence, $\|Y_n - y_0\|_X \leq b$ for all $n = 1, 2, \ldots$ and so by the triangle inequality,

$$\|Y_n\|_X \leq \|Y_n - y_0\|_X + \|y_0\| \leq b + \|y_0\|.$$

Hence, $\mathscr{H}$ is uniformly bounded, independently of $n = 1, 2, \ldots$.

(b) The map $\mathscr{H}$ is equicontinuous. For each $\varepsilon > 0$, there exists $\delta = \frac{\varepsilon}{M} > 0$ such that

$$\begin{aligned}\|Y_n(x) - Y_n(x')\|_X &\leq \left|\int_x^{x'} \|f(t, y_n(t))\|dt\right| \\ &\leq M|x - x'| \\ &\leq M\delta = \varepsilon,\end{aligned}$$

whenever $|x - x'| \leq \delta$. This completes the proof of the compactness of the map $T : X \longrightarrow X$.

By Schauder's fixed-point theorem, T has at least one fixed point $y = Ty \in X$, a solution y of the integral equation. Hence, y is defined on the interval $J = [x_0, x_0 + a] \subseteq I$ and is solution of problem (12.1). □

Theorem 12.4.2 (Peano's theorem, global existence). *Let $f : I \times \Omega \longrightarrow \mathbb{R}^n$ be a continuous and bounded function. Then problem* (12.1) *has at least one solution y defined on the whole interval J. y is called a global solution.*

Proof. Let $a > 0$ be such that $J_1 = [x_0, x_0 + a] \subseteq J$ and

$$M = \sup_{(t,x)\in I\times\Omega} \|f(t,x)\|$$

for f is bounded. For some positive constant b, let B be an open ball centered at y_0 with radius b. Then we follow exactly the same steps as in the proof of Theorem 12.4.1 using Schauder's fixed- point theorem for a fixed-point operator defined on the Banach space $X = \mathscr{C}(J_1, \overline{B})$. We obtain a solution defined on $[x_0, x_0+a]$ for all $a > 0$ such that $[x_0, x_0+a] \subseteq J$. Then the solution is defined over the entire interval J. □

12.4.2 Existence theory for boundary value problems

Consider the Dirichlet boundary value problem

$$\begin{cases} y''(x) = f(x, y(x), y'(x)), & 0 < x < 1, \\ y(0) = 0, \quad y(1) = 0, \end{cases} \tag{12.2}$$

where $f : I \times \mathbb{R}^2 \longrightarrow \mathbb{R}$ is continuous and satisfies the growth condition

($\mathscr{H}$) There exist $k > 0$ and $p > 0$ such that for all $(x, y, z) \in [0, 1] \times \mathbb{R}^2$, $|f(x, y, z)| \leq k(1 + |y|^p + |z|^p)$.

Theorem 12.4.3. *Further to Assumption* ($\mathscr{H}$), *assume that either*

$$\left(p = 1 \text{ and } 0 < k < \frac{8}{5}\right) \quad \text{or} \quad \left(p \neq 1 \text{ and } 0 < k < \frac{8}{2 + 4^p}\right).$$

Then BVP (12.2) *has at least one solution* $y \in \mathscr{C}^2([0,1];\mathbb{R})$.

Proof. Let G be the Green's function of the linear boundary value problem

$$\begin{cases} y''(x) = 0, & 0 < x < 1 \\ y(0) = y(1) = 0, \end{cases}$$

i. e.,

$$G(x,t) = \begin{cases} x(t-1), & 0 \le x \le t \le 1, \\ t(x-1), & 0 \le t \le x \le 1. \end{cases}$$

Then BVP (12.2) is equivalent to the integral equation

$$y(x) = \int_0^1 G(x,t) f(t, y(t), y'(t))dt.$$

This suggests to introduce the Banach space $X = \mathscr{C}^1([0,1];\mathbb{R})$ endowed with the norm

$$\|y\|_X = \max\left(\sup_{0 \le x \le 1} |y(x)|, \frac{1}{4} \sup_{0 \le x \le 1} |y'(x)| \right).$$

Define the operator $T : X \longrightarrow X$ by

$$Ty(x) = \int_0^1 G(x,t) f(t, y(t), y'(t))dt.$$

Then y is solution of BVP (12.2) if and only if y is a fixed point of T in X. First, notice that since f is a continuous function, Ty is a continuous function for each $y \in X$. We have

$$(Ty)'(x) = \int_0^1 \frac{\partial G}{\partial x}(x,t) f(t, y(t), y'(t))dt,$$

which is a continuous function, too. Hence, $Ty \in X$ for all $y \in X$, i. e., T is well-defined.

(1) The map T is continuous. Let $(y_n)_n \subset X$ be a sequence, which converges to some limit $y \in X$. We have $\lim_{n\to\infty} y_n(x) = y(x)$ and $\lim_{n\to\infty} y_n'(x) = y'(x)$ for all $x \in [0,1]$. Since f is continuous, we deduce that

$$\lim_{n\to\infty} f(x, y_n(x), y_n'(x)) = f(x, y(x), y'(x)), \quad \text{for all } x \in [0,1].$$

Since the interval $[0,1]$ is compact, the sequences $(Ty_n)_n$ and $(Ty_n)'_n$ converge uniformly to the limits Ty and $(Ty)'$, respectively. This shows that $(Ty_n)_n$ converges to Ty in X, i. e., T is sequentially continuous. Hence, T is continuous.

(2) A priori estimates. Since $\int_0^1 |G(x,t)|dt \le \frac{1}{8}$ for all $x \in [0,1]$,

$$\begin{aligned}
|Ty(x)| &\le \int_0^1 |G(x,t)| \cdot |f(t, y(t), y'(t)|dt \\
&\le k \int_0^1 |G(x,t)|(1 + |y(t)|^p + |y'(t)|^p)dt \\
&\le k \int_0^1 |G(x,t)|\Big(1 + \big(\sup_{0\le t\le 1} |y(t)|\big)^p + \big(\sup_{0\le t\le 1} |y'(t)|\big)^p\Big)dt.
\end{aligned}$$

Hence, for all $x \in [0,1]$,

$$|Ty(x)| \le \frac{k}{8}(1 + \|y\|_X^p + (4\|y\|_X)^p).$$

Also, for all $x \in [0,1]$

$$|(Ty)'(x)| \le \frac{k}{2}(1 + \|y\|_X^p + (4\|y\|_X)^p).$$

By definition of the norm $\|\cdot\|_X$, we deduce that

$$\begin{aligned}
\|Ty\|_X &= \max\Big(\sup_{0\le x\le 1} |Ty(x)|, \frac{1}{4} \sup_{0\le x\le 1} |(Ty)'(x)|\Big) \\
&\le \frac{k}{8}(1 + \|y\|_X^p + 4^p \|y\|_X^p).
\end{aligned}$$

(3) The map T self-maps a closed ball $B(0,R)$. Let $y \in B(0,R)$, where $R > 0$ is to be determined. Then $\|y\|_X \le R$ and

$$\|Ty\|_X \le \frac{k}{8}(1 + (1 + 4^p)R^p).$$

We distinguish between two cases:

(a) Case when $p = 1$. Then $\|Ty\|_X \le \frac{k}{8}(1 + 5R) \le R$ if and only if $\frac{k}{8} \le R\frac{8-5k}{8}$, i. e.,

$$R \ge R_0 = \frac{k}{8 - 5k}$$

because $0 < k < \frac{8}{5}$ by assumption.

(b) Case when $p \neq 1$. Then

$$\|Ty\|_X \leq \frac{k}{8}(1 + (1 + 4^p))R^p.$$

If we choose $0 < R < 1$, then

$$\|Ty\|_X \leq \frac{k}{8}(2 + 4^p),$$

which is less than 1 if $\frac{k}{8}(2 + 4^p) < 1$, that is if

$$0 < k < \frac{8}{2 + 4^p}.$$

In all cases, we obtain that

$$\|Ty\|_X \leq R, \quad \text{for all } y \in B(0, R),$$

which shows that the operator T self-maps the closed ball $B(0, R)$.

(4) The map $T : B(0, R) \longrightarrow B(0, R)$ is compact. The proof runs parallel to the one used to show the compactness of T in Theorem 12.4.1 (by means of the Ascoli–Arzéla lemma).

(5) Conclusion. By Schauder's fixed-point theorem, the mapping T has at least one fixed point $y \in X = \mathscr{C}^1([0, 1]); \mathbb{R})$ solution of BVP (12.2). Since f is continuous, y'' is continuous. Therefore, $y \in \mathscr{C}^2([0, 1]; \mathbb{R})$. □

12.5 Applications of homology theory

12.5.1 Degree of a map on the sphere

The definition of a degree presented in Section 6.2 is extended now to any dimension. It characterizes homotopic maps on the sphere. We consider a homology given by the Eilenberg–Steenrod axioms of the relative homology (Axiom 1–Axiom 7 of Definition 10.3.4).

Definition 12.5.1. Let $f : S^n \longrightarrow S^n$ be a continuous map. By Theorem 10.3.21, the induced homomorphism $f_* : H_n(S^n) \longrightarrow H_n(S^n)$ is isomorphic to a homomorphism on $\mathbb{Z}$. This homomorphism is completely determined by the image of 1, equivalently by the multiplication by some integer d, called the *degree of* f and denoted $d = f_*(1) = \deg f$.

Proposition 12.5.2. *We have*

(1) *The degree of a constant map is zero.*

(2) $\deg \mathrm{Id} = 1$.

Let $f, g : S^n \longrightarrow S^n$ be two continuous maps.

(3) $\deg f = \deg g$ *if and only if f and g are homotopic* (Hopf's classification theorem).
(4) $\deg(f \circ g) = (\deg f).(\deg g)$.
(5) *If f is right (or left) invertible, then* $\deg f = \pm 1$.

Proof. (1) Since f is constant, so is the induced homomorphism f_*. Since f_* is a homomorphism, $\deg f = f_*(1) = f_*(0) = 0$.

(2) By Axiom 1, Id_* is the identity on $H_n(S^n) \cong \mathbb{Z}$, i. e., $\mathrm{Id}_*(n) = n = n.1$, i. e., $\deg \mathrm{Id} = 1$.

(3) The sufficient condition follows from Axiom 5. Conversely, assume that $\deg f = \deg g$. By the definition of the degree, the homomorphisms f_* and g_* coincide. Then, for all $x \in S^n$,

$$f_*([x]) = g_*([x]) \iff [f(x)] = [g(x)] \iff f \simeq g.$$

(4) From Axiom 2, we have $(f \circ g)_* = f_* \circ g_*$. Hence,

$$d(f \circ g) = (f \circ g)_*(1) = f_*(g_*(1)) = f_*(\deg g) = (\deg g).(f_*(1)) = (\deg f).(\deg g).$$

(5) By properties (a) and (c), $(\deg f).(\deg f^{-1}) = \deg(f \circ f^{-1}) = \deg \mathrm{Id} = 1$. Since $\deg f$ is an integer, the result follows. □

The following result follows from Proposition 12.5.2 and the proof of part (1).

Corollary 12.5.3. (1) *A continuous maps $f : S^n \longrightarrow S^n$ is null-homotopic if and only if* $\deg f = 0$.
(2) *The composition of a continuous maps and a null-homotopic is a null-homotopic map.*

Proposition 12.5.4. *If $f : S^n \longrightarrow S^n$ is a homotopy equivalence (in particular a homeomorphism), then* $\deg f = \pm 1$.

Proof. Let $f : S^n \longrightarrow S^n$ and $g : S^n \longrightarrow S^n$ be such that $g \circ f \simeq \mathrm{Id}_{S^n}$ and $f \circ g \simeq \mathrm{Id}_{S^n}$. Then $1 = \deg \mathrm{Id} = \deg(f \circ g) = (\deg f).(\deg g)$. Then $\deg f = \deg g = \pm 1$. □

Proposition 12.5.5. *Let $R_\theta : S^1 \longrightarrow S^1$ be the rotation of angle θ given by $R_\theta(z) = ze^{i\theta}$. Then* $\deg R_\theta = 1$.

Proof. Let $R_\theta(z) = z_0 z$, where $z_0 = e^{i\theta}$, $\omega \in S^1$ is a square root of z_0 and $g(z) = \omega z$. Then g is also a rotation, hence a homeomorphism. By Proposition 12.5.4, $\deg g = \pm 1$. In addition, Proposition 12.5.2(4) implies that $\deg(g \circ g) = 1$. Since $g \circ g = R_\theta$, we infer $\deg R_\theta = 1$. □

Example 12.5.6. Consider the antipodal map $a : S^1 \longrightarrow S^1$ given by $a(z) = -z$. Since $a = R_\pi$, $\deg a = 1$. This result will be generalized to any sphere S^n in Proposition 12.5.25.

Corollary 12.5.7. *On the unit circle S^1, the identity, the antipodal map $a(z) = -z$ and the rotation are homotopic.*

Now, we return back again to Proposition 6.2.3 (see also Example 5.2.18(4)(b)). Using the induced map of f, we present a different proof.

Proposition 12.5.8. *Let $p(z) = z^n : S^1 \longrightarrow S^1$ be the n-fold covering map given by Theorem 5.1.15. Then $\deg f = n$.*

Proof. Consider the isomorphisms $\Psi : \pi_1(S^1) \longrightarrow \mathbb{Z}$ and $\Phi : \mathbb{Z} \to \pi_1(S^1)$ defined in Theorem 6.1.4 and Theorem 6.1.5 by $\Psi([f]) = \omega(f) = \widehat{f}(1)$ and $\Phi(n) = [e^{2ni\pi t},\ t \in I]$, respectively. Let the diagram

$$\mathbb{Z} \xrightarrow{\quad\Phi\quad} \pi_1(S^1) \xrightarrow{\quad p_*\quad} \pi_1(S^1) \xrightarrow{\quad\Psi\quad} \mathbb{Z}$$

Since by definition of the induced map $p_*([f]) = [p \circ f] = [f^n]$, we have $p_*([e^{2ik\pi t}]) = [e^{2ikn\pi t}]$. Then

$$(\Psi \circ p_* \circ \Phi)(k) = (\Psi \circ p_*)([e^{2ik\pi t}]) = \Psi([e^{2ikn\pi t}]) = nk. \qquad \square$$

Proposition 12.5.9. *Let $f : S^1 \longrightarrow S^1$ be a continuous map whose degree d is not a prime number and let $y_0 \in S^1$. Then there exist a positive integer n, a point $\widehat{x}_0 \in S^1$ and a continuous map $\widehat{f} : S^1 \longrightarrow S^1$ unique solution of the algebraic equation $(\widehat{f})^n = f$ subject to the condition $\widehat{f}(y_0) = \widehat{x}_0$.*

Proof. Let n be a divisor of d and $p(z) = z^n : S^1 \longrightarrow S^1$ the n-fold covering map given by Theorem 5.1.15. Since p is surjective, let $\widehat{x}_0 \in S^1$ be such that $p(\widehat{x}_0) = f(y_0)$. Notice that the condition involved in Theorem 5.2.16 holds. That is, $f_*(\mathbb{Z}) \subseteq p_*(\mathbb{Z})$, which is equivalent to $d\mathbb{Z} \subseteq n\mathbb{Z}$. Hence, f admits a unique lifting map $\widehat{f} : S^1 \longrightarrow S^1$ satisfying $p \circ \widehat{f} = f$ and $\widehat{f}(y_0) = \widehat{x}_0$, whence the claim. $\square$

Example 12.5.10. Let $f(z) = z^d$ be a power map, where d is not a prime number. Then the lifting map is also a power map $\widehat{f}(z) = z^k$, where $k = d/n$ and n is a divisor of d.

Proposition 12.5.11. *For $n = 1$, the degree $d_2 = \deg f$ and the degree d_1 defined in Definition 6.2.1 coincide.*

Proof. Let $d_2 = \deg f$ be the degree of the map $f : S^1 \longrightarrow S^1$ given by Definition 12.5.1 and let $p : S^1 \longrightarrow S^1$ be the power map given by $p(z) = z^{d_2}$. By Proposition 6.2.3, $\deg p = d_2$. Proposition 12.5.2 implies that the maps f and p are homotopic. By Proposition 6.1.2, we conclude that $d_1 = \deg f = \deg p = d_2$, as claimed. $\square$

Theorem 12.5.12. *Let $f : S^n \longrightarrow S^n$ be a continuous map. Then the following statements are equivalent:*
(1) *f is null-homotopic;*
(2) *The induced map f_* is the trivial homomorphism;*
(3) *$\deg f = 0$;*
(4) *f extends continuously to $\widetilde{f} : B^{n+1} \longrightarrow S^n$.*

Proof. By Theorem 6.4.1, statement (1) is equivalent to statement (4). The equivalence between (2) and (3) is an immediate consequence of the definition of the degree. Clearly, (1) is equivalent to (2) since for any constant map c, we have

$$f \simeq c \iff \deg f = \deg c = 0 \iff f_* = 0.$$

The equivalence between (1) and (3) is Corollary 12.5.3. □

Remark 12.5.13. (1) The equivalences (1) $\iff$ (2) $\iff$ (4) in Theorem 12.5.12 still hold for any target space X instead of S^n (see Theorem 6.4.1).
(2) However, one cannot replace S^n in both sides by a subspace $A \subset X$. Indeed, for $f = \mathrm{Id}_{|A}$, (1) means that A is contractible whereas (4) states that A is a retract of X.

In Theorem 12.5.12, the extension means that the target space S^{n-1} is kept the same. For a larger target space, $\widetilde{f}$ may exist even if f is not null-homotopic. This is shown by the following result known as Alexander's trick.

Proposition 12.5.14. *Let $h : S^n \longrightarrow S^n$ be a homeomorphism. Then:*
(1) *h is not null-homotopic;*
(2) *h has an "extension" $\widetilde{h} : B^{n+1} \longrightarrow B^{n+1}$, which is a homeomorphism;*
(3) *If h is the identity on S^n, $\widetilde{h}$ is homotopic to the identity on the ball B^{n+1}.*

Proof. (1) follows from Proposition 12.5.4.

(2) Define $\widetilde{h} : B^{n+1} \longrightarrow B^{n+1}$ by

$$\widetilde{h}(x) = \begin{cases} \|x\| h(x/\|x\|), & x \neq 0, \\ 0, & x = 0. \end{cases}$$

Since h assumes values in S^n, $\|\widetilde{h}(x)\| = \|x\|$. Hence, $\widetilde{h}$ is continuous. Moreover, $\widetilde{h}(x) = h(x)$ for all $x \in S^n$ and $\widetilde{h}$ is an inversion.

(3) A homotopy $H : B^{n+1} \times I \longrightarrow B^{n+1}$ is given by

$$H(x,t) = \begin{cases} t\widetilde{h}(x/t), & 0 < \|x\| \leq t, \\ x, & t \leq \|x\| \leq 1. \end{cases}$$

For $0 < \|x\| = t$, $x/t \in S^{n-1}$, and $\widetilde{h}(x/t) = h(x/t) = x/t$. Hence, $H(x,t) = x$, H is continuous at $x = 0$, both definition of H coincide and H is continuous on $B^{n+1} \times I$. Finally, $H(x,0) = x$ and $H(x,1) = \widetilde{h}(x)$ for all $x \in B^n$, as claimed. □

It turns out that Theorem 12.5.12 provides an equivalent form of Brouwer's fixed-point theorem. Statement (2) in the following formulation also bears Brouwer's name or Böhl's theorem.

Theorem 12.5.15. *The following statements are equivalent:*
(1) *Brouwer's fixed-point theorem holds for the closed ball B^n.*
(2) *The identity map $\mathrm{Id} : S^n \longrightarrow S^n$ is not null-homotopic.*
(3) *The inclusion map $i : S^n \hookrightarrow \mathbb{R}^{n+1} \setminus \{0\}$ is not null-homotopic.*

Proof. (1) $\Longrightarrow$ (2): Assume by contradiction that $\mathrm{Id} : S^n \longrightarrow S^n$ is null-homotopic. By Theorem 12.5.12, there is a continuous extension $\widetilde{\mathrm{Id}} : B^{n+1} \longrightarrow S^n \subset B^{n+1}$. Then $\widetilde{\mathrm{Id}}(x) = x$ and $-\widetilde{\mathrm{Id}}(x) = -x$ for all $x \in S^n$. Since $x \neq -x$, $-\widetilde{\mathrm{Id}}$ cannot have a fixed point (in S^n), violating Brouwer's fixed-point theorem.

(2) $\Longrightarrow$ (1): Suppose by contradiction that Brouwer's fixed-point theorem does not hold. By Remark 11.1.2, S^n is a retract of B^{n+1}. Hence, the identity map $\mathrm{Id} : S^n \longrightarrow S^n$ has a continuous extension $\widetilde{\mathrm{Id}} : B^{n+1} \longrightarrow S^n$, which means by Theorem 12.5.12 that Id is null-homotopic, a contradiction.

(2) $\Longrightarrow$ (3): Suppose by contradiction that (3) fails. By Theorem 6.4.1, the inclusion map i has a continuous extension $\tilde{i}$. Consider the following diagram, where $r(x) = x/\|x\|$ is the radial retraction:

$$B^{n+1} \xrightarrow{\quad \tilde{i} \quad} \mathbb{R}^{n+1} \setminus \{0\} \xrightarrow{\quad r \quad} S^n, \qquad f = r \circ \tilde{i} : B^{n+1} \longrightarrow S^n$$

Then $\tilde{i}(x) = x$ and $f(x) = x$ for all $x \in S^n$. Hence, f is a continuous extension of the identity on S^n, which violates Statement (2).

(3) $\Longrightarrow$ (2): Assume that (2) does not hold and let $\widetilde{\mathrm{Id}} : B^{n+1} \longrightarrow S^n \subset \mathbb{R}^{n+1} \setminus \{0\}$ be a continuous extension of the identity on S^n. Then $\widetilde{\mathrm{Id}}$ is also a continuous extension of the inclusion map i, contradicting (3). □

Remark 12.5.16. A proof of the equivalence between statements (1) and (2) of Theorem 12.5.15 has already been given in Remark 11.1.10 because (2) merely means that S^n is not contractible (see Theorem 7.5.13).

Since by Proposition 12.5.2, the degree of the identity map is 1, we deduce Brouwer's fixed- point theorem (Theorem 11.1.1).

Corollary 12.5.17. *Every continuous function $f : B^{n+1} \longrightarrow B^{n+1}$ has a fixed point.*

All equivalent formulations of Brouwer's fixed-point theorem proved in Chapter 11 and this chapter (Corollary 11.3.5 and Theorem 12.5.15) are collected in the following one.

Corollary 12.5.18. *The following statements, which are equivalent hold:*
(1) *Brouwer's fixed-point theorem for the closed ball B^n.*
(2) *Hartmann–Stampacchia theorem for nonempty compact convex sets.*
(3) *Knaster–Kuratowski–Mazurkiewicz theorem.*
(4) *Mini-max Ky Fan theorem.*

(5) *The boundary of a closed ball is not a retract of the ball.*
(6) *The boundary of a closed ball is not contractible.*
(7) *The identity map* Id : $S^n \longrightarrow S^n$ *is not null-homotopic.*
(8) *The inclusion map* $i : S^n \hookrightarrow \mathbb{R}^{n+1} \setminus \{0\}$ *is not null-homotopic.*
(9) *Corollary* 11.1.15 *holds.*

The following result extends the fundamental theorem of algebra Theorem 12.1.1 (see [25, Chapter XVI, 3.1]).

Proposition 12.5.19. *Let f be a continuous complex function, asymptotic to a power function, i. e.,* $\lim_{|z|\to\infty} \frac{f(z)}{z^n} = 1$, *for some* $n = 1, 2, \dots$. *Then f admits at least one zero.*

Proof. Assume to the contrary that $f(z) \neq 0$ for all $z \in \mathbb{C}$. The function $H(z,t) = \frac{f(tz)}{|f(tz)|}$ is a homotopy on $S^1 \times [0,1]$ joining $\frac{f(z)}{|f(z)|}$ and the constant map $\frac{f(0)}{|f(0)|}$. By Proposition 12.5.2(3), they have same degree, namely zero. By assumption, $\lim_{t\to\infty} \frac{H(z,t)}{z^n} = 1$, which implies $\|H(z,t_0) - z^n\| < 1$, for sufficiently large t_0. Let $z \in S^1$. If there is some $t \in [0,1]$ such that $tz^n + (1-t)H(z,t_0) = 0$, then $t \neq 0, 1$ and

$$\|H(z,t_0) - z^n\| = \frac{1}{1-t}\|z^n\| = \frac{1}{1-t} < 1,$$

which is impossible. Hence, the map G defined by

$$G(z,t) = \frac{tz^n + (1-t)H(z,t_0)}{\|tz^n + (1-t)H(z,t_0)\|}$$

is a homotopy between $H(z,t_0)$ and z^n. Appealing to Proposition 6.2.3, we conclude that $0 = \deg H(z,t_0) = \deg z^n = n$, a contradiction. □

12.5.2 Reflection and antipodal map

Definition 12.5.20. (1) The *antipodal map* is defined by $a(x) = -x$ for all $x \in S^n$, that is to say $a = -\mathrm{Id}_{|S^n}$.
(2) A continuous map $f : S^n \longrightarrow S^n$ is called a *reflection map* if $f(x_0, x_1, \dots, x_n) = (-x_0, x_1, \dots, x_n)$.

Remark 12.5.21. In Definition 6.4.4, a continuous map $f : S^n \longrightarrow S^n$ is antipode-preserving if it commutes with the antipodal map: $f \circ a = a \circ f$.

Remark 12.5.22. The reflection map can be defined for any coordinate x_i $(i = 0, 1, \dots n)$ instead of x_0. If $h : S^n \longrightarrow S^n$ is the homeomorphism, which commutes some coordinate with the first one, namely $h(x_0, x_1, \dots, x_i, \dots, x_n) = (x_i, x_1, \dots, x_0, \dots, x_n)$, then $h^{-1} \circ f_0 \circ h = f_i$, where f_0 is the reflection map in the above definition and f_i the reflection map for the coordinate x_i $(i = 0, 1, \dots, n)$.

Proposition 12.5.23. *If $f : S^n \longrightarrow S^n$ is a reflection map, then* $\deg f = -1$.

Proof. We proceed by induction on n. For $n = 0$, $S^0 = \{-1, 1\}$ and $f(-1) = 1$. Then the induced homomorphism satisfies $f_*(p) = q$, where $\{p, q\}$ generates $H_0(S^0) \cong \mathbb{Z} \oplus \mathbb{Z}$. Then $f_*(p - q) = q - p$ and thus $f_*(p) = -p$, proving the claim for $n = 0$. Assume the claim is true for $n - 1$, that is the degree of a reflection map $f : S^{n-1} \longrightarrow S^{n-1}$ is -1. Consider, by Axiom 3 of Definition 10.3.4, the commutative diagram

$$\begin{array}{ccc} H_n(S^n) & \xrightarrow{\quad f_* \quad} & H_n(S^n) \\ \partial_n \big\downarrow & & \big\downarrow \partial_n \\ H_{n-1}(S^{n-1}) & \xrightarrow{\quad g_* \quad} & H_{n-1}(S^{n-1}) \end{array}$$

where $g_* = (f_{|S^{n-1}})_*$ and ∂_n is the boundary homomorphism. By the induction assumption, we have

$$-\partial_n(p) = (g_* \circ \partial_n)(p) = (\partial_n \circ f_*)(p) = d\partial_n(p),$$

where $d = \deg f$ and f maps S^n in S^n. Hence, $d = -1$. □

Since $\deg \mathrm{Id} = 1$ (Proposition 12.5.2), from Proposition 12.5.4, we have the following.

Corollary 12.5.24. *Every homeomorphism $f : S^n \longrightarrow S^n$ is homotopic either to the identity or to the reflection map.*

For $n = 1$, the following result was considered in Example 12.5.6.

Proposition 12.5.25. *If $a : S^n \longrightarrow S^n$ is the antipodal map, then* $\deg a = (-1)^{n+1}$.

Proof. Let $f_i : S^n \longrightarrow S^n$ $(i = 0, 1, \ldots n)$ be the reflection maps defined in Remark 12.5.22. Then the map $a = f_0 \circ f_1 \circ \cdots \circ f_i \circ \cdots \circ f_n$ is antipodal. By Proposition 12.5.23 and Proposition 12.5.2(d), we have

$$\deg a = \overbrace{(-1)(-1)\ldots(-1)\ldots(-1)}^{n+1} = (-1)^{n+1}.$$ □

Corollary 12.5.26. *The antipodal map $a : S^n \longrightarrow S^n$ and the identity map on S^n are homotopic if and only if n is odd.*

Lemma 12.5.27. *Let $f : S^n \longrightarrow S^n$ be a continuous map. Then the Lefschetz number (Definition 11.6.1) and the degree of f (Definition 12.5.1) are related by the following formula:*

$$\Lambda(f) = 1 + (-1)^n \deg f. \tag{12.3}$$

Proof. The induced map $f_*^0 : H_0(S^n) \longrightarrow H_0(S^n)$ of f is homotopic to the identity map. Indeed, if $\sigma : \Delta_0 \longrightarrow S^n$ is a 0-simplex on S^n and $\varphi : \Delta_1 \longrightarrow S^n$ is a path joining σ to $f \circ \sigma$, then the boundary homomorphism is $\partial_1(\varphi) = f \circ \sigma - \sigma$. Then $[f \circ \sigma] = [\sigma]$. For $1 < i < n$, the induced homomorphisms $f_*^i : H_i(S^n) \longrightarrow H_i(S^n)$ are trivial but $f_*^n : H_n(S^n) \longrightarrow H_n(S^n)$ is determined by the degree of f over the one-dimensional vector space $H_n(S^n)$. Therefore,

$$\Lambda(f) = \sum_{i=0}^{i=n} (-1)^i \operatorname{tr}[f_*^i, H_i(S^n)] = 1 + 0 + \cdots + 0 + (-1)^n \deg f. \qquad \square$$

Proposition 12.5.28. *Let $f : S^n \longrightarrow S^n$ be a continuous map, which is not homotopic to the antipodal map. Then f has at least one fixed point $x \in S^n$.*

Proof. (1) *First proof.* Assume that f is fixed point free and define the homotopy $H : S^n \times [0,1] \longrightarrow S^n$ by

$$H(x,t) = \frac{tf(x) + (1-t)(-x)}{\|tf(x) + (1-t)(-x)\|}.$$

If $tf(x) + (1-t)(-x) = 0$, then equality $tf(x) = (1-t)x$ does not hold for $t = 1$ because $f(x) \neq 0$ for all $x \in S^n$. Hence, $\frac{t}{1-t} f(x) = x$. By taking the norms, we get $\frac{t}{1-t} =$, i. e., $t = \frac{1}{2}$ and then $f(x) = x$, which is a contradiction. Hence, H is well-defined and represents a continuous homotopy from f to the antipodal map a, contradicting the hypothesis of the proposition.

(2) *Second proof.* By Proposition 12.5.2(3) and Proposition 12.5.25, f not homotopic to the antipodal map is equivalent to saying that $\deg f \neq (-1)^{n+1}$. Owing to (12.3), $\Lambda(f) \neq 0$. Then a fixed point of f follows from Lefschetz fixed-point Theorem 11.6.2. $\square$

Corollary 12.5.29. *Let $f : S^n \longrightarrow S^n$ be a null-homotopic continuous map. Then f has at least a fixed point $x \in S^n$.*

Proof. (1) *First proof.* Just use Proposition 12.5.28 since f is not homotopic to the antipodal map.

(2) *Second proof.* By Theorem 12.5.12, f possesses a continuous extension $\tilde{f} : B^{n+1} \longrightarrow S^n$. By Brouwer's fixed-point theorem, $\tilde{f}$ has a fixed point $x \in B^{n+1}$, $x = \tilde{f}(x)$. Since $\tilde{f}$ takes values in S^n, $x \in S^n$ and by definition of the extension $f(x) = x$. $\square$

We end this section with the degree of the suspension map on the sphere. Let $f : S^n \longrightarrow S^n$ and $\Sigma(f) : \Sigma(S^n) \longrightarrow \Sigma(S^n)$ the suspension map of f (see Exercise 3, Sections 3.3–3.7). Since by Theorem 3.3.6 $\Sigma(S^n) = S^{n+1}$, $\Sigma(f)$ maps S^{n+1} into S^{n+1}. We have the following.

Proposition 12.5.30. *The maps f and $\Sigma(f)$ have the same degree.*

Proof. By Axiom 3 (Definition 10.3.4), the following diagram commutes:

$$\begin{array}{ccc} H_{n+1}(S^{n+1}) & \xrightarrow{\Sigma(f)_*} & H_{n+1}(S^{n+1}) \\ \Big\downarrow{\scriptstyle \partial_n} & & \Big\downarrow{\scriptstyle \partial_{n+1}} \\ H_n(S^n) & \xrightarrow{f_*} & H_n(S^n) \end{array}$$

where ∂_n is the boundary homomorphism. Let $d = \deg f$ and $\Sigma(d) = \deg \Sigma(f)$. Then, for every $m \in \mathbb{Z}$,

$$(f_* \circ \partial_n)(m) = d\partial_n(m) = \partial_{n+1}(\Sigma(d)m) = \Sigma(d)\partial_{n+1}(m).$$

Hence, both degrees are identical. □

Corollary 12.5.31. *For each $d \in \mathbb{Z}$ and all $n = 1, 2 \dots$, there exists a continuous map $f : S^n \longrightarrow S^n$ such that $\deg f = d$.*

Proof. We proceed by induction on n. For $n = 1$, $f(z) = z^d$ has degree d (Proposition 6.2.3). Assume the statement true for n, i. e., $\deg f = d$ for some $f : S^n \longrightarrow S^n$. By Proposition 12.5.30, the suspension $\Sigma(f) : \Sigma(S^n) = S^{n+1} \longrightarrow \Sigma(S^n) = S^{n+1}$ has degree equal to d. Therefore, the statement is true for $n + 1$. □

Corollary 12.5.32. *Let $h : \mathbb{Z} \longrightarrow \mathbb{Z}$ be a homomorphism. Then, for all $n = 1, 2 \dots$, there exists a continuous map $f : S^n \longrightarrow S^n$ whose induced homomorphism is h, $f_* = h$.*

Proof. By Corollary 12.5.31, there exists a continuous map $f : S^n \longrightarrow S^n$ such that $\deg f = h(1)$. By Definition 12.5.1, for all $n \in \mathbb{Z}$, $f_*(n) = h(1)n = h(n)$, which implies $f_* = h$. For $n = 1$, on may take $f(z) = z^{h(1)}$ since by Proposition 6.2.3, $\deg f = h(1)$. □

For $n = 1$, there exists also an obvious kind of converse to Corollary 12.5.31.

Proposition 12.5.33. *For every continuous function $f : S^n \longrightarrow S^n$, there exists a unique integer $d \in \mathbb{Z}$ such that f is homotopic to the power map z^d, namely $d = \deg f$.*

12.5.3 The hairy ball theorem

Definition 12.5.34. A continuous function $f : S^n \longrightarrow \mathbb{R}^{n+1}$ is a tangent vector field on S^n if the vectors x and $f(x)$ are perpendicular at x for every $x \in S^n$.

Theorem 12.5.35. *The following statements are equivalent:*

(1) *n is odd.*
(2) *The antipodal map a and the identity map on S^n are homotopic.*
(3) *There exists a tangent vector field $f : S^n \longrightarrow S^n$.*

Proof. By Corollary 12.5.26, (1) and (2) are equivalent.

(1) $\Longrightarrow$ (3): Define the mapping $f : S^{2n+1} \longrightarrow S^{2n+1}$ by

$$f(x_1, x_2, \ldots, x_{2n+1}, x_{2n+2}) = (x_2, -x_1, \ldots, x_{2n+2}, -x_{2n+1}).$$

Then $x \cdot f(x) = 0$, showing that $f(x)$ and x are perpendicular for all $x \in S^{2n+1}$.

(3) $\Longrightarrow$ (2): Let f be a tangent vector field on S^n and $H(x,t) = \cos(\pi t)x + \sin(\pi t)f(x)$. Since $f(x)$ and x are perpendicular for all $x \in S^n$, we have by Parseval's identity,

$$\|H(x,t)\|^2 = \|x\|^2 \cos^2(\pi t) + \|f(x)\|^2 \sin^2(\pi t) = 1.$$

Hence, $H : S^n \times I \longrightarrow S^n$ is a homotopy between the identity map Id_{S^n} and the antipodal map a. □

Corollary 12.5.36 (The hairy ball theorem). *Let $f : S^{2n} \longrightarrow \mathbb{R}^{2n+1}$ be a tangent vector field. Then f vanishes at least at one point.*

Proof. If not, for $x \in S^{2n}$, define $g(x) = \frac{f(x)}{\|f(x)\|}$. Then $g : S^{2n} \longrightarrow S^{2n}$ is a tangent vector field, violating Theorem 12.5.35. □

Remark 12.5.37. (1) Corollary 12.5.36 was first proved by Poincaré for $n = 1$ in 1885 and extended to higher dimensions by Brouwer in 1912.

(2) In [73, Theorem 1'], Milnor proved Corollary 12.5.36 in 1978 using only differential vector field arguments. Then he deduced Brouwer's fixed-point theorem [73, Theorem 2].

(3) Corollary 12.5.36 states that "curled" even-dimensional spheres cannot be combed. That is why Corollary 12.5.36 is also called in the literature the hairy dog theorem or the hedgehog combing theorem.

Remark 12.5.38. (1) Theorem 12.5.35 states that if n is even, then any continuous function $f : S^n \longrightarrow S^n$ is not a tangent vector field. As a consequence, a continuous function $f : S^{2n} \longrightarrow \mathbb{R}^{2n+1} \setminus \{0\}$ is not a tangent vector field for otherwise $g = f/\|f\| : S^{2n} \longrightarrow S^{2n}$ would be a tangent vector field.

(2) However, if n is odd, Theorem 12.5.35 states nothing about the existence of a continuous function $f : S^n \longrightarrow S^n$, which is not a tangent vector field. For example, constant functions on S^n are not tangent vector fields. There exist nontrivial examples as well. Let $f : B^{n+1} \longrightarrow \mathbb{R}^{n+1}$ be a continuous function such that $f(S^n) \subseteq \mathbb{R}^{n+1} \setminus \{0\}$. By Corollary 11.1.15, there exists $x_0 \in S^{n-1}$ such that $\langle x_0, f(x_0)\rangle > 0$. Hence, f is not a tangent vector field, whatever the dimension n is. For instance, one may consider the class of continuous functions $f : S^n \longrightarrow S^n$, which extends continuously to $\tilde{f} : B^{n+1} \longrightarrow S^n$. By Theorem 12.5.12, we infer that every null-homotopic function $f : S^n \longrightarrow S^n$ is not a tangent vector field for odd dimension (and of course for even dimension, too).

12.6 The Borsuk–Ulam theorem

In this section, we investigate some additional properties of continuous maps $f : S^n \longrightarrow \mathbb{R}^n$. The degree of a continuous map $f : S^n \longrightarrow S^n$ was studied in Subsection 12.5.1. For the class of antipode-preserving mapping, Borsuk proved the more general fact that the degree is odd (see Theorem 6.4.8). The latter result together with further properties of those functions f are proposed in some exercises of this chapter. A detailed discussion including more recent developments and some historical notes can be found in [69, Chapter 2]. After proving an equivalent version of the Borsuk–Ulam theorem for any dimension, we will consider some more equivalences, as we did in Section 6.4 for $n = 2$. Then we will present a further equivalent version, namely the Lyusternik–Schnirelmann theorem [65]. This is the aim of Subsection 12.6.1. Finally, some applications are given in Subsection 12.6.2. The Borsuk–Ulam theorem is further shown to imply the ham sandwich theorem as well as Brouwer's fixed-point theorem.

12.6.1 Equivalent versions

Theorem 12.6.1 (The Borsuk–Ulam theorem, version 1). *Let $f : S^n \longrightarrow \mathbb{R}^n$ be a continuous map, $n \geq 1$. Then there exists $x \in S^n$ such that $f(x) = f(-x)$.*

Note that for $n = 1$, the result is Exercise 6 of Section 1.5. When $n = 2$, Theorem 12.6.1 is just Theorem 6.4.6. The following result is a consequence of Theorem 12.6.1.

Corollary 12.6.2. *The sphere S^n cannot be embedded in the space $\mathbb{R}^n$.*

There exist several proofs of Theorem 12.6.1, which was first conjectured by Ulam in 1930 and then proved by Borsuk in 1933. The proofs employ homology, cohomology or degree theory. We refer the interested reader to [2, Theorem 3.8.15], [10, Theorem 20.2, Chapter IV], [70, Theorem 8.5.14], [83, Corollary 6.29], [102, Theorem 5.7.6], among other references.

We first check that Theorem 12.6.1 has the following equivalent version.

Theorem 12.6.3 (The Borsuk–Ulam theorem, version 2). *If $f : S^n \longrightarrow S^{n-1}$ is a continuous map, then f is not antipode-preserving.*

Proof. (1) Assume that Theorem 12.6.1 holds and by contradiction, let $f : S^n \longrightarrow S^{n-1}$ be a continuous map which is antipode-preserving. Then, by Assumption, there exists $x \in S^n$ such that $f(x) = f(-x) = -f(-x)$. Hence, $f(x) = 0$, a contradiction.

(2) Conversely, assume Theorem 12.6.3 and by contradiction suppose that we are given a continuous map $f : S^n \longrightarrow \mathbb{R}^n, f(x) \neq f(-x)$ for all $x \in S^n$. The map g defined by $g(x) = \frac{f(x)-f(-x)}{\|f(x)-f(-x)\|}$ is continuous and maps the sphere S^n into S^{n-1}. By Theorem 12.6.3, it is not antipode-preserving, which contradicts the definition of g. □

Inspired by [42, Theorem (26.15)], [83, Theorem 12.39] and [94, Corollary 8, Section 8, Chapter 5], we prove a more general result using the fundamental groups.

Theorem 12.6.4 (The Borsuk–Ulam theorem, general version). *If $f : S^n \longrightarrow S^m$ is a continuous map, then f is not antipode-preserving whenever $n > m \geq 1$.*

Since we have already investigated the case $n = 2, m = 1$ in Theorem 6.4.14, we restrict our attention to the case $n > m > 1$.

Proof. We use the reduction to absurdity and suppose that f is antipode-preserving. Consider the diagram

$$\begin{array}{ccc} S^n & \xrightarrow{\quad f \quad} & S^m \\ {\scriptstyle \pi_{(n)}}\big\downarrow & & \big\downarrow{\scriptstyle \pi_{(m)}} \\ \mathbb{RP}^n & \xrightarrow{\quad g=\overline{f} \quad} & \mathbb{RP}^m \end{array}$$

where the canonical projections $\pi_{(k)} : S^k \longrightarrow \mathbb{RP}^k$ $(k = n, m)$ are the covering maps given by Proposition 5.1.7. Denote by $\sim_k$ and $\pi_{(k)}(x) = [x]_k$ the equivalence relation and the equivalence class of x in the sphere S^k, respectively $(k = n, m)$. Since $\overline{f}([x]_n) = [f(x)]_m$, $g = \overline{f}$ is well-defined and the diagram commutes. Indeed, if $x, y \in S^n$ and $x \sim_n y$, then $x = \pm y$. Hence,

$$f(x) = f(\pm y) = \pm f(y) \quad \Longleftrightarrow \quad f(x) \sim_m f(y).$$

In other words,

$$\pi_{(n)}(x) = \pi_{(n)}(y) \quad \Longrightarrow \quad \pi_{(m)}(f(x)) = \pi_{(m)}f(y).$$

Let $X = \mathbb{RP}^m$, $\widehat{X} = S^m$, $p = \pi_m$, $Y = \mathbb{RP}^n$ and $g : \mathbb{RP}^n \longrightarrow \mathbb{RP}^m$. For $n > m \geq 2$, by [83, Lemma 12.38] or [94, Corollary 7, Section 8, Chapter 5], g has a unique lifting map $\widehat{g} : \mathbb{RP}^n \longrightarrow S^m$ such that $\pi_m \circ \widehat{g} = g$. By Theorem 5.2.16, this is achieved if the induced homomorphism g_* is proved to be trivial. Hence, $\pi_m \circ (\widehat{g} \circ \pi_n) = g \circ \pi_n = \pi_m \circ f$. Then $\widehat{g} \circ \pi_n$ and f lift the same map. Since they have same image under π_m, either $(\widehat{g} \circ \pi_n)(x) = f(x)$ or $(\widehat{g} \circ \pi_n)(x) = -f(x) = f(-x)$. In the second case, $(\widehat{g} \circ \pi_n)(-x) = f(-x)$ because $\pi_n(x) = \pi_n(-x)$. Consequently, $\widehat{g} \circ \pi_n$ and f agree at either x or $-x$. By uniqueness of the lifting map, we conclude that $\widehat{g} \circ \pi_n = f$. However,

$$(\widehat{g} \circ \pi_n)(x) = (\widehat{g} \circ \pi_n)(-x) = f(x) = f(-x),$$

which is a contradiction. □

Corollary 12.6.5. *For $n \geq 2$, let $f : S^n \longrightarrow S^1$ be a continuous map. Then f is null-homotopic and there exists $n \in S^n$ such that $f(-x) \neq -f(x)$.*

The proof of the following corollary is analogous to the case $n = 2$ treated in Section 6.4. The equivalence between (1) and (3) is Theorem 12.6.3. To deal with statement (4), we apply Theorem 6.4.1 instead of Theorem 6.2.9. In [20, Theorem (10.6.3)], [25, Theorem 6.1, Chapter XVI], [45, Proposition 2B.6], it is shown that if $f : S^n \longrightarrow S^n$ is antipode-preserving, then deg(f) is an odd number (Borsuk–Ulam theorem, Exercise 10, this chapter) and so f is not null-homotopic, that is to say Statement (5) of the corollary holds true. The reader can also refer to [20, Theorem (10.6.1)] for the equivalences. Here, f is a continuous map.

Corollary 12.6.6. *The following statements, which are equivalent, hold true:*
(1) *If $f : S^n \longrightarrow \mathbb{R}^n$ is antipode-preserving, then there exists $x \in S^n$ such that $f(x) = f(-x)$.*
(2) *If $f : S^n \longrightarrow \mathbb{R}^n$ is antipode-preserving, then there exists $x \in S^n$ such that $f(x) = 0$.*
(3) *If $f : S^n \longrightarrow S^{n-1}$, then f is not antipode-preserving.*
(4) *If $f : B^n \longrightarrow S^{n-1}$, then $f_{|S^{n-1}}$ is not antipode-preserving.*
(5) *If $f : S^{n-1} \longrightarrow S^{n-1}$ is antipode-preserving, then f is not null-homotopic.*

Example 12.6.7. By Example 9.1.18, the sphere S^n is $(n-1)$-connected. However, S^n is not n-connected for otherwise the antipodal map $a : S^n \longrightarrow S^n$ would have a continuous extension $\tilde{a} : B^{n+1} \longrightarrow S^n$, which contradicts Corollary 12.6.6(4). This is an alternative proof to Example 9.2.14, which appeals to the higher fundamental groups.

Theorem 12.6.8 (The Lyusternik–Schnirelmann theorem). (1) *Let $\{F_k\}_{1\le k\le n}$ be a family of closed sets that cover S^{n-1}. Then there exist $k \in \{1, n\}$ and $x_k \in F_k$ such that $(-x_k) \in F_k$.*
(2) *This formulation is equivalent to the Borsuk–Ulam theorem.*

Proof. (1) The Borsuk–Ulam theorem implies the Lyusternik–Schnirelmann theorem. The result is trivial for $n = 1$. Assume that $F_k \cap (-F_k) = \emptyset$, for $k \in \{1, \dots, n-1\}$ otherwise we are done. We claim that $F_n \cap (-F_n) \neq \emptyset$. Define the functions f_k $(1 \le k \le n-1)$ by

$$\begin{cases} 0, & x \in F_k, \\ 1, & x \in -F_k. \end{cases}$$

By the Tietze extension theorem (Theorem A.5.1), the f_k functions can be continuously extended to $\tilde{f}_k : S^{n-1} \longrightarrow [0,1]$. By applying the Borsuk–Ulam theorem to the function

$$\tilde{f} = (\tilde{f}_1, \dots, \tilde{f}_{n-1}) : S^{n-1} \longrightarrow \mathbb{R}^{n-1},$$

we get some $x_0 \in S^{n-1}$ such that $\tilde{f}(x_0) = \tilde{f}(-x_0)$. Then $\tilde{f}_k(x_0) = \tilde{f}_k(-x_0)$, for all $k \in \{1, 2, \dots, n-1\}$. Then, by definition of the functions f_k, we have $x_0 \notin F_k \cup (-F_k)$. Otherwise,

$$\begin{aligned} x_0 \in F_k \quad &\Longrightarrow \quad f_k(x_0) = 0 \quad \Longrightarrow \quad f_k(-x_0) = 0 \\ &\Longrightarrow \quad -(x_0) \in F_k \quad \Longrightarrow \quad x_0 \in (-F_k) \\ &\Longrightarrow \quad f_k(x_0) = 1, \end{aligned}$$

a contradiction. Since the families $\{F_k\}_{1\leq k\leq n}$ and $\{-F_k\}_{1\leq k\leq n}$ cover the sphere S^{n-1}, we obtain that $x_0 \in F_n \cap (-F_n)$.

(2) The Lyusternik–Schnirelmann theorem implies the Borsuk–Ulam theorem. By [69, p. 24], the sphere S^{n-1} can be covered by $(n+1)$ closed sets $F_1, \dots, F_{n+1}$ such that no F_i contains a pair of antipodal points. If $f : S^n \longrightarrow S^{n-1}$ is a continuous antipode-preserving map, the sets $f^{-1}(F_1), \dots, f^{-1}(F_{n+1})$ would contradict Lyusternik–Schnirelmann theorem. □

12.6.2 Applications

An alternative proof to Corollary 10.4.20 is first presented.

Corollary 12.6.9. *The spaces $\mathbb{R}^m$ and $\mathbb{R}^n$ are homeomorphic if and only if $m = n$.*

Proof. Let $f : \mathbb{R}^n \longrightarrow \mathbb{R}^m$ be a homeomorphism with $m < n$, i. e., $m \leq n-1$. The restriction $g = f_{|S^{n-1}} : S^{n-1} \longrightarrow \mathbb{R}^m \subset \mathbb{R}^{n-1}$ is continuous and injective. By Borsuk–Ulam Theorem 12.6.1, there exists $x \in S^{n-1}$ such that $g(x) = g(-x)$. Then g is not injective, a contradiction. The converse is obvious. □

Theorem 12.6.10 (The ham sandwich theorem). *Let $(B_i)_{1\leq i\leq n}$ be a family of measurable and bounded subsets of $\mathbb{R}^n$. Then there exists a hyperplane H of $\mathbb{R}^n$, which divides each of the sets B_i in two parts of same measure.*

Recall the following.

Definition 12.6.11. An hyperplane H of $\mathbb{R}^n$ is a set of points $(x_1, x_2, \dots, x_n)$ such that

$$a_1x_1 + a_2x_2 + \cdots a_nx_n = c,$$

where the constants $c, a_1, a_2, \dots, a_n$ are such that $a_1, a_2, \dots, a_n$ are not all equal to zero. In vector notation,

$$H = \{x \in \mathbb{R}^n : \langle a, x\rangle = c\},$$

where $a = \langle a_1, a_2, \dots, a_n\rangle$ and $x = \langle x_1, x_2, \dots, x_n\rangle$.

Remark 12.6.12. (1) Theorem 12.6.10 is obvious in the case B_i are balls of Euclidean plane $\mathbb{R}^2$ or Euclidean space $\mathbb{R}^3$.

(2) Consider a family B_i $(1 \leq i \leq 3)$, which consists of three chunks of bread, kosher and cheese. Then the hyperplane H is represented by a knife, which is used to bisect the sandwich with its ingredients.

Proof. Fix some $a = (0, 0, \dots, 1) \in \mathbb{R}^{n+1}$. For $1 \leq k \leq n$, set

$$E_k(x) = \{y \in B_k : (y - a).x \geq 0\}$$

and consider the function $f : S^n \longrightarrow \mathbb{R}^n, f = (f_1, \dots, f_n)$ given by

$$f_k(x) = \text{meas}(E_k(x)), \quad \text{for all } k \in [0, 1],$$

where an element $y \in \mathbb{R}^n$ is identified with $y' = (y, 0) \in \mathbb{R}^{n+1}$ and where . refers to the inner product in $\mathbb{R}^{n+1}$.

To show that f is continuous, let $(x_j)_j$ be a sequence, which converges to some limit $x \in S^n$, as $j \to \infty$. Since the sets B_k are bounded, as $j \to \infty$, the sequences $\chi_{E_k(x_j)}$ converge to $\chi_{E_k(x)}$ for all $k \in [1, n]$. Here, the function χ_A is the characteristic function of the set A. By the Lebesgue dominated convergence theorem, as $j \to \infty$, the sequence of functions $f_k(x_j)$ converges to $f_k(x)$ for $1 \le k \le n$. Then the Borsuk–Ulam theorem yields some $x_0 \in S^n$ such that $f(x_0) = f(-x_0)$. For all $k \in [1, n]$, we have

$$\text{meas}\{y \in B_k : (y - a).x_0 \ge 0\} = \text{meas}\{y \in B_k : (y - a).x_0 \le 0\}. \tag{12.4}$$

Reasoning by contradiction, suppose that the first n components of x_0 are zero. Since x_0 belongs to the sphere, we deduce that $x_0^{n+1} = \pm 1$, and thus for all $k \in [1, n]$,

$$\text{meas}\{y \in B_k : (y_{n+1} - 1).x_0 \ge 0\} = \text{meas}\{y \in B_k : (y_{n+1} - 1) \le 0\}.$$

However, $y_{n+1} = 0$ and then meas $B_k = 0$ for all $k \in [1, n]$, contradicting the supposition. By (12.4), for all $k \in [1, n]$,

$$\text{meas}\{y \in B_k : y.x_0 \ge x_0^{n+1}\} = \text{meas}\{y \in B_k : y.x_0 \le x_0^{n+1}\}.$$

To conclude, just take the hyperplane $H = \{y \in \mathbb{R}^n\colon\ y.x_0 = x_0^{n+1}\}$. □

Theorem 12.6.13. *Borsuk–Ulam theorem implies Brouwer's fixed-point theorem.*

Proof. By contradiction, let $f : B^n \longrightarrow B^n$ be a continuous fixed-point free map. Consider the retraction $r : B^n \longrightarrow S^{n-1}$ as defined in the proof of Theorem 7.6.2. Then $r(x) = x$ for all $x \in S^{n-1}$, and thus r is antipode-preserving on S^{n-1}, contradicting Corollary 12.6.6(4). □

Exercises (Chapter 12)

1. Let $U \subset \mathbb{R}^n$ and $V \subset \mathbb{R}^m$ be two open sets with $n \neq m$. Prove that U and V are not homeomorphic (invariance of domain theorem).

2. Let $A \subseteq \mathbb{R}^n$. Prove that A is not homeomorphic to S^n.

3. Let $f, g : S^n \longrightarrow S^n$ be continuous maps. Suppose that $f(x) \neq g(x)$ for all $x \in S^n$. Prove that

(1) f is homotopic to $a \circ g$, where a is the antipodal map;

(2) $\deg f = (-1)^{n+1} \deg g$;
(3) Let $g(x) = -x$ for all $x \in S^n$. Prove that $f \simeq \mathrm{Id}_{S^n}$.

4. Let $f, g : S^n \longrightarrow S^n$ be continuous maps.
(1) Suppose that $f(x) \neq -g(x)$ for all $x \in S^n$. Prove that f and g are homotopic.
(2) Suppose that for some $0 < k < 2$, $\|f(x) - g(x)\| \leq k\|g(x)\|$ for all $x \in S^n$. Prove that f and g are homotopic.
(3) Suppose that there exists $0 < k < 2$ such that $\|f(x) - x\| \leq k$ for all $x \in S^n$. Prove that either f has a fixed point or there exists a vector field on S^n. (*Hint:* Use Exercise 3(1) both with Theorem 12.5.35).

5. Let $f : S^n \longrightarrow S^n$ be given by $f(x) = Ax$, where $A \in \mathscr{M}^n$ is a $n \times n$ nonsingular matrix. Prove that $\deg f = \det A$, the determinant of A. (*Hint:* See [10, Proposition 7.1, Chapter IV] or [52, Proposition 10.2.4] for an orthogonal matrix A.)

6. Let $f : S^n \longrightarrow S^n$ be a continuous map such that $\deg f \neq 1$. Prove that f sends some point $x \in S^n$ to its antipode, i. e., $f(x) = -x$. (*Hint:* Compose f with the antipodal map, compute the degree and use Proposition 12.5.28.)

7. Let $f : S^n \longrightarrow S^n$ be a continuous map such that $\deg f = 0$. Prove that there exist $x, y \in S^n$ such that $f(x) = x$ and $f(y) = -y$. (For $n = 1$, this is Exercise 7, Chapter 6). (*Hint:* Use Proposition 12.5.28, Corollary 12.5.29 and Exercise 6.)

8. (1) Let $f : S^n \longrightarrow S^n$ be a continuous map such that $\deg f \neq 0$. Prove that f is surjective (for $n = 1$, this is Exercise 9, Chapter 6). (*Hint:* Argue by contradiction and use Corollary 4.9.9 and Theorem 12.5.12.)
(2) Construct a continuous function $f : S^n \longrightarrow S^n$, which is surjective and null-homotopic. Conclude. (*Hint:* By induction, continue Question (3) of Exercise 9, Chapter 6 using the suspension of a sphere.)

9. Let $f : S^n \longrightarrow S^n$ be a continuous map antipode-preserving, which is not homotopic to the antipodal map. Prove that f has a least two antipodal fixed points. (*Hint:* Apply Proposition 12.5.28.)

10. Assume that $f : S^n \longrightarrow S^n$ is antipode-preserving.
(1) Prove that $\deg f$ is odd (this is Borsuk–Ulam theorem—an odd map has an odd degree) (the case $n = 1$ is Theorem 6.4.8). (*Hint:* See, e. g., [10, Theorem 20.6, Chapter IV], [25, Chapter XVI, Theorem 6.1] or [45, Exercise 14, Section 2.2].)
(2) Deduce that f is not null-homotopic.
(3) Deduce that f is surjective. (Use Question 8(1)).

11. Let $f : S^n \longrightarrow S^n$ be a continuous map such that $\deg f$ is even. Prove that:
(1) there exists $x \in S^n$ such that $f(-x) \neq -f(x)$. (*Hint:* Apply Borsuk–Ulam theorem),

(2) there exists $x \in S^n$ such that $f(-x) = f(x)$. (*Hint:* Argue by contradiction, consider the map given by $g(x) = \frac{f(x)-f(-x)}{\|f(x)-f(-x)\|}$, construct a homotopy and apply Borsuk–Ulam theorem.)

12. Let $f : S^n \longrightarrow S^n$ be a continuous map. Prove that:

(1) (a) if f is antipode collapsing (i. e., $f = f \circ a$), then $\deg f$ is even (an even map has an even degree).

(b) If n is even, then f is null-homotopic.

(2) If $\deg f$ is odd, then there exists $x \in S^n$ such that $f(x) = -f(-x)$ (partial converse of the Borsuk–Ulam theorem). (*Hint:* Argue by contradiction, consider the map given by $g(x) = \frac{f(x)+f(-x)}{\|f(x)+f(-x)\|}$, and apply part (1).)

13. Let $f : S^{2n+1} \longrightarrow S^{2n+1}$ be a continuous map such that $\deg f \neq 1$. Prove that f has a fixed point and sends some point to its antipode. (*Hint:* Evaluate $\deg(a \circ f)$ and make use of Proposition 12.5.28.)

14. Let $f : S^{2n} \longrightarrow S^{2n}$ be a continuous map.

(1) Prove that $f^2 = f \circ f$ has a fixed point. (*Hint:* Evaluate $\deg f^2$ and make use of Proposition 12.5.28.)

(2) Show that there exists $x \in S^{2n}$ such that $\langle f(x), x\rangle \neq 0$. (*Hint:* Use Corollary 12.5.36 or Question (1).)

15. Let $f : S^{2n} \longrightarrow S^{2n}$ be a continuous map.

(1) Prove that either f has a fixed point or f sends some point to its antipode. More precisely, there exists $x \in S^{2n}$ such that either $f(x) = x$ or $f(x) = -x$. (*Hint:* Argue by contradiction and use Exercise 3.)

(2) Deduce Question (2) of Exercise 14 above.

16. Let $f : S^{2n} \longrightarrow S^{2n}$ be a continuous map.

(1) Suppose that $\deg f \neq -1$. Prove that f has a fixed point. (*Hint:* Use Proposition 12.5.28.)

(2) Suppose that $\deg f \neq 1$. Prove that f sends some point to its antipode. (*Hint:* Use Proposition 12.5.28.)

(3) Conclude when $\deg f \neq \pm 1$.

17. Let $f : S^{2n} \longrightarrow \mathbb{R}^{2n+1} \setminus \{0\}$ be a continuous function.

(a) Prove that there exist $x \in S^{2n}$ and $\lambda \neq 0$ such that $f(x) = \lambda x$. (*Hint:* Use Corollary 12.5.36.)

(b) Deduce Question (2) of Exercise 14 above.

18. Let $f : \mathbb{R}^n \longrightarrow \mathbb{R}^n$ be a continuous map such that for all $x \in S^{n-1}$, the scalar product $\langle f(x), x\rangle$ has constant sign. Prove that there exists $x \in B^n$ such that $f(x) = 0$. (*Hint:* Use Corollary 11.1.17 and consider $g = -f$.)

19. Let $A \subset S^n$ be a closed proper subset. Prove the existence of a homeomorphism $h : S^n \longrightarrow S^n$ such that $\deg h = \deg h^{-1} = 1$ and $h(A) \subset S^n_+$. (*Hint:* See [25, Chapter XVI, 7.1].)

20. Let $S^k \subset \mathbb{R}^{n+1}$ and $1 \leq k \leq n$. Prove that the sphere S^k separates $\mathbb{R}^{n+1}$ if and only if $k = n$. (*Hint:* Use Exercise 21 of Chapter 6 and [25, Theorem 2.1, Chapter XVII].) This is a further version of Jordan–Brouwer separation theorem.

21. (1) Let $X \subset \mathbb{R}^{n+1}$ be compact and $A \subset X$ a retract of X. Prove that $\mathbb{R}^{n+1} \setminus A$ cannot have more components than $\mathbb{R}^{n+1} \setminus X$. (*Hint:* See [25, Chapter XVII, 4.4].)
(2) Deduce that S^k is not a retract of S^n for $1 \leq k < n$ (and recapture Corollary 9.2.15).

22. Assume that A is a deformation retract of X. Prove that $\mathbb{R}^{n+1} \setminus A$ and $\mathbb{R}^{n+1} \setminus X$ have the same number of components. (*Hint:* See [25, Chapter XVII, Exercise 4.2].)

23. Prove that every continuous map $f : \mathbb{RP}^{2n} \longrightarrow \mathbb{RP}^{2n}$ has a fixed point.

(*Hint:* Using Theorem 5.2.16, prove the existence of the lifting map $g = \widehat{f \circ \pi} : S^{2n} \longrightarrow S^{2n}$ of the composite map $f \circ \pi$, where $\pi : S^{2n} \longrightarrow \mathbb{RP}^{2n}$ is the projection map and then apply Exercise 15.)

For $n = 1$, one as well argue by contradiction and use the map $h : S^2 \longrightarrow S^2$ defined by $h(x) = \frac{g(x) \times x}{\|g(x) \times x\|}$, where $\times$ is the vector product in $\mathbb{R}^3$. Then a contradiction with the hairy ball Theorem 12.5.35 is reached.)

24. If n is odd, prove that the real projective space $\mathbb{RP}^n$ and the complex projective space $\mathbb{CP}^n$ have not the fixed-point property (see [101, Exercices 38.18, 38.19, p. 255]).

25. (Kakutani's fixed-point theorem). Let $C \subset \mathbb{R}^n$ be a nonempty compact convex subset and $F : C \multimap \mathscr{P}_{cl,cv}(C)$ an upper semicontinuous multivalued map with nonempty closed convex values (F is upper semicontinuous if for every open set $V \subset C$, the inverse image $F^{-1}(U)$ is open in C). Prove that F has a fixed point $x \in F(x)$. (*Hint:* Start with a simplex C and use Brouwer's fixed-point theorem together with the fact that an upper semicontinuous multivalued map F has a closed graph $\{(x, y) \in C \times C : y \in F(x)\}$.)

Appendices

A Auxiliary results in analysis

A.1 Carathéodory's lemma and Mazur's lemma

Concerning the results of this section, we refer the reader to [7, 54, 84].

Definition A.1.1. A subset A of a vector space X is convex if for all points x, y in A, the segment

$$[x,y] = \{\lambda x + (1-\lambda)y,\ \lambda \in [0,1]\}$$

lies entirely in A.

Proposition A.1.2. *Let A be a convex subset of a normed space.*
(1) *The closure $\overline{A}$ and the interior set $\mathring{A}$ are convex sets.*
(2) *For every $x \in \mathring{A}$, $y \in \partial A$, and $\lambda \in (0,1]$, $z = \lambda x + (1-\lambda)y \in \mathring{A}$.*

Proof. (1)(a) Let $x, y \in \overline{A}$ and $\lambda \in [0,1]$. Then there exist two sequences $(x_n)_n \subset A$ and $(y_n)_n \subset A$ such that $x = \lim_{n\to\infty} x_n$ and $y = \lim_{n\to\infty} y_n$. By convexity of A, $\lambda x_n + (1-\lambda)y_n \in A$ for all $n = 1, 2, \ldots$ and $\lambda \in [0,1]$. Taking the limit as $n \to \infty$, we find $\lambda x + (1-t)\lambda \in \overline{A}$, as claimed.

(b) Let $x, y \in \mathring{A}$ and $\lambda \in [0,1]$. By convexity of A, $\lambda x + (1-\lambda)y \in A$, i. e., $\lambda\mathring{A} + (1-\lambda)\mathring{A} \subset A$. It is easily checked that $\lambda\mathring{A} + (1-\lambda)\mathring{A}$ is an open set (assume for instance that $0 \in \mathring{A}$ and check that λA is open for all λ nonzero scalar number whenever A is). Since $\mathring{A}$ is the largest open set contained in A, we deduce that $\lambda\mathring{A} + (1-\lambda)\mathring{A} \subset \mathring{A}$. Therefore, $\lambda x + (1-\lambda)y \in \mathring{A}$.

(2) Let $\lambda \in (0,1)$. By assumptions, there exists $\delta > 0$ such that $\mathring{B}(x,\delta) \subset A$ and for all $\varepsilon > 0$, there exists $y_\varepsilon \in A_\varepsilon \cap \mathring{B}(y,\varepsilon)$. Let $z_\varepsilon = \lambda x + (1-\lambda)y_\varepsilon$. We claim that $z \in \mathring{B}(z_\varepsilon, r) \subset A$ for some $r > 0$. We have

(a) $z \in \mathring{B}(z_\varepsilon, r)$. Indeed,

$$\|z - z_\varepsilon\| = (1-\lambda)\|y - y_\varepsilon\| < (1-\lambda)\varepsilon \le r,$$

whenever $0 < \varepsilon \le \frac{r}{1-\lambda}$.

(b) Let $z' \in \mathring{B}(z_\varepsilon, r)$. Then

$$\begin{aligned}\|z' - z_\varepsilon\| < r \quad &\Longleftrightarrow \quad \|z' - \lambda x - (1-\lambda)y_\varepsilon\| < r\\ &\Longleftrightarrow \quad \left\|\frac{z'}{\lambda} - \lambda x - \frac{(-\lambda)}{\lambda}y_\varepsilon - x\right\| < r/\lambda \le \delta,\end{aligned}$$

whenever $0 < r \le \lambda\delta$ in which case

$$\frac{1}{\lambda}(z' - (1-\lambda)y_\varepsilon) \in \mathring{B}(x,\delta) \subset A.$$

https://doi.org/10.1515/9783111517384-013

Since A is convex and $y_\varepsilon \in A$, we conclude that

$$\lambda\left(\frac{1}{\lambda}(z' - (1-\lambda)y_\varepsilon)\right) + (1-\lambda)y_\varepsilon = z' \in A.$$

This shows that $\mathring{B}(z_\varepsilon, r) \subset A$ and concludes the proof. □

Lemma A.1.3. *Let A be a subset of a vector space X. The following statements are equivalent:*
(1) *A is convex.*
(2) *For all $k = 1, 2, \ldots$, $\sum_{i=1}^{i=k} \lambda_i x_i \in A$ whenever $x_i \in A$, $0 \le \lambda_i \le 1$ for all $i = 1, 2 \ldots, k$ and $\sum_{i=1}^{i=k} \lambda_i = 1$.*

Proof. (2) $\Longrightarrow$ (1): $k = 2$ is Definition A.1.1.

(1) $\Longrightarrow$ (2): (2) is already true for $k = 1, 2$. By induction, assume it is true for k and let $x_i \in A$, $0 \le \lambda_i \le 1$ for all $i = 1, 2 \ldots, k+1$ with $\sum_{i=1}^{i=k+1} \lambda_i = 1$. We may assume that all λ_i are nonzero for otherwise we are done by the induction hypothesis. Let $\lambda = \sum_{i=1}^{i=k} \lambda_i > 0$ and $\mu_i = \frac{\lambda_i}{\lambda} > 0$, $i = 1, 2 \ldots, k$. Then $\sum_{i=1}^{i=k} \mu_i = 1$. By the induction hypothesis, $\sum_{i=1}^{i=k} \mu_i x_i \in A$. Since A is convex,

$$A \ni \lambda \sum_{i=1}^{i=k} \mu_i x_i + (1-\lambda)x_{k+1} = \lambda \sum_{i=1}^{i=k} \frac{\lambda_i}{\lambda} x_i + \lambda_{k+1} x_{k+1} = \sum_{i=1}^{i=k+1} \lambda_i x_i$$

because $\lambda_{k+1} = 1 - \sum_{i=1}^{i=k} \lambda_i = 1 - \lambda$. Hence, (2) is true for $k+1$. □

Definition A.1.4. (1) The convex hull of a set A is the smallest convex set that contains A. It is denoted $\mathrm{Co}(A)$.
(2) The closed convex hull of A is the smallest closed convex set containing A. It is denoted $\overline{\mathrm{Co}}(A)$.

Remark A.1.5. If a set is convex, then so is its closure. Hence, $\overline{\mathrm{Co}(A)}$ is closed, convex and contains A. This implies that $\overline{\mathrm{Co}}(A) \subseteq \overline{\mathrm{Co}(A)}$. In addition by definition of $\mathrm{Co}(A)$, we have $A \subseteq \mathrm{Co}(A) \subseteq \overline{\mathrm{Co}}(A)$. By taking the closures, $\overline{\mathrm{Co}(A)} \subseteq \overline{\mathrm{Co}}(A)$. Therefore,

$$\overline{\mathrm{Co}}(A) = \overline{\mathrm{Co}(A)}.$$

The following lemma is known as Carathéodory's characterization of the convex hull in vector spaces.

Lemma A.1.6. *Let A be a subset of a vector space X. Then*

$$\mathrm{Co}(A) = \left\{\sum_{i=1}^{k} \lambda_i x_i : \ x_i \in A,\ \lambda_i \ge 0,\ \text{for all } i = 1, 2, \ldots, k \text{ and } \sum_{i=1}^{k} \lambda_i = 1,\ k = 1, 2, \ldots\right\}.$$

Proof. (1) Let B denote the right-hand set and C any convex set, which contains A. By Lemma A.1.3, $B \subset C$. In particular, for $C = \mathrm{Co}(A)$, $B \subset \mathrm{Co}(A)$.

(2) Conversely, since $A \subset B \subset \mathrm{Co}(A)$, it suffices to check that B is convex. Let x, y belong to B. Then there exists $k = 1, 2, \ldots$ such that $x = \sum_{i=1}^{k} \lambda_i x_i$ and $y = \sum_{i=1}^{k} \mu_i y_i$. Hence, $x_i \in A$, $y_i \in A$, $\lambda_i \geq 0$, $\mu_i \geq 0$ for all $i = 1, 2, \ldots, k$, and $\sum_{i=1}^{k} \lambda_i = \sum_{i=1}^{k} \mu_i = 1$. For any $\lambda \in [0, 1]$, we have

$$\lambda x + (1 - \lambda y) = \sum_{i=1}^{k} \lambda \lambda_i x_i + \sum_{i=1}^{k} (1 - \lambda) \mu_i y_i \in B$$

because $\sum_{i=1}^{k} \lambda \lambda_i + (1 - \lambda) \mu_i = 1$. □

Proposition A.1.7. *Let S be the boundary of the closed unit ball $B = B(0, 1)$ in a normed space X. Then $\mathrm{Co}(S) = B$.*

Proof. Since $S \subset B$ and B is convex, $\mathrm{Co}(S) \subseteq B$. Conversely, let $x \in B$ and $\lambda = \frac{1+\|x\|}{2}$. Then $\lambda \in (0, 1]$ and $x = \lambda x/\|x\| + (1 - \lambda)(-x)/\|x\|$, where $x/\|x\|$ and $-x/\|x\|$ lie in B. Hence, $x \in \mathrm{Co}(S)$ which completes the proof. □

We are now in position to deal with the finite-dimensional case.

Proposition A.1.8. *Let $A \subset \mathbb{R}^n$. Then*

$$\mathrm{Co}(A) = \left\{ \sum_{i=1}^{n+1} \lambda_i x_i : x_i \in A,\ \lambda_i \geq 0,\ \textit{for all } i = 1, 2, \ldots, n+1 \textit{ and } \sum_{i=1}^{n+1} \lambda_i = 1, k = 1, 2, \ldots, n+1 \right\}.$$

Proof. If B denotes the right-hand set, then by Lemma A.1.6, $B \subset \mathrm{Co}(A)$. To prove the converse, let $x \in \mathrm{Co}(A)$. Then, again by Lemma A.1.6, there exist $k = 1, 2, \ldots$, $x_1, \ldots, x_k$ points of A, and scalars $\lambda_i \geq 0$ with $\sum_{i=1}^{k} \lambda_i = 1$ such that

$$x = \sum_{i=1}^{k} \lambda_i x_i = \sum_{i=1}^{k-1} \lambda_i (x_i - x_k) + x_k.$$

If $k \leq n + 1$, we are done. If $k > n + 1$, then $k - 1 > n$, and thus the $(k - 1)$ vectors $x_1 - x_k, \ldots, x_{k-1} - x_k$ are necessarily linearly dependent. Then there exist scalars $\alpha_1, \alpha_2, \ldots, \alpha_{k-1}$ not all zero such that $\sum_{i=1}^{k-1} \alpha_i (x_i - x_k) = 0$. Let $\alpha_k = -\sum_{i=1}^{k-1} \alpha_i$. Then $\sum_{i=1}^{k} \alpha_i x_i = \sum_{i=1}^{k} \alpha_i = 0$. Since we can suppose that at least one α_i is positive, let $\mu = \min\{\frac{\lambda_i}{\alpha_i}, i = 1, 2, \ldots, k, \alpha_i > 0\} \geq 0$ and $\beta_i = \lambda_i - \mu \alpha_i \geq 0$. Then

$$x = \sum_{i=1}^{k-1} (\lambda_i - \mu \alpha_i)(x_i - x_k) + x_k = \sum_{i=1}^{k} \beta_i x_i$$

and $\sum_{i=1}^{i=k} \beta_i = \sum_{i=1}^{i=k} \lambda_i - \mu \sum_{i=1}^{i=k} \alpha_i = 1$. This shows that x is a convex combination of the vectors x_i, $1 \leq i \leq k$. By definition of μ, there is some $i_0 = 1, 2, \ldots, k$ such that $\beta_{i_0} = 0$.

Thus, x is a convex combination of the $k-1$ points of A. Repeating the process, we obtain that x is a convex combination of the $k-1, k-2, \dots, n+1$ points of A, as claimed. □

Remark A.1.9. When $e_1, \dots, e_n$ are basis vectors of $\mathbb{R}^n$, the convex hull $\mathrm{Co}(\{e_1, \dots, e_n\})$ is a $(n-1)$-simplex of dimension $n-1$ (see Definition 10.1.1).

Corollary A.1.10. *Assume that X is a finite-dimensional normed space and $K \subset X$ a subset. Then*

$$K \text{ compact} \quad \Longrightarrow \quad \mathrm{Co}(K) \text{ compact}.$$

Proof. Let $n = \dim X$. Define the map:

$$\varphi : S \times K^{n+1} \longrightarrow X$$

by

$$\varphi(\lambda, x) = \sum_{i=1}^{n+1} \lambda_i x_i,$$

where $x = (x_1, x_2, \dots, x_{n+1}) \in K^{n+1}$ and $\lambda = (\lambda_1, \lambda_2, \dots, \lambda_{n+1}) \in S$, with

$$S = \left\{ (\lambda_1, \lambda_2, \dots, \lambda_{n+1}) \in \mathbb{R}^{n+1} : \lambda_i \geq 0 \text{ and } \sum_{i=1}^{n+1} \lambda_i = 1 \right\}.$$

We can see that φ is continuous and $S \times K^{n+1}$ is compact. Since by Proposition A.1.8, $\mathrm{Co}(K) = \varphi(S \times K^{n+1})$, $\mathrm{Co}(K)$ is a compact subset of X as the continuous image of a compact set. □

In general, the convex hull of a closed set is not closed (even in finite dimension) and the convex hull of a compact set is not necessarily compact. However, we have the following.

Lemma A.1.11 (Mazur's theorem). *Assume that X is a Banach space and $K \subset X$. Then*

$$K \text{ relatively compact} \quad \Longrightarrow \quad \mathrm{Co}(K) \text{ relatively compact}.$$

Proof. The proof uses the following characterization of compact sets in metric spaces (see Proposition 1.3.34 and Remark 1.3.35)

$$\text{A set } A \text{ is compact} \quad \Longleftrightarrow \quad A \text{ is complete and totally bounded}.$$

Since X is a Banach space, the closed subset $\overline{\mathrm{Co}}(K)$ is complete. Consequently, it is sufficient to check that $\overline{\mathrm{Co}}(K)$ is totally bounded (see Definition 1.3.27), which can be proved using the fact that $\overline{K}$ is totally bounded. □

A.2 The pasting (or gluing) lemma

The following result from point-set topology concerns the extension of continuous functions. The pasting lemma was extensively referred to throughout this book.

Lemma A.2.1. *Let $X = A \cup B$, where A and B are closed (or open) subsets of a topological space $(X, \mathscr{T})$. Let $f : A \longrightarrow Y$ and $g : B \longrightarrow Y$ be two continuous maps such that $f(x) = g(x)$ for all $x \in A \cap B$. Then the extended function*

$$h(x) = \begin{cases} f(x), & x \in A, \\ g(x), & x \in B \end{cases}$$

is continuous on X.

Proof. Suppose that A and B are closed subsets of X and let $C \subset Y$ be a closed subset. Since f and g agree on the intersection $A \cap B$, h is well-defined. Then $h^{-1}(C) = f^{-1}(C) \cup g^{-1}(C)$, where $f^{-1}(C)$ is closed in A and $g^{-1}(C)$ is closed in B. In addition, $f^{-1}(C)$ and $g^{-1}(C)$ are closed in X for A and B are closed in X. Therefore, $h^{-1}(C)$ is closed in X. □

Remark A.2.2. Lemma A.2.1 still holds for a finite numbers of closed subsets B_i that cover X or a possibly an infinite number of open subsets A_i that cover X.

A.3 Lebesgue number

Lemma A.3.1. *Let E be a compact metric space and $(U_\alpha)_{\alpha\in J}$ an open covering of E. Then there exists $\lambda > 0$, called the Lebsegue number, such that for every subspace $F \subset E$ whose diameter is less than λ, there exists $\alpha_0 \in J$ such that $F \subset U_{\alpha_0}$.*

Proof. Suppose to the contrary that there exists no such Lebesgue number. Then, for $\lambda = 1/n$, we obtain a sequence of subsets $(F_n)_n \subset E$ such that $\operatorname{diam} F_n < 1/n$ for all positive integer n and $F_n \not\subset U_\alpha$ for all $\alpha \in J$. Define a sequence of elements $(x_n)_n$ by picking some $x_n \in F_n$. Suppose that some subsequence has a limit $x = \lim_{k\to\infty} x_{n_k}$ and let $U_\alpha \ni x$ for some $\alpha \in J$. Since U_α is open and the balls $(B(x, 1/k))_k$ form a basis for the topology of the metric space E, there exists $k_0 = 1, 2, \ldots$ such that $B(x, 1/k_0) \subset U_\alpha$. Since x is the limit of the subsequence, there exists $k_1 \in \{1, 2, \ldots\}$ such that $x_{n_k} \in B(x, 1/k_0)$ for all $k \geq k_1$. Choose k sufficiently large such that $n_k > n_{k_1} > 2k_0$ and $x_{n_k} \in B(x, 1/2k_0)$. To show that $F_{n_k} \subset U_\alpha$, let $y \in F_{n_k}$. By the triangle inequality, we get

$$\begin{aligned} d(y, x) &\leq d(y, x_{n_k}) + d(x_{n_k}, x) \\ &\leq \operatorname{diam} F_{n_k} + 1/2k_0 \\ &\leq 1/n_k + 1/2k_0 \\ &\leq 1/2k_0 + 1/2k_0 = 1/k_0. \end{aligned}$$

Hence, $y \in B(x, 1/k_0)$, and thus $y \in U_\alpha$, a contradiction to $F_n \not\subset U_\alpha$ for all $\alpha \in J$. We have proved that $(x_n)_n$ has no convergent subsequence, infringing the sequential compactness of the space E. □

A.4 Urysohn's lemma

The following result of concerns the separation of disjoint closed subsets of a normal space (see Definition 1.1.18). A proof may be found, e. g., in [27, Theorem 1.5.11].

Theorem A.4.1. *Let A and B be two disjoint closed subsets of a normal space X. Then there exists a continuous function $f : X \longrightarrow [0,1]$ such that $f(x) = 0$ for all $x \in A$ and $f(x) = 1$ for all $x \in B$.*

A.5 Dugundgi's extension theorem

For the sake of completeness, we begin with an extension theorem valid in the finite-dimensional case and proved by Tietze in 1915 (see [98]).

Theorem A.5.1. *Let $K \subset \mathbb{R}^n$ be a compact subset. Then every continuous function $f : K \longrightarrow \mathbb{R}^n$ can be extended to the whole space $\mathbb{R}^n$.*

Proof. Since K is compact, there exists a dense subset $\{a_1, \dots, a_N\}$ in K. For $x \notin K$, define

$$\phi_i(x) = \max\left\{2 - \frac{\|x - a_i\|}{d(x,K)}, 0\right\}, \quad i = 1, \dots, N.$$

Then we can consider the following extension of f:

$$\tilde{f}(x) = \begin{cases} (\sum_{i=1}^N 2^{-i}\phi_i(x))^{-1} \sum_{i=1}^N 2^{-i}\phi_i(x) f(a_i), & x \notin K, \\ f(x), & x \in K. \end{cases}$$ □

A detailed proof of the following lemma can be found in [21, Theorem 5.7], [25, Theorem 6.1, Chapter IX], [85, Theorem 6.1.1].

Lemma A.5.2. *Let E be a metric space, $A \subset E$ a closed subset and Y a normed space. Then every continuous function $f : A \longrightarrow Y$ has a continuous extension $\tilde{f} : E \longrightarrow Y$ such that $\tilde{f}(E) \subset \mathrm{Co}(f(A))$, where $\mathrm{Co}(A)$ is the convex hull of A.*

Corollary A.5.3. *Let X and Y be two normed spaces, X Banach, $A \subset X$ a closed subset of X and $f : A \longrightarrow Y$ a compact map. Then f has a compact extension $\tilde{f} : X \longrightarrow Y$ such that $\tilde{f}(X) \subset \mathrm{Co}(f(A))$.*

A.6 Stampacchia's lemma

The following result from functional analysis is known as Stampacchia's projection lemma or the Hilbert projection theorem (see, e. g., [54, Theorems IV.5.1, V.3.2]).

Lemma A.6.1. *Let $(H, \langle\cdot\rangle)$ be a Hilbert space and $C \subset H$ a nonempty closed convex subset. Then:*
(1) *For all $x \in H$, there exists a unique $x_0 \in C$ such that $\|x - x_0\| = \mathrm{d}(x, C)$. The point x_0 is denoted $x_0 = P(x)$ and defines a map $P : H \longrightarrow C$.*
(2) *$\langle x - P(x), y - P(x)\rangle \leq 0$ for all $x \in F$ and $y \in C$.*
(3) *$P^2(x) = P(x)$ for all $x \in H$.*
(4) *$\|P(x) - P(y)\| \leq \|x - y\|$ for all $x, y \in H$ (the map P is 1-Lipschitz continuous, Definition 1.2.6).*

The point x_0 is called the projection of x on C.

Proof. (1) For $x \in H$, let $\delta = \mathrm{d}(x, C) = \inf_{y \in C} \|x - y\|$. Then there exists a sequence $(x_n)_n \subset C$ such that $\lim_{n\to\infty} \|x - x_n\| = \delta$. By the parallelogram identity,

$$\|x + y\|^2 + \|x - y\|^2 = 2(\|x\|^2 + \|y\|^2).$$

We can easily derive the median identity, written for x, x_p, x_q, i. e.,

$$\|x_p - x_q\|^2 = 2(\|x - x_p\|^2 + \|x - x_q\|^2) - 4\left\|x - \frac{x_p + x_q}{2}\right\|^2.$$

Since C is convex, $\frac{1}{2}(x_p + x_q) \in C$. Then $\lim_{p,q\to\infty} \|x_p - x_q\| = 0$, proving that (x_n) is a Cauchy sequence. Since C is closed, and hence complete in H, there exists $x_0 = \lim_{n\to\infty} x_n$ such that $\|x_0 - x\| = \delta$. The uniqueness of x_0 follows from the median identity or from (2).

(2) Since $\|x - x_0\| \leq \|x - y\|$ for all $y \in C$,

$$\|x - x_0\|^2 \leq \|x - x_0 - t(y - x_0)\|^2, \quad \text{for all } y \in C \text{ and all } t \in [0, 1],$$

i. e.,

$$2\langle x - x_0, y - x_0\rangle \leq t\|y - x_0\|^2, \quad \text{for all } t \in [0, 1],$$

The claim follows by taking the limit as $t \to 0^+$.

(3) follows from (1).

(4) Let $x' = P(x)$ and $y' = P(y)$. We have

$$x' - y' = (x' - x) + (y - y') + (x - y)$$

and

$$\begin{aligned}\|x'-y'\|^2 &= \langle x'-y', x'-y'\rangle \\ &= \langle x'-x, x'-y'\rangle + \langle y-y', x'-y'\rangle + \langle x-y, x'-y'\rangle.\end{aligned}$$

Using (2) and the Cauchy–Schwartz inequality, we get

$$\begin{aligned}\|x'-y'\|^2 &\leq \langle x-y, x'-y'\rangle \\ &\leq |\langle x-y, x'-y'\rangle| \\ &\leq \|x-y\|.\|x'-y'\|,\end{aligned}$$

proving the claim. □

A.7 Schauder's approximation theorem

Let $A = \{a_1, a_2, \dots, a_N\}$ be a finite set in a normed space X. For some positive ε, the Schauder approximation map $p_\varepsilon : X \longrightarrow X$ is defined by

$$p_\varepsilon(x) = \frac{1}{\sum_{i=1}^N m_i(x)} \sum_{i=1}^N m_i(x) a_i,$$

where $m_i(x) = \max(0, \varepsilon - \|x - a_i\|)$, $i = 1, 2, \dots, N$. Note that $m_i(x) > 0$ for all x in the open ball $B(a_i, \varepsilon)$ and $p_\varepsilon(x) \in \mathrm{Co}(A)$. Let $\mathscr{B}_\varepsilon = \bigcup_{i=1}^N B(a_i, \varepsilon)$. The following two results can be found either in [40, Proposition (2.2) and Theorem (2.3), Section 6] or [103, Proposition 2.12, p. 55].

Lemma A.7.1. *Assume $A \subset C$, where $C \subset X$ is a convex subset. Then:*
(1) *$p_\varepsilon : \mathscr{B}_\varepsilon \longrightarrow \mathrm{Co}(A) \subset C$ is compact;*
(2) *$\|p_\varepsilon(x) - x\| < \varepsilon$ for all $x \in \mathscr{B}_\varepsilon$;*

Proof. (1) follows from the fact that $\mathrm{Co}(A)$ is compact.

(2) We have the estimates

$$\begin{aligned}\|p_\varepsilon(x) - x\| &= \left\| \frac{\sum_{i=1}^N m_i(x) a_i}{\sum_{i=1}^N m_i(x)} - \frac{\sum_{i=1}^N m_i(x) x}{\sum_{i=1}^N m_i(x)} \right\| \\ &= \frac{1}{\sum_{i=1}^N m_i(x)} \left\| \sum_{i=1}^N m_i(x)(x - a_i) \right\| \\ &\leq \frac{1}{\sum_{i=1}^N m_i(x)} \sum_{i=1}^N m_i(x) \|x - a_i\| < \varepsilon.\end{aligned}$$ □

Theorem A.7.2. *Let X be a normed space and $C \subset X$ a convex subset. Let $f : X \longrightarrow C$ be a compact map. Then, for each positive ε, there exists a finite set $A = \{a_1, a_2, \dots, a_N\} \subset$*

$f(X) \subset C$ and a compact map $f_\varepsilon : X \longrightarrow C$ whose range is contained in a finite-dimensional space $X_\varepsilon \subset C$ such that:
(1) $\|f_\varepsilon(x) - f(x)\| < \varepsilon$ *for all* $x \in X$;
(2) $f_\varepsilon(X) \subset \mathrm{Co}(A) \subset C$.

Proof. Let $\varepsilon > 0$. We have

$$\overline{f(X)} \subset \bigcup_{a \in f(X)} B(a, \varepsilon).$$

Since f is compact, there exist finite number of points $A = \{a_1, a_2 \dots, a_N\} \in f(X)$ such that

$$\overline{f(X)} \subset \mathscr{B}_\varepsilon = \bigcup_{i=1}^{N} B(a_i, \varepsilon).$$

By Lemma A.7.1, we can take $f_\varepsilon = p_\varepsilon \circ f$, where p_ε is the Schauder approximation map. Clearly, f_ε is compact and since $p_\varepsilon(X) \subset \mathrm{Co}(A)$ and C is convex, we deduce part (2). □

As a consequence, we deduce the following.

Corollary A.7.3. *Let X be a normed space and $K \subset X$ a compact subset. Then, for each $\varepsilon > 0$, there exist a finite-dimensional subspace $X_\varepsilon \subset X$ and a continuous map $p_\varepsilon : K \longrightarrow X_\varepsilon$ such that*

$$\|p_\varepsilon(x) - x\| < \varepsilon, \quad \textit{for all } x \in K.$$

In addition, $p_\varepsilon(K) \subset \overline{\mathrm{Co}}(K)$.

Proof. Apply Theorem A.7.2 by taking the inclusion map $f = i : K \hookrightarrow C = \mathrm{Co}(K)$. □

A direct proof of Corollary A.7.3 can be found in [56, Lemma 5.2.2].

A.8 Minkowski's functional

Let $0 \in \mathring{C}$, where $C \subset X$ is a bounded convex subset of a normed space X. The gauge functional (or Minkowski's functional) (see, e. g., [54]) $j = j_C : X \longrightarrow [0, \infty)$ is defined by

$$j(x) = \inf\left\{t > 0 : \frac{x}{t} \in C\right\}.$$

Let $0 < r < R$ be such that $B(0, r) \subset C \subset B(0, R)$.

Lemma A.8.1. *j satisfies the following properties:*
(1) $\frac{\|x\|}{R} \le j(x) \le \frac{\|x\|}{r}$ *for all* $x \in X$;

(2) $j(x) = 0 \iff x = 0$;
(3) $\frac{\|x\|}{R} \le j(x) \le 1$ *for all* $x \in C$;
(4) $j(\lambda x) = \lambda j(x)$ *for all* $x \in X$ *and for all* $\lambda \ge 0$ *(j is positively homogeneous);*
(5) $j(x_1 + x_2) \le j(x_1) + j(x_2)$ *for all* $x_1, x_2 \in X$ *(j is subadditive);*
(6) *j is continuous;*
(7) $\{x \in X : j(x) < 1\} = \mathring{C}$, $\{x \in X : j(x) \le 1\} = \overline{C}$ *and* $\{x \in X : j(x) = 1\} = \partial C$.

Proof. (1) Since $B(0,r) \subset C$ for all $0 < \varepsilon < 1$, $\frac{\varepsilon r}{\|x\|} x \in B(0,r) \subset C$. Hence, $\frac{\|x\|}{\varepsilon r} \ge j(x)$ for all $0 < \varepsilon < 1$. Letting $\varepsilon \to 1$, we get $0 \le j(x) \le \frac{\|x\|}{r}$ for all $x \in X$. Since $C \subset B(0,R)$, for all positive t, if $\frac{x}{t} \in C$, then $\frac{x}{t} \in B(0,R)$. Hence, $\frac{\|x\|}{R} \le t$, for all $t > 0$. Taking the infimum over $t > 0$, we find $j(x) \ge \frac{\|x\|}{R}$ for all $x \in X$.

(2) follows from (1).

(3) For $x \in C$, $\frac{x}{1} \in C$ implies $j(x) \le 1$.

(4) Notice that for positive λ, if $\frac{\lambda x}{t} \in C$ then $\frac{x}{t} = \frac{1}{\lambda}(\frac{\lambda x}{t}) \in \frac{1}{\lambda}C$. As a consequence, for all $x \in X$,

$$\begin{aligned} j(\lambda x) &= \inf\left\{t > 0 : \frac{\lambda x}{t} \in C\right\} \\ &= \inf\left\{t > 0 : \frac{x}{t} \in \frac{1}{\lambda}C\right\} \\ &= \inf\left\{\lambda\frac{t}{\lambda} > 0 : \frac{x}{t} \in \frac{1}{\lambda}C\right\} \\ &= \inf\left\{\lambda t' > 0 : \frac{x}{\lambda t'} \in \frac{1}{\lambda}C\right\} \\ &= \inf\left\{\lambda t' > 0 : \frac{x}{t'} \in C\right\} \\ &= \lambda \inf\left\{t' > 0 : \frac{x}{t'} \in C\right\} \\ &= \lambda j(x). \end{aligned}$$

(5) Let $t_1, t_2 > 0$ and $x_1, x_2 \in X$. We have

$$\frac{x_1 + x_2}{t_1 + t_2} = \frac{t_1}{t_1 + t_2}\frac{x_1}{t_1} + \frac{t_2}{t_1 + t_2}\frac{x_2}{t_2}.$$

If $\frac{x_1}{t_1}$ and $\frac{x_2}{t_2}$ lie in C, then by convexity of C, $\frac{x_1+x_2}{t_1+t_2} \in C$. Taking the infimum over elements t_1, t_2 such that $\frac{x_1}{t_1}, \frac{x_2}{t_2} \in C$ and using the definition of j, we conclude that $j(x_1 + x_2) \le j(x_1) + j(x_2)$.

(6) By property (5), we have

$$j(x) = j(x + h - h) \le j(x + h) + j(-h).$$

Hence,

$$j(x) - j(-h) \leq j(x+h) \leq j(x) + j(h),$$

i. e.,

$$-j(-h) \leq j(x+h) - j(x) \leq j(h).$$

Equivalently, using property (1),

$$|j(x+h) - j(x)| \leq \max(|j(h)|, |j(-h)|) \leq \frac{\|h\|}{r}, \quad \text{for all } h \in X.$$

As a consequence $\lim_{h \to 0} j(x+h) = j(x)$, for all $h \in X$.

(7) Let $x \in X$ be such that $j(x) < 1$. By definition of j, there is some $0 < t_0 < 1$ such that $x/t_0 \in C$. Then $x = t_0 x/t_0 + (1 - t_0)0 \in C$. This with property (3) yield

$$j^{-1}(-\infty, 1) \subseteq C \subseteq j^{-1}(-\infty, 1].$$

Then, taking the interior sets (closure sets, respectively) and using property (6) of the continuity of j, we obtain the characterization of the interior of C (closure of C, respectively). The last characterization then follows from the definition of the boundary of a set. □

Remark A.8.2. To define Minkowski's function, it is not necessary for C to be convex or bounded. Convexity is needed to show the subadditivity property. In the general case, j may be infinite but it is finite if C is absorbing, i. e., for all $x \in C$, there exists $\lambda_0 > 0$ such that $x \in \lambda C$ for all $|\lambda| > \lambda_0$ (see [22] or [93, Lemma 4.2.5]). The boundedness of C is used to show properties (1), (2) and (3). The space X needs only to be a vector space [54].

A.9 Ascoli–Arzéla lemma

Definition A.9.1. Let E and F be two metric spaces and $H \subset C(E, F)$ a family of continuous functions from E to F. H is called equicontinuous if for all $\varepsilon > 0$, there exists $\delta > 0$ such that for all $x, y \in E$,

$$d_E(x, y) \leq \delta \quad \Longrightarrow \quad d_F(f(x), f(y)) \leq \varepsilon, \quad \text{for all } f \in H.$$

Lemma A.9.2. *Let E be a compact metric space and F a complete metric space. Let $\mathscr{H} \subset \mathscr{C}(E, F)$ be a bounded family subset. Then*

$$\mathscr{H} \text{ relatively compact} \quad \Longleftrightarrow \quad \begin{cases} \mathscr{H} \text{ is equicontinuous} \\ \text{and} \\ \text{for all } x \in E, \text{ the set } \mathscr{H}(x) \text{ is relatively compact,} \end{cases}$$

where $\mathscr{H}(x) = \{f(x) : f \in \mathscr{H}\}$.

Proof. For the sake of simplicity of presentation, let $F = (F, \|\cdot\|)$ be a Banach space. For the general case, we refer the reader to, e. g., [76, Corollary 45.5]. Let $X = \mathscr{C}(E, F)$.

(1) Necessity.

(a) If $\mathscr{H}$ is relatively compact, then for all $\varepsilon > 0$, there exist a finite number of elements $\{f_i\}_{1\le i\le n}$ in X such that $\mathscr{H} \subset \bigcup_{i=1}^{n} B(f_i, \varepsilon/3)$, i. e., symbolically

$$\forall f \in \mathscr{H}, \exists i \in \{1, \dots, n\}, \quad \|f - f_i\|_X \le \varepsilon/3.$$

Then

$$\forall f \in \mathscr{H}, \forall x \in E, \exists i \in \{1, \dots, n\}, \quad \|f(x) - f_i(x)\| \le \varepsilon/3.$$

Hence,

$$\mathscr{H}(x) \subset \bigcup_{i=1}^{n} B(f_i(x), \varepsilon/3).$$

Therefore, $\mathscr{H}(x)$ is relatively compact in F.

(b) We claim that $\mathscr{H}$ is equicontinuous. For each $i = 1, 2, \dots, n$, the function f_i is continuous. As a consequence, for all positive ε, there exists $\delta_i > 0$, such that for all $x, y \in E$, we have

$$d(x,y) \le \delta_i \quad \Longrightarrow \quad \|f_i(x) - f_i(y)\| \le \varepsilon/3.$$

Let $\delta = \min_{1\le i\le n} \delta_i$ and $f \in \mathscr{H}$. Then there exists $i \in \{1, \dots, n\}$ such that $f \in B(f_i, \varepsilon/3)$ and for all $x, y \in E$,

$$\begin{aligned} d(x,y) \le \delta \quad \Longrightarrow \quad \|f(x) - f(y)\| &\le \|f(x) - f_i(x)\| + \|f(y) - f_i(y)\| \\ &\quad + \|f_i(x) - f_i(y)\| \\ &\le \varepsilon/3 + \varepsilon/3 + \varepsilon/3 = \varepsilon, \end{aligned}$$

proving the equicontinuity of $\mathscr{H}$.

(2) Sufficiency. Since $X = \mathscr{C}(E, F)$ is complete, it is sufficient to prove that $\mathscr{H}$ is totally bounded (see Definition 1.3.27, Proposition 1.3.34 and Remark 1.3.35). Let $\varepsilon > 0$. Since $\mathscr{H}$ is equi-continuous for every $x \in E$, there exists some $\delta > 0$ such that for all $y \in E$ and $f \in \mathscr{H}$,

$$d(x,y) \le \delta \quad \Longrightarrow \quad \|f(x) - f(y)\| \le \varepsilon/4.$$

Since E is compact, it can be covered by a finite number of balls $B_{x_i} = B(x_i, r), 1 \le i \le m$. By supposition, each subset $\mathscr{H}(x)$ is relatively compact in F and so is their finite union $\mathscr{H} = \bigcup_{i=1}^{m} \mathscr{H}(x_i)$. Then we can cover $\mathscr{H}$ by a finite number of ball centered at c_j $(1 \le j \le p)$ and with radius $\frac{\varepsilon}{4}$. Let $J_1 = \{1, 2, \dots, m\}, J_2 = \{1, 2, \dots, p\}$ and Φ the set of all mappings

$\varphi : J_1 \longrightarrow J_2$. For all $\varphi \in \Phi$, denote by L_φ the set of all mappings $f \in \mathscr{H}$ such that for all $i \in J_1$,

$$\|f(x_i) - c_{\varphi(i)}\| \leq \varepsilon/4.$$

Some of the sets L_φ may be empty, but $\mathscr{H}$ is covered by the union of L_φ. We need to prove that the diameter of each L_φ is less than or equal to ε. Let $f, g \in L_\varphi$. For every $y \in E$, there exists $i \in J_1$ such that $y \in B_{x_i}$. Hence,

$$\|f(y) - f(x_i)\| \leq \varepsilon/4 \quad \text{and} \quad \|g(y) - g(x_i)\| \leq \varepsilon/4.$$

For all $y \in E$, we have the estimates

$$\begin{aligned} \|f(y) - g(y)\| &\leq \|f(y) - f(x_i)\| + \|g(y) - g(x_i)\| \\ &\quad + \|f(x_i) - c_{\varphi(i)}\| + \|g(x_i) - c_{\varphi(i)}\| \\ &\leq \varepsilon/4 + \varepsilon/4 + \varepsilon/4 + \varepsilon/4 = \varepsilon. \end{aligned}$$

Hence, $\|f - g\|_{\mathscr{C}(E,F)} \leq \varepsilon$ and the claim follows. □

As a consequence of the Ascoli–Arzéla lemma, we have the following.

Corollary A.9.3. *Let E be a compact metric space and F a Banach space with finite dimension. Then a bounded family $\mathscr{H} \subset \mathscr{C}(E, F)$ is relatively compact if and only if $\mathscr{H}$ is equicontinuous.*

This is due to the fact that if A is a subset of a finite-dimensional space, then

$$A \text{ bounded} \quad \Longleftrightarrow \quad A \text{ relatively compact}.$$

B Basics in algebra

The essential of algebra concepts we needed in this book is recalled in this section. More details can be found in most textbooks of algebra, e. g., [4].

B.1 Group

Definition B.1.1. A group is a pair $(G, *)$ such that the internal operation $*$ is associative, there exists an identity element $e \in G$ and each element x of G has an inverse x', i. e., $x * x' = x' * x = e$. The group is called Abelian if the binary operation $*$ is commutative.

With an additive group notation, $* = +$, $e = 0$, the inverse of $x \in G$ is $-x$ and $nx = \overbrace{x + x + \cdots x}^{n}$. With a multiplicative group notation, $* = \cdot$, (the dot may be omitted), $e = 1$, the inverse of $x \in G$ is x^{-1} and $x^n = \overbrace{x \cdot x \cdot \ldots x}^{n}$.

A subset $H \subset G$ is called subgroup if $(H, *)$ is a group. It is easy to check that H is a subgroup if and only if $x * y^{-1} \in H$ for all $x, y \in H$.

Let $(G, *)$ and $(G', \circ)$ be two groups and $f : G \longrightarrow G'$ a map.

Definition B.1.2. (1) f is called a homomorphism if for all $x, y \in G, f(x * y) = f(x) \circ f(y)$.
(2) A monomorphism is an injective homomorphism.
(3) An epimorphism is a surjective homomorphism.
(4) An isomorphism is a bijective homomorphism.

Remark B.1.3. If f is a homomorphism, then $f(x') = (f(x))'$, where x' is the inverse of x.

Definition B.1.4. The kernel and the image (range) of a homomorphism f are the sets

$$\operatorname{Ker} f = \{x \in G : f(x) = e'\} \quad \text{and} \quad \operatorname{Im} f = \{y \in G' : \exists x \in G, f(x) = y\},$$

respectively, where e' is the identity element of the group G'.

Example B.1.5. Let $\mathbb{R}$ be the additive group of real numbers and $\mathbb{C} \setminus \{0\}$ the multiplicative group of nonzero complex numbers. Then the map $f : (\mathbb{R}, +) \longrightarrow (\mathbb{C} \setminus \{0\}, \cdot)$ given by $f(t) = e^{2i\pi t}$ is a homomorphism and $\operatorname{Ker} f \cong \mathbb{Z}$.

It is easily verified that

Proposition B.1.6. (1) *The sets* $\operatorname{Ker} f$ *and* $\operatorname{Im} f$ *are subgroups of* G *and* G'*, respectively;*
(2) f *is one-to-one if and only if* $\operatorname{Ker} f = \{e\}$*;*
(3) f *is onto if and only if* $\operatorname{Im} f = G'$.

B.2 Quotient group

Definition B.2.1. Let $g \in G$, where G is a multiplicative group and let H be a subset of G. The left coset and the right coset of H by an element $g \in G$ are defined as

$$gH = \{gh : h \in H\} \quad \text{and} \quad Hg = \{hg : h \in H\},$$

respectively.

Remark B.2.2. For every $g \in G$, the left coset gH and the right coset Hg of H are equipotent to the subset H (i. e., with a bijection between them) with the bijections

$$h \in H \longmapsto gh \quad \text{and} \quad h \in H \longmapsto hg,$$

respectively.

Definition B.2.3. Let H be *a subgroup* of G. Define on G an equivalence relation by $g \sim_l g'$ if and only if there exists $h \in H$ such that $g = g'h$. Equivalently, $(g')^{-1}g \in H$.

Remark B.2.4. (1) Any equivalence class is a left coset. Indeed, for all $g \in G$,

$$\begin{aligned}[g]_l &= \{g' \in G : g^{-1}g' \in H\}\\ &= \{g' \in G : \exists\, h \in H : g^{-1}g' = h\}\\ &= \{g' \in G : \exists\, h \in H : g' = gh\}\\ &= gH.\end{aligned}$$

(2) The equivalence relation given by "$g \sim_r g'$ if and only if there exists $h \in H$ such that $g = hg'$" defines the right coset as an equivalence class.
(3) Since gH and Hg are equivalence classes, the left cosets (the right cosets, respectively) are either disjoint or identical. In other words,

$$g \sim_l g' \iff gH = g'H \quad \text{and} \quad g \sim_r g' \iff Hg = Hg'.$$

(4) If G is an Abelian group, the left coset and the right coset are identical.
(5) The quotient set $G/_{\sim_l}$ ($G/_{\sim_r}$, respectively) is described as the set of left cosets gH, $g \in G$ (right cosets $Hg, g \in G$, respectively) of H. In general, these quotient sets have not group structures.

Definition B.2.5. A subgroup $H \subset G$ is said to be normal if $g^{-1}hg \in H$ for all $g \in G$ and $h \in H$.

If G is an Abelian group, then all subgroups are normal. From the definition, we have the following.

Proposition B.2.6. *The following statements are equivalent:*
(1) *$H \subset G$ is normal.*
(2) *For all $g \in G$ and $h \in H$, there exists $h' \in H$ such that $gh' = hg$.*
(3) *For all $g \in G$, $g^{-1}Hg = H$.*

A normal subgroup is denoted $H \trianglelefteq G$. We have the following.

Proposition B.2.7. *H is a normal subgroup of G if and only if the left coset and the right coset of H by any element $g \in G$ are equal, i. e., $gH = Hg$ for all $g \in G$.*

Proof. (1) Assume that the condition is satisfied and let $g \in G$ and $h \in H$. Then $gH = Hg$ yields some $h' \in H$ such that $ghg^{-1} \in H$. By Proposition B.2.6, H is normal.

(2) Conversely, let $h' = gh \in gH$ for some $h \in H$. Then $h'g^{-1} = ghg^{-1}$. Since H is normal, $h'g^{-1} \in H$, and thus $h' = h''g \in Hg$ with $h'' = h'g^{-1} \in H$. The proof of the converse inclusion is the same. □

Example B.2.8. If $f : G \longrightarrow G'$ is a homomorphism, then the kernel subgroup $\operatorname{Ker} f \subset G$ is normal.

Proposition B.2.9. *Let $H \subset G$ be a normal subgroup. Then:*
(1) *The set of all cosets has a group structure under the operation $(gH)\cdot(g'H) = (gg')H$ for all $g, g' \in G$. It is called the quotient group of G modulo H and denoted G/H.*
(2) *The quotient map $\pi : G \longrightarrow G/H$ given by $\pi(g) = gH$ is a homomorphism.*
(3) *We have* $\operatorname{Ker}\pi = H$.

Proof. (1) (a) The internal operation is clearly well-defined. Indeed, if $g' \sim g''$, then $(g'')^{-1}g' \in H$, which is equivalent to $(gg'')^{-1}(gg') \in H$, i. e., $gg' \sim gg''$ and so

$$(gg')H = (gg'')H \iff (gH)\cdot(g'H) = (gH)\cdot(g''H).$$

Note here that we have not used the normality of H. However, $g \sim g''$ means $(g'')^{-1}g \in H$. Then $(gg')H = (g''g')H$ if and only if $(g''g')^{-1}(gg') \in H$. By associativity,

$$(g')^{-1}((g'')^{-1}g)g' = (g')^{-1}(g'')^{-1}(gg') \in H,$$

which holds since H is normal.

(b) The associativity of the internal operation follows from the associativity of the dot operation in the group G.

(c) The identity element is $eH = H$, where $e = e_G$ is the identity element in G (and also in H). The symmetric of the element gH is $g^{-1}H$.

(2) follows from the definition of the internal operation in G/H.

(3) (a) If $g \in \operatorname{Ker}\pi$, then $gH = \pi(g) = H$, by definition of the identity element. In particular, $g = ge \in H$, i. e., $\operatorname{Ker}\pi \subset H$.

(b) Conversely, let $g \in H$. Then, for any $h \in H$, $h' = g^{-1}h \in H$ and so $h = gh' \in gH$, proving that $H \subset gH$. For the converse inclusion, if $h \in H$, then $gh \in H$ because H is a subgroup. Hence, $gH \subset H$, and consequently equality $gH = H$ holds. Equivalently, $\pi(g) = H$. Hence, $g \in \operatorname{Ker}\pi$, proving that $H \subset \operatorname{Ker}\pi$, which completes the proof of (3). □

Example B.2.10. Let $G = \mathbb{Z}$ be the additive group of integers and $H = n\mathbb{Z}$ for some $n \in \{1, 2, \ldots\}$ the subgroup of all multiples of n. Define a relation of equivalence in G by $m \sim m'$ if and only if $m - m' \in n\mathbb{Z}$, i. e., $m \cong m'[n]$. Then $H = \operatorname{Ker} h$, where the homomorphism h is $h(k) \cong k[n]$ for all $k \in \mathbb{Z}$. Owing to example B.2.8, H is a normal subgroup. By Proposition B.2.9, $\mathbb{Z}/n\mathbb{Z}$ is a quotient group and $H = \operatorname{Ker}\pi$, where the quotient map is given by $\pi(k) = kn\mathbb{Z}$ for all $k \in \mathbb{Z}$.

Example B.2.11. (1) The quotient group G/G is isomorphic to the trivial group with the identity element $\{e\}$ because $\operatorname{Ker}\pi = G$.
(2) The quotient $G/\{e\}$ is isomorphic to G because $\pi(g) = \{g\}$, for $g \in G$.

Remark B.2.12. (1) Example B.2.8 and Proposition B.2.9(3) show that a subgroup is normal if and only if it is the kernel of some homomorphism.

(2) We have $\operatorname{Ker} \pi = H$. However, H is the identity element in G/H but not in G. So, π is not an isomorphism.

A special situation of a normal subgroup holds in a topological group. The following result can be found, e. g., in [6, Theorem (4.10)].

Proposition B.2.13. *Let G be a topological group (Definition 4.7.1) and let K denote the connected component of G, which contains the identity element. Then K is a closed normal subgroup of G.*

Proof. The set K is closed as a connected component. To check that K is normal, note that for every $x \in K$, the set $\{x^{-1}y : y \in K\} = x^{-1}K$ is connected as the continuous image of K under the map $\phi(y) = x^{-1}y$. By definition of K, $x^{-1}K \subset K$. Hence, $x^{-1}y \in K$ for all $(x, y) \in K^2$. Also, $x^{-1}yx \in K$ for all $(x, y) \in G \times K$, proving the claim. □

The general formulation of the first theorem of homomorphism is the following.

Theorem B.2.14. *Let $f : G \longrightarrow G'$ be a group homomorphism. Then:*
(1) *The quotient $G/\operatorname{Ker} f$ is a group.*
(2) *The quotient group $G/\operatorname{Ker} f$ is isomorphic to $\operatorname{Im} f$.*
(3) *If f is surjective, $G/\operatorname{Ker} f$ and G' are isomorphic.*

Proof. Part (1) is a consequence of Example B.2.8 and Proposition B.2.9(1).

(2) Define the map $\phi : G/\operatorname{Ker} f \longrightarrow G'$ by $\phi([x]) = f(x)$, i. e., $\phi \circ \pi = f$, where π is the quotient map given in Proposition B.2.9 as shown in the diagram:

$$\begin{array}{ccc} G & \xrightarrow{\quad f \quad} & G' \\ {\scriptstyle \pi_{\operatorname{Ker} f}} \big\downarrow & & \big\uparrow {\scriptstyle i} \\ G/_{\operatorname{Ker} f} & \xrightarrow{\quad \phi \quad} & \operatorname{Im} f \end{array}$$

Then:

(a) ϕ is well-defined. Indeed, if $[x] = [x']$, i. e., $x \sim x'$, then

$$\begin{aligned} (x')^{-1}x \in \operatorname{Ker} f \quad &\Longleftrightarrow \quad f((x')^{-1}).f(x) = 1 \\ &\Longleftrightarrow \quad (f(x'))^{-1}f(x) = 1 \\ &\Longleftrightarrow \quad f(x) = f(x'), \end{aligned}$$

because f is a homomorphism.

(b) In addition, part (a) also shows that ϕ is one-to-one.

(c) The map ϕ is a homomorphism because

$$\phi([x].[x']) = \phi([xx']) = f(xx') = f(x)f(x') = \phi([x])\phi([x']).$$

□

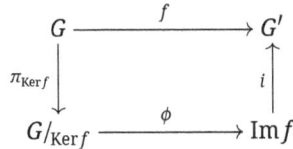

The following result allows factorization of homomorphism through the projection map.

Theorem B.2.15. *Let $f : G \longrightarrow G'$ be a group homomorphism and $K \subset \operatorname{Ker} f$ a normal subgroup. Then there exists a unique homomorphism $\tilde{f} : G/K \longrightarrow G'$ such that the following diagram commutes:*

$$\begin{array}{ccc} G & \xrightarrow{\pi} & G/K \\ & \searrow^{f} & \downarrow \tilde{f} \\ & & G'. \end{array}$$

Proof. It is immediate to check that $\tilde{f}$ is defined as the map ϕ in the proof of Theorem B.2.14. □

Definition B.2.16. The direct sum $G \oplus G'$ of two additive groups G, G' is the Cartesian product $G \times G'$ with the operation

$$(g_1, g_1') \oplus (g_2, g_2') = (g_1 + g_2, g_1' + g_2').$$

B.3 Cyclic group

Let $(G, \cdot)$ be a multiplicative group. For $g \in G$, by g^n it is meant $g.g \dots .g$, n times. For $n < 0$, $g^{-n} = (g^{-1})^n$, where g^{-1} is the inverse of g. The set of all elements of the form g^n, $n \in \mathbb{Z}$, is denoted $[g]$. It forms a subgroup of G.

Definition B.3.1. An element g of a group G has a finite order if there is some positive integer n such that $g^n = 1$. The smallest such an n is called the order of g. If g has no finite order, we say that it has infinite order.

Example B.3.2. Let $f : \mathbb{Z} \longrightarrow G$ be given by $f(k) = g^k$, for some $g \in G$, where G is a multiplicative group. Then f is a homomorphism. Additionally, $\operatorname{Ker} f \cong n\mathbb{Z}$ if g has finite order n and $\operatorname{Ker} f \cong 0$ if g has infinite order.

Definition B.3.3. (1) The group G is said to be cyclic if there exists some $g \in G$ such that $G = [g]$. g is called the generator of the group G.
(2) A cyclic group is finite if the generator g has a finite order (Definition B.3.1). Otherwise, it is infinite.

Remark B.3.4. Any two finite cyclic groups of the same order are isomorphic.

Example B.3.5. (1) The set of integers $\mathbb{Z}$ is an infinite cyclic additive group with generator 1.
(2) The set of integers modulo n is an additive finite cyclic group of order n generated by $[1]$. It is denoted $\mathbb{Z}_n = \mathbb{Z}/n\mathbb{Z}$ (Example B.2.10).

Example B.3.6. The multiplicative group of the complex nth roots of the unity is a finite cyclic group generated by $e^{\frac{2i\pi}{n}}$ since $(e^{\frac{2i\pi}{n}})^k = e^{\frac{2ik\pi}{n}}$ $(k = 1, 2, \ldots, n)$ and has order n for $(e^{\frac{2i\pi}{n}})^n = 1$.

Proposition B.3.7. (1) *Every finite cyclic group is isomorphic to* $\mathbb{Z}_n$ *for some positive* n.
(2) *Every infinite cyclic group is isomorphic to* $\mathbb{Z}$.

More generally, we have the following.

Definition B.3.8. Let $S \subset G$ be any subset of a multiplicative group G. We say that G is generated by S if every element of G can be written as a product of positive and negative powers of elements of S. It is denoted $G = \langle S \rangle$.

Remark B.3.9. It can be checked that $\langle S \rangle$ is the smallest subgroup that contains the subset S.

Example B.3.10. If S is a one-element set, then G is a cyclic group.

Remark B.3.11. The generator is not unique. For example, in $\mathbb{Z}_3$, $\{1\}$ and $\{2\}$ are both generators of the cyclic group of order 3.

B.4 Abelianization of a group

Definition B.4.1. Let G be a multiplicative group and $x, y \in G$.
(1) The commutator of x and y is the element

$$[x, y] = xyx^{-1}y^{-1}.$$

(2) The subgroup of G generated by the set of all commutators in G is called the commutator subgroup of G and denoted $[G, G]$.

Remark B.4.2. (1) Clearly, $[x, y] = 1$ if and only if x and y commute.
(2) The commutator $[G, G]$ is the normal subgroup of G generated by the elements $xyx^{-1}y^{-1}$, for $x, y \in G$.

Remark B.4.3. Let $xyx^{-1}y^{-1}$ be an element of the commutator $[G, G]$. Then,

$$g[x, y]g^{-1} = [gxg^{-1}, gyg^{-1}]$$

which implies $x[G, G] \cong [G, G]x$ for every $x \in G$. Hence, the quotient group $G/[G, G]$ is Abelian.

Definition B.4.4. The quotient group $G/[G, G]$ is called the Abelianization of the group G.

In the general case, for a group G and a normal subgroup H, we have the following.

Proposition B.4.5. *The quotient group G/H is Abelian if and only if $[G, G] \subseteq H$.*

Proof. (1) If $[G, G] \subseteq H$, then $[g, g'] = gg'g^{-1}g'^{-1} \in H$ for all g, g' in G. By Definition B.2.3, $gg' \sim_l g'g$, i. e., $[gg'] = [g'g]$ or $gg'H = g'gH$, for all g, g' in G. Hence, G/H is Abelian.

(2) Conversely, if G/H is Abelian, then for all g, g' in G, $[gg'] = [g'g]$. Hence, $[g, g'] = g'g^{-1}g'^{-1} \in H$, proving the claim. $\qquad\square$

B.5 Free product

Let $(G, \cdot)$ be a multiplicative group and $G_i \subset G$, $i \in J = [1, n]$ a family of subgroups.

Definition B.5.1. The group G is called a free product of subgroups G_i and is denoted

$$G = \prod_{i \in J}^* G_i = G_1 * G_2 * \cdots * G_n$$

if there are homomorphisms $\phi_i : G_i \longrightarrow G$, $i \in J$ such that for any group H and any homomorphisms $\psi_i : G_i \longrightarrow H$, there exists a unique homomorphism $f : G \longrightarrow H$ such that the following diagram commutes:

$$\begin{array}{ccc} G_i & \xrightarrow{\phi_i} & G \\ & \searrow{\psi_i} & \downarrow{f} \\ & & H \end{array}$$

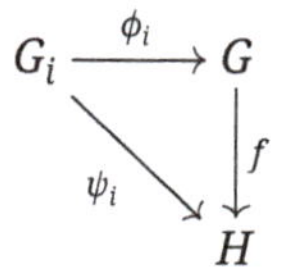

Then the groups $G_1, G_2, \ldots, G_n$ are called the generators of G.

Remark B.5.2. (1) If ϕ_i are monomorphisms (for instance in the case of inclusion maps), each subgroup G_i can be identified with its image. Thus, every element of G can be written as a finite product of elements of the groups G_i, $i \in J = [1, n]$.

(2) In the case of additive groups, G is called direct sum and is denoted $G = \bigoplus_{i \in J} G_i$.

The existence of a free product is assured, e. g., by [67, Theorem 4.2], where one can find the following.

Example B.5.3. Let $G_1 = \langle x_1 \rangle = \{1, x_1\}$ and $G_2 = \langle x_2 \rangle = \{1, x_2\}$ be two cyclic groups, each of order 2, where $\{1\}$ is the identity element.Then their free product $G_1 * G_2$ consists of the following elements:

$$x_1, x_1 x_2, x_1 x_2 x_1, \ldots \quad \text{or} \quad x_2, x_2 x_1, x_2 x_1 x_2, \ldots.$$

The elements $x_1 x_2$ and $x_2 x_1$ are different and each one has infinite order. The free product $G_1 * G_2$ is denoted $\{1, x_1, x_2\}$ or $\langle x_1, x_2 \rangle$.

Example B.5.4. More generally, let $G = \langle x \mid x^m = 1\rangle$, and $H = \langle y \mid y^n = 1\rangle$ be two cyclic groups of orders m and n, respectively. Then their free product is

$$G * H = \langle x, y \mid x^m = y^n = 1\rangle.$$

In the case when the free product in Definition B.5.1 consists of only one set (not necessarily a group), we have the following.

Definition B.5.5. A free group G on a set S is a group such that there exists a map $\phi : S \longrightarrow G$ satisfying the following property: for every group H and every map $\psi : S \longrightarrow H$, there exists a unique morphism f such that $\psi = f \circ \phi$:

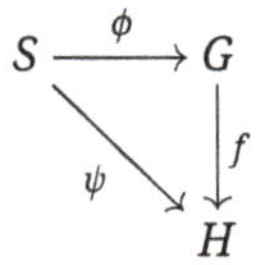

B.6 Free product with amalgamation

Let G_1, G_2, H be groups and $\phi_i : H \longrightarrow G_i$ two group homomorphisms ($i = 1, 2$):

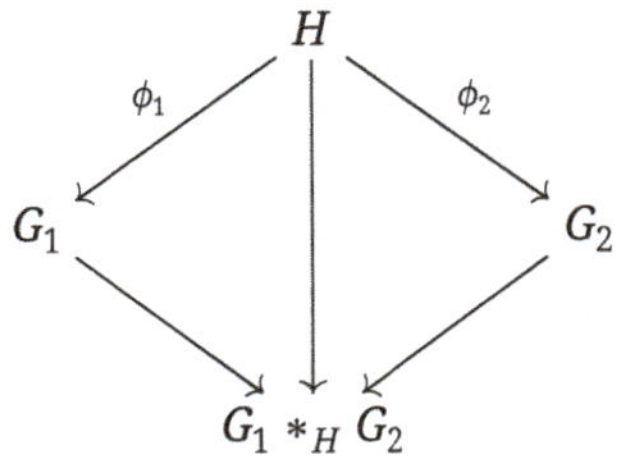

Definition B.6.1. The free product of G_1, G_2 with amalgamation H (or amalgamated on H), denoted $G_1 *_H G_2$, is defined as the quotient group $(G_1 * G_2)/N$, where N is the least normal subgroup of $G_1 * G_2$ that contains the elements $\phi_1(h)\phi_2(h)^{-1}$, $h \in H$.

Example B.6.2. If $H \subset G_i$ and ϕ_i are the inclusion maps, for $i = 1, 2$, then N is the trivial group and thus $G_1 *_H G_2 \cong G_1 * G_2$.

Example B.6.3. If $H \cong \{e\}$, then $G_1 *_H G_2 \cong G_1 * G_2$.

Example B.6.4. If $G_2 \cong \{e\}$, then $G_1 *_H G_2 \cong G_1/N$, where N is the least normal subgroup of G_1 that contains the elements $\phi_1(h)$, $h \in H$.

Let N_1, N_2 be normal subgroups of two groups G_1, G_2, respectively. By [76, Theorem 68.7], we have the following.

Proposition B.6.5. *If N is the least normal subgroup of $G_1 * G_2$ that contains N_1 and N_2, then*

$$(G_1 * G_2)/N \cong (G_1/N_1) * (G_2/N_2).$$

Example B.6.6. We have

$$(\mathbb{Z} * \mathbb{Z})/(2\mathbb{Z} * 2\mathbb{Z}) \cong (\mathbb{Z}/2\mathbb{Z}) * (\mathbb{Z}/2\mathbb{Z}).$$

Taking N_1 the identity element subgroup and $N_2 = G_2$, we get the following.

Corollary B.6.7. *If N is the least normal subgroup of $G_1 * G_2$ that contains G_2, then*

$$(G_1 * G_2)/N \cong G_1. \tag{B.1}$$

Remark B.6.8. Let $h_1 : G_1 \longrightarrow G_1$ be the identity map, $h_2 : G_2 \longrightarrow G_1$ be the trivial one and $h : G_1 * G_2 \longrightarrow G_1$ the extension given by

$$h(x) = \begin{cases} h_1(x), & x \in G_1, \\ h_2(x), & x \in G_2. \end{cases}$$

Then h is surjective with kernel G_2. Hence,

$$(G_1 * G_2)/G_2 \cong G_1, \tag{B.2}$$

proving directly a particular case of Corollary B.6.7.

B.7 Free group

Definition B.7.1. (1) An Abelian group G is called free if there exists a subset $A \subset G$ (called basis of G) such that every element $g \in G$ has the unique representation

$$g = \sum_{x \in A} n_x x,$$

where $n_x \in \mathbb{Z}$ and the sum is zero for all but finitely many x. In other words, a free Abelian group is a free $\mathbb{Z}$-module, i. e., a free module over the integers.

(2) The cardinality of A is called the rank of the group G (or dimension of G) and is denoted $\dim G$.

(3) Let G be a free Abelian group with basis $\{e_1, e_2, \ldots, e_n\}$ and $f : G \longrightarrow G$ a homomorphism. Then, for all $1 \leq i \leq n$, $f(e_i) = \sum_{i=1}^{i=n} a_{ji} e_j$ and the trace of f, denoted $\mathrm{tr}[f, G]$, is the trace of the matrix $(a_{ij})_{1 \leq i,j \leq n}$.

Remark B.7.2. (1) Condition (1) in Definition B.7.1 is equivalent to saying that A generates G and

$$\sum_{n_i \in \mathbb{Z}, x_i \in A, i=1,\dots,N} n_i x_i = 0 \iff n_i = 0, \quad \text{for all } i = 1, \dots, N.$$

(2) Nielsen–Schreier theorem states that if H is a subgroup of a free Abelian group G, then H is a free Abelian group (see [75, Lemma 11.2]).

Example B.7.3. A multiplicative group G generated by its elements $\{x_1, x_2, \dots, x_N\}$ is an Abelian free group. Every element $x \in G$ is uniquely expressed as

$$x = x_{i_1}^{\delta_1} x_{i_2}^{\delta_2} \dots x_{i_N}^{\delta_N},$$

where $\delta_j \in \mathbb{Z}$, for $j = 1, 2, \dots, N$. Rank of G is N.

Remark B.7.4. By Remark B.8.7, a free Abelian group G is isomorphic to a finite direct sum of the ring $\mathbb{Z}$.

Example B.7.5. (1) The rank of the group $\mathbb{Z}$ is 1 for $n = n.1$ for all $n \in \mathbb{Z}$.
(2) The rank of the group $\mathbb{Z} \oplus \mathbb{Z}$ is $1 + 1 = 2$.

Example B.7.6. (1) The rank of the group $\mathbb{Z} \oplus \mathbb{Z}/2$ is $1 + 0 = 1$.
(2) The rank of the group $\mathbb{Z}/2 = \mathbb{Z}/2\mathbb{Z} =$ is 0. Indeed, $[1]$ is a generator of $\mathbb{Z}/2 = \{[0], [1]\}$ for $[1] = 1.[1]$ and $[0] = [2] = [1] + [1] = 2[1]$ but $[0]$ is not a generator. Yet, the rank cannot be 1 because $[0] = 2[1] = \cdots = 2n[1]$ for all $n \in \mathbb{Z}$. So, the representation is not unique. $\mathbb{Z}/2$ is not a free Abelian group.

Example B.7.7. Let E be a nonempty set and G the additive group of all functions $f : E \longrightarrow R$, which take the value zero for all but only a finite number of elements of E, say $f(e_1), f(e_2), \dots, f(e_N)$. For each $e \in E$, define $f_e(x) = 1$ if $x = e$ and $f_e(x) = 0$ otherwise. Then every element f of G is uniquely represented as

$$f = \sum_{j=1}^{N} f(e_j) f_{e_j}.$$

Hence, G is a free Abelian group with basis $\mathscr{B} = \{f_e : e \in E\}$ and rank N.

Example B.7.8. Let E be a space and R a ring. For each $x \in E$, define the function

$$e_x(y) = \begin{cases} 1, & y = x, \\ 0, & y \neq x. \end{cases}$$

Then the injective map $x \mapsto e_x$ identifies the space E with the set $\mathscr{B} = \{e_x, \ x \in E\}$. Let $L(E; R)$ denote the free Abelian group generated by the basis $\mathscr{B}$. Each element $f \in L(X; R)$

can be expressed as a sum $f = \sum_{x \in E} \lambda_x e_x$, where the $\lambda_x \in R$ are zero but for a finite number of elements $x \in E$.

B.8 Module

Definition B.8.1. A module M over a ring R is an Abelian group $(M, +)$ with an operation $R \times M \longrightarrow M$ such that for all $r, s \in R$ and all $x, y \in M$,

$$\begin{cases} r(x+y) = rx + ry, \\ (r+s)x = rx + sx, \\ (rs)x = r(sx), \\ ex = x, \end{cases}$$

where e is the identity element of the ring R.

Remark B.8.2. When $R = \mathbb{Z}$, the module M is an Abelian group. If R is a field, then M is a vector space.

Definition B.8.3. Let M and N be modules. A morphism of modules is a map $f : M \longrightarrow N$ such that for all $r, s \in R$ and all $x, y \in M$,

$$f(rx + sy) = rf(x) + sf(y).$$

Definition B.8.4. A module is (finitely) generated if there exist finitely many elements $x_1, x_2, \ldots, x_n$ in M such that every element of M is a linear combination of $x_1, x_2, \ldots, x_n$ with coefficients in the ring M.

Definition B.8.5. A module is cyclic if it is generated by one element.

Definition B.8.6. A free module is a module with a basis $\mathscr{B} = \{b_i,\ i = 1, \ldots, n\}$ (R-linearly independent and finitely generator).

Remark B.8.7. Thus every element m of the module M can uniquely written as $\sum_{i=1}^{n} \sigma_i b_i$, where for each $i = 1, \ldots, n$, $\sigma_i \in R$ and $b_i \in \mathscr{B}$. By considering the module isomorphism $m \mapsto (\sigma_1, \sigma_2, \ldots, \sigma_n)$, we obtain that a free module is a module isomorphic to a direct sum $\bigoplus_{1 \le i \le n} R_i$, where $R_i = R$ for each index i, that is the direct sum of a finite copies of the ring R.

Definition B.8.8. (1) A chain complex is a sequence of morphisms of modules

$$\cdots \xrightarrow{f_3} M_2 \xrightarrow{f_2} M_1 \xrightarrow{f_1} M_0 \cdots$$

such that $f_n \circ f_{n+1} = 0$, i. e., $\operatorname{Im} f_{n+1} \subseteq \operatorname{Ker} f_n$.

(2) A chain is called exact if $\operatorname{Im} f_{n+1} = \operatorname{Ker} f_n$.

Example B.8.9. A short exact sequence is an exact sequence

$$0 \longrightarrow M_2 \xrightarrow{f} M_1 \xrightarrow{g} M_0 \longrightarrow 0,$$

where f is injective and g is surjective.

Example B.8.10. A morphism of modules $f : M \longrightarrow N$ defines a short exact sequence

$$0 \longrightarrow K \longrightarrow M \xrightarrow{f} N \longrightarrow C \longrightarrow 0,$$

where $K = \operatorname{Ker} f$ and $C = \operatorname{coKer} f = N / \operatorname{Im} f$.

Subject Index

https://doi.org/10.1515/9783111517384-014

Author Index

https://doi.org/10.1515/9783111517384-015

Bibliography

[1] ADAMS, COLIN; FRANZOSA, ROBERT. *Introduction to Topology: Pure and Applied*. Pearson, 2008.

[2] ADHIKARI, MAHIMA RANJAN. *Basic Algebraic Topology and its Applications*. Springer, [New Delhi], 2016.

[3] ALTMAN, MIECZYSŁAW. *On the extension of linear transformations in Banach spaces*. Bull. Acad. Polon. Sci. Sér. Sci. Math. Astr. Phys. 6 (1958), 241–248 (unbound insert).

[4] ALUFFI, PAOLO. *Algebra: Chapter 0*. Graduate Studies in Mathematics, 104. American Mathematical Society, Providence, RI, 2009.

[5] ARKOWITZ, MARTIN. *Introduction to Homotopy Theory*. Universitext. Springer, New York, 2011.

[6] ARMSTRONG, MARK ANTHONY. *Basic Topology*. Corrected reprint of the 1979 original. Undergraduate Texts in Mathematics. Springer-Verlag, New York–Berlin, 1983.

[7] BARBU, VIOREL; PRECUPANU, TEODOR. *Convexity and Optimization in Banach Spaces*. Fourth edition. Springer Monographs in Mathematics. Springer, Dordrecht, 2012.

[8] BORSUK, KAROL. *Über gewisse Invarianten der ε-Abbildungen* (German). Math. Ann. 108 (1933), no. 1, 311–318.

[9] BORSUK, KAROL. *Theory of Retracts*. Monografie Matematyczne, 44. Państwowe Wydawnictwo Naukowe, Warsaw, 1967.

[10] BREDON, GLEN E. *Topology and Geometry*. Corrected third printing of the 1993 original. Graduate Texts in Mathematics, 139. Springer-Verlag, New York, 1997.

[11] BROWN, RONALD. *Elements of Modern Topology*. McGraw-Hill Book Co., New York–Toronto–London, 1968.

[12] BROWN, RONALD. *Topology and Groupoids*. Third edition of Elements of Modern Topology [McGraw-Hill, New York, 1968; MR0227979]. With 1 CD-ROM (Windows, Macintosh and UNIX). BookSurge, LLC, Charleston, SC, 2006.

[13] BROUWER, L. E. J. *Über Abbildungen von Mannigfaltigkeiten* (German). Math. Ann. 71 (1911), no. 3, 320–327.

[14] BROUWER, L. E. J. *Über Abbildungen von Mannigfaltigkeiten* (German). Math. Ann. 71 (1911), no. 1, 97–115.

[15] BROUWER, L. E. J. *Über Abbildungen von Mannigfaltigkeiten* (German). Math. Ann. 71 (1912), no. 4, 598.

[16] CROOM, FRED H. *Basic Concepts of Algebraic Topology*. Undergraduate Texts in Mathematics. Springer-Verlag, New York–Heidelberg, 1978.

[17] CROSSLEY, MARTIN D. *Essential Topology*. Springer Undergraduate Mathematics Series. Springer-Verlag London, Ltd., London, 2005.

[18] DEIMLING, KLAUS. *Nonlinear Functional Analysis*. Springer Verlag, Berlin, New York. Oxford, 1985.

[19] DEO, SATYA. *Algebraic Topology. A Primer*. Second edition. Texts and Readings in Mathematics, 27. Hindustan Book Agency, New Delhi, 2018.

[20] TOM DIECK, TAMMO. *Algebraic Topology*. EMS Textbooks in Mathematics. European Mathematical Society (EMS), Zürich, 2008.

[21] DJEBALI, SMAÏL; GÓRNIEWICZ, LECH; OUAHAB, ABDELGHANI. *Solution Sets for Differential Equations and Inclusions*. De Gruyter Series in Nonlinear Analysis and Applications, 18. Walter de Gruyter & Co., Berlin, 2013.

[22] Djebali, Smaïl. *Fixed point theory for* 1*-set contractions: a survey*. Applied Mathematics in Tunisia, 53–100, Springer Proc. Math. Stat., 131, Springer, Cham, 2015.

[23] DODSON, C. T. J.; PARKER, PHILLIP E. *A User's Guide to Algebraic Topology*. Mathematics and its Applications, 387. Kluwer Academic Publishers Group, Dordrecht, 1997.

[24] DOLD, ALBRECHT. *Lectures on Algebraic Topology*. Reprint of the 1972 edition. Classics in Mathematics. Springer-Verlag, Berlin, 1995.

[25] DUGUNDJI, JAMES. *Topology*. Allyn and Bacon, Inc., Boston, Mass., 1966.

[26] EILENBERG, SAMUEL; STEENROD, NORMAN E. *Axiomatic approach to homology theory*. Proc. Natl. Acad. Sci. USA 31 (1945), 117–120.

https://doi.org/10.1515/9783111517384-016

[27] Engelking, Ryszard; Sieklucki, Karol. *Topology: A Geometric Approach*. Translated from the Polish original by Adam Ostaszewski. Sigma Series in Pure Mathematics, 4. Heldermann Verlag, Berlin, 1992.

[28] Evens, Len; Thompson, Rob. *Algebraic Topology*. Textbook Lecture Notes, Northwestern University City University of New York, 2014.

[29] Fan, Ky. *Fixed-point and minimax theorems in locally convex topological linear spaces*. Proc. Natl. Acad. Sci. USA 38 (1952), 121–126.

[30] Fan, Ky. *A minimax inequality and applications*. Inequalities, III (Proc. Third Sympos., Univ. California, Los Angeles, Calif., 1969; dedicated to the memory of Theodore S. Motzkin), pp. 103–113. Academic Press, New York, 1972.

[31] Félix, Yves; Tanré, Daniel. *Topologie Algébrique. Cours et Exercises Corrigés* (French). Dunod, Paris, 2010.

[32] Fernández, T. *A note on the Borsuk non-retraction theorem*. Monatshefte Math. 145 (2005), no. 2, 95–96.

[33] Fomenko, Anatoly; Fuchs, Dmitry. *Homotopical Topology*. Second edition. Graduate Texts in Mathematics, 273. Springer, [Cham], 2016.

[34] Fox, R. H. *On homotopy type and deformation retracts*. Ann. Math. (2) 44 (1943), 40–50.

[35] Gallier, Jean; Xu, Dianna. *A Guide to the Classification Theorem for Compact Surfaces*. Geometry and Computing, 9. Springer, Heidelberg, 2013.

[36] Gamelin, Theodore W.; Greene, Robert Everist. *Introduction to Topology*. Second edition. Dover Publications, Inc., Mineola, NY, 1999.

[37] Geoghegan, Ross. *Topological Methods in Group Theory*. Graduate Texts in Mathematics, 243. Springer, New York, 2008.

[38] Godbillon, Claude. *Éléments de Topologie Algébrique* (French). Hermann, Paris, 1971.

[39] Gramain, André. *Topology of Surfaces*. Translated from the French by Leo F. Boron, Charles O. Christenson and Bryan A. Smith. BCS Associates, Moscow, ID, 1984.

[40] Granas, Andrzej; Dugundji, James. *Fixed Point Theory*. Springer Monographs in Mathematics. Springer-Verlag, New York, 2003.

[41] Gray, Brayton. *Homotopy Theory. An Introduction to Algebraic Topology*. Pure and Applied Mathematics, 64. Academic Press [Harcourt Brace Jovanovich, Publishers], New York–London, 1975.

[42] Greenberg, Marvin J.; Harper, John R. *Algebraic Topology. A First Course*. Mathematics Lecture Note Series, 58. Benjamin/Cummings Publishing Co., Inc., Advanced Book Program, Reading, Mass., 1981.

[43] Hadamard, J. *Sur quelques questions du calcul des variations* (French). Ann. Sci. Éc. Norm. Supér. (3) 24 (1907), 203–231.

[44] Hartman, Philip; Stampacchia, Guido. *On some non-linear elliptic differential-functional equations*. Acta Math. 115 (1966), 271–310.

[45] Hatcher, Allen. *Algebraic Topology*. Cambridge University Press, Cambridge, 2002.

[46] Hocking, John G.; Young, Gail S. *Topology*. Addison-Wesley Publishing Co., Inc., Reading, Mass.–London, 1961.

[47] Hofmann, Karl H.; Martin, John R. *Möbius manifolds, monoids, and retracts of topological groups*. Semigroup Forum 90 (2015), no. 2, 301–316.

[48] Hopf, Heinz. *Über die Drehung der Tangenten und Sehnen ebener Kurven* (German). Compos. Math. 2 (1935), 50–62.

[49] Hu, Sze-tsen. *Homotopy Theory*. Pure and Applied Mathematics, Vol. VIII. Academic Press, New York–London 1959.

[50] Hu, Sze-tsen. *Theory of Retracts*. Wayne State University Press, Detroit 1965.

[51] Hu, Sze-tsen. *Homology Theory: A First Course in Algebraic Topology*. Holden-Day, Inc., San Francisco, Calif.–London–Amsterdam, 1966.

[52] Jeanneret, A.; Lines, D. *Invitation à la Topologie Algébrique. Tome I: Homologie* (French). Cépadués Editions, 2014.

[53] Kalajdzievski, Sasho. *An Illustrated Introduction to Topology and Homotopy*. CRC Press, Boca Raton, FL, 2015.

[54] KANTOROVICH, L. V.; AKILOV, G. P. *Functional Analysis*. Translated from the Russian by Howard L. Silcock. Second edition. Pergamon Press, Oxford–Elmsford, N.Y., 1982.

[55] KERVAIRE, MICHEL A.; MILNOR, JOHN W. *Groups of homotopy spheres. I.* Ann. Math. (2) 77 (1963), 504–537.

[56] KESAVAN, S. *Topics in Functional Analysis and Applications*. John Wiley & Sons, Inc., New York, 1989.

[57] KLEE, V. L., JR. *A note on topological properties of normed linear spaces*. Proc. Am. Math. Soc. 7 (1956), 673–674.

[58] KNASTER, B.; KURATOWSKI, C.; MAZURKIEWICZ, S. *Ein Beweis des fixpunktsatzes für n-dimensionale simplexe* (German). Fundam. Math. 14 (1929), 132–137.

[59] KOSNIOWSKI, CZES. *A First Course in Algebraic Topology*. Cambridge University Press, 1980.

[60] LAHIRI, B. K. *A First Course in Algebraic Topology*. Alpha Science International Ltd, UK, 2000.

[61] LAWSON, TERRY. *Topology: A Geometric Approach*. Oxford Graduate Texts in Mathematics, 9. Oxford University Press, Oxford, 2003.

[62] LEE, JOHN M. *Introduction to Topological Manifolds*. Second edition. Graduate Texts in Mathematics, 202. Springer, New York, 2011.

[63] LERAY, JEAN; SCHAUDER, JULES. *Topologie et équations fonctionnelles* (French). Ann. Sci. Éc. Norm. Supér. (3) 51 (1934), 45–78.

[64] LUNDELL, ALBERT T.; WEINGRAM, STEPHEN. *The Topology of CW Complexes*. The University Series in Higher Mathematics. Van Nostrand Reinhold Co., New York, 1969.

[65] LYUSTERNIK, L.; ŠNIREL'MAN, L. *Topological methods in variational problems and their application to the differential geometry of surfaces* (Russian). Usp. Mat. Nauk 2, (1947). no. 1(17), 166–217.

[66] MANETTI, MARCO. *Topology*. Translated from the 2014 Italian edition by Simon G. Chiossi. Unitext, 91. La Matematica per il 3+2. Springer, Cham, 2015.

[67] MASSEY, WILLIAM S. *Algebraic Topology: An Introduction*. Harcourt, Brace & World, Inc., New York, 1967.

[68] MASSEY, WILLIAM S. *A Basic Course in Algebraic Topology*. Graduate Texts in Mathematics, 127. Springer-Verlag, New York, 1991.

[69] MATOUŠEK, JIŘÍ. *Using the Borsuk–Ulam Theorem. Lectures on Topological Methods in Combinatorics and Geometry*. Written in cooperation with Anders Björner and Günter M. Ziegler. Universitext. Springer-Verlag, Berlin, 2003.

[70] MAUNDER, C. R. F. *Algebraic Topology*. Reprint of the 1980 edition. Dover Publications, Inc., Mineola, NY, 1996.

[71] MAY, J. P. *A Concise Course in Algebraic Topology*. Chicago Lectures in Mathematics. University of Chicago Press, Chicago, IL, 1999.

[72] MAYER, JOERG. *Algebraic Topology*. Prentice-Hall, Inc., Englewood Cliffs, N.J., 1972.

[73] MILNOR, JOHN. *Analytic proofs of the "hairy ball theorem" and the Brouwer fixed-point theorem*. Am. Math. Mon. 85 (1978), no. 7, 521–524.

[74] MÖNCH, HARALD. *Boundary value problems for nonlinear ordinary differential equations of second order in Banach spaces*. Nonlinear Anal. 4 (1980), no. 5, 985–999.

[75] MUNKRES, JAMES R. *Elements of Algebraic Topology*. Addison-Wesley Publishing Company, Menlo Park, CA, 1984.

[76] MUNKRES, JAMES R. *Topology*. Second edition. Prentice Hall, Inc., Upper Saddle River, NJ, 2000.

[77] NEWMAN, W. M.; SPROULL, R. F. *Principles of Interactive Computer Graphics*. McGraw-Hill, New York, 1979.

[78] POINCARÉ, HENRI. *Analysis situs*. J. Éc. Polytech. (2), (1895), no. 1, 1–123.

[79] REES, ELMER G. *Topology. Notes on geometry*. Universitext. [University Textbook]. Springer-Verlag, Berlin–New York, 1983.

[80] ROGERS, C. A. *A less strange version of Milnor's proof of Brouwer's fixed-point theorem*. Am. Math. Mon. 87 (1980), no. 7, 525–527.

[81] ROLFSEN, DAL. *Knots and Links*. Corrected reprint of the 1976 original. Mathematics Lecture Series, 7. Publish or Perish, Inc., Houston, TX, 1990.

[82] ROTHE, ERICH. *Über Abbildungsklassen von Kugeln des Hilbertschen Raumes* (German). Compos. Math. 4 (1937), 294–307.

[83] ROTMAN, JOSEPH J. *An Introduction to Algebraic Topology*. Graduate Texts in Mathematics, 119. Springer-Verlag, New York, 1988.

[84] ROYDEN, H. L.; FITZPATRICK, P. M. *Real Analysis*. Fourth edition. Pearson Education, 2010.

[85] SAKAI, KATSURO. *Geometric Aspects of General Topology*. Springer Monographs in Mathematics. Springer, Tokyo, 2013.

[86] SÁNCHEZ-GRANERO, M. A. *Retraction of a compact subspace onto its boundary*. Monatshefte Math. 139 (2003), no. 2, 169–172.

[87] SCHAEFER, HELMUT. *Über die Methode der a priori-Schranken* (German). Math. Ann. 129 (1955), 415–416.

[88] SCHAUDER, J. *Der Fixpunktsatz in Funktionalräumen* (German). Stud. Math. 2 (1930), 171–180.

[89] SEIFERT, HERBERT; THRELFALL, WILLIAM. *Seifert and Threlfall: A Textbook of Topology*. Translated from the German edition of 1934 by Michael A. Goldman. With a preface by Joan S. Birman. With "Topology of 3-dimensional fibered spaces" by Seifert. Translated from the German by Wolfgang Heil. Pure and Applied Mathematics, 89. Academic Press, Inc. [Harcourt Brace Jovanovich, Publishers], New York–London, 1980.

[90] SHASTRI, ANANT R. *Basic Algebraic Topology*. With a foreword by Peter Wong. CRC Press, Boca Raton, FL, 2014.

[91] SIERADSKI, ALLAN J. *An Introduction to Topology and Homotopy*. The Prindle, Weber & Schmidt Series in Advanced Mathematics. PWS-KENT Publishing Co., Boston, MA, 1992.

[92] SINGH, TEJ BAHADUR. *Introduction to Topology*. Springer, Singapore, 2019.

[93] SMART, D. R. *Fixed Point Theorems*. Cambridge Tracts in Mathematics, 66. Cambridge University Press, London–New York, 1974.

[94] SPANIER, EDWIN H. *Algebraic Topology*. Corrected reprint. Springer-Verlag, New York–Berlin, 1981.

[95] SPIVAK, MICHAEL. *A Comprehensive Introduction to Differential Geometry*, Vol. V. Second edition. Publish or Perish, Inc., Wilmington, Del., 1979.

[96] STONE, A. H. *Paracompactness and product spaces*. Bull. Am. Math. Soc.. 54 (1948), 977–982.

[97] SWITZER, ROBERT M. *Algebraic Topology – Homotopy and Homology*. Reprint of the 1975 original [Springer, New York;] Classics in Mathematics. Springer-Verlag, Berlin, 2002.

[98] TIETZE, HEINRICH. *Über Funktionen, die auf einer abgeschlossenen Menge stetig sind* (German). J. Reine Angew. Math. 145 (1915), 9–14.

[99] VASSILIEV, V. A. *Introduction to Topology*. Translated from the 1997 Russian original by A. Sossinski. Student Mathematical Library, 14. American Mathematical Society, Providence, RI, 2001.

[100] VICK, JAMES W. *Homology Theory. An Introduction to Algebraic Topology*. Second edition. Graduate Texts in Mathematics, 145. Springer-Verlag, New York, 1994.

[101] VIRO, O. YA.; IVANOV, O. A.; NETSVETAEV, N. YU.; KHARLAMOV, V. M. *Elementary Topology. Problem Textbook*. American Mathematical Society, Providence, RI, 2008.

[102] WEINTRAUB, STEVEN H. *Fundamentals of Algebraic Topology*. Graduate Texts in Mathematics, 270. Springer, New York, 2014.

[103] ZEIDLER, EBERHARD. *Nonlinear Functional Analysis and its Aplications. Vol. I: Fixed Point Theorems*. Springer Verlag, New York, 1986.

[104] ZHANG, GUOWEI; SUN, JINGXIAN. *A generalization of the cone expansion and compression fixed point theorem and applications*. Nonlinear Anal. 67 (2007), no. 2, 579–586.

www.ingramcontent.com/pod-product-compliance
Lightning Source LLC
Chambersburg PA
CBHW080651220326
41598CB00033B/5170
* 9 7 8 3 1 1 1 5 1 7 3 6 0 *